S0-CKS-239

Contributions to Current Research in Geophysics (CCRG)

3

Stress in the Earth

Editor: Max Wyss
Cooperative Institute for Research in Environmental Sciences, University of Colorado, Boulder, Colorado, U.S.A.

Reprint from PAGEOPH

1977 Birkhäuser Verlag Basel und Stuttgart

Reprinted from Pure and Applied Geophysics (PAGEOPH), Volume 115 (1977), No. 1–2

CIP-Kurztitelaufnahme der Deutschen Bibliothek

Stress in the earth: reprint from PAGEOPH/
ed.: Max Wyss. - 1. Aufl. - Basel, Stuttgart:
Birkhäuser, 1977.
(Contributions to current research in geophysics; 3)
ISBN 3-7643-0952-0

NE: Wyss, Max [Hrsg.]

CONTENTS Page

Editor's Note

Knowledge of the stress in the earth is fundamental to the understanding of tectonic processes. Yet stress is an elusive variable which is difficult to obtain for even one point in the earth. Because stress measurements are costly they are scarce, and one is tempted to assume that local values may have regional significance which may not be the case. Of the six independent values in the stress tensor, usually only one or two, sometimes four, and seldom six can be estimated depending on the method used. Stress is generally not measured directly but inferred from a wide variety of observations often involving specialized techniques. As a result, scientists using one approach may not be familiar with recent results obtained by different methods. The Chapman Conference on 'Stress in the Lithosphere' at Aspen, Colorado, in September, 1976 made an important contribution toward eliminating the communication gap between workers in various disciplines concerned with stresses in the earth. Stress determinations obtained by the different methods are not usually published in the same journals. Therefore, I thought it was a good idea to publish a special issue on stresses, which would include articles describing various stress measuring techniques. Using the Chapman Conference participants as a nucleus, I also tried to include work from as many continents as possible.

The objective of the special issue is to assemble current research articles describing new approaches and results, with some review for the reader unfamiliar with special techniques. The articles cover a variety of studies involving overcoring, hydraulic fracturing, geological mapping, grain-size paleopiezometry, flow laws of minerals, velocity anisotropy, seismicity, earthquake focal mechanisms, earthquake stress-drops, and theoretical analysis of earthquake faulting and plate motions. I hope that the reader will find these topics informative and useful, and that earth scientists of various special fields will become aware of new work relevant to their problems.

The results presented in this volume indicate that we are on the way to being able to compile regional and even global stress maps, similar to the slip-vector maps from which the plate motions were derived in the 1960's. Such stress maps are especially needed as a base for theoretical studies, which attempt to solve the question of the driving mechanism of plate motions.

I wish to express my thanks to all who helped to create this special issue. I am deeply indebted to the organizers of the Chapman Conference for allowing me to solicit papers from among the participants. I am also very grateful for the speedy reviews which made the timely appearance of this issue possible.

MAX WYSS

Pageoph, Vol. 115 (1977), Birkhäuser Verlag, Basel

Principal Horizontal Stresses in Southern Africa

By N. C. GAY[1]

Abstract – A review of *in situ* stress measurements at eight regional localities in Africa south of the 15°S parallel shows that average directions of the horizontal principal stresses are N–S and E–W. These directions agree with principal stress orientations deduced from earthquake fault plane solutions. However, the maximum and minimum principal horizontal stresses are not consistently oriented parallel to either the N–S or E–W direction; they may vary within an individual region because of local geological structures and from region to region. At the sites within the Witwatersrand sediments (all at depths greater than 500 m) the maximum stress tends to lie NW–SE but at three of the four sites outside the Witwatersrand sediments (all at depths less than 500 m) this stress is oriented approximately N–S.

The data reported here are compared with horizontal stresses predicted for Southern Africa by SOLOMON, SLEEP and RICHARDSON (1975) from various plate tectonic driving force models. The agreement between orientations is fair for all sites but only the deep sites in the Witwatersrand sediments have comparable stress magnitudes.

Key words: Stress *in-situ*; Intraplate stress field; Tectonics of Africa.

1. Introduction

Information on the state of stress in eight regional localities in Africa south of the 15°S parallel is now available from strain relief measurements carried out in near-surface engineering sites and deep mines. At all but two of these localities measurements have been made at more than one site. Most of these data have been discussed in detail by Gay (1975). However, two sets of previously unpublished data have since been made available. These are the results obtained by VAN HEERDEN (1974) and VAN HEERDEN and ORR (1974) at the Ruacana Power Station, South West Africa, and those obtained by VAN HEERDEN (R. KERSTEN, personal communication, 1976) at the Prieska Copper Mine, Cape Province, South Africa. At both of these localities, measurements were made using the CSIR triaxial strain cell (LEEMAN 1968). The detailed results and regional geology around the sites are given in the Appendix.

In the earlier paper, GAY (1975) showed that at shallow depths (<500 m) horizontal stresses tended to be larger than vertical stresses but that at greater depths vertical stresses were nearly twice as large as the horizontal ones. This variation in the relative magnitudes of horizontal and vertical stresses agrees well with VOIGHT's (1966) denudation theory. The new results from the Prieska Copper Mine and Ruacana

[1]) Bernard Price Institute of Geophysical Research, University of the Witwatersrand, Johannesburg, South Africa.

Power Station are consistent with the curve calculated by GAY (1975, Fig. 2) illustrating this point. The axes of principal stress are normally inclined to the vertical and horizontal (GAY 1975, Fig. 3) suggesting that the observed states of stress include components of early tectonic and gravitational forces in addition to present day gravitational and, possibly, tectonic forces. At some sites both the orientations and magnitudes of the principal stresses appeared to be influenced by local geologic structures and no dominant trend in the orientation of the stress ellipsoids over Southern Africa was observed.

The purpose of this note is to report the magnitudes and orientations of the principal horizontal stresses at each regional locality and at individual sites within each region. These stresses were calculated from the three-dimensional states of stress at each site, as given in Table I of GAY (1975) and the Appendix to this paper, using the elementary stress relationships given in JAEGER and COOK (1969, Chap. 2).

At all sites strain relief overcoring methods were used to determine the state of stress. All measurements were made in boreholes at depths sufficiently far into the rock for local stress concentrations due to the excavation to be minimal; i.e. they had negligible influence on the orientations and magnitudes of the ambient or virgin principal stresses. In the Witwatersrand mines, the only site sufficiently close to an

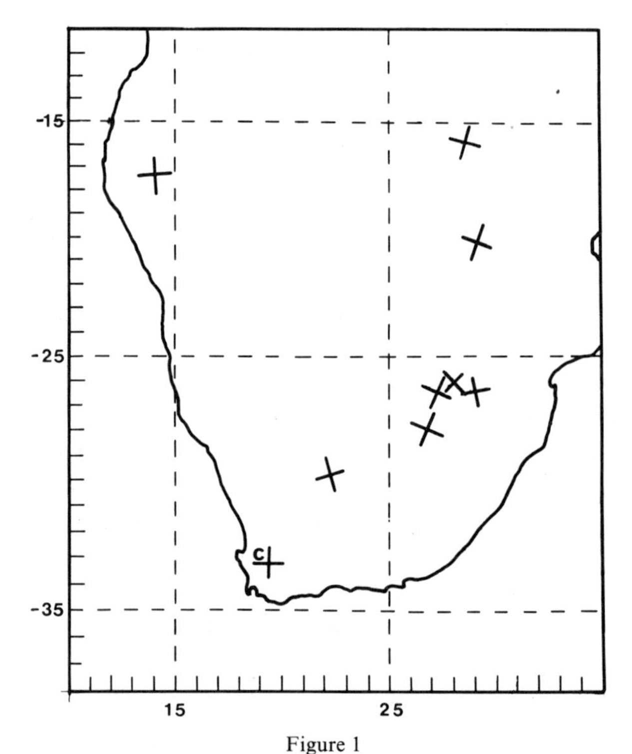

Figure 1
Average principal horizontal stress directions in Southern Africa as determined from *in-situ* stress measurements and the 1969 Ceres earthquake (C).

area of active mining for the stress field to be affected was site 8, in the Harmony Gold Mine. PALLISTER (1974, private communication) corrected for these mining induced stresses by means of an elastic analogue solution (SALAMON *et al.* 1964). These corrected results were used to obtain the data reported here.

2. *Results*

Table I lists the magnitudes and orientations of the principal horizontal stresses, σ_1 and σ_2, at each measuring site as well as the average stress directions for each region containing more than one site. With the possible exception of the mean directions for the Central Witwatersrand region, these average stress orientations are remarkably consistent, the overall average being 001° (s.d 25°) and 271° (s.d 25°). Figure 1 is a map of Southern Africa on which the average horizontal principal stress directions are plotted at each regional locality.

The relative magnitudes of the principal stresses are not shown on the map because the orientations of σ_1 and σ_2, respectively, are not necessarily consistent within any particular region. This is particularly so for the sites within the Witwatersrand Group. For example, σ_1 at Bracken Mine (site 1) is oriented at nearly 90° to the σ_1 directions for the other two mines in the Evander region. Similarly, sites 7(i) and 7(ii) in the Doornfontein Gold Mine, which are 5 km apart, also have their respective σ_1 directions perpendicular to each other and at Durban Roodepoort Deep Gold Mine the stress orientations change by 37° between sites 5(i) and (ii), which are only 3 km apart. In shallower sites outside the Witwatersrand sediments, the respective orientations of σ_1 and σ_2 are far more stable, and only site 12(iv), at Ruacana, has the maximum principal stress oriented at a high angle to the regional trend.

Also listed in Table 1 are the magnitudes of the vertical stress at each site. In the Witwatersrand sites, the vertical stress is always larger than either of the principal horizontal stresses except at sites 1 and 5(i) where there is geologic evidence of large components of horizontal residual stresses (c.f. GAY 1975, CAHNBLEY 1970). By contrast, the vertical stress is less than the maximum horizontal principal stress at sites 9(i) and (ii) at the Prieska Copper Mine and smaller than both horizontal principal stresses at all the other sites outside the Witwatersrand Group.

3. *Discussion*

The data presented here and by GAY (1975) cover a significant portion of the sub-continent of Southern Africa and they provide a fairly good picture of the state of stress in the upper levels of a large block of continental crust. The most striking aspect of the results is the consistent N–S (001°; s.d 25°) and E–W (271°; s.d 25°) directions of the horizontal principal stresses. These directions do not appear to change with depth of measurement nor are they greatly influenced by regional geologic structures.

They coincide remarkably well with the orientations of the principal stresses deduced from the fault plane solution for the 1969 Ceres earthquake by Green and McGarr (1972). These stress directions are plotted at point C on Fig. 1; the maximum compressive stress is oriented approximately E–W and the minimum N–S. They also agree with stress orientations deduced from fault plane solutions for earthquakes in Zambia and Rhodesia which indicated that the least compressive stress was oriented approximately east–west (MAASHA and MOLNAR, 1972, Table I, events 7, 8, 9, 10). These events were the result of normal faulting with the maximum compressive stress near vertical.

The fault plane solutions quoted above show a change through 90° in the orientation of the least compressive stress over a large part of Southern Africa. Similar changes in the orientations of the maximum and minimum stresses within individual regions also occur, normally due to local structures. Also, there appears to be a distinct difference between the average stress orientations at sites within the Witwatersrand sedimentary basin and those outside it. The mean orientation of the average principal stress directions for each region within the basin is: σ_1 307° (s.d 12°) and σ_2 037° (s.d 11°). For those localities outside the Witwatersrand, the stresses at Kafue Gorge are approximately parallel to those in the Witwatersrand sediments but at Prieska, Shabani and Ruacana, σ_1 lies approximately N–S and σ_2 E–W. Note that the Kafue Gorge orientation is opposite to those deduced by MAASHA and MOLNAR (1972) from the earthquake fault plane solutions mentioned above, and that for Prieska also contradicts the Ceres earthquake solution. The Prieska data may be influenced by the regional structural geology. The mine is situated adjacent to a major NW–SE trending regional lineament known as the Brakbos fault zone (VILJOEN *et al.* 1975). Movement along this zone is right lateral and any residual components of stress associated with the structure, would have a compressive stress oriented N–S.

Recently, SOLOMON *et al.* (1975) have tried to deduce the orientations and relative magnitudes of principal horizontal stresses within lithospheric plates for different driving-force models. To do this they assume that the lithosphere is a thin elastic shell in a membrane state of stress so that the magnitudes and orientations of the stresses are constant through the thickness of the membrane. In the Southern African subcontinent, their analysis predicts tensional stresses with σ_1, the least tensile stress, striking approximately 075° and σ_2 striking 345°, for all models. The magnitudes of the principal stresses are of the order of tens of megapascals.

The results reported here cover a depth range of 2.5 km and consistent directions of the horizontal principal stresses over this depth range suggest that they can be compared with Solomon *et al.*'s predictions. The comparison between the observed NS–EW average orientations and the predicted orientations is quite good. Clearly the assumption of constant stress magnitudes does not hold over the depth range covered here. However, some comparison with the theoretical magnitudes can be obtained from the 'deviatoric' stress (σ_m) at each site, calculated according to the equation $\sigma_m = \sigma_H - \sigma_V$ where σ_H is the measured horizontal principal stress and σ_V the vertical stress.[2] From the values of σ_H and σ_V given in Table 1, σ_m is generally compressive at the near surface sites outside the Witwatersrand system and ranges in value from −6 to 12 MPa. At the shallowest site in the Witwatersrand system, σ_m is small (2.5 to −2.8 MPa) and with increasing depth it becomes tensile and has magnitudes as large as −35 MPa. These magnitudes are comparable to those predicted by SOLOMON *et al.* (1975).

The compressive stresses at the shallower sites are undoubtedly due, in part at least, to the increase in horizontal stresses relative to the vertical stress during denudation (VOIGHT 1966; GAY 1975). This effect is not allowed for in SOLOMON *et al.*'s (1975) analysis and so one would expect the data from the deeper sites to be more suitable for assessing the validity of the models. Significantly, the data from the deep sites is more akin to the predicted stress state than that from the shallower sites.

Table 1

Vertical and principal horizontal stresses at localities in Southern Africa

A. SITES WITHIN THE WITWATERSRAND GROUP SEDIMENTS

Evander region (Latitude: 26°23′S; Longitude: 29°00′E)

(1) Bracken Gold Mine
Depth 508 m, σ_v − 13.9 MPa
σ_1 − 16.4 MPa, strike 344°
σ_2 − 11.1 MPa, strike 074°

(2) Winkelhaak Gold Mine
Depth 1226 m, σ_v − 38.4 MPa
σ_1 − 32.1 MPa, strike 084°
σ_2 − 31.2 MPa, strike 354°

(3) Kinross Gold Mine
Depth 1577 m, σ_v − 49.5 MPa
σ_1 − 37.2 MPa, strike 081°
σ_2 − 26.4 MPa, strike 351°
Average directions of principal stresses: 350°, 080°

Central Witwatersrand Region (Latitude: 26°11′S; Longitude: 28°04′E)

(4) East Rand Proprietary Mines, Boksburg
Depth 2400 m, σ_v − 37.4 MPa
σ_1 − 31.7 MPa, strike 323°
σ_2 − 22.2 MPa, strike 053°

(5) Durban Roodepoort Deep Gold Mine, Roodepoort
(i) Depth 2500 m, σ_v − 59.0 MPa
σ_1 − 84.9 MPa, strike 330°
σ_2 − 35.9 MPa, strike 060°
(ii) Depth 2300 m, σ_v − 68.5 MPa
σ_1 − 53.7 MPa, strike 293°
σ_2 − 38.7 MPa, strike 023°
Average directions of principal stresses · 315°, 045°

[2]) In SOLOMON *et al.*'s (1975) paper, the term 'deviatoric stress' is used in a sense other than that of classical continuum mechanics and can be interpreted as the equivalent of σ_m defined here. This was pointed out to me by an anonymous reviewer.

Table 1—*continued*

Carletonville region (Latitude: 26°22′S; Longitude: 27°25′E)
(6) Western Deep Levels Gold Mine
Depth 1770 m, σ_v – 47.4 MPa
σ_1 – 33.9 MPa, strike 284°
σ_2 – 21.4 MPa, strike 014°

(7) Doornfontein Gold Mine
(i) Depth 2320 m, σ_v – 58.5 MPa
σ_1 – 40.4 MPa, strike 034°
σ_2 – 23.1 MPa, strike 304°
(ii) Depth 1320 m, σ_v – 39.0 MPa
σ_1 – 25.1 MPa, strike 298°
σ_2 – 12.4 MPa, strike 028°
Average directions of principal stresses: 025°, 295°

Welkom region (Latitude: 28°00′S; Longitude: 26°45′E)
(8) Harmony Gold Mine
Depth 1500 m, σ_v – 33.1 MPa
σ_1 – 19.0 MPa, strike 292°
σ_2 – 14.2 MPa, strike 022°

B. Sites not in the Witwatersrand Group Sediments

(9) Prieska Copper Mine, Copperton (Latitude: 29°57′S; Longitude: 22°10′E)
(i) Depth 279 m, σ_v – 8.8 MPa
σ_1 – 16.4 MPa, strike 359°
σ_2 – 2.7 MPa, strike 089°
(ii) Depth 410 m, σ_v – 9.6 MPa
σ_1 – 13.0 MPa, strike 330°
σ_2 – 6.4 MPa, strike 060°
Average directions of principal stresses: 345°, 075°

(10) Kafue Gorge Hydroelectric scheme (Latitude: 15°55′S; Longitude: 28°45′E)
(i) Depth 400 m, σ_v – 13.8 MPa
σ_1 – 25.6 MPa, strike 280°
σ_2 – 19.1 MPa, strike 010°
(ii) Depth 160 m, σ_v – 8.9 MPa
σ_1 – 16.9 MPa, strike 301°
σ_2 – 12.5 MPa, strike 031°
Average direction of principal stresses 290°, 020°

(11) Shabani Asbestos Mine (Latitude: 20°20′S; Longitude: 30°05′E)
Depth 350 m, σ_v – 10.7 MPa
σ_1 – 16.5 MPa, strike 020°
σ_2 – 13.4 MPa, strike 290°

(12) Ruacana Hydroelectric scheme (Latitude: 17°20′S; Longitude: 14°00′E)
(i) Depth 135 m, σ_v – 4.4 MPa
σ_1 – 11.7 MPa, strike 336°
σ_2 – 6.2 MPa, strike 066°
(ii) Depth 128 m, σ_v – 3.9 MPa
σ_1 – 9.0 MPa, strike 012°
σ_2 – 6.4 MPa, strike 282°

Table 1—*continued*

(iii) Depth 113 m, σ_v – 4.85 MPa
σ_1 – 11.1 MPa, strike 350°
σ_2 – 5.7 MPa, strike 080°
(iv) Depth 109 m, σ_v – 3.2 MPa
σ_1 – 8.0 MPa, strike 271°
σ_2 – 5.6 MPa, strike 001°
Average direction of principal stresses: 355°, 085°

Mean of average directions of principal stresses: 001° (s.d 25°); 271° (s.d 25°)

Appendix

Details of stress measuring sites at Prieska Copper Mine and Ruacana Hydroelectric Scheme

Prieska

Sites at depths of 279 m and 410 m; measurements made using the CSIR triaxial strain cell.

Results: 279 m $\sigma_1 = 22.4$ MPa, plunges at 22° to 004°; $\sigma_2 = 8.8$ MPa, plunges at 48° to 123°; $\sigma_3 = 2.5$ MPa, plunges at 33° to 260°. Vertical stress $\sigma_v = 8.8$ MPa; mean horizontal stress $\sigma_h = 12.45$; $\sigma_h/\sigma_v = 1.41$. *410 m* $\sigma_1 = 13.04$ MPa, plunges at 6° to 330°; $\sigma_2 = 9.65$ MPa, plunges at 78° to 098°; $\sigma_3 = 6.35$ MPa, plunges at 6° to 239°; $\sigma_v = 9.63$ MPa; $\sigma_h = 9.7$ MPa; $\sigma_h/\sigma_v = 1.01$.

Geology: Folded quartz amphibolite schists of the Marydale meta-sediments. The mine is bounded by the Brakbos fault zone, a right lateral shear zone which is one of many similar fault zones striking NW–SE over a large area of the NW Cape Province.

Ruacana

Sites at depths of 109, 113, 128 and 135 m; measurements made using the CSIR triaxial strain cell by Van Heerden (1974) and Van Heerden and Orr (1974).

Results: 109m $\sigma_1 = 8.9$ MPa, plunges at 22° to 076°; $\sigma_2 = 6.1$ MPa, plunges at 18° to 338°; $\sigma_3 = 2.25$ MPa, plunges at 30° to 032°. Vertical stress $\sigma_v = 3.23$ MPa; mean horizontal stress $\sigma_h = 6.78$ MPa; $\sigma_h/\sigma_v = 2.10$. *113m* $\sigma_1 = 11.4$ MPa, plunges at 11° to 171°; $\sigma_2 = 6.9$ MPa, plunges at 34° to 269°; $\sigma_3 = 3.36$ MPa, plunges at 53° to 068°; $\sigma_v = 4.85$ MPa; $\sigma_h = 8.42$ MPa; $\sigma_h/\sigma_v = 1.74$. *128m* $\sigma_1 = 9.2$ MPa, plunges at 12° to 018°; $\sigma_2 = 7.65$ MPa, plunges at 7° to 286°; $\sigma_3 = 2.5$ MPa, plunges at 76° to 172°; $\sigma_v = 3.9$ MPa; $\sigma_h = 7.79$ MPa; $\sigma_h/\sigma_v = 1.98$. *135m* $\sigma_1 = 11.8$ MPa, plunges at 7° to 160°; σ_2 = 8.4 MPa, plunges at 36° to 255°; σ_3 = 2.0 MPa, plunges at 53° to 061°; σ_v = 4.38 MPa; σ_h = 8.95 MPa; σ_h/σ_v = 2.04.

Results are the mean of more than one measurement; the discrepancy between individual measurements was large at the 128 m site.

Geology: Pink-grey, coarse grained porphyroblastic gneiss with minor amphibolite bands and pegmatite or aplite dykes. The gneisses have a foliation striking 060° and dipping 70°–80° SE; they are cut by two orthogonal sets of vertical joints and one flat-lying set. A 40 m wide mylonite zone which strikes 060° and dips 65°–80° SE also traverses the area.

4. *Acknowledgements*

I am grateful to Mr. G. Chunnet of Hydroconsults, Pretoria, South Africa, and Mr. R. W. D. Kersten of Anglo-Transvaal Investment Co. Ltd., Johannesburg, South Africa, for permission to use the Ruacana and Prieska Copper Mine results. Art McGarr reviewed the manuscript. In addition, I am grateful to an anonymous reviewer for pointing out errors in my understanding of Solomon *et al.*'s paper and for suggesting many improvements.

References

Cahnbley, H. (1970), *Grundlagenuntersuchungen über das Entspannungsbohrverfahren während des praktischen Einsatzes in grosser Tiefe*. Dissertation, Technical University of Clausthal, Germany.

Gay, N. C. (1975), *In-situ stress measurements in Southern Africa,* Tectonophysics, *29*, 447–459.

Green, R. W. E. and McGarr, A. (1972), *A comparison of the focal mechanism and aftershock distribution of the Ceres, South African earthquake of September 29, 1969*, Bull. Seism. Soc. Amer. *62*, 869–871.

Jaeger, J. C. and Cook, N. G. W., *Fundamentals of Rock Mechanics* (Methuen, London 1969), 513 pp.

Leeman, E. R. (1968), *The determination of the complete state of stress in rock in a single borehole – Laboratory and underground measurements,* Int. J. Rock Mech. Mining Sci. *5*, 31–56.

Maasha, N. and Molnar, P. (1972), *Earthquake fault parameters and tectonics in Africa,* J. Geophys. Res. *77*, 5731–5743.

Salamon, M. D. G., Ryder, J. A., and Ortlepp, W. D. (1964), *An analogue solution for determining the elastic response of strata surrounding tabular mining excavations,* J. S. Afr. Inst. Min. Metall. *65*, 115–137.

Solomon, S. C., Sleep, N. H., and Richardson, R. N. (1975), *On the forces driving plate tectonics: Inferences from absolute plate velocities and intraplate stress,* Geophys. J. R. astr. Soc. *42*, 769–801.

Van Heerden, W. L. (1974), *Additional rock stress measurements at the Ruacana Power Station, S.W.A.*, Rep. Counc. Sci. Ind. Res. S. Afr., no. ME 1324.

Van Heerden, W. L. and Orr, C. M. (1974), *Rock stress measurements at the Ruacana Power Station, S.W.A.,* Rep. Counc. Sci. Ind. Res. S. Afr. no. ME 1311.

Viljoen, R. P., Viljoen, M. J., Grootenboer, J. and Longshaw, T. G. (1975), *ERTS – 1 imagery: applications in geology and mineral exploration,* Min. Sci. Engng. *7*, 132–168.

Voight, B. (1966), *Beziehung zwischen grossen horizontalen spannungen im Gebirge und der Tektonik und der Abstragung*, Ist. Congr. Int. Soc. Rock Mech., Lisbon, 1966, Proc. *2*, 51–56.

(Received 21st September 1976)

Pageoph, Vol. 115 (1977), Birkhäuser Verlag, Basel

Central Europe: Active or Residual Tectonic Stresses

By Gerhard Greiner and J. Henning Illies[1])

Abstract – The regional stress field in the Western Alps and their northern foreland has been investigated by *in situ* stress determinations. More than 600 strain relief measurements were made with resistance strain gages in boreholes carried out in mines, tunnels and quarries. The stresses calculated and data obtained from other papers were used to get a detailed idea of the stress conditions in Central Europe.

The measurements confirm a continuous flux of compressive stress from the Alps to the northern foreland east of the Rhinegraben. The largest stresses are observed in the Central Alps, the lowest in the Rhinegraben rift system. The horizontal stresses exceed at nearly all places the vertical ones. Evidently the excess of horizontal stress is generated by active plate tectonics in the Alps. A tectonic model to explain the observed stress pattern is presented.

Key words: Stress *in-situ*; Intraplate stress field; Tectonics of central Europe.

1. Method

At the test sites the strain relief technique was used to determine the absolute stresses in the interior of the rock masses. This method was primarily developed in South Africa and is known as the 'doorstopper' strain gage device (Leeman 1969; Cahnbley 1970); it is nowadays one of the most frequently used techniques to determine rock stresses *in situ* (Leeman 1970).

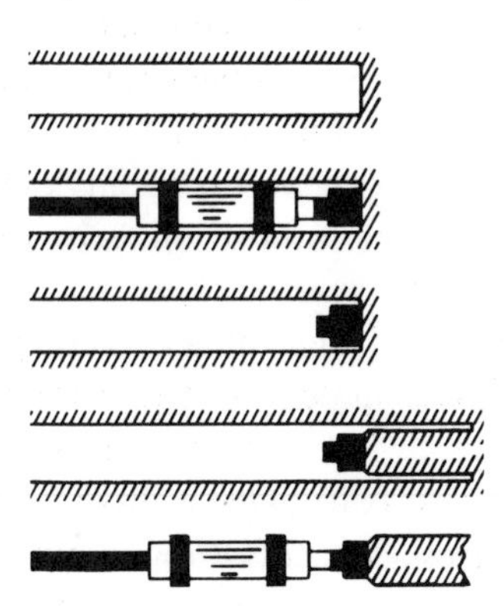

Figure 1
Schematic illustration of the overcoring method for the determination of *in situ* stress by strain relief. Five sequential steps are shown from top to bottom.

[1]) Geological Institute, University, D-7500 Karlsruhe, Kaiserstr. 12, BRD.

The principle of this overcoring method is shown in Fig. 1: A hole is drilled into the rock *in situ* and its end is smoothed, polished and dried up. Then an oriented rosette of resistance strain gages is glued against the end, i.e., in a plane normal to the axis of the borehole. At this time the applied and residual stresses are still present. After initial readings the installation tool is taken away. The strain gage then is overcored. As a consequence of this overcoring procedure the rock element with the strain cell is separated from the surrounding rock, i.e., it is separated from the stresses *in situ*. The result of this operation is a radial straining of the core proportional to the amount of the stress release in the rock. These radial strains are determined by difference gage readings outside the borehole. Using Hooke's law $\sigma = E \cdot \varepsilon$, where σ is the stress, ε the measured strain and E Young's modulus of the rock, it is possible to calculate the stresses acting at the end of the borehole. Considering the stress concentrations at the flattened end of the borehole (Bonnechere, 1972) the virgin rock stresses in the rock mass around the borehole are to be determined.

2. *Locations of measurements – geological setting of the test sites*

Table 1 shows all available data of the test sites:

'A' – 'K' represent measurements done by our working group in a variety of rocks of different tectonic levels, e.g., 'C' and 'D' are measurements in the folded Hercynian basement, 'K' is a series of measurements in the Alpine fold belt. The other test sites are situated in the sedimentary cover of the Rhinegraben flanks or in the graben fill itself.

In order to obtain sufficient data for statistical evaluations, measurements have been performed at 10 different locations between the Alps and the region of Luxembourg. At each location more than 60 strain readings have been taken so that the following investigations are based upon more than 600 separate strain values. The test sites were situated in mines, tunnels, and quarries which were carefully examined to have the optimal conditions with reference to position, geometric feature, rock type, and geologic structure.

The measurements from tunnels were performed continuously in boreholes with increasing depths. The readings have been taken at least up to distances of about 7–8 times the radius of the underground openings, i.e., sometimes 25 m away from the walls of the tunnels. This was necessary for escaping the stress concentrations around these technical operation places. Fig. 2 shows the variation of the stresses with increasing depth in a borehole performed in dioritic rocks: The directions do not vary more than 10 degrees at different measuring points with one exception in the vicinity of a fracture zone at a depth of 6.5 m. The standard deviation of the amounts is about 12% when neglecting the extremely high values near

Table 1

	Location of measurement	Longitude	Latitude	Overburden (m)	Type of rock	Direction of greatest principal stress	Excess of horizontal stress (bars)
A	Nennig	6°23′01″	49°31′23″	—	Middle Triassic limestone	126° ± 6°	23.0
B	Oppenheim	8°21′08″	49°51′51″	—	Tertiary limestone	126° ± 6°	−3.0
C	Auerbach	8°39′06″	49°42′30″	ca. 140	Hercynian diorite	125° ± 12°	24.5
D	Albersweiler	8°01′19″	49°13′21″	—	Precambrian gneiss	75° ± 5°	6.5
E	Wössingen	8°37′00″	49°01′03″	—	Middle Triassic limestone	140° ± 11°	21.5
F	Onstmettingen	9°01′00″	48°17′00″	ca. 25	Upper Jurassic limestone	130° ± 14°	16.5
G	Straßberg	9°05′00″	48°10′27″	ca. 35	Upper Jurassic limestone	152° ± 12°	17.0
H	Bollschweil	7°46′39″	47°55′24″	—	Middle Jurassic limestone	153° ± 19°	8.0
I	Kleinkems	7°31′53″	47°40′40″	—	Upper Jurassic limestone	176° ± 13°	19.0
K	Grimsel	ca. 8°19′	ca. 46°34′	ca. 265	Granodioritic massif	171° ± 11°	160.0
L	Luzern	ca. 8°17′	ca. 47°04′	ca. 90	Molasse sandstone	ca. 104°	ca. 50.0
M	Hochkönig Mts.	ca. 13°06′	ca. 47°26′	ca. 750	Paleozoic greywacke	ca. 160°	ca. 60.0
N	Mont Blanc	ca. 6°54′	ca. 45°53′	ca. 1800	Cryst. schists, Granite	ca. 40°	ca. 360.0
O	Lago Delio	ca. 8°47′	ca. 46°06′	ca. 180	Gneiss polymetamorphic	ca. 140°	ca. 250.0
P	Val Camonica	ca. 10°19′	ca. 46°05′	ca. 230	Phyllite Alpine basement	ca. 165°	ca. 208.0
Q	Gotthard	ca. 8°35′	ca. 46°35′	ca. 550	Granitic massif		ca. 149.0
R	Waldeck	ca. 9°05′	ca. 51°12′	270	Hercynian greywacke		ca. 77.0

A–I, R Alpine foreland; K–Q Alpine fold belt

the fracture zone and 15% when considering these values. Exactly the same results were obtained from boreholes in marbles at the same location.

In the other cases, several measuring boreholes up to depths of more than 10 m were drilled at the floors of the quarries to obtain enough strain data for statistic calculations.

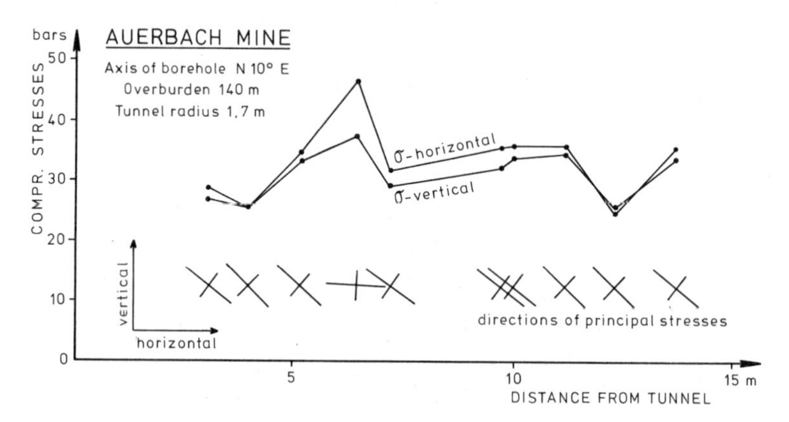

Figure 2
Variation of stresses in a borehole with increasing depth.

'L' represents a series of measurements done by GYSEL (1975) in a tunnel excavated in rocks of the folded subalpine molasse near Lucerne, Switzerland.

'M' From the Eastern Alps stress values are reported by BRÜCKL *et al.* (1975) taken in greywackes of a mine south of Salzburg, Austria.

'N' HAST (1973) measured in the Mont Blanc road tunnel in crystalline schists and granites.

'O' and P' In the Italian Alps measurements were performed by DOLCETTA (1972) in two galleries of hydroelectric power stations. The values here had been obtained in gneisses and phyllites of the polymetamorphic Alpine basement.

'Q' In the Gotthard region (Swiss Alps) KOVÁRI *et al.* (1972) measured in granites around a highway tunnel.

'R' Finally LÖGTERS (1974) reports stress data from the Hercynian basement of the eastern part of the Rhenish Massif. At this location a specially designed measuring instrument in a large borehole of 900 mm diameter was used to record continuously the strain during the drilling operations.

Summarizing it may be stated that in the described area data from surely more than a thousand strain measurements are available.

3. *Results*

(*a*) *Directions of maximum horizontal compression* (*Fig. 3 and Table 1*)

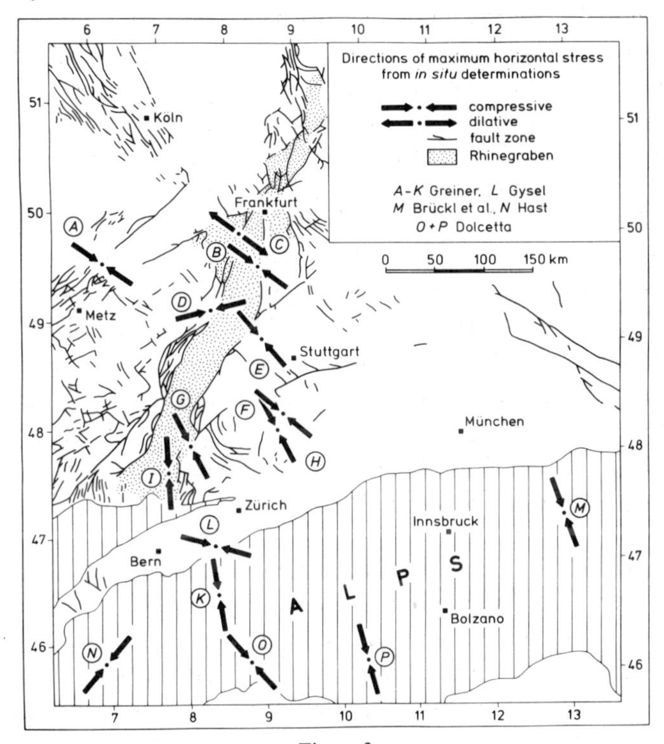

Figure 3

Tectonic map of the Central Alps and the Rhinegraben with stress directions obtained by overcoring.

The results of the series of measurements are plotted on a tectonic map of Central Europe:

First of all there is the Rhinegraben rift system traversing this area in a NNE to SSW direction over a distance of about 300 km from Basel in Switzerland up to Frankfurt in the north. Rhinegraben rifting is still active as indicated by seismic activity (AHORNER 1975), high accumulation rates of Quaternary sediments (BARTZ 1974) and geodetic observations (SCHWARZ 1976). The Rhinegraben probably acts as an active subplate boundary in the regional stress conditions, i.e., as a vertical discontinuity in the sense of continuum-mechanics. The fractures of this rift system go down to a depth of about 7–8 km (DOHR 1957).

The Alpine collision front forms the southern rim of the area investigated. Alpine folding and nappe formation had evolved in Tertiary times. In Recent times a regional updoming is observed up to 1.7 mm/year (Schweizerisches Landesnivellement). Moreover, seismic activity and seismotectonic observations demonstrate an active advance of Alpine orogeny.

Between the southern end of the Rhinegraben and the northern rim of the Alpine fold belt the Jura Mts. intervene. The Pliocene folded anticlines of this mountain range intrude the southern mouth of the Rhinegraben with a $\pm$ NS directed movement.

As mentioned before the measurements had been performed in different rock types of different ages and of different tectonic structure. On this account it is surprising to see that the map of maximum compressive stresses (Fig. 3) shows at most of the test sites a preferred NW–SE to NNW–SSE direction of σ_1.

In the Alps, the maximum compressive stress direction is mostly NNW to SSE. Different from that, HAST (1973) reports a NE–SW trend of σ_1 from the Mont Blanc tunnel. This direction seems to have not only local character for it is confirmed by the values obtained from fault plane solutions of this region (BLUM *et al.* 1976).

More to the north GYSEL (1975) reports a WNW to ESE direction for the maximum compressive stress near Lucerne, Switzerland. These measurements were performed at two test sites in a tunnel crossing the southern flank of an anticline, about 90 m below surface. It is remarkable that the directions of the largest horizontal stresses, that are in both cases the largest principal stresses σ_1 also, are equal, whereas their magnitudes differ widely. The extraordinary σ_1 directions may be influenced by local effects, e.g., release of stress by the topography around the test sites (GYSEL 1975).

The region north of the Alps is characterized by a relatively uniform distribution of the σ_1 directions also. In the southernmost part of the Rhinegraben σ_1 directions are observed in the range of about 160–180°, similar to the results in the adjacent parts of the Alps. Going more northward a systematical anticlockwise rotation of the σ_1 directions may be recognized. Regardless the values in the southernmost parts of the Rhinegraben the σ_1 directions demonstrate the same general trends as calculated before by fault plane solutions of earthquakes (AHORNER 1975).

At the western rim of the northern segment of the Rhinegraben fluctuating values had been stated. Obviously the largest amounts of the Alpine stresses are consumed by stress release along the active rift of the Rhinegraben. The nearly EW trend as measured in Albersweiler (location 'D') is in good agreement with active dip slip faulting observed in the same area (ILLIES and GREINER 1976).

(*b*) *Amounts of horizontal stresses* (*Fig. 4 and Table 1*)

Contrary to the approximately uniform distribution of the directions the amounts of the stresses show larger variations. As reported in the literature, the scatter of the total amounts is much wider than that of the directions (BONNECHERE 1969; CAHNBLEY 1970). The accuracy of the measurements is normally about $\pm 15\%$. This fact has to be considered when interpretating *in situ* stress data.

Nevertheless it is possible to point out some systematic features in the stress distributions of Central Europe. In order to compare values of horizontal stresses measured at different depths, the excess of horizontal stress σ_{ex} was calculated as the

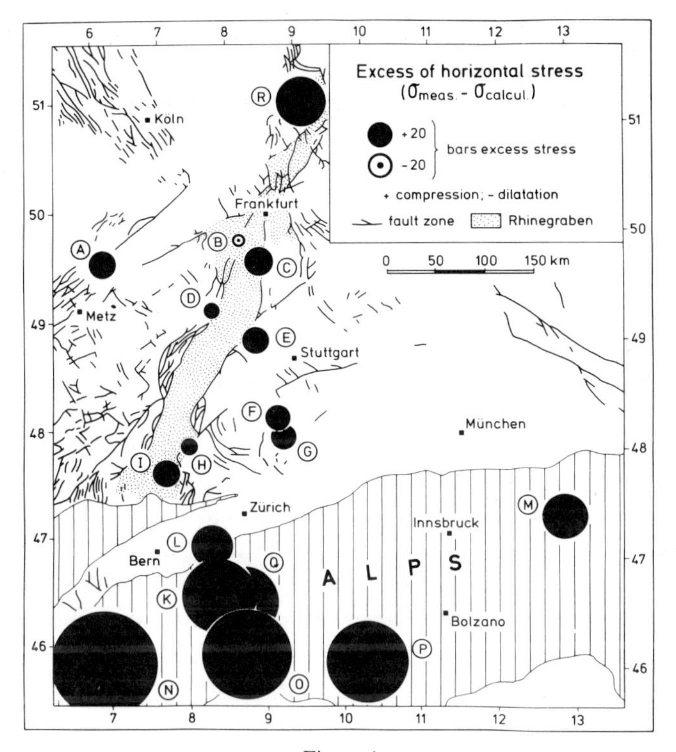

Figure 4
Tectonic map of the Central Alps and Rhinegraben with the magnitudes of the excess horizontal stress.

difference of the measured horizontal stress σ_{meas} and the horizontal stress σ_{hor} as calculated by the following equations:

$$\sigma_{ex} = \sigma_{meas} - \sigma_{hor} \tag{1}$$

and

$$\sigma_{hor} = \sigma_{vert} \frac{\mu}{1 - \mu} \tag{2}$$

hence

$$\sigma_{ex} = \sigma_{meas} - \sigma_{vert} \frac{\mu}{1 - \mu} \tag{3}$$

Where σ_{hor} is the theoretical horizontal stress (Poisson's stress), σ_{vert} is the vertical stress as caused by the weight of the overburden (gravitational stress or lithostatic stress) and μ is Poisson's ratio.

The validity of equation (2) is sometimes questioned, because the results of calculations using it are sometimes not in agreement with the stresses measured. The superposition of tectonic and residual stresses on the gravitational stresses produces this difference and hence this is a criterion for the excess of horizontal stress. Figure 4 shows the excess horizontal stress plotted on the tectonic map. The sizes of the black circles are proportional to the excesses of horizontal stress. It can clearly be seen that a systematic geographical distribution of the stresses is present. In the northern foreland of the Alps there are relatively small stresses observed. Nevertheless it has to be stated that the stresses are (with one exception) all compressive and at all places higher than expected. A minimum is observed in the region of the Rhinegraben – away from its flanks the values tend to increase. The graben acts as a vertical discontinuity where there is a stress consumption to be noted. From the northern rim of the Alpine fold belt up to the Central Alps a systematic increase of the horizontal stresses is observed. A similar increase of the stresses towards the northern rim of the Alpine fold belt may be documented by a systematic increase of the seismic velocities in the molasse trough north of the Alps (LOHR 1969), by extremely strong deformations of the galleries of lignite mines south of Munich (HEISSBAUER 1975) and by abnormal high formation pressures in some oilfields in the molasse basin (LEMCKE 1973).

It is worth mentioning that the regions with the highest rates of seismicity e.g., Rhinegraben and Hohenzollerngraben (about 60 km south of Stuttgart) are not characterized by the presence of abnormally high compressive horizontal stresses. This points to the fact that in the region investigated, areas of high local seismicity act in the sense of stress release and not of stress accumulation.

4. Residual or active tectonic stresses?

The *in situ* state of stress consists in most cases of a superposition of several different kinds, e.g., gravitational stresses (lithostatic stresses) due to the weight of the overburden, topographical stresses as created by the morphology, stresses induced by works of man (stress concentrations around underground openings), tectonic stresses as a result of recent tectonics and residual stresses due to events in geologic history. When interpreting *in situ* stress data, the first three kinds mentioned above may easily be filtered out from the *in situ* data. The separation of residual and tectonic stresses is more complicated and needs specific investigations.

Residual stresses are usually defined as stresses which exist in the Earth's crust as residues from former tectonic deformations (MÜLLER 1963). Measured residual stresses are known to be related to Tertiary, Mesozoic, Palaeozoic and even Precambrian tectonic events (e.g., EISBACHER and BIELENSTEIN 1971; LEE and NICHOLS 1972; FRIEDMAN 1972). VOIGHT (1966) gives a more general definition of the residual stresses: 'Residual stresses are systems of stresses on the inside of a body which are in equilibrium, or approach to equilibrium, when neither normal nor shear stresses are transmitted through its exterior surfaces.' Residual stresses are therefore stresses that remain in a body, when all external load is taken away. They superimpose stresses conditioned by external forces. Voight's definition implies the complication to fix the limit of such a geologic domain where the residual stresses exist: Is such a domain in the scale of crystals as shown by X-ray diffraction studies of FRIEDMAN (1972), is it in the scale of a quarry as reported by LEE and NICHOLS (1972) or is it finally in the scale of larger geologic units as proved by ENGELDER and SBAR (1976) in the northeastern part of the United States?

For determining residual stresses two special investigations have been performed:

(a) Measurements in different rock types, of different ages and tectonic development, all of the same region. As mentioned above the *in situ* state of stress consists of a superposition of applied and residual stresses (FRIEDMAN 1972). The stress field didn't remain constant in amount and direction in geologic time (e.g., HERGET 1972; ILLIES 1974; GREINER 1975). Rocks from different tectonic levels within one and the same region are expected to show a wide scatter of directions and amounts due to different amounts of residual stresses. The applied tectonic stresses operate uniformly to all rocks in a region and will not produce formation-specific differences in amounts and directions.

The results of the measurements of the area in question do not show substantial variations in orientations and amounts: They are in the range of $\pm 10°$ respectively to the directions and of about 15% in the amounts what is the specific scattering of the results when using the strain relief technique. This implies that there is no significant influence of residual stresses recorded although the ages of the rocks vary from Tertiary to Precambrian. If residual stresses are present they have to be very small compared with the applied tectonic stresses. Clearly the majority of the observed stresses in Central Europe are derivable from the Alpine fold belt: Here the 'generator' of the active stresses is situated as seen by the large amounts of the horizontal stresses (Fig. 4).

(b) 'Double overcoring' (Fig. 5). A second experiment for detecting residual stresses is 'double overcoring':

An overcoring procedure with a large bit is performed to unbound a rock element with a strain gage from the surrounding rocks. This is eliminating applied stresses and a part of the residual stresses in the rock element. Inside of the core another smaller one is drilled to disturb the equilibrium of the internal residual stresses. What is expected is an additional second strain caused by the second overcoring procedure. ENGELDER and SBAR (1976) used a similar technique for measurements in the floor of quarries. They concluded that a major component of the *in situ* strains was residual. These residual strains had different orientations than the primarily measured superimposed *in situ* stresses. NICHOLS (1975) performed a similar experiment to prove the significance of the residual stresses. In a large block of granite cut from the floor of a quarry to remove boundary loads he found very high residual stresses (several hundreds of bars) that varied considerably in orientations and amounts.

A typical time-strain diagram of a double overcoring procedure in Triassic limestones at test site 'A' is presented in Fig. 6. The strain recorded by four strain gages is plotted to the time of the overcoring procedure. As being observed most of the strain is relieved at the first stage of the recording time, a short time later a total

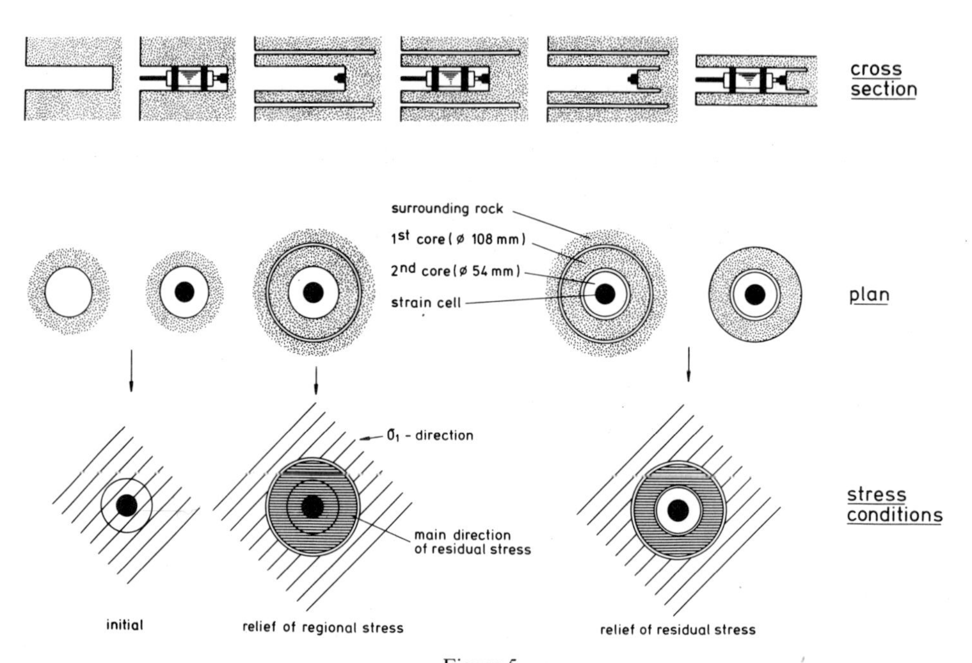

Figure 5
Schematic illustration of the double overcoring techniques used for the identification of the residual and tectonic components of the *in situ* stress.

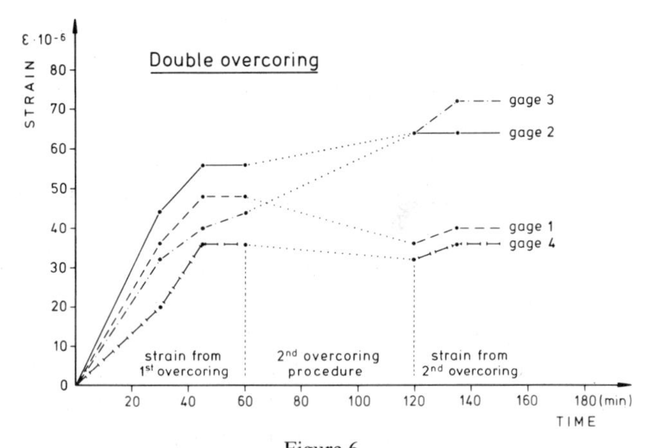

Figure 6
Strain relief resulting from the double overcoring technique at test site 'A'.

strain relief is reached (documented by the horizontal branch of the curve). A second overcore procedure does not furnish a significant second part of strain. This is evidence for the absence of residual stresses in the rocks measured (gage 3 was defect).

Following the considerations above it seems to be proved that the dominant parts of the excess stresses measured in Central Europe are not residual stresses but of active tectonic origin. Possibly future measurements in the basement with the double overcoring method may bring more evidence for residual stresses. The dominant active tectonic stresses are responsible for the recent tectonic activity and seismicity of the region (ILLIES, 1975).

5. *Examples of geologic data interpretation*

Western Alps and Rhinegraben (*Fig. 7*)

From the Central Alps where the excess of horizontal stress culminates, down to the northern foreland a decreasing gradient of tectonic stresses may be observed, similar to the morphological slope of the mountain range. The amounts of horizontal stresses in the foreland appear comparatively low. As shown by geodetic observations, wide parts of the mountain belt are actually under regional uplift producing annual rates up to about 1.7 mm. In eastern Switzerland the isobases of recent updoming (PAVONI 1975) are oriented about normal to the directions of the greatest principal stress. Most of the measured σ_1 directions follow the NW–SE trend as calculated before by fault plane solutions of earthquakes (AHORNER 1975). South of the Rhinegraben this general trend appears modified by a convergent turn towards the southern end of the graben. The crustal unit north of the Alps is provided with a strongly consolidated basement of Hercynian age. The Rhinegraben forms a weakness point within the rigid abutment of Alpine plate collision, predestinated to attract strain release of Alpine stresses. This regional effect of the Rhinegraben subplate boundary superimposes the general stress field as generated by Alpine plate tectonics its direction of maximum compression trends about NW–SE over wide parts of Central Europe.

Alpine folding and nappe formation came to cease in Pliocene times. The tectonosphere of the mountain range as being in a state of substantial consolidation reacted upon the recent advance of plate collision and lithospheric thickening by the described dome-like upbuckling. Instead of the consolidated fold belt still mobile zones of the foreland absorbed now the progression of crustal shortening. The Rhinegraben which had been primarily formed as a tensional rift valley in mid-Eocene to Lower Miocene times, had been reactivated to a shear belt since fluctuating Alpine stresses met this zone as fitting the sinistral shear component of the regional stress field (ILLIES and GREINER 1976). An Upper Pliocene to Recent fault pattern allocated to this and the conjugate dextral shear mediates between the southern segment of the Rhinegraben and the adjacent part of the Alps in Switzerland. The so localized

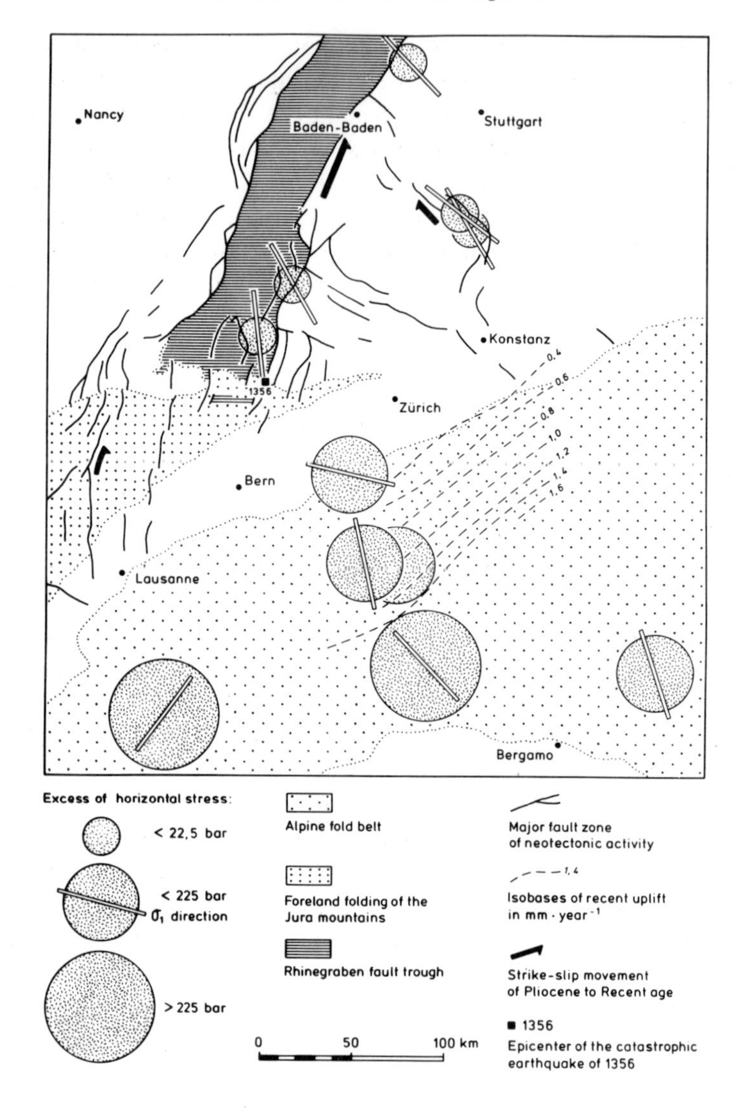

Figure 7
Tectonic map of the Central Alps and the southern Rhinegraben with measured stress directions and magnitudes.

yielding of the foreland north of the collision front may be illustrated by the distribution of earthquake epicenters. Dense clusters of historical seismic events accentuate the crustal mobility at the Baden-Baden – Lausanne – Konstanz triangle. The most catastrophic earthquake in Europe north of the Alps was that of Basel in 1356, located in the very center of this unit.

Rhinegraben and Lower Rhine embayment (Fig. 8)

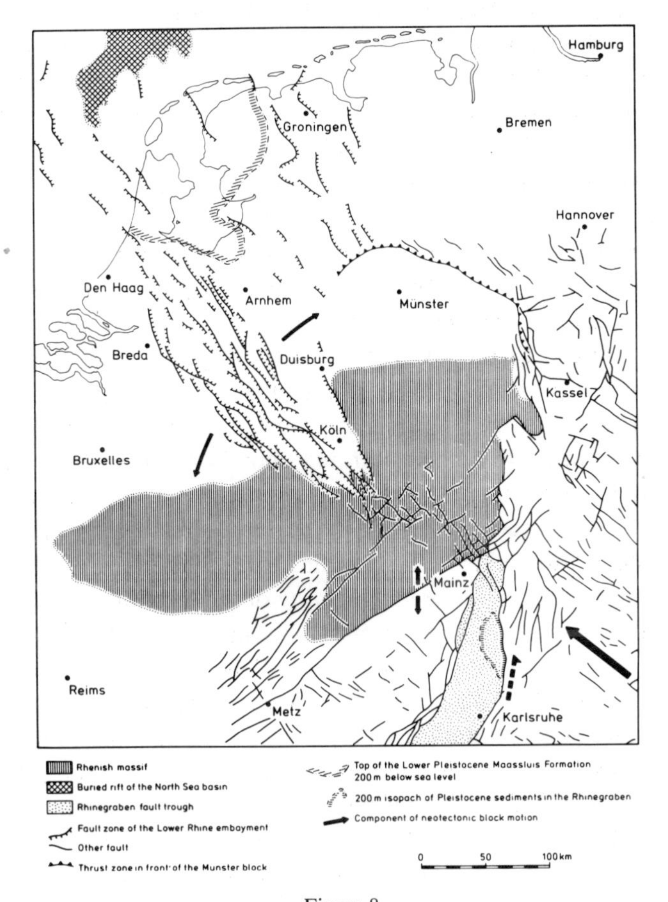

Figure 8
Tectonic map of the northern Rhinegraben and the Lower Rhine embayment with directions of stress induced block motions.

With an about 38 km wide overlap Fig. 8 describes the neotectonic behavior of the block pattern in the area attached northward to Fig. 7. Active subsidence and faulting of the Rhinegraben will be mainly controlled by a sinistral shear motion that appertains to the regional stress conditions measured along the eastern frame of the graben. Since the appropriate shift of the South German block unit is conditioned by Alpine plate tectonics, the mobility and seismicity of the rift belt modifies the stress flux by strain release. Consequently, the western rim of this part of the graben appears nearly free of horizontal compression; it is found to be in the shade of stresses generated by Alpine plate tectonics (ILLIES and GREINER 1976). AHORNER (1975) calculated the seismic slip rate of the graben to be about 0.05 mm/year. Near Frankfurt the active slide-bar mechanism is interrupted since the Hercynian block of the Rhenish massif had conditioned a lateral offset of the primary northward continuation of the Paleogene rift valley, From this point onward the horizontal motion of the South German block will be transmitted immediately to the adjacent Rhenish massif (ILLIES and GREINER 1976). As conditioned by the individual displacements of the block units concerned old lineaments along the southern rim of the Rhenish massif broke up. This is shown by fault plane solutions of local earthquakes (AHORNER and MURAWSKI 1975) and by dislocations of Pleistocene river terrasses.

A continuous belt of seismic epicenters mediates between the Rhinegraben and the fault trough of the Lower Rhine embayment, traversing the Rhenish massif between Mainz and Köln. An about 135° average trend of σ_1 has been calculated by fault plane solutions of earthquakes (AHORNER 1975). A framework of stress controlled neotectonic ruptures illustrates how geologically the fault action came about (STENGEL-RUTKOWSKI, 1976). Volcanic eruptions of Quaternary age in the Eifel and the Neuwied area appear to be concentrated upon NW–SE striking lines. The axis of Pleistocene to Recent rifting of the Lower Rhine embayment, cutting the northern part of the massif, is in accordance with the regional stress regime. The rift segment exhibits a fan-shaped widening with spreading rates that increase towards the NW. This implies a corresponding divergent yielding of the framing parts of the massif.

The observed decay of the Rhenish massif by active rift valley propagation could be explained by the role of the Central graben of the North Sea basin which sets in immediately north of the Lower Rhine embayment (HEYBROEK 1974). This unit remained in a state of an extinct rift valley since Paleocene times. But its weakened basement as well as the underconsolidation of its thick sedimentary infill had turned this belt into a crushable zone which is introduced into the northern abutment of the Rhenish massif. Hereby the massif, since being under a NW–SE directed compression, has been forced to give way by the observed outward distortion of the rims of the Lower Rhine embayment. The downwarped thickness contours of Pleistocene strata surrounding the southern end of the Central graben illustrate, how this buried rift had been affected. Below the island of Terschelling the base of the Pleistocene is found at a maximum depth of about 600 m. The contours of the top layer of the Lower Pleistocene Maassluis Formation (VAN MONTFRANS 1975) demonstrate a funnel-shaped basin that intervenes between the passive (northern) and the active (southern) segment of a quasi continuous rift belt.

REFERENCES

AHORNER, L. (1962), *Untersuchungen zur quartären Bruchtektonik der Niederrheinischen Bucht*, Eiszeitalter und Gegenwart, *13*, 24–105.

AHORNER, L. (1975), *Present-day stress field and seismotectonic block movements along major fault zones in Central Europe*, Tectonophysics *29*, 233–249.

AHORNER, L. and MURAWSKI, H. (1975), *Erdbebentätigkeit und geologischer Werdegang der Hunsrück-Südrand-Störung*, Z. deutsch. geol. Ges. *126*, 63–82.

BARTZ, J., *Die Mächtigkeit des Quartärs im Oberrheingraben* in *Approaches to Taphrogenesis* (ed. H. Illies and K. Fuchs) (Schweizerbart, Stuttgart, 1974), p. 78–87.

BLUM, R., BOCK, G., MERKLER, G. and FUCHS, K. (1976), *Beobachtung von Seismizität an Stauseen und Untersuchung ihrer Ursachen*, Sonderforschungsbereich 77 Felsmechanik, Universität Karlsruhe, Jahresbericht 1975, p. 135–152.

BONNECHERE, F. (1969), *A comparative field study of rock stress determination techniques*, U.S. Army Corps of Engineers, Missouri River Division, Omaha, Nebraska, Technical Rep. 1–69, p. 1–56.

BONNECHERE, F. (1972), *Contrainte dans la région centrale d'un fond plate de forage*, Luzern, Int. Sympos. für Untertagebau, p. 447–456.

BRÜCKL, E., ROCH, K. H. and SCHEIDEGGER, A. E. (1975), *Significance of stress measurements in the Hochkönig Massif in Austria*, Tectonophysics *29*, 315–322.

CAHNBLEY, H., *Grundlagenuntersuchungen über das Entspannungsbohrverfahren während des praktischen Einsatzes in großer Teufe*, (TU Clausthal, Diss. 1970), 166 p.

DOHR, G. (1957), *Zur reflexionsseismischen Erfassung sehr tiefer Unstetigkeitsflächen*, Erdöl und Kohle *10*, 278–281.

DOLCETTA, M. (1972), *Rock load on the support structures of two large underground hydroelectric power stations*, Luzern, Int. Sympos. für Untertagebau, p. 406–446.

EISBACHER, G. H. and BIELENSTEIN, H. U. (1971), *Elastic strain recovery in Proterozoic rocks near Elliot Lake, Ontario*, J. Geophys. Res. *76*, 2012–2021.

ENGELDER, J. T. and SBAR, M. L. (1976), *Evidence for uniform strain orientation in the Potsdam Sandstone, northern New York, from in situ measurements*, J. Geophys. Res. *81*, 3013–3017.

FRIEDMAN, L. (1972), *Residual elastic strain in rocks*, Tectonophysics *15*, 297–330.

GREINER, G. (1975), *In situ stress measurements in southwest Germany*, Tectonophysics *29*, 265–274.

GYSEL, M. (1975), *In situ stress measurements of the primary stress state in the Sonnenberg tunnel in Lucerne, Switzerland*, Tectonophysics *29*, 301–314, and *31*, 307.

HAST, N. (1973), *Global measurements of absolute stress*, Phil. Trans. R. Soc. London *A274*, 409–419.

HEISSBAUER, H. (1975), *Die Gebirgsmechanik beim Abbau in großer Teufe des Kohlenbergwerkes Peißenberg und ihre Auswirkungen auf die Bergtechnik*, Geologica Bavarica *73*, 37–53.

HERGET, G. (1972), *Tectonic fabric and current stress field at an iron mine in the Lake Superior region*, 24th IGC Montreal, Sect. 13, p. 241–248.

HEYBROEK, P. (1974), *Explanation to tectonic maps of the Netherlands*, Geologie en Mijnbouw *53*, 43–50.

ILLIES, H. (1974), *Intra Plattentektonik in Mitteleuropa und der Rheingraben*, Oberrhein. geol. Abh. *23*, 1–24.

ILLIES, J. H. (1975), *Intraplate tectonics in stable Europe as related to plate tectonics in the Alpine system*, Geol. Rundsch. *64*, 677–699.

ILLIES, H. and GREINER, G. (1976), *Regionales stress-Feld und Neotektonik in Mitteleuropa*, Oberrhein. geol. Abh. *25*, 1–40.

KOVÁRI, K., AMSTAD, CH. and GROB, H. (1972), *Ein Beitrag zum Problem der Spannungsmessung im Fels*, Luzern, Int. Sympos. für Untertagebau, p. 501–512.

LEE, F. T. and NICHOLS, T. C., JR. (1972), *Some effects of geologic structure, engineering operations and thermal changes on rock-mass behavior*, 24th IGC Montreal, Sect. 13, p. 261–272.

LEEMAN, E. R. (1969), *The 'Doorstopper' and triaxial rock stress measuring instruments developed by the CSIR*, J. S. Afr. Inst. Min. Metall. *69*, 305–339.

LEEMAN, E. R. (1970), *Experience throughout the world with the CSIR 'Doorstopper' rock stress measuring equipment*, Proc. 2nd Cong. Int. Soc. Rock Mech. *3*, 419–425.

LEMCKE, K. (1973), *Zur nachpermischen Geschichte des nördlichen Alpenvorlandes*, Geologica Bavarica, *69*, 5–48.

LÖGTERS, G. (1974), *Interfels-Meßstern zur Bestimmung der Spannungsverteilung in situ*, Interfels Messtechnik Information, 1974, p. 33–35.

LOHR, J. (1969), *Die seismischen Geschwindigkeiten in der jüngeren Molasse im ostschweizerischen und deutschen Alpenvorland*, Geophys. Prosp. *17*, 111–125.

van MONTFRANS, H. M., *Toelichting bij de ondiepe breukenkaart met diepteligging van de Formatie van Maassluis* in *Toelichting bij geologische overzichtskaarten van Nederland*, (ed. W. H. Zagwijn and C. J. van Staalduinen, Rijks Geologische Dienst, Haarlem, 1975), p. 103–108.

MÜLLER, L., *Der Felsbau*, Enke, Stuttgart, 624 p.

NICHOLS, T. C. (1975), *Deformations associated with relaxation of residual stresses in a sample of Barre granite from Vermont*, U.S. Geol. Surv. Prof. Pap. 875, p. 1–35.

PAVONI, N. (1975), *Working group 1: Recent Crustal Movements*, International Geodynamics Project, First Report of Switzerland, p. 3–17.

SCHWARZ, E. (1976), *Präzisionsnivellement und rezente Krustenbewegung dargestellt am nördlichen Oberrheingraben*, Z. f. Vermessungswesen *101*, 14–25.

SCHWEIZERISCHES LANDESNIVELLEMENT (1976), *Rezente Krustenbewegungen*, Eidg. Landestopographie, Sekt. Nivellement.

STENGEL-RUTKOWSKI, W. (1976), *Idsteiner Senke und Limburger Becken im Licht neuer Bohrergebnisse und Aufschlüsse (Rheinisches Schiefergebirge)*, Geol. Jb. Hessen, 104, p. 183–224.

VOIGHT, B. (1966), *Restspannungen im Gestein*, Proc. 1st Congr. Int. Soc. Rock Mech. *2*, 45–50.

(Received 17th November 1976)

Pageoph, Vol. 115 (1977), Birkhäuser Verlag, Basel

A Mechanism for Strain Relaxation of Barre Granite: Opening of Microfractures[1])

By TERRY ENGELDER,[2]) MARC L. SBAR[3]) and ROBERT KRANZ[2])

Summary – Strain relaxation in the Barre Granite and surrounding metasediments in Vermont, was measured by overcoring strain gauge rosettes bonded to outcrop surfaces. The average maximum expansion upon relieving 15.2 cm diameter cores trends N55°W, while the average maximum expansion of 7.6 cm diameter cores coaxial with 15.2 cm cores trends N70°W. The maximum strain relief of the internal overcores is normal to the microfracture fabric. Therefore, the mechanism of strain relaxation is attributed to the opening of microfractures either parallel to the rift direction of the Barre Granite or parallel to the foliation of the metasediment. The lack of parallelism between the normal to the rift plane and the maximum expansion of the initial overcore suggests an externally applied strain superimposed on the strain caused by opening of microfractures.

Key words: Stress *in-situ*; Microfracture strain; Strain relaxation; Residual strain.

1. Introduction

The strain relaxation data presented in this paper come from one of nine suites gathered in various areas of the northeastern United States (Fig. 1). In addition to a companion paper (ENGELDER and SBAR [1]) and ENGELDER and SBAR [2] other data suites will be published in the near future. In gathering these data we hope to evaluate surface strain relief techniques as a means of measuring stress near the surface of the crust. An initial step in the evaluation of surface strain relief data is to understand the rheological processes occurring during overcoring. Out data from Barre, Vermont, are relevant to this problem.

The relaxation of *in situ* strain which can be observed during overcoring experiments amounts to more than the simple rebound of an isotropic elastic body cut free from its boundary loads. Several relaxation mechanisms contribute either to components of instantaneous or time-dependent recovery of *in situ* strain (VARNES and LEE [3], NICHOLS and SAVAGE [4]). These mechanisms may be either elastic or inelastic; the latter may occur instantaneously or as a function of time. An important characteristic of *in situ* strain is that it can also be recovered from a rock free of boundary loads. This phenomenon can be attributed to 'residual' stresses locked in a rock during its burial. VOIGHT [5] recognized that rocks as well as metals contain

---

[1]) Lamont-Doherty Geological Observatory Contribution No. 2455.

[2]) Lamont-Doherty Geological Observatory, Palisades, New York 10964.

[3]) Now at: Department of Geosciences, University of Arizona, Tucson, Arizona 85721, USA.

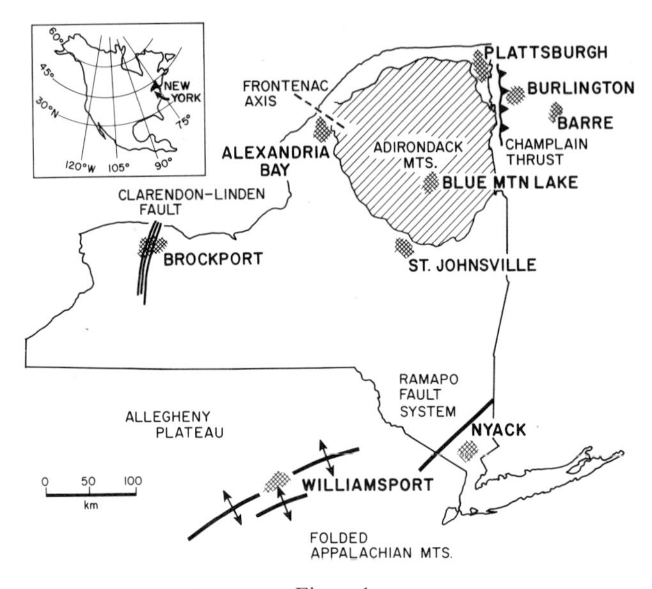

Figure 1
Nine localities, indicated by hatch marks, where *in situ* strain has been measured.

these residual stresses or 'systems of stresses on the inside of a body which are in equilibrium, or approach equilibrium, when neither normal nor shear stresses are transmitted through its exterior surfaces'.

Complications arise from using the strain relaxation of rocks for the purpose of stress measurements because of the difficulty of distinguishing between residual strain and strains caused by tectonic forces which are boundary loads. Moreover, residual strain may mimic elastic strains caused by boundary loads. However, residual strains may contain useful information concerning the tectonic history of the rock (VOIGHT and ST. PIERRE [6]).

Residual strains have been detected using techniques for measuring biaxial strain parallel to free faces of rock (HOSKINS *et al.* [7], FRIEDMAN [8], SWOLFS *et al.* [9], [10], NICHOLS [11], ENGELDER and SBAR [2]). GALLAGHER *et al.*'s [12] photoelastic experiments and SWOLFS *et al.*'s [9], [10] experiments in Cedar City quartz diorite suggest the complex nature of residual strains in rocks and demonstrate that upon cutting a body containing residual strains the largest strain changes occur near the fresh surface. Likewise, NICHOLS' [11] experiments with a block of Barre Granite demonstrate that surficial strains occur on all surfaces during the cutting of fresh surfaces into a rock containing residual strains.

Residual strain is not a universally observed phenomenon in rocks which are free at their boundaries. Using the borehole deformation gauge, HOOKER and DUVALL [13] report an absence of any stress field in a freed slab of porphyritic granite gneiss at Arabia Mountain, Georgia. With the same gauge, MORGAN *et al.* [14] failed to detect any relaxation in a free block of trona from Westvaco, Wyoming. In such cases it is significant that an attempt was made to recover a strain inside a freed block rather than at or near the surface of the block. Using the CSIR strain cell which detects surface strains at the bottom of a borehole, GREINER [personal communication, 1976] did not measure a residual strain in Mesozoic limestones north of the Alps.

In this paper, we present further data from the strain relaxation of Barre Granite and surrounding metasediments in the vicinity of Barre, Vermont. HOOKER and JOHNSON [15] used the Bureau of Mines borehole deformation gauge to measure stress in the Barre Granite. Experience with Barre Granite by NICHOLS [11] suggested that we might detect a residual strain at the surface of Barre Granite.

2. *The Barre Granite*

The Barre Granite intrudes a quartzite-phyllite containing Middle Silurian to Lower Devonian fossils. A standard view is that the phyllite is the Gile Mountain Formation after WHITE and JAHNS [16]. This phyllite was subjected to staurolite-zone metamorphism and folded during the Middle Devonian Acadian orogeny. The Barre Granite belongs to the Late Devonian Plutonic Series (PAGE [17]) which includes a group of late post-tectonic plutons, differentiated at depth and emplaced in part by shouldering aside their walls and in part by stoping. In the vicinity of the Barre Granite quarries, the Gile Mountain Formation has a N20°E striking foliation which dips steeply to the west. The Barre Granite's rift direction strikes about N30°E, almost parallel to the foliation of the surrounding metasediments.

The modal composition of Barre Granite is: quartz – 25%; potash feldspar – 20%; plagioclase – 35%; biotite – 9%; muscovite – 9%; and accessories – 2% (CHAYES [18]). The granite has marked mechanical anistropy which parallels the rift plane and correlates with the preferred orientation of microfractures and fluid inclusions [DOUGLAS and VOIGHT [19]). Barre Granite has an orthorhombic symmetry delineated by sonic velocities; the fastest direction is parallel to the rift plane (BUR *et al.* [20]). Residual strains have been observed on the microscopic scale by X-ray techniques (FRIEDMAN [8]) and on a macroscopic scale by overcoring relaxation techniques (NICHOLS [11]). The granite contains large strains relieved during quarry operations with maximum expansion normal to the rift plane (WHITE [21]) and microfracture fabric.

3. *Strain relief of Barre Granite*

Experimental procedure

In situ strain was measured by overcoring three component (60°) foil-resistance strain gauge rosettes bonded to the horizontal rock surface (FRIEDMAN [8], BROWN [22], SWOLFS *et al.* [10], ENGELDER and SBAR [2]). Strain relaxation accompanied the cutting of a vertical 15.2 cm diameter core to a depth of 20 cm using a diamond masonry bit. A 7.6 cm diameter core coaxial to the 15.2 cm core was cut to further relax residual strain (SWOLFS *et al.* [9], NICHOLS [11]).

Strain relaxation following relief of the rock and thermal effects are primarily responsible for total strain changes recorded over a period of days. Thermal effects can occur because the self compensation of a strain gauge for a temperature change is rarely perfect. However, by using a switching box, a number of strain gauges can be balanced against the same compensation gauge in the second arm of a half bridge. When bonded to a rock, these gauges have approximately the same thermal drift. To further reduce thermal contamination of the strain data, we subtract the thermal strain changes of a compensated, but unovercored, gauge (or the average of several gauges) from the strain changes of each overcored gauge. To obtain enough readings for a reliable correction we monitor the gauges for several days before and after overcoring. In practice this correction for thermal drift is not perfect but minimizes thermal effects so that we can resolve a strain relaxation of $\pm 10\ \mu\varepsilon$ ($1\ \mu\varepsilon$ = a strain of 10^{-6}).

Three outcrops of Barre Granite were selected for strain relief measurements (Fig. 2). The outcrop labelled Hilltop was located at a distance of more than one

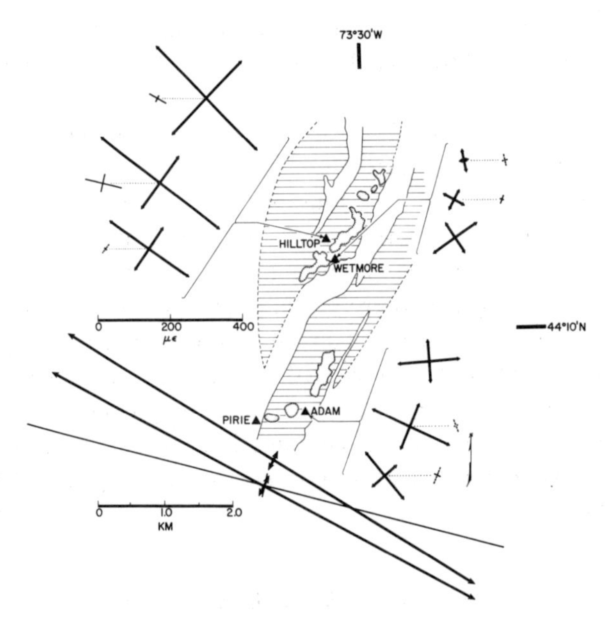

Figure 2

Geology and *in situ* strain in the vicinity of the Barre Granite quarries, Vermont. Outcrops of Barre Granite are indicated by hatch marks. A Paleozoic metasediment surrounds the granite. Four sample sites are named. The magnitude and orientation of the strain relieved by the initial overcore are indicated by dark arrows. The magnitude and orientation of the strain relieved by an internal overcore are indicated by lighter lines. Solid lines represent expansion and dashed lines represent contractions. A scale for the magnitude of the relieved strain is given in microstrain ($\mu\varepsilon$). Quarries are shown within hatch marks.

quarry depth (80 m) from the lip of the Smith quarry, whereas the one labelled Wetmore was 20 m from the lip. The outcrop labelled Adam was located 150 meters from the lip of the Adam quarry.

The strain relaxation of Barre Granite

The greatest strain relaxation occurred at Hilltop where the orientation of maximum expansion for three measurements clustered within 5° of N50°W. At Adam, the orientation of maximum expansion varied up to 28° about N68°W. The lowest strain was recovered from Wetmore where the orientation of maximum expansion seemed random. The orientation of maximum expansion was more consistent among measurements at places where larger strains were relieved. In areas of smaller strain relief the magnitude of the relieved strain varied considerably.

At Hilltop the orientations of maximum expansion obtained by the external and internal overcore were within 25° of each other, but the magnitude of the internal strain relief was at least 75% smaller than the initial overcore. With the exception of one measurement from Hilltop, all internal overcores were performed within 3 days of the initial external overcore. At Adam, the orientation of maximum expansion

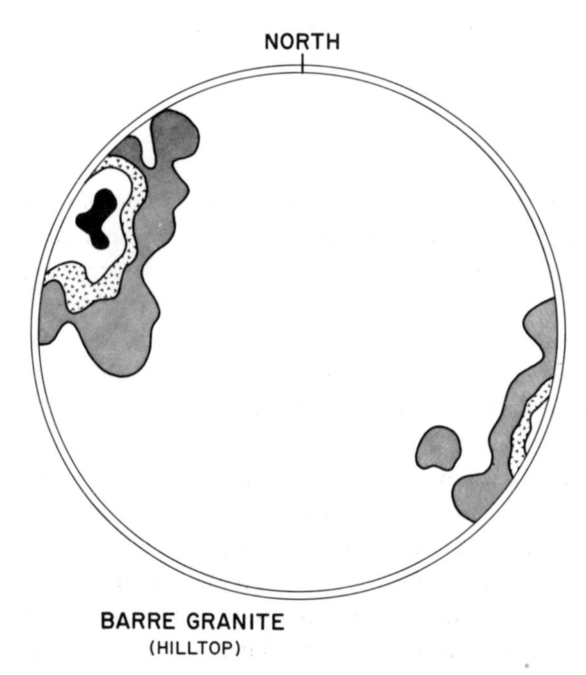

Figure 3

Orientation of 100 poles to microfractures within quartz grains of Barre Granite from Hilltop. Plane of each diagram is the horizontal surface of the outcrop on which *in situ* strain was measured; data plotted in equal-area, lower hemisphere projection. Microfractures are measured only in a thin section cut parallel with the horizontal plane of the outcrop. Contours are at 15%, 10%, 5%, 1% per 1% area.

following the internal overcore differed as much as 41° from that of the initial overcore and the strain was either a small expansion or contraction. The orientation of maximum expansion after the internal overcore at Wetmore was almost normal to the initial maximum expansion and the magnitudes of the axes of the strain ellipse were all less than 25 $\mu\varepsilon$ in either expansion or contraction.

The rift of the Barre Granite strikes N30°E (DOUGLAS and VOIGHT [19]). The average of three initial overcores at Hilltop shows a maximum expansion oriented approximately N50°W, which is 10° more northerly than the normal to the preferred orientation to microfractures (Fig. 3). However, the bisector of the angle of the two most divergent orientations of maximum expansion of the internal overcore, N63°W, is within 3° of the normal to the strike of the rift plane. At Adam, the average of the directions of maximum expansion was about normal to a poorly defined rift plane oriented between N30°E and N39°E. Finally, we observed no relation between the orientation of maximum expansion at Wetmore to the rift plane cited in the literature.

Strain relief of two cores at Pirie in the Gile Mountain Formation consisted of a maximum expansion of 1292 and 1342 $\mu\varepsilon$ with an orientation of N69°W and N73°W (Fig. 2). Relief following one internal overcore was larger (1500 $\mu\varepsilon$) than the initial overcore with axes of the strain ellipse oriented 10° counterclockwise from those of the initial strain relief. The maximum expansion of the internal overcore was approximately normal to the strike of the foliation of this metasediment.

4. *Laboratory analysis*

We subjected our cores to mechanical tests and petrographic analyses to define a mechanical anisotropy in the plane in which strain relaxation was measured. Information on the mechanical anisotropy and data from the internal overcore yield clues concerning the mechanisms contributing to the strain relaxation of the Barre Granite.

Open microfractures and closed microfractures outlined by fluid inclusions are strongly oriented parallel to the rift plane of the Barre Granite (Fig. 3). At Hilltop the maximum concentration of poles to microfractures indicates that most microfractures strike within a few degrees of N30°E. At Hilltop the internal overcore for two samples gave maximum expansions within 2° of the normal to the preferred strike of the microfractures. Microfractures are also strongly oriented at Wetmore and Adam. However, the latter has fewer open microfractures.

To define the mechanical anisotropy across the diameter of our 7.6 cm cores in the laboratory we used both dynamic and static tests. The former consisted of measuring the compressional wave velocity using the pulse matching technique described by MATTABONI and SCHREIBER [23]. In order to mate the barium titanate transducers to the core, we ground twelve flats parallel to the axis, thus allowing us to measure the pulse velocity in six directions at 30° intervals parallel to the plane of strain relief.

The compressional wave velocities for one sample from each of the four *in situ* strain sites is illustrated in Fig. 4. The compressional wave velocity is related to the orientation of the rift plane and preferred orientation of microfractures in Barre Granite. The velocity is fastest parallel to the rift or foliation planes and slowest normal to these planes. The compressional wave velocity anisotropy varies among the *in situ* strain sites. The greatest difference, 53%, occurs in the metasediment and the least, 3%, is found at Adam. Samples taken from approximately the same position in the pluton, Wetmore and Hilltop, have nearly the same compressional wave velocities and velocity variation with direction whereas the sample, Adam, is different. In all instances the maximum expansion upon overcoring was sub-parallel to the direction of the lowest compressional wave velocity.

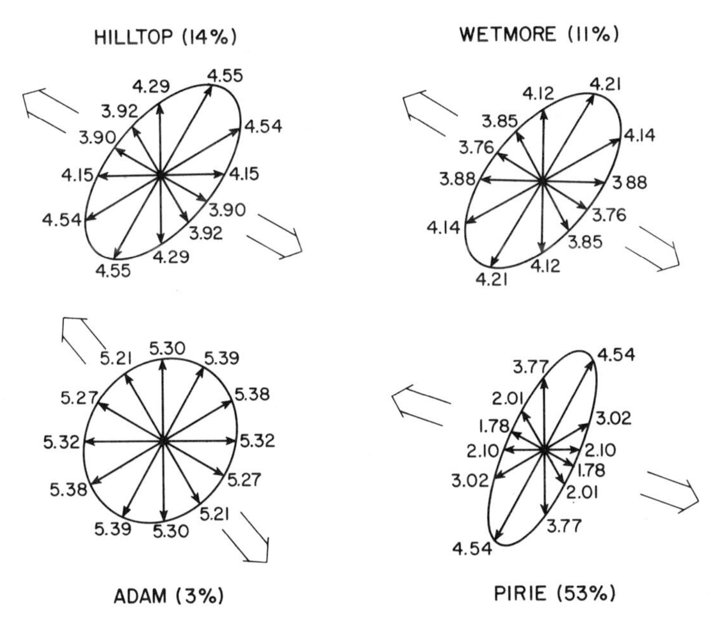

Figure 4
Laboratory measurements of compressional wave velocities in the horizontal plane of samples taken near Barre, Vermont. Units for velocity are km/sec. The orientation of the maximum expansion following initial overcoring is shown with arrows outside each velocity ellipse. Velocity anisotropy is listed next to the name of the sample site. Samples from Hilltop, Adam, and Wetmore are Barre Granite. Sample from Pirie is a metasediment.

To determine the static moduli, uniaxial compression tests were used. Because the moduli across the diameter of the cores were of greatest interest, the cores were loaded in each of six directions normal to the core's axes. The twelve flats ground on the sides of the cores are also necessary to accommodate the loading anvils. The load was monitored with a calibrated beryllium–copper load cell; shortening across the diameter of the cores was measured with a linear variable differential transformer transducer. Strain gauge rosettes with components at 60° were monitored to measure strain at the center of the core. Although flats were ground parallel to the core axis, we assume as an approximation that we are loading a finite cylindrical core across its diameter.

The stresses and strains inside a cylinder loaded radially may be determined using the Airy stress function expressed in terms of two analytic functions of a complex variable $z = x + iy$ (JAEGER and COOK [24]). In this boundary value problem, the normal and tangential stresses applied to the boundary can be represented by a Fourier series. The solution of the Airy stress function by HONDROS [25] indicates a complicated stress and strain distribution inside the cylinder (Fig. 5). Because of the

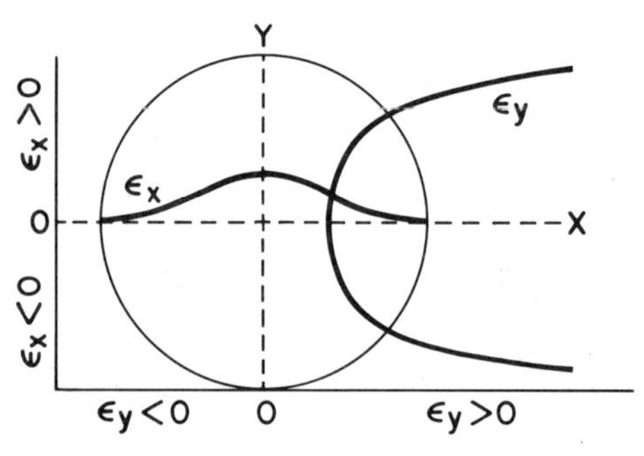

Figure 5
Theoretical strain distribution along the vertical (OY) and horizontal (OX) diameters of a cylinder loaded along the vertical diameter (HONDROS [25]). ε_x and ε_y – radial strain at points on the OX and OY axes.

complicated strain pattern inside the cylinder the ratio of load to diametrical shortening is not an accurate measure of the elastic moduli. However, calculation of the Young's modulus and Poisson's ratio is possible using center strains and the equations derived by HONDROS [25]:

$$\nu = -\left[\frac{\varepsilon_{ry} + 3\varepsilon_{\theta y}}{2(\varepsilon_{ry} - \varepsilon_{\theta y})}\right]$$

and:

$$E = \frac{2P(1 + \nu)(1 - 2\nu)}{\pi Dt[\nu\varepsilon_{ry} + (1 - \nu)\varepsilon_{\theta y}]}$$

where ε_{ry} = radial normal strain at a point on OY axis
$\varepsilon_{\theta y}$ = tangential normal strain at a point situated on the OY axis
D = cylinder diameter
P = applied load
ν = Poisson's ratio
E = Young's modulus
t = cylinder length

For a comparison with the dynamic properties of the rocks the compressional wave velocity (V_p) may be estimated using the relation:

$$V_p^2 = \frac{E(1 - v)}{(1 + v)(1 - 2v)\zeta}$$

where ζ is the density of the rock.

Three assumptions necessary for the calculation of v and E are: (1) each component of the strain gauge rosette measures strain at the center of the core, (2) strain in some directions can be computed by measuring strain along other directions, and (3) E and v are the same for small strains in compression and tension. The first is necessary because the components of a strain gauge rosette are not centered, but all are no more than $\frac{1}{5}$th of a radius away from the axis of the core. In addition, radial strain on the x axis is actually measured rather than tangential strain on the y axis. At the center of the core the two are equal. The second assumption is made because both the radial normal strain (ε_{ry}) and the tangential normal strain ($\varepsilon_{\theta y}$) are necessary in order to calculate Poisson's ratio in a particular direction. With a strain gauge rosette of three components bonded to the end of our cylinder, it is possible to measure just three of six components of both ε_{ry} and $\varepsilon_{\theta y}$. The other three components of each are estimated by averaging the two components at 30° from the unknown component.

Data from diametrical loading of cores at 30° intervals around the axes are listed in Table 1. Cores from Hilltop and Adam were tested. Tests on the metasediment from Pirie failed because of the cores' tendency to split along foliation when loaded

Table 1

Barre granite (Hilltop) (17,703 NT)

Direction	N0°E	N30°E	N60°E	N90°E	N60°W	N30°W
$\varepsilon_{ry} \times 10^{-6}$	−331	−298*)	−249	−373*)	−498	−414*
$\varepsilon_{\theta y} \times 10^{-6}$	+201.5*)	+269	+232.5*)	+196	+165*)	+134
v	0.26	0.46	0.96	0.18	~0.00	~0.00
E	0.31 mb	0.32 mb	0.36 mb	0.26 mb	0.20 mb	0.25 mb
V_p	3.82 km/sec	v too high for calculation		3.30 km/sec	2.77 km/sec	3.10 km/sec

Barre granite (Adam) (17,703 NT)

Direction	N0°E	N30°E	N60°E	N90°E	N60°W	N30°W
$\varepsilon_{ry} \times 10^{-6}$	−171	−160*)	−148	−162*)	−175	−173*)
$\varepsilon_{\theta y} \times 10^{-6}$	77.5*)	81	77.5*)	74	74*)	74
v	0.12	0.17	0.18	0.13	0.09	0.10
E	0.60 mb	0.65 mb	0.69 mb	0.65 mb	0.58 mb	0.60 mb
V_p	4.96 km/sec	5.12 km/sec	5.31 km/sec	5.04 km/sec	4.71 km/sec	4.80 km/sec

*) Strains estimated by average of strains at 30° on either side of direction in question.

parallel to the foliation. Each direction was stressed to 174 bars which was necessary to restore the strain released by overcoring.

General comments concerning the static tests are (1) an equivalent load in the N60°W direction on Hilltop, Adam and Pirie cores restored strain nearly equal to that relieved at each of the three sites; (2) when loaded in certain directions cores from Hilltop yielded unreasonable Poisson's ratios (>0.5) and (3) when it is possible to calculate a compressional wave velocity from static data, the calculated velocity varies with direction much like the measured velocity but the calculated velocity is slightly slower than that measured.

5. Residual strain

For strain relaxation measurements conducted on Potsdam Sandstone, any strain recovered by an internal overcore into a body without boundary loads is defined as residual strain, regardless of the mechanism of strain relaxation or how the strain was imposed on the rock body (ENGELDER and SBAR [2]). This notion of residual strain in rock is more general than the standard definition as outlined by FRIEDMAN [8] and VOIGHT and ST. PIERRE [6]. A standard notion of residual strain comes from the metals literature where MCCLINTOOK and ARGON [26] define residual strains as recoverable elastic distortions of constituent crystals or grains that satisfy internal equilibrium conditions and that exist in a given volume of rock with no external boundary loads. FRIEDMAN [8] envisions a two component system: (1) elastically distorted crystals, reflecting previous loads, that are locked into the aggregate by (2) other elastically distorted crystals or cement. We suggest that this hypothetical system be modified to include non-elastic distortions and further expanded so that microfractures within crystals or cement be considered as a low modulus component. FRIEDMAN points out [8, page 299] that if one part of the residual strain system consists of low modulus elements, these low modulus elements will yield relatively large strains and dominate the total strain relaxation, even though the internal forces causing the strains in high and low modulus elements are opposite and equal. Furthermore, if the low modulus element has a non-random fabric, such as the microfractures in Barre Granite, then that fabric will control the orientation of strain relaxation.

Residual strains are imposed on a rock when crystals become elastically distorted in response to paleotectonic, paleotopographic, and thermal and chemical histories of the rock (see page 301; FRIEDMAN [8]). Because our measurements, were made at the surface, we cannot exclude the possibility that recoverable strains (residual strains) were generated by physical and chemical processes associated with weathering and may reflect a non-random fabric.

6. Mechanisms of strain relaxation

We suggest that the opening of microfractures in the Barre Granite is one major mechanism of strain relaxation. This is the same conclusion reached by NORMAN [27] for the strain relaxation of crystalline rocks near Atlanta, Georgia. Our interpretation is based on the similarity of orientation among the strain relief, the microfractures, dynamic properties and static properties of the Barre Granite.

A load of 17,703 N on our cores is equivalent to a stress of about 100 bars at the center of the core. For the same stress loaded axially on cylinders of Barre Granite, DOUGLAS and VOIGHT [19] found that Young's modulus varied with direction between 0.23 and 0.41 mb. These values are similar to those obtained with our cores from Hilltop. Douglas and Voight conclude that the mechanical anisotropy of Barre Granite is produced by small preferentially oriented cracks parallel to the rift.

The diametrical loading behavior of Barre Granite is compatible with a rock containing a set of microfractures which influence the mechanical properties of the rock. A load normal to the microfractures will close them with comparatively little strain normal to the load, thus giving the granite a very low v in that direction. Likewise, their closure during loading gives the granite a low E perpendicular to the microfractures because the microfractures have a low stiffness. In contrast a positive load parallel to the plane of the microfractures will tend to open the fractures. This behavior is manifest as an anisotropic v. E and v are both relatively higher for loading parallel to the microfractures.

A compelling reason for suggesting that a mechanism for the strain relaxation of the internal overcore in Barre Granite is the opening of cracks is that the orientation of maximum expansion is at right angles to the strike of the predominant orientation of microfractures. Two additional observations support our contention that microfractures relax upon overcoring.

We have identified a relationship between the numbers of open microfractures per unit volume and the magnitude of strain components upon internal overcoring. Thin sections from Adam had fewer open microfractures than observed in thin sections from Hilltop. The magnitude of the strain relieved by the internal overcores at Adam is much smaller than that relieved at Hilltop (Fig. 2). The rock with the larger number of microfractures per unit volume has the larger component of strain relaxation.

Second, the orientation of *in situ* strain relief at Adam is not as consistent as at Hilltop (Fig. 2). In samples from Adam we measured fewer microfractures per unit volume which were more randomly oriented compared to microfractures in Hilltop samples. This is confirmed by laboratory velocity measurements. The larger compressional wave anisotropy at Hilltop is a manifestation of a stronger preferred crack orientation whereas the higher velocities at Adam are manifestations of a smaller number of microfractures per unit volume. If the mechanism for strain relaxation is the opening of microfractures, we would expect the orientation of total strain relaxation to vary from sample to sample at sites which lack a strong preferred orientation of microfractures.

Samples from Wetmore and Hilltop have similar microfracture fabric, and dynamic and static characteristics, yet their *in situ* strain measurements do not correlate.

These data illustrate the inconsistency which may occur between outcrops of rocks which have similar mechanical properties. Unlike data from the other three sites there is no pattern among relaxation of the three 15.2 cm cores or between the 15.2 cm core and the subsequent internal overcore at Wetmore. Because there is no pattern we suspect that the data from Wetmore is of poor quality, although an explanation for this is qualitative. The lip of a 50 m deep quarry was only 20 m from the sample site and quarry operations will have relieved some of the strain prior to our measurement. A major sheeting fracture parallel to the surface or extreme surface weathering could also alter the strain boundary conditions. Finally, there may be a real difference between *in situ* strain or strain relaxation mechanisms from point to point in the Barre Granite as was suggested by NICHOLS [11].

Thin sectioned samples from Hilltop generally contain one open microfracture per mm and can have as many as five open microfractures per mm. In order for the Barre Granite to relax by 400 $\mu\varepsilon$, one microfracture per mm would have to open by 0.4 μm. Cracks opened to 0.4 μm or more are commonly observed in rocks such as a granite (BRACE *et al.* [28]). Thus a strain of 4×10^{-4} could be reasonably accounted for by the opening of microfractures.

X-ray diffraction studies of fresh Barre Granite indicate that quartz crystals in the granite are elastically distorted with their greatest elongation normal to the rift plane (FRIEDMAN [8]). Individual grains removed from the rock (say by overcoring) will therefore have a maximum contraction normal to the rift plane. Total strains produced in this manner will be on the order of 10^{-4}. NICHOLS [11] found the strain relief associated with internal overcoring of an unweathered block of Barre Granite to be compatible with the strain locked into the quartz crystals. The maximum contraction of one of his cores was normal to the rift plane of the block. Since we relieved strains which caused expansion of the core, and since we attribute this expansion to the opening of microfractures, it is quite possible that the internal system of stresses prior to overcoring was altered by weathering, so that prior to overcoring, individual grains were already relieved to some extent.

Weathering may have several effects, among which are the chemical and physical alteration of crack surfaces to decrease their tensile strength[4]) and/or elastic stiffness. The cracks may start to open in response to the residual elastic elongation of the quartz grains and further open when boundary restraints are removed by overcoring. The stiffness of cracks or grains with cracks may be orders of magnitude less than the stiffness of whole grains. This difference in stiffness would result in a net expansion normal to the rift plane counteracting the tendency for unfractured crystals to contract.

Even in the absence of boundary loads internal stresses prior to overcoring are necessary to produce a net strain upon overcoring. By definition these stresses are residual and may be manifest in Barre Granite by elastic distortion. In Barre Granite

[4]) Many cracks in Barre Granite are healed and may have a certain tensile strength.

exposed to weathering this residual stress may be partially relieved by the opening of microfractures.

Upon internal overcoring, Barre Granite at the Hilltop site expanded perpendicular to its microfracture fabric in response to the relaxation of a residual strain. However, there was a lack of parallelism between the normal to the rift plane or microfracture fabric and the maximum expansion direction of the initial overcore. One explanation for this is the presence of an externally applied strain of undefinable magnitude superimposed on the strain relieved by the opening of microfractures, and having principal axes incongruent with the strain relieved by the internal overcore. Alternatively, the residual strain axes may become reorientated after the initial overcoring, making it impossible to separate applied and residual strains.

7. Conclusion

The measured direction of maximum expansion at the surface of Barre Granite is normal to the rift plane originating in the preferred orientation of microfractures. These measurements are in agreement with similar measurements made by SWOLFS [29] in the floor of the Wetmore Quarry. The dominant mechanism of strain relaxation at the surface of Barre Granite is the opening of microfractures.

Acknowledgements

Thomas Chen, Philip Lelyveld, and Kevin Powell assisted in data gathering and reduction. Paul Pomeroy and Warren Prell assisted in administration of the project. We wish to acknowledge many discussions with Hans Swolfs and T. C. Nichols on the subject of residual strain. The Rock of Ages Monument Company, Barre, Vermont granted permission to use their land and facilities. The work was supported by a contract with the New York State Energy Research and Development Authority, the Nuclear Regulatory Commission and National Science Foundation Grant EAR-74-07923. Roger Bilham and A. B. Watts reviewed the manuscript.

[1] ENGELDER, T. and SBAR, M. L. (1977), *The relationship between* in situ *strain relaxation and outcrop fractures in the Potsdam sandstone, Alexandria Bay, New York*, Pure appl. Geophys. (this volume).

[2] ENGELDER, T. and SBAR, M. L. (1976), *Evidence for uniform strain orientation in the Potsdam sandstone, northern New York, from* in situ *measurements*, J. Geophys. Res. *81*, 3013–3017.

[3] VARNES, D. J. and LEE, F. T. (1973), *Hypothesis of mobilization of residual stress in rock*, Geol. Soc. Am. Bull. *83*, 2863–2866.

[4] NICHOLS, T. C. and SAVAGE, W. Z. (1977), *Rock strain recovery-factor in foundation design*, Am. Soc. Civil Eng., Specialty Conf. on Rock Eng. for Foundations and Slopes (in press).

[5] VOIGHT, V. (1966), *Restspannungen im Gestein*, Int. Soc. Rock. Mech., Proc., 1st Cong. Lisbon *1*, 45–50.

[6] VOIGHT, B. and ST. PIERRE, B. H. P. (1974), *Stress history and rock stress*, Advances in Rock Mechanics, Proc. 3rd Cong. ISRM *II*, 580–582.

[7] HOSKINS, E. R., RUSSELL, J. E., BECK, K. and MOHRMAN, D. (1972), In situ *and laboratory measurements of residual strain in the coast ranges north of San Francisco Bay*, EOS *53*, 1117.

[8] FRIEDMAN, M. (1972), *Residual elastic strain in rocks*, Tectonophysics *15*, 297–330.

[9] SWOLFS, H. S., PRATT, H. R. and HANDIN, J. (1973), In situ *measurements of strain relief in a tectonically active area*, Terra Tek Report TR 74–5, 38 pp.

[10] SWOLFS, H. S., HANDIN, J. and PRATT, H. R. (1974), *Field measurements of residual strain in granitic rock masses*, Advances in Rock Mechanics, Proc. 3rd Cong. ISRM *II*, 563–568.

[11] NICHOLS, T. C. (1975), *Deformations associated with relaxation of residual stresses in a sample of Barre Granite from Vermont*, U.S. Geological Survey Prof. Paper 875, 32 pp.

[12] GALLAGHER, J. J., FRIEDMAN, M., HANDIN, J. and SOWERS, G. (1974), *Experimental studies relating to microfracture in sandstone*, Tectonophysics *21*, 203–247.

[13] HOOKER, V. E. and DUVALL, W. I. (1966), *Stresses in rock outcrops near Atlanta, Ga.*, Bur. of Mines R. I. 6860, 18 pp.

[14] MORGAN, T. A., FISCHER, W. G. and STURGIS, W. J. (1965), *Distributions of stress in the Westvaco trona mine, Westvaco, Wyo.*, U.S. Bur. Mines Rept. Inv. 6675, 58 pp.

[15] HOOKER, V. E. and JOHNSON, C. F. (1969), *Near-surface horizontal stresses, including the effects of rock anisotropy*, U.S. Bur. Mines Rept. Inv. 7224, 29 pp.

[16] WHITE, W. S. and JAHNS, R. H. (1950), *Structure of central and east-central Vermont*, J. Geol. *58*, 179–220.

[17] PAGE, L. R. *Devonian plutonic rocks in New England*, 371–383 in *Studies of Appalachian Geology: Northern and Maritime* (eds. E. Zen, W. S. White, J. J. Hadley and J. B. Thompson) (Interscience Pub., New York 1968), 475 pp.

[18] CHAYES, R. (1952), *The finer-grained calcalkaline granites of New England*, J. Geol. *60*, 207–254.

[19] DOUGLAS, P. M. and VOIGHT, B. (1969), *Anisotropy of granites – a reflection of microscopic fabric*, Geotechnique *19*, 376–398.

[20] BUR, T. R., HJELMSTAD, K. E. and THILL, R. E. (1969), *An ultrasonic method for determining the attenuation symmetry of materials*, U.S. Bur. Mines Rept. Inv. 7335, 8 pp.

[21] WHITE, W. S. (1946), *Rock-bursts in the granite quarries at Barre, Vermont*, U.S. Geol. Survey Circ. 13, 15 pp.

[22] BROWN, A. (1974), *Photoelastic measurements of recoverable strain at four sites*, Tectonophysics *21*, 135–164.

[23] MATTABONI, P. and SCHREIBER, E. (1967), *Method of pulse transmission measurements for determining sound velocities*, J. Geophys. Res. *72*, 5160–5163.

[24] JAEGER, J. C. and COOK, N. G. W., *Fundamentals of Rock Mechanics* (Methuen, London 1969), 513 pp.

[25] HONDROS, G. (1959), *The evaluation of Poisson's ratio and the modulus of materials of low tensile resistance by the Brazilian (indirect tensile) test with particular reference to concrete*, Aust. J. Appl. Sci. *10*, 243–264.

[26] MCCLINTOOK, F. A. and ARGON, A. S., *Mechanical Behavior of Materials* (Addison–Wesley, Reading, Massachusetts 1966), 770 pp.

[27] NORMAN, C. E. (1976), *Geometric relationships between geologic structure and ground stresses near Atlanta, Ga.*, U.S. Bur. Mines Rept. Inv. 7365, 24 pp.

[28] BRACE, W. F., SILVER, E., HADLEY, K. and GOETZE, C. (1972), *Cracks and pores: a closer look*, Science *178*, 162–164.

[29] SWOLFS, H. S. (1976), *Field investigation of strain relaxation and sonic velocities in Barre Granite, Barre, Vermont*, Terra Tek Report TR 76–13, 14 pp.

(Received 20th December 1976)

Pageoph, Vol. 115 1977), Birkhàuser Verlag, Basel

The Relationship Between *In Situ* Strain Relaxation and Outcrop Fractures in the Potsdam Sandstone, Alexandria Bay, New York[1])

By Terry Engelder[2]) and Marc L. Sbar[3])

Summary – *In situ* strain was measured by overcoring foil-resistance strain gauge rosettes bonded to five outcrops of Potsdam Sandstone near Alexandria Bay, New York. Strain relaxation magnitude and orientation correlated with the area of the intact outcrop outlined by intersecting vertical fractures. The maximum expansion occurred at the outcrop with the largest area between intersecting fractures. Outcrops with more than one set of longer, open fractures or more complicated fracture patterns have lower recoverable strains. Strain relaxation was lowest next to a postglacial pop-up. The orientation of the pop-up indicated relief of an ENE directed compression, the direction also observed as the maximum expansion at the outcrop yielding the largest strain relaxation.

Key words: Stress in situ; Strain relaxation; Intraplate stress field.

1. Introduction

To complement the companion paper discussing the significance of strain relaxation of the Barre Granite we describe the strain relaxation of the Potsdam Sandstone near Alexandria Bay, New York (Fig. 1 of Engelder *et al.* [1]). This is a region presumably affected by ENE maximum horizontal compressive stress (Sbar and Sykes [2]). Strain relaxation measurements in the Potsdam Sandstone provide an opportunity to test Sbar and Sykes' observation of uniformly oriented stress in the lithosphere and the extent of transmission of this stress to the surface. In addition we could sample *in situ* strain associated with different fracture patterns within outcroppings of one rock unit in order to further clarify the relationship between *in situ* strain and outcrop fractures. These data also supplement strain relaxation measurements in the Potsdam Sandstone near Plattsburgh, New York (Engelder and Sbar [3]; Fig. 1 of Engelder *et al.* [1]).

Many *in situ* strain measurements suggest a relationship between the orientations of a dominant fracture set and local strain relaxation (Preston [4]; Eisbacher and Bielenstein [5]; Brown [6]; Swolfs *et al.* [7]). This relationship exists at all depths from exposed bedding plane surfaces down to 1 km near underground mines. The fact that there seems to be a relationship suggests that caution is necessary when using strain

[1]) Lamont-Doherty Geological Observatory Contribution No. 2456.

[2]) Lamont-Doherty Geological Observatory, Palisades, New York 10964.

[3]) Now at: Department of Geosciences, University of Arizona, Tucson, Arizona 85721, USA.

relief measurements to detect a regional tectonic stress. In some instances it may be difficult to delineate strain fields whose local orientations are perturbed by fractures. Using solid-inclusion probes, LEE *et al.* [8] demonstrated that conspicuous structures such as fractures and foliation do influence the redistribution of strain around an actively excavated opening.

The relationship of *in situ* strain and locally dominant fracture sets to a regional tectonic stress is further complicated by residual strains. FRIEDMAN and LOGAN [9] demonstrated that residual strains influence the mechanical properties of sandstone so that induced fractures propagate parallel to the maximum compressive residual strain. This implies that fracture orientation may be controlled by a residual stress. Likewise, PRICE [10] presents a model for the downwarp and uplift of a basin in which residual strain and fractures are controlled by the shape of the basin rather than through transmission of more deeply seated stresses.

Methodology

In situ strain was measured by overcoring foil-resistance three component strain gauge rosettes epoxied on a polished horizontal outcrop (FRIEDMAN [11]; BROWN [12]; SWOLFS *et al.* [7]; ENGELDER and SBAR [3]). Strain relaxation accompanied the cutting of a vertical 15.2 cm diameter core using a diamond masonry bit with its axis centered on the strain gauge rosette (overcoring). After four or more days a 7.6 cm diameter core coaxial with the 15.2 cm core was cut to further relax residual strain (SWOLFS *et al.* [13]; NICHOLS [14]). The larger is the initial overcore whereas the smaller is an internal overcore.

Three rosettes were bonded to each outcrop chosen for our experiment. Spacing between rosettes was 1 to 2 m with vertical fractures between each rosette. The gauges were monitored for several days before and after overcoring to establish a base line from which to subtract thermal fluctuations.

2. Geology of Alexandria Bay vicinity

Outcrops of lower Paleozoic sediments and Precambrian basement lace the countryside in the vicinity of Alexandria Bay. Many outcrops of the basal Cambrian Potsdam Sandstone are still bare following recent glacial scouring. Generally a quartz arenite but locally a subarkose to arkose, the sandstone was deposited on an erosional surface with less than 100 m of relief (CUSHING *et al.* [15]). Silica overgrowths generally cement the sand, but calcareous cement appears in the uppermost 5 m of the formation (CUSHING *et al.* [15]).

Reactivation of high angle faults in the Precambrian basement followed the deposition of the Potsdam Sandstone. These reactivated faults, whose general strike is N50°E, are expressed as monoclines in the lower Paleozoic sediments. A DAMES and

MOORE Technical Report [16] suggests that the Potsdam sands were unconsolidated at the time of fault motion and that the motion on the reactivated faults was absorbed by soft sediment deformation in the Potsdam sand and by subsequent monoclinal flexing of overlying lithified carbonates. Some NE striking fractures within the Potsdam Sandstone are attributed to stresses associated with minor slip in basement faults following lithification of the sand. This faulting was quite local and did not affect any of the outcrops sampled in our study.

The most recent tectonic features in the Potsdam Sandstone are a number of post-glacial folds commonly called 'pop-ups'. These pop-ups consist of slabs as much as 3–4 m thick which have buckled upward as much as 4 m following the most recent glaciation (Fig. 1). The pop-ups apparently buckled upward in response to a compression until the fold fractured along its crest and flanks, furnishing relief to the bent flanks and permitting them to straighten (CUSHING *et al.* [15]). The orientation of pop-ups has been used to infer the direction of maximum compressive stress in the crust (SBAR and SYKES [2]). The Alexandria Bay pop-ups have a fold axis of NNW which implies that the crustal stress is most compressive in the ENE direction.

The five outcrops selected for our measurements are labelled Baker, Frazier, Kirkey, Nelson and Ostrander (Fig. 2). The Potsdam Sandstone is a quartz arenite at Baker, Frazier, Nelson and Ostrander but a subarkose at Kirkey (Table 1). Kirkey is located near the top of the formation where calcareous cement is abundant. The grain size distribution of this subarkose is bimodal whereas the quartz arenites are unimodal with silica cement.

Figure 1
Postglacial pop-up viewed looking NNW along its axis. The strain relaxation measurements at Nelson were located 30 m ENE from the axis, which is exposed for 100 m.

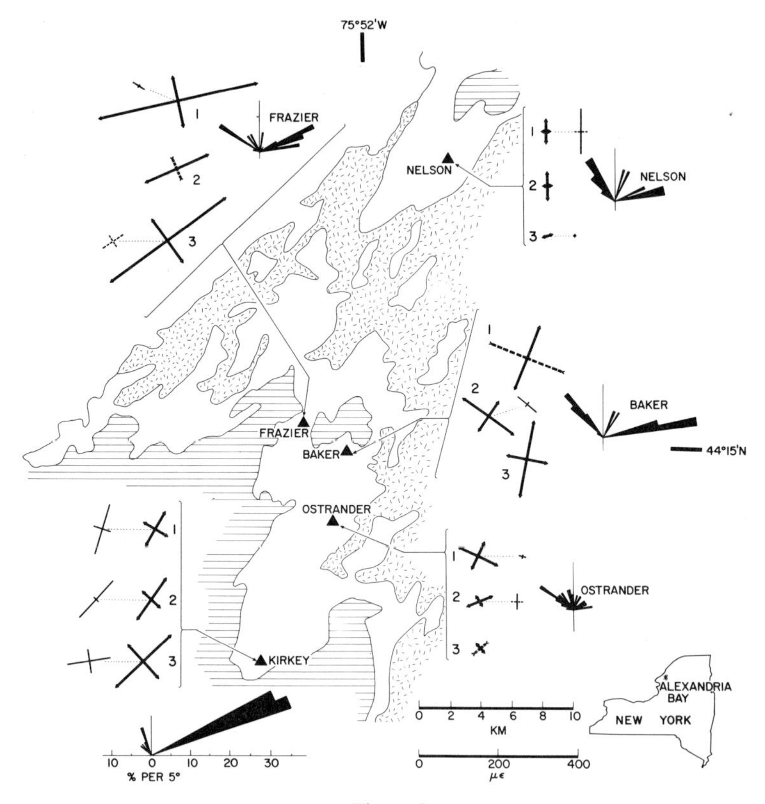

Figure 2
Geology and *in situ* strain in the vicinity of Alexandria Bay, New York. Outcrops of the Precambrian basement are stippled; outcrops of the Potsdam Sandstone are the blank; and outcrops of the lower Paleozoic carbonates are indicated by horizontal lines. The azimuth of the intersection of fractures with outcrop surface are indicated by rose diagrams. This data is divided into 5° intervals and displayed in percent of fractures per 5° interval. Five sample sites are named. The magnitude and orientation of the strain relieved by the initial overcore are indicated by dark arrows. The magnitude and orientation of the strain relieved by an internal overcore are indicated by lighter lines. Solid arrows represent expansion and dashed arrows represent contractions. The magnitude of the relieved strain is represented by a scale in microstrain (με). Four internal overcores cut in the lab four months after initial overcore are: Frazier – 3, Kirkey – 3, Nelson – 3, and Ostrander – 2.

Table 1
Modal analysis of Potsdam Sandstone, Alexandria Bay

	Quartz	Feldspar	Carbonate cement	Silica cement	Rock fragment	Pores	Accessory minerals
1. Baker	86	—	—	13	1	—	—
2. Ostrander	83	—	—	16	1	—	—
3. Nelson	82.3	—	—	6.3	2.3	8.3	0.8
4. Frazier	76	—	—	24	—	—	—
5. Kirkey	66	13.5	15.5	3.5	0.5	0.5	0.5

3. *Relation between macrofractures and in situ strain*

The orientation of the intersection of fractures with the surface of each outcrop is illustrated by means of rose diagrams (Fig. 2). In the vicinity of Alexandria Bay the most prominent fracture sets strike within 15° of either N65°E or N35°W. Locally a third fracture set appears at N20°E. The majority of fractures at Frazier extend less than 2 m on the outcrop face (Fig. 3), whereas the majority of fractures at Kirkey can be traced up to 50 m (Fig. 4). Fracture lengths are between these extremes at Nelson, Baker and Ostrander. Fracture spacing is between 50 and 70 cm at Baker, 5 and 70 cm at Ostrander, 2 and 100 cm at Nelson, 20 and 100 cm at Frazier, and 50 and 100 cm at Kirkey.

To characterize the state of an outcrop with regard to fractures, some quantitative

Figure 3

View along 0.3 to 2 m long fractures trending ENE at Frazier. The NW trending fracture in the foreground forms one edge of a mechanically continuous portion of the outcrop, which extends back to the vicinity of the geologist. Since the short ENE trending fractures do not outline discrete blocks, they form short discontinuities without interrupting the mechanical continuity of the outcrop.

Figure 4
ENE trending fractures at Kirkey. Many of these fractures are open and thus transmit no normal stress.

combination of fracture length, distance between parallel fractures, orientation of fractures and separation of walls of individual fractures should be considered. In practice it is difficult to combine all four parameters in one function that is meaningful. We feel that it is significant to characterize the outcrops using the area outlined by vertical fractures. This defines the size of mechanically continuous portions of the outcrop. Usually long-intersecting fractures define the boundaries of mechanically continuous portions as illustrated in Fig. 5 where each continuous portion is about 1 m^2. Short but widely spaced fractures form discontinuities without completely interrupting the mechanical continuity of the outcrop (Fig. 3). To be concise we shall refer to the degree of fracturing in each outcrop as fracture density where a higher fracture density is characterized by smaller areas between vertical fractures.

In each outcrop the area outlined by vertical fractures is variable (Fig. 6). Most outcrops of Potsdam Sandstone have areas between 1 and 10 m^2 as is the case at Frazier, Kirkey, and Baker (Figs. 3, 4, and 5). Exceptions occur in the vicinity of postglacial pop-ups such as Nelson (Fig. 2) and on outcrops containing glacial striations. Parts of the outcrop at Nelson were shattered either before or during the formation of the pop-up and, therefore, are plotted in Fig. 6 as very small areas outlined by vertical fractures. Ostrander is a knob of Potsdam Sandstone with 10 m of relief over an area of 2×10^4 m^2. During glaciation many boulders were dragged over the knob forming glacial striations (Fig. 7). Fractures generated during glaciation are responsible for the many orientations of fractures at Ostrander (Fig. 2).

Figure 5
Three fracture sets at Baker. The view is toward the SSW looking along the least dense of the three sets.

The magnitude of the strain relief following the 15.2 cm overcore varies with the fracture density within the Potsdam Sandstone (Figs. 2 and 6). Frazier had the largest strain relaxation and lowest fracture density. Relaxation of intermediate magnitude occurred at Baker and Kirkey each of which has one or more open fracture sets. Small changes occurred at Ostrander, an outcrop characterized by intense fracturing in many orientations, and Nelson, an outcrop undoubtedly affected by a postglacial pop-up.

The relation between orientation of strain relaxation and fractures is not consistent. The maximum expansion at Frazier parallels a major fracture set whereas the maximum expansion of the different gauges trends 15° to 30° from the strike of the same fracture set at Kirkey. At Baker and Ostrander there is no obvious relationship between maximum expansion and fracture orientation primarily because neither fractures nor the strain relief have a single orientation. All three strain measurements at Baker have maximum expansions within a few degrees of the strike of one of the three fracture sets. On the intensely fractured outcrop at Ostrander, two of the maximum expansion directions are within a few degrees of the most preferred orientation for fractures. At Nelson a small amount of strain relief was measured; two rosettes recorded a change on just the north component whereas the third rosette only recorded small changes on the other two components. Here the strain relief was not oriented with respect to outcrop fractures.

Expansion during the internal overcoring of subarkose from Kirkey was more than 100 $\mu\varepsilon$ compared with relaxations of less than 100 $\mu\varepsilon$ for all of the quartz arenites. Internal overcoring of the subarkose yielded strains comparable in magnitude and

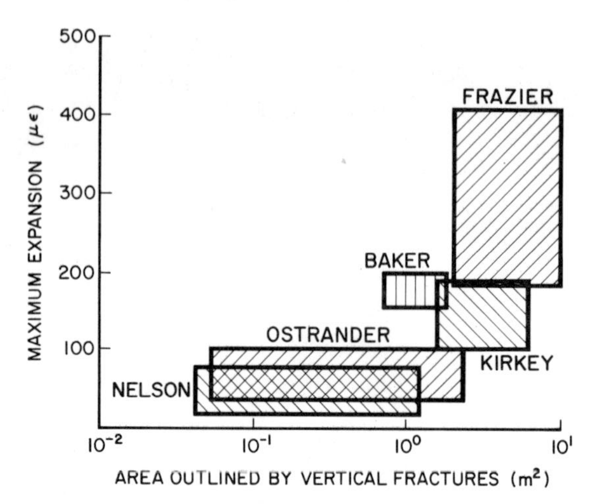

Figure 6
Maximum elongation of strain relaxation following initial overcore plotted as a function of area outlined by vertical fractures (m^2) for five outcrops of Potsdam Sandstone on which *in situ* strain was measured. Hatched areas indicate the extent of variation of both strain relaxation and area outlined by vertical fractures at each outcrop.

Figure 7
Fractures associated with a glacial wear track at Ostrander. The concave shape of the fractures indicates that the boulder causing this wear track moved from right to left. Other short fractures are seen in the top portion of the photo.

orientation to that following the initial overcore whereas three quartz arenites all yielded small strains with either maximum expansions or minimum contractions oriented in the northwest quadrant. Only the quartz arenite from Nelson yielded internal overcoring strains comparable in magnitude and orientation to strains relieved during initial overcoring. Most internal overcores were cut within four days of the initial overcore but four samples listed in Figure Caption 2 were cut four months later. There is no apparent correlation between outcrop fractures and the orientation of the maximum expansion of internal overcores at Kirkey and Nelson where significant strain was relieved.

4. *Laboratory analyses*

To define the mechanisms of relaxation of the Potsdam Sandstone, we subjected our samples to petrographic analyses and mechanical tests similar to those described in ENGELDER *et al.* [1].

In thin section, observable characteristics which might contribute to mechanical anisotropy or strain relaxation are not present or are subtle. The only sample with any microfracture fabric is the subarkose from Kirkey (Fig. 8). Samples from Frazier and Kirkey have slight long grain axis fabrics which may be attributed to deposition in a fluvial environment (Fig. 9). All of the quartz arenites (Baker, Frazier, Nelson, and Ostrander) have serrated grain boundaries characteristic of pressure solution.

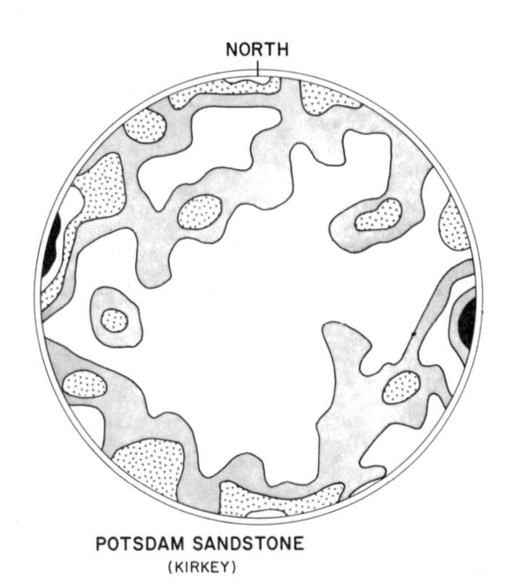

Figure 8
Orientation of poles to microfractures within quartz grains of Potsdam Sandstone at Kirkey. Plane of each diagram is the horizontal surface of the outcrop on which *in situ* strain was measured; data plotted in equal-area, lower hemisphere projection. Microfractures are measured only in a thin section cut parallel with the horizontal plane of the outcrop. Contours are at 6 %, 4 %, 2 %, 1 % per 1 % area. This diagram is for gauge 3 at Kirkey. Similar microfracture orientations were obtained at Kirkey gauge 2.

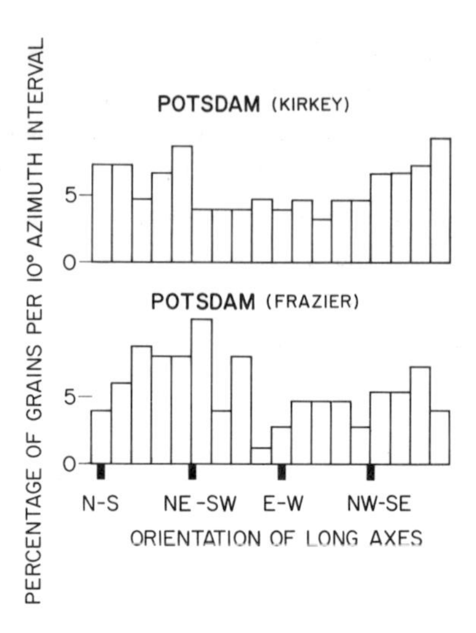

Figure 9
Orientation of long axes of quartz grains in sandstone. Long axes of 150 grains have been measured from each thin section cut parallel to the horizontal plane of the outcrop. Azimuths have been divided into 10° intervals.

The compressional wave velocity and the amount of velocity anisotropy vary from sample to sample (Fig. 10). The highest anisotropy for Potsdam Sandstone occurs in samples from Kirkey which is the only subarkose in which an *in situ* strain measurement was made. Two samples of the subarkose illustrate the variation in velocity between samples. In both cases the anisotropy has the same orientation. The lowest velocity occurred in the Nelson sample which was noticeably more porous than other samples from Alexandria Bay. The fast direction in Nelson is normal to the axis of the postglacial pop-up. One sample from Frazier has a 3% anisotropy whereas another sample has an anistropy of less than 1% (not plotted in Fig. 10).

A core from Frazier with a 3% velocity anisotropy was stressed diametrically in each of six directions to 174 b (Table 2). This load was necessary to achieve radial center strains equivalent to those relieved in the field. From these data compressional wave velocities for each direction in the sample may be estimated using a density of 2.4 gm/cm^2. The sample is elastically anisotropic with the stiffest direction parallel to the direction of maximum expansion upon initial relief. Here the calculated compressional wave velocity anisotropy estimated from static tests is 16% and does not have the same orientation as the measured anisotropy (Fig. 10).

The dynamic tests are probably more indicative of the true anisotropy of Frazier. The static tests are very sensitive to misalignment of the loading anvils and nonlinearity of the rock. Any error in measuring strain (Table 2) is magnified in the calculation of the compressional wave velocity. The orientation of a slight velocity anisotropy of 3% is difficult to find using static tests because of the possibilities for experimental error.

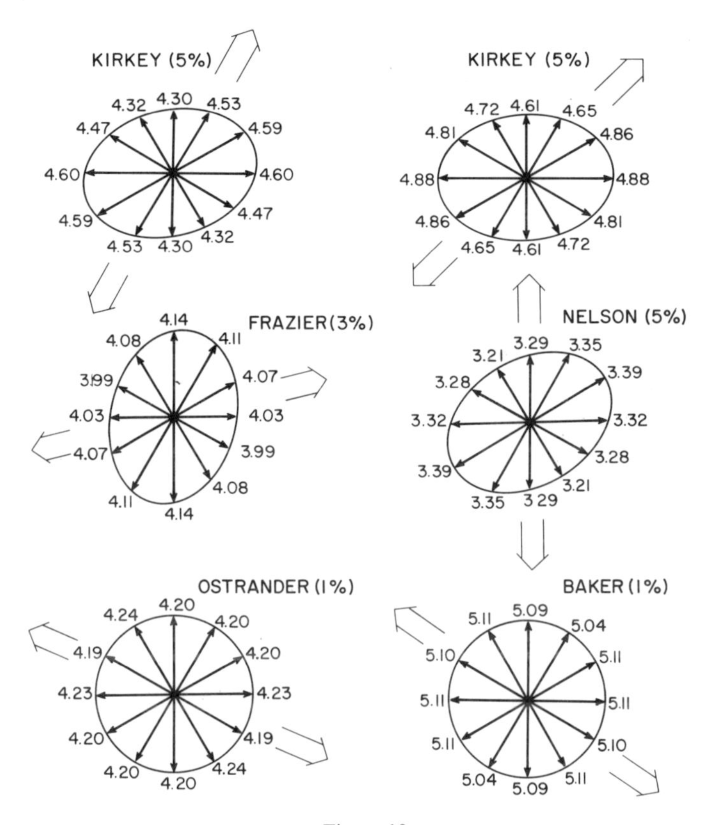

Figure 10
Compressional wave velocities in the horizontal plane of samples of Potsdam Sandstone taken near Alexandria Bay, New York. Units for velocity are km/sec. The orientation of the maximum expansion following initial overcoring is shown with arrows outside each velocity ellipse. Velocity anisotropy is listed next to the name of the sample site. The dimensions and shape of the ellipses are only diagrammatic and do not precisely reflect the velocity anisotropy

Table 2
Potsdam Sandstone (Frazier) (17703 NT)

Direction	N0°E	N30°E	N60°E	N90°E	N60°W	N30°W
ε_{ry}	−248	−215*	−182	−200.2*	−219	−233.5*
$\varepsilon_{\theta y}$	102.5*	104	88.5*	73	87*	101
ν	0.08	0.15	0.15	0.03	0.07	0.10
E	0.41 mb	0.48 mb	0.56 mb	0.50 mb	0.47 mb	0.44 mb
V_p	4.16 km/sec	4.59 km/sec	4.97 km/sec	4.56 km/sec	4.45 km/sec	4.33 km/sec

*) Strains estimated from average of strains at 30° on either side of direction in question.

However, the only fabric of Frazier is a preferred orientation of long grain axes which are more closely aligned with the orientation of the maximum for the static modulus.

5. *Discussion*

Residual strains

As indicated in the companion paper (ENGELDER *et al.* [1]), strain relaxation is influenced by residual strain for some rocks such as the Barre Granite, so care must be exercised in relating strain relaxation to tectonic stresses. The presence of residual strain is indicated by relaxation following internal overcoring. We concluded in the companion paper that residual strains observed during internal overcoring contributed significantly to the relaxation of the initial overcore.

Unlike the Barre Granite, the mechanism associated with the relaxation of residual strain in the Potsdam Sandstone is not obvious. In some cases the strain relieved by the internal overcore is not significantly greater than our experimental error of $\pm 10\ \mu\varepsilon$.

Significant strains were recovered at Kirkey following double overcoring of a subarkose which has an anisotropy (Fig. 10) and a petrographic fabric (Figs. 8 and 9); however, a 60° variation in maximum expansion following internal overcoring and the difference in orientation between the anisotropy and petrographic fabric make it difficult to suggest a unique mechanism for relaxation of residual strain. Opening of microfractures is one possibility but with the preferred strike of about N10°E, the microfractures are not suitably oriented to be the exclusive cause for the relaxation of any of the three internal cores. Another feature unique to the subarkose is a calcite cement which might also contribute to residual strain relaxation in some yet unknown manner.

We observed little, if any, anisotropy which correlates with a relaxation mechanism for residual strain in the quartz arenites (Fig. 10). Evidence for a small anisotropy comes from samples taken at Frazier where quartz grains were deposited with their long axes trending northeast (Fig. 9). The dynamic moduli of samples from Frazier do not reflect this grain shape fabric. However, the static tests indicated that these samples have a high modulus trending N60°E or almost parallel to most preferred orientation of long grain axes (Table 2). Maximum expansion following internal overcoring was normal to the long grain axes or parallel to the low static modulus direction. Yet, because expansion following the initial overcore was parallel to the high modulus direction of the rock, the behavior of cores from Frazier is not comparable to cores taken from the Barre Granite where maximum expansion during both initial and internal overcoring was parallel to the low modulus direction (ENGELDER *et al.* [1]).

One quartz arenite core sampled at Nelson did behave like the Barre Granite relaxing upon internal overcoring with a large component of strain parallel to the maximum expansion during initial overcoring. Although this sample did have a velocity anisotropy its minimum was 30° from the maximum expansion of 'residual' strain, as was the case for Kirkey.

Our conception of the relationship among residual strain, a velocity anisotropy, and a petrographic fabric comes from the behavior of Barre Granite. Basically we found that residual strain occurs in rocks with a strong anisotropy which is the manifestation of a petrographic fabric. This relationship appears to be the same for the Potsdam Sandstone. A subarkose from Kirkey and a quartz arenite from Nelson both have the largest velocity anisotropy and yield significant residual strains which contribute a major component of strain to the initial strain relaxation. We also conclude that residual strain contributed only a small component to the initial strain relaxation of the quartz arenites at Frazier, Baker, and Ostrander.

Outcrop fractures

In the Alexandria Bay region we observed a correlation between magnitude of strain relief and the area of outcrop outlined by vertical fractures (Fig. 6). Outcrops with sets of intersecting open fractures or more complicated fracture patterns have lower recoverable strains. In addition, the relation between maximum expansion and fracture orientation is complicated.

Our hypothesis is that outcrop fractures interrupt tectonic strain transmitted from depth in the lithosphere by reducing the strain magnitude and reorienting the principal axes of this *in situ* strain. These fractures may also partially relieve residual strain, but if residual strain relaxation arises from an interaction with some mechanical element of the rock fabric, as was the case for Barre Granite, a partial relief of residual strain by fracture propagation should not cause its reorientation. Our data support this hypothesis. Although the subarkose at Kirkey was cut by large, open fractures, residual strain as measured after internal overcoring in three different cores had about the same magnitude and orientation. Likewise, large components of residual strain relaxed during initial overcoring; these strains also had about the same magnitude and orientation. The small components of residual strain measured in the quartz arenites from Baker, Frazier, and Ostrander all have maximum expansions in the northwest quadrant. The scatter in magnitude and orientation was about the same regardless of the outcrop fracture length and spacing. The greater intensity of fracturing at Baker or Ostrander did not affect the residual strain relative to that at Frazier.

Considering the magnitude of the relaxation following the initial overcore, there is a direct relationship between the area outlined by vertical fractures and the magnitude at Frazier, Baker, and Ostrander. We interpret this to indicate that more closely spaced and longer fractures interrupt tectonic strain. The spacing of intersecting fractures at Frazier is as much as three meters. This spacing seems to be about the minimum fracture spacing necessary for transmission of a large enough tectonic strain so that strain in blocks separated by vertical fractures is similar in magnitude and orientation. The closer spacing at Baker and Ostrander correlates with an increase in scatter of *in situ* strain orientation. This scatter may be caused by open fractures relieving strain normal to the fracture. Outcrops with three or more fracture sets such as Baker, and Ostrander have three or more directions along which tectonic strain could have been relieved near the surface.

Little, if any, tectonic strain was measured at Kirkey. This outcrop contains the longest and most open fractures observed in the vicinity of Alexandria Bay. Although the outcrop contains comparatively large areas outlined by vertical fractures, the fracture sets are apparently long enough, deep enough, or open enough to isolate the outcrop from tectonic strains.

Likewise, the sandstone in the vicinity of the post-glacial pop-up at Nelson contains no component of tectonic strain. Apparently a fairly large volume (10^5 m^3) of sandstone sheared along a bedding plane when the sandstone popped up, thus isolating this volume from tectonic loading. Based on the orientation of the pop-up, the maximum compressive strain prior to its formation was N58°E or parallel to the maximum expansion of the strain relieved at Frazier.

Assuming that the pop-up was an elastic instability, the Euler formula for determining the critical stress for buckling can be used to estimate the ENE directed stress at Nelson (R. PLUMB, personal communication). The thickness of the elastic member which can be measured in the outcrop is 0.5 m (Fig. 1). From laboratory tests we measured a Young's modulus of 0.3 mb. We estimate that the elastic member extends at least 30 m on either side of its axis based on our measurements indicating that the sandstone is detached. Using a length of 60 m and assuming that the member has fixed ends and no side restraints, we calculate a maximum critical buckling stress of 260 bars. The critical stress decreases for longer members. We assume that after buckling the sandstone member collapsed by breaking along vertical fractures.

Regional stress

Our data suggest that tectonic strain at depth in the lithosphere is transmitted to the surface. The strain relief at Frazier appears to be influenced the least by outcrop fractures and residual strain. The ENE trend of the maximum elongation for strain at Frazier is the same as the orientation of the maximum compressive stress as inferred from post-glacial pop-ups and fault plane solutions in New York State (SBAR and SYKES [2]). This maximum compressive stress is believed to be a manifestation of a uniformly oriented regional stress pervading the lithosphere in the northeastern United States.

Strain in the N60°E direction was 200 $\mu\varepsilon$ at Frazier. If Young's Modulus in that direction is on the order of 0.5 mb (Table 2), we have relieved a stress of 100 b at the surface. Many authors have presented graphs showing horizontal stress as a function of depth. In curves such as that shown in HERGET [17] the horizontal stress is about 100 bars at the surface.

Near Alexandria Bay only one outcrop of five sampled appears to contain a significant tectonic strain. Thus we conclude that fractures decouple outcrops from tectonic strain.

Acknowledgement

This work was supported by contracts with the New York State Energy Research and Development Authority and Nuclear Regulatory Commission plus National Science Foundation Grants EAR-74-07923 and DES-75-03640. JOHN BEAVAN and TRACY JOHNSON reviewed the manuscript.

REFERENCES

[1] ENGELDER, T., SBAR, M. L. and KRANZ, R. (1977), *Strain relaxation of Barre Granite: Residual strain attributed to opening of microfractures*, Pure and Appl. Geophys. (this volume).

[2] SBAR, M. L. and SYKES, L. R. (1973), *Contemporary compressive stress and seismicity in eastern North America: An example of intra-plate tectonics*, Geol. Soc. Am. Bull. *84*, 1861–1882.

[3] ENGELDER, J. T. and SBAR, M. L. (1976), *Evidence for uniform strain orientation in the Potsdam sandstone, northern New York, from in situ measurements*, J. Geophys. Res. *81*, 3013–3017.

[4] PRESTON, D. A. (1968), *Photoelastic measurement of elastic strain recovery in outcropping rocks*, Trans. AGU *49*, 302.

[5] EISBACHER, G. H. and BIELENSTEIN, H. U. (1971), *Elastic strain recovery in Proterozoic rocks near Elliot Lake, Ontario*, J. Geophys. Res. *76*, 2012–2021.

[6] BROWN, A. (1974), *Photoelastic measurement of recoverable strain at four sites*, Tectonophysics *21*, 135–164.

[7] SWOLFS, H. S., HANDIN, J. and PRATT, H. R. (1974), *Field measurements of residual strain in granitic rock masses*, Advances in Rock Mechanics, Proc. 3rd Cong. ISRM *II*, 563–568.

[8] LEE, F. T. and NICHOLS, T. C. (1972), *Some effects of geologic structure, engineering operations and thermal changes on rock-mass behavior*, Proc. 24th Int. Geol. Cong., Montreal, Sec. *13*, 261–272.

[9] FRIEDMAN, M. and LOGAN, J. M. (1970), *The influence of residual elastic strain on the orientation of experimental fractures in three quartzose sandstones*, J. Geophys. Res. *75*, 387–405.

[10] PRICE, N. J. (1974), *The development of stress systems and fracture patterns in undeformed sediments*, Advances in Rock Mechanics, Proc. 3rd Cong. ISRM *I*, 487–496.

[11] FRIEDMAN, M. (1972), *Residual elastic strain in rocks*, Tectonophysics *15*, 297–330.

[12] BROWN, A. (1973), *In situ strain measurement by photoelastic gauges, I: The method*, Tectonophysics *19*, 383–397.

[13] SWOLFS, H. S., PRATT, H. R. and HANDIN, J. (1973), *In situ measurements of strain relief in a tectonically active area*, Terra Tek Report TR 74–5, 38 p.

[14] NICHOLS, T. C. (1975), *Deformations associated with relaxation of residual stresses in a sample of Barre Granite from Vermont*, U.S. Geological Survey Prof. Paper 875, 32 p.

[15] CUSHING, H. P., FAIRCHILD, H. L., RUEDEMANN, R., and SMYTH, C. H. (1910), *Geology of the Thousand Islands region*, New York State Museum, Bulletin 145, 1–194.

[16] DAMES and MOORE Consulting Engineers (1974), *Seismo-tectonic conditions in the St. Lawrence River Valley region*, Report to NYS Atomic and Space Development Authority under Job. No. 7465-020.

[17] HERGET, G. (1974), *Ground stress determination in Canada*, Rock Mechanics *6*, 53–64.

(Received 11th November 1976)

Pageoph, Vol. 115 (1977), Birkhäuser Verlag, Basel

Reflections on Measurement of Residual Stress in Rock

By TERRY E. TULLIS[1])

Abstract – Consideration of the behaviour of elastic bodies shows that it is not possible for a set of overcoring measurements that are made within isolated blocks to show residual strains or stresses that have a non-zero average, unless the size of the equilibrium volume over which the residual stresses balance is both considerably larger than the volume of the overcore and smaller than the size of the isolated block. Since some results have been reported that do not match these constraints, non-elastic behaviour must have occurred during overcoring. A possible explanation is that oriented microcracks are somehow opened by the overcoring. In some cases, stress fields induced by microcracking near the overcoring cut may explain measured strain changes far from the cut. Consideration of various reported measurements in terms of relative sizes of overcoring and equilibrium volumes shows that care is necessary when interpreting residual stress measurements.

Key words: Stress residual; Stress *in-situ*; Microfracture strain.

1. *Introduction*

Numerous measurements have been made of strain changes in rocks that have been attributed to residual stresses. Gaining an understanding of these measurements is important both in order to determine what tectonic significance to attribute to them and to determine to what extent corrections must be made for them when measuring applied *in situ* stress by overcoring or other similar stress relief techniques. This note solves neither of these problems, but instead sets out the types of measurement results that are physically possible with elastic behaviour. A comparison with some of the measurements reported in the literature suggests that non-elastic processes dominate in some of these measurements.

2. *Definition of terms*

Residual stress will be defined as those stresses that are contained within an isolated body that is free from external tractions and that are not attributable to the action of body forces, body torques, or thermal gradients. Applied stresses will be defined as those stresses that arise within a body as a result of surface tractions on its boundaries, body forces and body torques upon its volume, or thermal gradients.

---

[1]) Department of Geological Sciences, Brown University, Providence, Rhode Island, 02912, USA.

Residual stresses in a body must balance one another. The equilibrium volume will be defined as the smallest rectangular parallelpiped for which each of the Cartesian components of residual stress (1) averages to zero over its interfaces, and (2) exerts no net moment, i.e., shows no average correlation between magnitude and position on its interfaces (BISHOP and HILL 1951, p. 422). The actual residual stress at any point on the interfaces of the equilibrium volume will not be zero in general since these interfaces are contained within the body and the body contains residual stresses. The equilibrium volume has the property that the volume average of the stresses within it is zero. A body containing residual stresses may be thought of as being built up of a collection of equilibrium volumes. The equilibrium volume is essentially the same as the locking domain or equilibrium volume of VARNES and LEE (1972, p. 2865), the residual stress domain or equilibrium volume of SWOLFS, *et al.* (1974, p. 564), or the self-equilibrating unit of RUSSELL and HOSKINS (1973, p. 3). Equilibrium volumes can exist on any scale from the submicroscopic to the megascopic (RUSSELL and HOSKINS 1973, p. 4–8).

Initial overcoring will be used to refer to any operation that introduces stress-free boundaries around some portion of a body previously containing applied stresses. Second overcoring will be used to refer to any such stress relief operation that is made within a body previously isolated from applied stresses by initial overcoring.

3. *Behaviour of elastic bodies*

The question of interest in this discussion is what strains can occur as a result of second overcoring of a perfectly elastic body that contained residual stresses. Since the new surfaces of the second overcore previously had stresses existing on them, there will necessarily be a rearrangement of the stresses in the second overcore. Whether or not the stress will change a measurable amount at a given point of observation depends upon the distance of that point from the new surface relative to the characteristic scale of the residual stresses. According to St. Venant's principle (cf. LOVE 1934, p. 132) the influence of local stress concentrations falls off rapidly away from such concentrations so that it is unimportant beyond a few characteristic dimensions of the stress concentration. Consequently, one would not expect, for example, to be able to measure a change in stress or strain several grain diameters from a newly created surface in a polycrystalline aggregate possessing only grain scale residual stresses as long as the average stress existing at the site of the surface was zero prior to the cut being made.

In order to illustrate and quantify this application of St. Venant's principle, consider the stress field within a two dimensional elastic half-space arising from tractions upon its surface. These tractions will be such that their average over the surface is zero. Consider sinusoidal variations in either normal or shear stress. Any arbitrary variation of surface tractions can be constructed from these two cases by

Fourier analysis, since solutions may be superposed in linear elasticity. The solution for the stress distribution within the elastic half-space is obtained from TIMOSHENKO and GOODIER (1951, p. 47, equation (e)) by choosing appropriate boundary conditions. The x coordinate lies along the surface and the y coordinate increases into the half-space.

For the sinusoidally varying normal stress case the boundary conditions are that at $y = 0$, $\sigma_y = A \sin \pi x/l$ and $\tau_{xy} = 0$ where l is a half wave-length of the variations and A is their amplitude. Also, as $y \to \infty$, $\sigma_y \to 0$ and $\tau_{xy} \to 0$. Using these boundary conditions to find the values of the four constants in the solutions for the stresses, substituting these constants into the equations and simplfying, we obtain the result:

$$\begin{aligned} \sigma_x &= A \sin \frac{\pi x}{l} e^{-\pi y/l} \left(1 - \frac{\pi y}{l}\right) \\ \sigma_y &= A \sin \frac{\pi x}{l} e^{-\pi y/l} \left(1 + \frac{\pi y}{l}\right) \\ \tau_{xy} &= -A \cos \frac{\pi x}{l} e^{-\pi y/l} \left(\frac{\pi y}{l}\right) \end{aligned} \tag{1}$$

For the sinusoidally varying shear stress case the boundary condition as $y \to \infty$ is the same whereas at $y = 0$, $\sigma_y = 0$ and $\tau_{xy} = A \cos \pi x/l$. Using the same procedure for this case yields:

$$\begin{aligned} \sigma_x &= A \sin \frac{\pi x}{l} e^{-\pi y/l} \left(2 - \frac{\pi y}{l}\right) \\ \sigma_y &= A \sin \frac{\pi x}{l} e^{-\pi y/l} \left(\frac{\pi y}{l}\right) \\ \tau_{xy} &= A \cos \frac{\pi x}{l} e^{-\pi y/l} \left(1 - \frac{\pi y}{l}\right) \end{aligned} \tag{2}$$

In all of these cases the stress components fall off exponentially with the ratio y/l of distance from the surface to half wave-length of the surface variations. The value of these stresses as a function of y/l is shown in Fig. 1. If the stresses also vary along the surface in the third dimension, they will fall off even more rapidly as a function of y/l.

Imagine a situation in which a second overcoring is to be made in a body containing residual stresses with equilibrium volumes much smaller than the size of the overcore. Along the future surface of the second overcore the average stresses must be zero, but the actual stresses are assumed to vary with half wave-lengths equal to the grain diameter and to have an amplitude of 100 MPa. Upon overcoring these amplitudes are reduced to zero and the stresses near the surface change accordingly. Equations (1) and (2) as illustrated in Fig. 1 show that the elastic stress changes resulting from this overcoring are less than 0.1 MPa within 3 grain diameters of the surface. The stress changes will be only 10^{-4} MPa within 5 grain diameters and 10^{-10} MPa within 10 grain diameters. The above calculations give the maximum changes that occur as a function of y/l. However, the changes are periodic as a function of x/l and hence a measuring technique that averaged over several wave-lengths would show no change, even very near the surface.

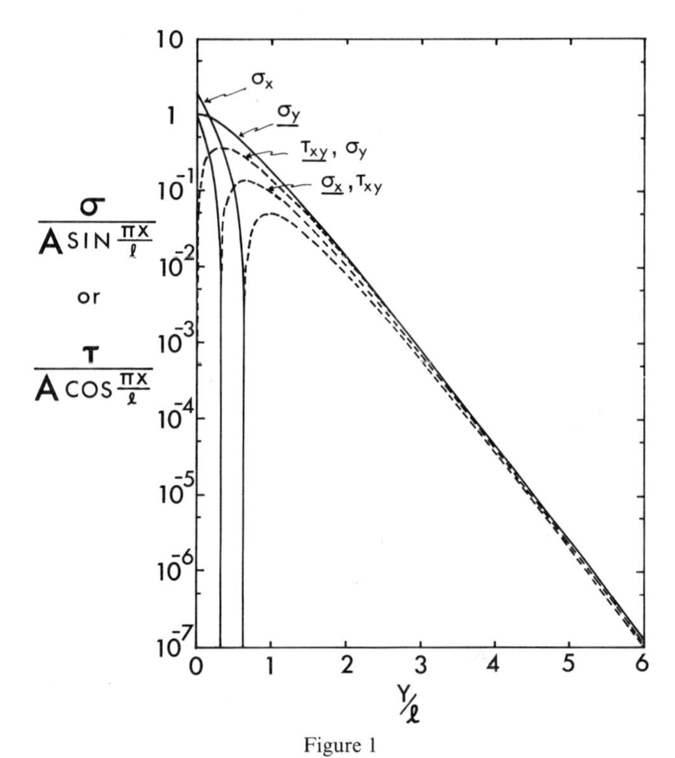

Figure 1

Decrease of stress magnitudes with distance into an elastic half-space that has sinusoidally varying normal or shear stress on its surface. Distance y is normalized by half wave-length l of the sinusoidal variations. Underlined stress components are for situation with sinusoidally varying normal stress σ_y and zero shear stress τ_{xy} on surface; non-underlined components are for opposite situation. Lines are solid for positive stresses and dashed for negative ones.

The above calculations suggest that it is helpful to identify three situations that might arise in attempting to measure residual stresses by second overcoring. These situations arise from the three possible sizes of the volume isolated by the second overcore relative to the equilibrium volume for the residual stress. In the first situation, the volume of the second overcore is quite large relative to the equilibrium volume. In the second situation, the two volumes are of approximately the same order of magnitude. In the third situation, the volume of the second overcore is quite small relative to the equilibrium volume.

In situation one, we can directly use the results of the stress distribution calculations presented above to say that anywhere away from the newly created free surfaces there will be no measurable change in stress or strain upon creation of those surfaces. The important point is that although there are residual stresses and strains at the measurement point, the second overcoring cannot release them as long as all behaviour is elastic.

In situation two, it is impossible to find a place within the second overcore that is free from the influence of the stress changes that occurred at its surface. If the second overcore isolated at least an equilibrium volume then the average of the stress changes measured throughout the overcore will be zero both before and after the cut. If the size of the second overcore is somewhat smaller than the equilibrium volume, then the average stress changes within a single overcore will not be zero. However, in either of these cases although a measurement made at only one point in one overcore will show a stress change, an average of such measurements over enough overcores will yield zero stress or strain change. This is a particularly important and fundamental point, namely that although for the situation in which the equilibrium and second overcore volumes are of the same order it is definitely possible to measure residual stresses, no overall direction or pattern of residual stress can emerge from a statistically adequate number of second overcoring measurements in a body behaving elastically.

In situation three in which the volume of the second overcore is much smaller than the equilibrium volume, the stress within the isolated volume will be essentially homogeneous. If the initial overcoring freed a volume of the same order of size as the volume of the second overcore, then the residual stresses within that first volume were also essentially homogeneous. Thus, those residual stresses were relieved along with the applied stresses by the first overcore and, hence, no residual stress remains to be relieved by the second overcore. If the volume relieved by the initial overcore was much bigger than that relieved by the second overcore then there would be a non-zero residual stress to be relieved by the second overcore. Whether the average of a number of such measurements is zero or non-zero depends on their spatial distribution relative to the equilibrium volume and to the volume of the initial overcore.

4. *Implications for previously reported measurements*

Measurements incompatible with elastic behaviour

HOSKINS *et al.* (1972) report making a series of measurements of residual strain by second overcoring in which they consistently found the same orientation of residual strains. Of their 21 measurements, 6 were undercore measurements made in the field on outcrops and 15 were undercore measurements made in the lab on isolated oriented blocks. For the purposes of this discussion their undercoring can be regarded as equivalent to overcoring. The lab measurements should clearly be regarded as second overcoring. Because the field measurements were made on highly fractured, essentially loose rock, they are also regarded to be second overcoring measurements (HOSKINS, personal communication, 1976). Because of the incompatibility between these results and elastic behaviour as outlined above, the only conclusion possible is that the rock is not behaving elastically. Since microcracks are known to be important in the low pressure mechanical behaviour of rocks (c.f. BRACE 1965; PENG and JOHNSON 1972; SIMMONS *et al.* 1975) it seems likely that they may be playing an important role here (VARNES and LEE 1972, p. 2864–2865; FRIEDMAN 1972, p. 289; RUSSELL and HOSKINS 1973, p. 10; VOIGHT and ST. PIERRE 1974, p. 581–582). It is notable that the strains reported in this study were nearly all expansions rather than contractions, probably resulting from permanent opening of microcracks. The explanation for the strain upon overcoring to be different in one direction than another could be that pre-existing cracks had a preferred orientation resulting from the stresses existing during their formation. It is also possible that residual stresses on a grain scale could contribute to opening of preferrentially oriented microcracks (FRIEDMAN and LOGAN 1970; VOIGHT and ST. PIERRE 1974, p. 581–582). Even if the strains in this study were due to the opening of microcracks, it is not clear why they opened during second overcoring. It could be in response to the static or high frequency mechanical stress the rock experiences during overcoring, to temperature changes and differential thermal expansion cracking, or to the interaction of cooling water with the microcracks. It would be desirable to study this situation further because the unknown operative processes may also be important in numerous other overcoring measurements of *in situ* stress, but be less obvious because in most cases the strains can be interpreted in terms of elastic recovery of applied or residual stresses. Presumably some correction to such measurements should be made for the influence of such non-elastic processes, but at present no basis exists for doing this.

Another example of strains resulting from non-elastic processes accompanying a stress relief operation in an isolated rock containing residual stresses has been given by FRIEDMAN (1972, p. 311–315). In this case, the strains are measured by X-ray determination of the *d*-spacing of $\{32\bar{5}4\}$ in quartz and the second overcore operation is simply the controlled propagation of a macroscopic tensile fracture. Significant strain changes occurred up to 15 grain diameters of the fracture surface. Although the X-ray measured strain changes within the quartz grains themselves are elastic, FRIEDMAN (1972, p. 314–315) suggests that the non-elastic process of microcracking along grain boundaries adjacent to the macroscopic fracture is responsible for relaxing the strains within the grains. This interpretation is supported by measurements of fracture–surface energy (FRIEDMAN *et al.* 1972).

Measurements possibly explained by non-elastic behaviour

Interesting measurements of residual stresses having some features in common have been reported by NICHOLS (1975) and by SWOLFS *et al.* (1974). A sequence of concentric second overcoring was performed in both cases, the work of NICHOLS

(1975) having been done by actual overcoring on a 220 mm cube of Barre granite in the laboratory and the work of SWOLFS *et al.* (1974) having been done by cutting sets of four slots in the field on a 2.5 meter cube of Cedar City quartz diorite that had previously been isolated on its sides by similar slot cutting. In Nichols' work the concentric overcores were successively larger. In SWOLFS' *et al.* (1974) work the concentric overcores were successively smaller. In both studies, the strain in every annulus that was isolated consisted of radial contractions and circumferential expansions. The similarity and symmetry of these results suggests that they might be produced by some artifact of the measuring process and thus might not accurately represent the residual stresses in the rock. A tentative explanation is offered below for these patterns, although it is not able to explain all the measurements, particularly the strains observed in the centrally located block in each case.

NICHOLS (1975, p. 19–20) reports that there is a zone of microcracking about 1 mm thick produced adjacent to the surfaces created by overcoring. Since microcracking is a process that creates dilatancy it seems possible that the tendency for this layer to expand could create stresses in the adjacent part of the rock. In fact, the situation is formally identical to the thermal stresses that arise upon heating the surface of a body. The analysis of thermal stresses in solid and hollow cylinders of TIMOSHENKO and GOODIER (1951, p. 408–412) may be used directly. In using the equations of Timoshenko and Goodier we will interpret their linear thermal expansion $\propto T$ to be the linear strain of an isotropic microcrack dilatancy that would occur in the absence of the stresses produced by that dilatancy. The actual strains that occur in the dilatant layer are not equal to $\propto T$, but are those given by equation (239) of TIMOSHENKO and GOODIER (1951, p. 408). It is artificial to assume that dilatancy tends to be uniform throughout the thickness of the 1 mm layer and zero elsewhere, but it simplifies the analysis and it will give the same stress and strain fields far from the layer as would a more realistic decay of dilatancy away from the surface.

Three cases need to be analyzed to compare with the results of NICHOLS (1975) and of SWOLFS *et al.* (1974). The first is the expansion of a surface layer on the inner surface of a hollow cylinder, the second is the expansion of a surface layer on the outer surface of a hollow cylinder, and the third is the expansion of a surface layer on the outer surface of a solid cylinder. In the first case, all of the stresses in the layer will be compressive. The strains in the cylinder itself will be radial contractions and circumferential expansions, the magnitudes of which decrease with radial distance outward from the inner surface. In the second and third cases, the radial stress in the layer is tensile while its circumferential and axial stresses are compressive. In both these cases the strains within the cylinder in the radial and circumferential directions are equal to one another. These strains are expansions in the case of the solid cylinder and are homogeneous within the cylinder. For the hollow cylinder they are expansions in the outer part and contractions in the inner part, the transition point depending on Poisson's ratio.

Because the radial stresses are tensile in both cases of dilatancy on the outer surface, the solutions for the strain within the cylinders are probably irrelevant. Since the surface layer is full of microcracks it will not be able to support tensile stress and thus cannot strain the cylinder as assumed in the thermal stress calculations.

However, the compressive stresses in the case of the expanded layer on the inner surface of the hollow cylinder will be able to strain the hollow cylinder. The equations for the radial and circumferential strains in this case in which a layer of thickness d is assumed to undergo a uniform temperature increase of T are found, following TIMOSHENKO and GOODIER (1951, equations 239, 244, 245 and 246), to be:

$$\varepsilon_r = \frac{\propto Td}{(1-v)}\left(a + \frac{d}{2}\right)\left\{\frac{1}{(b^2 - a^2)}\left[\frac{(r^2 - a^2)(1-v)}{r^2} - 2v\right] - \frac{(1+v)}{r^2}\right\}$$

$$\varepsilon_\theta = \frac{\propto Td}{(1-v)}\left(a + \frac{d}{2}\right)\left\{\frac{1}{(b^2 - a^2)}\left[\frac{(r^2 - a^2)(1-v)}{r^2} - 2v\right] + \frac{(1+v)}{r^2}\right\}$$

where v is Poisson's ratio and a and b are the inner and outer radii respectively. These equations apply for any radius r greater than $a + d$ and less than or equal to b; different expressions must be used for the strains within the layer of thickness d.

These expressions can be applied to many of the second overcoring operations reported by NICHOLS (1975, p. 6–16). Shown in Fig. 2 are the predicted strains according to this mechanism that may be compared with the measured strains of his Fig. 6 (NICHOLS, 1975, p. 9; FRIEDMAN, 1972, p. 320, Fig. 4). The parameters used for calculating this figure have been chosen to match Nichols' situation: $a = 25.5$ mm, $b = 110$ mm, $v = 0.14$ and $d = 1$ mm. The magnitude of $\propto T$ was chosen so as to give strain magnitudes similar to the ones he measured. Its value is 0.0053 which means that the linear strain in the dilatant region would be 0.5% if it were not under stress. The circumferential and axial compressive stresses are about 200 MPa in the dilatant layer (using $E = 34$ GPa) which means that the circumferential and axial strain in this layer would be only about 5×10^{-6} and the radial strain would be about 0.7%. Thus, within the 1 mm thick layer, radial opening of cracks oriented parallel to the inner surface would total only 7 microns. These values do not seem excessive when one looks at Nichols' photograph of the surface (NICHOLS, 1975, p. 20, Fig. 14B). If this mechanism is the correct explanation for the observed strains, local point stresses between diamonds and rock during drilling must exceed 200 MPa in order to cause the fractures in the presence of the resulting stress. This does not seem unreasonable, especially since the fractures are observed.

Quantitative application of this model to the study of SWOLFS *et al.* (1974) is not possible since their geometry was square rather than circular. Ignoring that difference the dimensions of their block and the larger strains measured by them (400×10^{-6} in the isolated wall) requires a larger thickness for the cracked layer (17 mm) and larger $\propto T$ (0.02) although the stresses need be no larger than 200 MPa since the modulus of the rock is low (8 GPa). Since the slots were made by drilling over-lapping drill holes these values may be possible although they seem excessive.

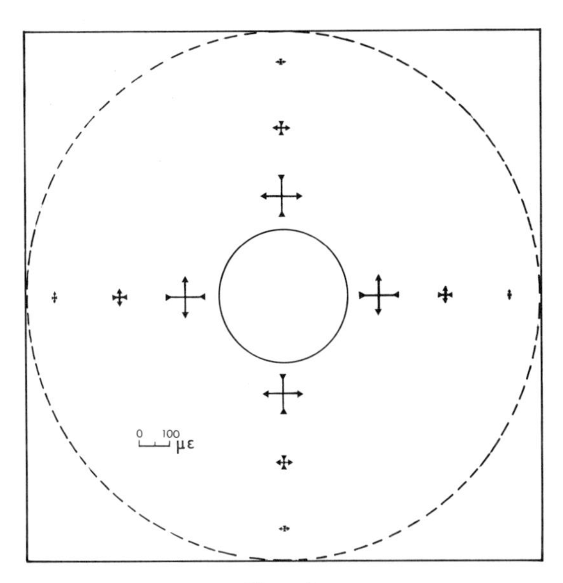

Figure 2
Strains in a 220 mm diameter infinite cylinder due to a 51 mm diameter inner hole lined with a 1 mm thick layer of dilatant (microcracked) material in which compressive stresses reach about 200 MPa. See text for details. Since strains are insensitive to outer diameter, results for indicated cube will be similar. Strains are similar to those measured by NICHOLS (1975, Fig. 6; FRIEDMAN 1972, Fig. 4) although effects of his free measurement surface are neglected here.

This process is unable to explain the contraction of the central core of NICHOLS (1975, p. 9, Fig. 6) or the decaying pattern of radial contraction and circumferential expansion toward the centre of the central block of SWOLFS *et al.* (1974, p. 567, Fig. 6). No strains would be predicted by this mechanism since the tensile stresses on the outer surfaces of the blocks would result in failure rather than stressing of the interior. Since the behaviour of the central block was different in the two studies, it may be that its behaviour comes closer to representing the residual stresses in the two studies than does the systematic behaviour I have attempted to explain here.

Other measurements

A few other representative measurements of residual stress reported in the literature will be briefly discussed to illustrate how they may fit into the scheme of relative sizes of equilibrium and second overcore volumes.

RUSSELL and HOSKINS (1973, p. 10–15) report measuring changes in residual strain by second overcoring in a 152 mm diameter, 41 mm thick disk. Upon drilling a kerf only 5 mm deep into one side of the disk with an 83 mm diameter bit, contraction strains of order 60×10^{-6} occurred on the opposite face. The maximum grain diameter was 4 mm so at least 100 grains must have been cut by the 5 mm deep kerf. If the residual stresses were balanced on the grain scale, St. Venant's principle shows us that elastic strain changes could not be measured on the opposite face of the disk. Thus, either these measurements represent a case of situation two in which the equilibrium and second overcore volumes are of the same order, or non-elastic processes occurred during overcoring. Since the strains were contractions, crack opening cannot be offered as the explanation. Perhaps crack closing could occur, but this seems unlikely without the continued existence of a compressive stress to hold them closed.

At Rangely, Colorado *in situ* stresses were measured by a number of different techniques (DE LA CRUZ and RALEIGH, 1972; FRIEDMAN, 1972, p. 306–310). The measured stresses are felt to be residual by FRIEDMAN (1972, p. 309–310) because of the proximity of the horizontal measurement surface to a free vertical face and because of the agreement in stress directions between the overcoring results and the X-ray results. However, if the overcoring did elastically release only residual stresses, and should therefore be thought of as second overcoring (with the initial overcoring having been due to erosion and macrofracturing), then the measurements represent a case of situation three in which the equilibrium volume is much bigger than the second overcore volume. This is because the measurements all give similar results indicating that only a small part of an equilibrium volume is being sampled. On the other hand the residual stresses measured using X-rays on isolated samples must be balanced on a grain scale. In other words, although it is possible that both the field and the X-ray measurements are measuring residual stresses, the stresses are on two different scales and cannot both be locked in by the same process. However, if non-elastic processes are occurring during overcoring then the overcoring measurements could be related to the grain scale stresses. The stresses of both scales are similar in direction to the stresses at depth that are responsible for the earthquakes at Rangely (RALEIGH *et al.* 1972; HAIMSON 1973). It does not seem helpful to consider stresses on this larger scale to be residual since it is not possible to show that they are not due to tractions on the boundaries of the region, to body forces, or to thermal gradients. A reasonable explanation for similarity in stress directions would be that tectonic stresses in the area at present are similar to what they were at the time the residual stresses measured with X-rays were locked in.

A somewhat similar situation exists in the overcoring done in northeastern New York by ENGELDER and SBAR (1976). Here, initial overcoring measurements systematically show directions of expansion about 90° to each other in different aged members of the Potsdam sandstone. They interpret this result to mean that the relieved stresses are residual rather than applied and believe the stresses to be locked in on a grain scale (ENGELDER and SBAR 1976, p. 3015–3016). If this is true then, following arguments presented above, non-elastic processes must be releasing the grain scale stresses. Expansions do dominate so crack opening could be important, but significant contractions are also measured. If the response is only elastic then the relieved stresses are either applied stresses, residual stresses with equilibrium volumes much larger than the area covered by their measurements, or some combination. ENGELDER and SBAR (1976, p. 3015) also report the results of second overcoring, but no statistically significant directions of strain were found so it is not necessary to attribute those strains to non-elastic behaviour.

SWOLFS *et al.* (1974, p. 565–566) report a second overcoring measurement on a sample in the laboratory that shows the same orientation of maximum expansion as was found by initial overcoring in the field. They conclude from this agreement that the equilibrium volume is larger than the overcore volume. This reasoning cannot be correct for elastic behaviour if the second overcore measurement is assumed to be statistically significant, since it would be a case of situation three and no second overcoring strain could occur. Since all the strains were expansions in both the lab and field overcoring, crack opening could be important. However, the majority of the microcracks in the rock are oriented in the wrong direction for further opening of them to explain the directionality of the expansions.

5. Conclusion

Both elastic and non-elastic behaviour apparently occurs during at least some overcoring measurements. It is not clear what non-elastic processes operate, although permanent opening of microcracks seems one reasonable possibility. That permanent *closing* of microcracks could occur in the absence of permanent compressive macrostresses is difficult to visualize, but this may be possible and it deserves further consideration. Additional careful observations are needed to aid in understanding the types of non-elastic processes that can occur during overcoring and the situations in which they occur. With our present poor understanding it is not possible to assess adequately the influence these processes may have on measurements of both residual and applied stress.

If one assumes perfectly elastic behaviour, then, as long as a statistically significant number of measurements are made, it is not possible for residual stresses to introduce error into the average values of applied stresses measured by initial overcoring, except in the case where the residual stress equilibrium volume is larger than the volume over which the measurements are averaged.

Acknowledgement

I thank the Research School of Earth Sciences at the Australian National University for supporting me as a Visiting Fellow during the time this was written.

REFERENCES

BISHOP, J. F. W. and HILL, R. (1951), *A theory of the plastic distortion of a polycrystalline aggregate under combined stresses*, Phil. Mag. *42*, 414–427.

BRACE, W. F. (1965), *Some new measurements of linear compressibility of rocks*, J. Geophys. Res. *70*, 391–398.

DE LA CRUZ, R. and RALEIGH, C. B. (1972), *Absolute stress measurements at the Rangeley anticline, northwestern Colorado*, Int. J. Rock Mech. Min. Sci. *9*, 625–634.

ENGELDER, J. T. and SBAR, M. L. (1976), *Evidence for uniform strain orientation in the Potsdam sandstone, northern New York, from in situ measurements*, J. Geophys. Res. *81*, 3013–3017.

FRIEDMAN, M. (1972), *Residual elastic strain in rocks*, Tectonophysics, *15*, 297–330.

FRIEDMAN, M., HANDIN, J. and ALAN, G. (1972), *Fracture-surface energy of rocks*, Int. J. Rock Mech. Min. Sci. *9*, 757–766.

FRIEDMAN, M. and LOGAN, J. (1970), *The influence of residual elastic strain on the orientation of experimental fractures in three quartzose sandstones*, J. Geophys. Res. *75*, 387–405.

HAIMSON, B. C. (1973), *Earthquake related stresses at Rangeley, Colorado*, in *New horizons in rock mechanics, Proc. Fourteenth Symposium on Rock Mechanics* (ed. H. R. Hardy, Jr. and R. Stefanko) (Am. Soc. Civil Engineers, New York, 1973), pp. 689–708.

HOSKINS, E. R., RUSSELL, J. E., BECK, K. and MOHRMAN, D. (1972), *In situ and laboratory measurements of residual strain in the Coast Ranges north of San Francisco Bay*, EOS *53*, 1117.

LOVE, A. E. H., *A treatise on the mathematical theory of elasticity*, (Cambridge University Press, 1934), 643 p.

NICHOLS, T. C. (1975), *Deformations associated with relaxation of residual stresses in a sample of Barre granite from Vermont*, U.S. Geol. Surv. Prof. Pap. 875, 32p.

PENG, S. and JOHNSON, A. M. (1972), *Crack growth and faulting in cylindrical specimens of Chelmsford granite*, Int. J. Rock. Mech. Min. Sci. *9*, 37–86.

RALEIGH, C. B., HEALY, J. H. and BREDEHOEFF, J. D. (1972), *Faulting and crustal stress at Rangely, Colorado*, Am. Geophys. Union Menograph *16*, 275–284.

RUSSELL, J. E. and HOSKINS, E. R. (1973), *Residual stresses in rock*, in *New horizons in rock mechanics, Proc. Fourteenth Symposium on Rock Mechanics* (ed. H. R. Hardy Jr. and R. Stefanko) (Am. Soc. Civil Engineers, New York, 1973), p. 1–24.

SIMMONS, G., TODD, T. and BALDRIDGE, W. S. (1975), *Toward a quantitative relationship between elastic properties and cracks in low porosity rocks*, Am. J. Sci. *275*, 318–345.

SWOLFS, H. S., HANDIN, J. and PRATT, H. R. (1974), *Field measurements of residual strain in granitic rock masses, advances in rock mechanics*, Proc. of 3rd Congress of ISRM, *2*, 563–568, U.S. Committee for Rock Mechanics, National Academy of Sciences, Washington, D.C.

TIMOSHENKO, S. and GOODIER, J. N., *Theory of elasticity* (McGraw-Hill, New York, 1951), 506 p.

VARNES, D. J. and LEE, F. T. (1972), *Hypothesis of mobilization of residual stress in rock*, Geol. Soc. America Bull. *83*, 2863–2866.

VOIGHT, B. and ST.-PIERRE, H. P. (1974), *Stress history and rock stress: Advances in Rock Mechanics*, Proc. of 3rd Congress of ISRM, *2*, p. 580–582, U.S. Committee for Rock Mechanics, National Academy of Sciences, Washington, D.C.

(Received 1st November 1976)

Pageoph, Vol. 115 (1977), Birkhäuser Verlag, Basel

The Stress State Near Spanish Peaks, Colorado Determined From a Dike Pattern

By OTTO H. MULLER[1]) and DAVID D. POLLARD[2])

Abstract – The radial pattern of syenite and syenodiorite dikes of the Spanish Peaks region is analysed using theories of elasticity and dike emplacement. The three basic components of Odé's model for the dike pattern (a pressurized, circular hole; a rigid, planar boundary; and uniform regional stresses) are adopted, but modified to free the regional stresses from the constraint of being orthogonal to the rigid boundary. Dike areal density, the White Peaks intrusion, the strike of the upturned Mesozoic strata, and the contact between these strata and the intensely folded and faulted Paleozoic rocks are used to orient the rigid boundary along a north–south line. The line of dike terminations locates the rigid boundary about 8 km west of West Peak. The location of a circular plug, Goemmer Butte, is chosen as a point of isotropic stress. A map correlating the location of isotropic stress points with regional stress parameters is derived from the theory and used to determine a regional stress orientation (N82E) and a normalized stress magnitude. The stress trajectory map constructed using these parameters mimics the dike pattern exceptionally well. The model indicates that the regional principal stress difference was less than 0.05 times the driving pressure in the West Peak intrusion. The regional stress difference probably did not exceed 5 MN/m^2.

Key words: Stress field regional; Volcanic dike pattern; Paleopiezometer.

1. Introduction

The geology of the Spanish Peaks region and the petrology of its igneous rocks have been described by ENDLICH (1878), HILLS (1901), KNOPF (1936), and JOHNSON (1960, 1961, 1968). Sedimentary rocks of the Raton basin, which range in age from Cretaceous to Eocene and are predominantly shales, sandstones, and conglomerates, form the broad asymmetric La Veta syncline (Fig. 1a). The steep western limb of the syncline abuts the intensely folded and faulted Paleozoic rocks of the Sangre de Cristo Mountains, while the eastern limb merges into the horizontal strata of the plains. Two prominent topographic features, West and East Spanish Peak, are about 11 km and 15 km east of the Paleozoic–Mesozoic contact and rise almost 2 km above the surrounding plains. The peaks are composed predominantly of igneous rock as well as some metamorphosed and deformed sedimentary rock. The principal intrusive form associated with each peak is described by Johnson as a stock. The rocks of the peaks, being more resistant to weathering and erosion than the surrounding sedimentary rocks, produce the dramatic relief. Numerous nearly vertical dikes crop

[1]) Colgate University, Hamilton, New York 13346.
[2]) U.S. Geological Survey, Menlo Park, California 94025.

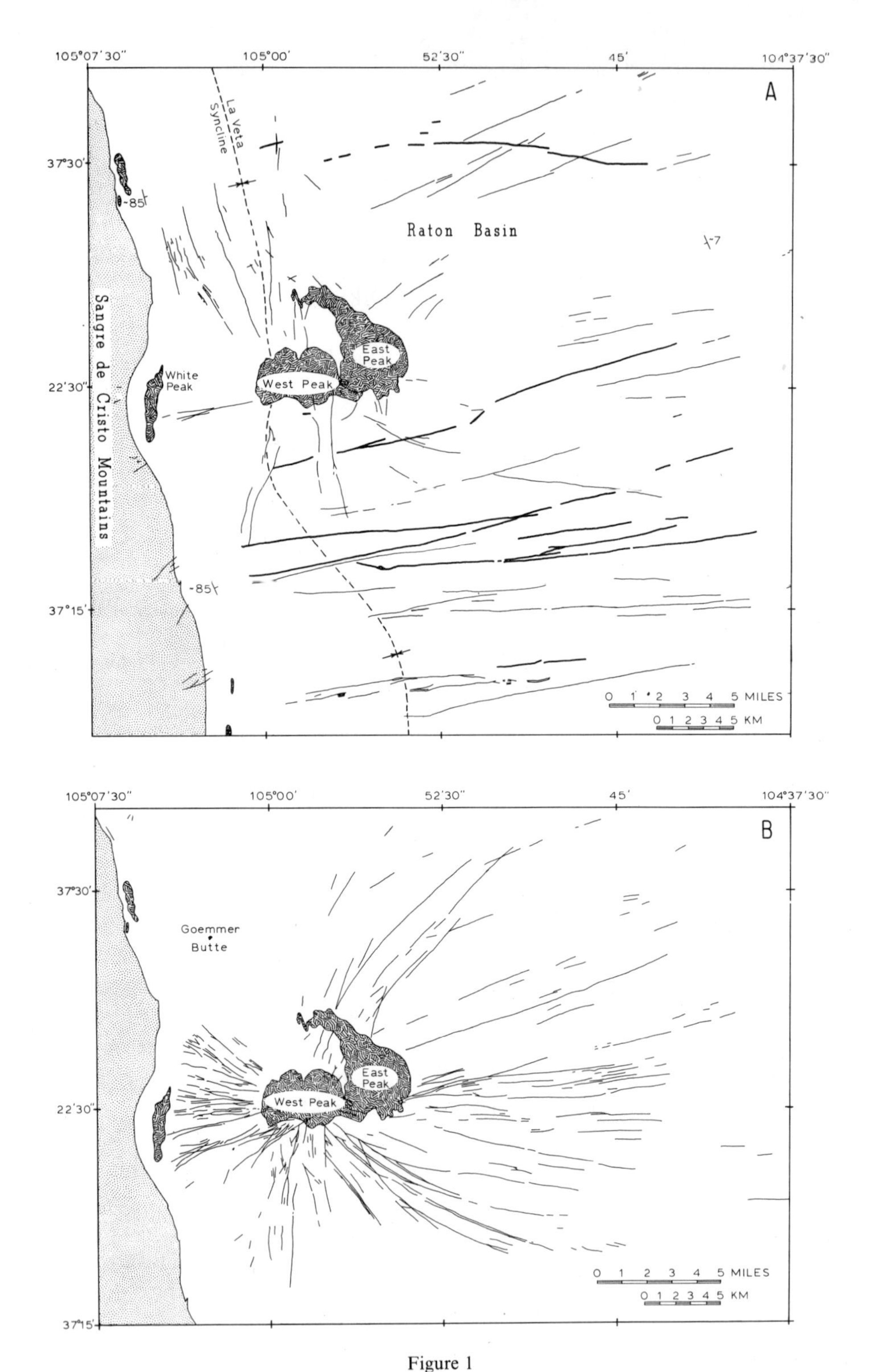

Figure 1

Maps of dikes (solid lines) in the Spanish Peaks region of south central Colorado. Stippled pattern is intensely folded and faulted Paleozoic rock. Hatched pattern is igneous rock. From JOHNSON (1968). (A) Dikes present in Johnson's 1968 map, but not used in our analysis. Thicker lines are syenodiorite dikes of the subparallel swarm. Thinner lines are dikes having compositions other than syenite or syenodiorite. (B) Syenite and syenodiorite dikes used in our analysis.

out around the peaks and some of these form a radial pattern focused approximately on West Peak (Fig. 1b).

The pattern formed by the intersection of the ground surface and these dikes (i.e., the map pattern) was the subject of papers by ODÉ (1957) and JOHNSON (1961). Odé pointed out the similarity between the pattern of maximum principal stress trajectories for a particular elasticity theory model, and the dike pattern as mapped by KNOPF (1936). Odé used all of the dikes on Knopf's map regardless of petrology or orientation. His model consisted of a pressurized circular hole adjacent to a rigid, planar boundary in a two dimensional elastic plate. On this stress system he superimposed a homogeneous far field stress having principal stresses parallel and perpendicular to the rigid boundary. The hole was believed analogous to West Peak stock, the rigid boundary represented the Sangre de Cristo mountain front (i.e., the Paleozoic–Mesozoic contact), and the far field stresses accounted for the regional stress state in the horizontal plane at the time of intrusion. Odé's work suggests that the dikes were emplaced along stress trajectories having orientations which were locally controlled by the stock and Sangre de Cristo range, while far from these features the dikes curved into the trend of the maximum principal regional stress. This work has been cited in several textbooks as an excellent example of the application of theoretical work to structural problems in geology (HILLS, 1963, pp. 377–380; A. JOHNSON, 1970, pp. 400–428). We agree with these citations, but suggest that further development of the model is warranted in light of recent geologic work in the region.

JOHNSON (1961, 1968) presented a new map of the igneous rocks and separated the dikes on a geometric basis into a radial set and a subparallel set trending roughly east–west. We presume that the subparallel set is not related to West Peak and should therefore be excluded from the analysis of the radial pattern. Johnson's map includes many radial dikes not on Knopf's map, particularly to the west of West Peak, which should be included in the analysis. Because none of the radial dikes contact the stock and many are discontinuous along their outcrop trend, Johnson argued that the magma intruded from below the present erosion surface. This contention is supported by SMITH (1975), who used elongated vesicles, and by POLLARD *et al.* (1975), who used contact offsets, to suggest an upward and outward propagation direction for some of the dikes. On the basis of the diverse petrology of different dikes and their cross-cutting relationships, Johnson delineated several (at least seven) different magmatic phases. We cannot justify the assumption that the same stress field existed during all of these phases. Johnson also suggested that the dikes selectively intruded along joints that formed prior to igneous activity. These new field data and interpretations may cause some readers to question the appropriateness of Odé's stress analysis. However, we present arguments below which support the continued use of the basic components of the analysis.

Our first objective is to review the basic postulates of the theory and discuss its applicability to the origin of the dike pattern. Next we modify the model of Odé to account for an arbitrary orientation of the far field stresses and point out some new deductions from this model. Geologic criteria based in part upon Johnson's map are used to select dikes for analysis and to fix the model parameters which are used to derive a trajectory pattern. Finally, we interpret the pattern of selected dikes in the Spanish Peaks region in light of the modified model and present a stress field which accounts for the dike pattern.

2. *Evaluation of the Theory*

For the purpose of discussion we will briefly review the fundamental postulates of elasticity theory without presenting the governing equations which may be found in standard texts (TIMOSHENKO and GOODIER, 1951; MUSKHELISHVILI, 1975). In formulating the two dimensional differential equations of equilibrium for stresses and of compatibility for strains and displacements, one postulates that the model is a static continuum. A linear relation between stress and strain is postulated and the 21 independent elastic constants of the generalized Hooke's Law are reduced to two familiar elastic moduli (Young's modulus; Poisson's ratio) by postulating the model is homogeneous and isotropic with respect to these moduli. The resulting stress-strain equations can be modified further by postulating conditions of plane strain or generalized plane stress and the result used to write the compatibility equation in terms of stresses. Then, neglecting body forces in the plane of interest, the equilibrium and compatibility equations are combined to derive the biharmonic differential equation for the Airy stress function. This equation must be satisfied along with the boundary conditions to find a suitable solution for a particular problem.

Upon first comparing the model postulates with the complexity of the geology in the Spanish Peaks region, it may seem remarkable that a good correlation exists between the stress trajectory pattern derived by Odé and the dike pattern on the map of Knopf. We have insufficient data on the rock properties and physical conditions at the time of intrusion to evaluate differences between the idealized model and the real situation precisely, but we feel that a qualitative evaluation is necessary.

In order to compare the dike pattern with a stress trajectory pattern, we need not resolve stresses within areas less than about 1 km^2. On this scale a horizontal slab of the earth's crust does not contain voids which would make the continuum model inappropriate. Because of the relatively great magma viscosity and small driving pressures, we conclude that inertial forces related to intrusion were negligible and a static model is appropriate. Temperatures and confining pressures in the upper 5 km of the crust over most of the Spanish Peaks region probably did not exceed 250°C and 150 MN/m^2. Under such conditions many sedimentary rocks are adequately described as linear elastic materials (HANDIN, 1966). Within several kilometers of the West Peak intrusive complex or within several meters of the periphery of advancing dikes, elevated temperatures and large differential stresses resulted in non-linear and irrecoverable deformation not described by the model.

Lithologic variation in the sedimentary rock of the Spanish Peaks region forms a distinct geologic inhomogeneity which may also represent a mechanical inhomogeneity. For clastic sedimentary rock, Young's modulus varies from 6×10^{10} to 10^{12} N/m^2 and Poisson's ratio varies from 0.03 to 0.3 (BIRCH, 1966). The ranges of these moduli over much of the Raton Basin were probably smaller because the strata are nearly horizontal (sub-parallel to the plane of the model) and have similar histories of compaction, cementation, and deformation. Another geologic inhomogeneity exists west of West Peak, where the strata of the Raton Basin are abruptly upturned, thrust faulted, and juxtaposed against the intensely folded and faulted rocks of the Sangre de Cristo mountains. The dike pattern was apparently influenced by one or more of these structures because dikes extend less than 10 km to the west, but up to 30 km to the east of the peak. The model treats this inhomogeneity by imposing special boundary conditions along a line paralleling the trend of these structures. Any differences in elastic moduli between rocks on either side of this line are not accounted for directly, but rather through the choice of boundary conditions.

The possibility of significant directional dependency (anisotropy) of elastic moduli is minimized by considering a horizontal slab which is sub-parallel to strata over much of the region. The rocks do contain tectonic joints; however, a joint pattern described by JOHNSON (1961) as complex, perhaps random, may not produce a significant anisotropy. Locally a dike might propagate along a joint, but it would tend to select joints oriented parallel to the maximum principal stress. A significant moduli anisotropy in the horizontal plane may exist in the upturned beds west of West Peak; however, its effect is treated by the boundary conditions.

A two dimensional analysis is justified in that most dikes of the region are nearly vertical and some can be traced over a 1 km change in elevation (JOHNSON, 1961). We represent the stock by a vertical cylindrical body although little supportive evidence is available. Choice between a plane strain or generalized plane stress case is not clear cut. The plane stress case may be a good description of conditions near the ground surface, but probably does not apply at depth. Near the mid-depth of the intrusive structures the plane strain case should provide a reasonable description of natural conditions, but it does not apply near the ground surface. Nevertheless, the in-plane stress distribution and stress trajectories are identical for both plane stress and plane strain. The stress state is independent of elastic moduli if the in-plane body forces are constant or zero, a likely condition for a horizontal plane.

The postulates of sheet intrusion theory also must be examined. ANDERSON (1938, 1972) proposed that the wedging effect of magma in dikes will be greatest along fissures which are perpendicular to the 'smallest rock pressure,' and that these fissures will offer the 'least resistance' to intrusion of the magma. STEVENS (1911) made a similar proposal. The above statements are easily corroborated, if one accepts a very eccentric, pressurized elliptical hole in a two dimensional elastic body as a good model for a dike; one interprets 'smallest rock pressure' to mean least principal far field stress; and one specifies uniform far field stresses (POLLARD, 1973). The tensile stress (wedging action) acting across the end of the long axis of the hole is maximized when the length of the hole is perpendicular to the minimum principal far field stress. Also, a given dilation of the hole is accomplished by the least internal pressure (least resistance) with this geometrical arrangement. Because of symmetry, one trajectory of maximum principal stress extends from the ends of the hole parallel to its long axis. Upon tensile failure, the hole should propagate along this trajectory maintaining a linear form.

The situation for the Spanish Peaks dikes is more complex than this simple model for a single dike. Because a uniform regional stress field is believed to have been perturbed by the stock and mountain front prior to dike intrusion, stress trajectories were not straight lines. How a dike would respond to such a non-uniform stress field is not known, but several studies bear on the problem. Theoretical work by ERDOGAN and SIH (1963), POLLARD (1973), and SIH (1974) suggest that if the maximum principal far field stress is not parallel to the dike, the initial propagation direction should be out of the original plane of the dike and generally toward the direction of maximum far field stress. Experiments by BRACE and BOMBOLAKIS (1963). ERDOGAN and SIH (1963), and HOEK and BIENIAWSKI (1965) confirm these tendencies. Thus a qualitative argument can be made that in a non-uniform stress field a dike should adjust its course of propagation to follow a maximum principal stress trajectory. This adjustment may result in the division of a dike into closely-spaced, *en echelon* segments (POLLARD *et al*, 1975). Such details of dike form are ignored in this analysis although SMITH (1975) has clearly shown that many dikes south-east of West Peak are made up of *en echelon* segments. We use only the general trend of dikes as mapped by JOHNSON (1968).

We will propose a single stress trajectory pattern to compare with the dike pattern. However, a detailed analysis by SMITH (1975) identified two foci for radial dikes south-east of West Peak. In addition, emplacement of a dike may modify the original stress field up to distances of about twice the dike length (POLLARD, 1973). The discontinuous nature of dikes would reduce the area of stress modification, still, after several dikes intruded near West Peak, we would expect the original stress pattern to have changed, and we would not expect the closely-spaced dikes to follow a unique pattern of trajectories. The good correlation between dikes and trajectories found by Odé leads us to suggest that the region around the peak continued to feel the effects of the original stress field while stresses induced by each dike relaxed. The pattern of dikes seems to reflect this unique stress field due to the stock, boundary, and regional stresses without significant alteration by dike intrusion or movement of the focus.

Geologic evidence indicates that dike propagation was not restricted to a horizontal plane. Dikes probably intruded upward and outward from the West Peak area through the horizontal section considered by the model. Because dikes are nearly vertical over 1 km depths, we conclude that the horizontal stress trajectories did not vary substantially over this depth. Thus, the fact that propagation was not horizontal is not a substantive criticism of the model.

3. Derivation of the Model

The model combines the several solutions suggested by Odé and having parameters shown in Fig. 2. Solutions are presented in terms of Airy stress functions which may be differentiated to derive the stress field (TIMOSHENKO and GOODIER, 1951, p. 56). Because the stress trajectory pattern depends upon the deviatoric stress and not the mean stress, an arbitrary pressure may be added over the entire region without changing the results.

One of the solutions treats the stock at West Peak as a pressurized cylindrical hole (source) of radius, r_0, in an infinite two dimensional sheet with no stress at infinity. A constant normal stress and zero shear stress are applied on the hole boundary. The stress function for this geometry is $\Phi_s = A_1 \ln r_1$, where r_1 is the radial coordinate centered inside the hole (TIMOSHENKO and GOODIER, 1951, pp. 55–60). Because the radial stress equals the pressure, P, at $r_1 = r_0$ we must have $A_1 = Pr_0^2$ to satisfy the boundary condition. The mechanical effects produced by structures paralleling the Sangre de Cristo mountain front are treated by superimposing another pressurized cylindrical hole (image) at a distance of 2 d, twice that from West Peak to the mountain front. The stress function for the image is $\Phi_i = A_2 \ln r_2$ where r_2 is centered at the image hole. If the two holes have the same radius and pressure. $A_2 = Pr_0^2$. For the source-image system with uniform internal pressure the stress function is $\Phi_{si} = Pr_0^2 \ln r_1 r_2$.

Superposition of the image stresses in this manner results in only an approximate solution to the problem. The desired boundary conditions on the source are not satisfied. The shear stress will not be zero; the normal stress will not be a constant equal to P; and the stress trajectories will not be radial. However, if the distance between source and image is large compared to the radii of the holes, these alterations are small and can be ignored. For example, the radial stress induced by uniform pressure in the image is $P(r_0/r_2)^2$. Using distances appropriate for West Peak stock ($r \leq 2.6$ km; $d = 8.1$ km) the pressure on the boundary of the source is altered by less than 4%. With equal hole radii and equal positive pressure in both source and image, all displacements perpendicular to the mirror plane between the source and image cancel. Displacements parallel to the mirror plane do not vanish, but double due to the superposition of the image. The mirror plane is a principal stress trajectory with zero shear stress and thus acts as if it were a lubricated rigid boundary.

We judge from Odé's work that equal pressures in the source and image result in a more satisfactory trajectory pattern than equal magnitudes but opposite sign. To further improve the correspondence between the stress trajectories and the dike pattern, Odé superimposed a homogeneous far field stress system (C, B in Odé's paper; S_1, S_3 in the present paper) which ignores the presence of the source and image. The stress function for a homogeneous stress in an infinite region without holes is $\Phi_f = \frac{1}{2}r^2 S_1 \cos^2 \theta + \frac{1}{2}r^2 S_3 \sin^2 \theta$. The S_1 direction corresponds to $\theta = 0°$ and is the maximum principal stress. Odé observed that the symmetry axis of the dike pattern was nearly perpendicular to the long range trend of the Sangre de Cristo range and assumed this indicated that the principal regional stresses acted perpendicular and parallel to the mountain front. He therefore superimposed the far field stresses orthogonal to the rigid boundary by centering the polar coordinates midway between the source and image and letting $\theta = 90°$ correspond to the rigid boundary.

The stress function for the far field stress used by Odé may be modified by replacing θ with $\theta - \phi$ where ϕ is an arbitrary angle between the direction of the maximum principal far field stress, S_1, and the perpendicular to the rigid boundary (Fig. 2). The stress function is

$$\Phi_f = \tfrac{1}{2}S_1 r^2 \cos^2(\theta - \phi) + \tfrac{1}{2}S_3 r^2 \sin^2(\theta - \phi).$$

This modification allows us to vary the orientation of the far field stresses and rigid boundary.

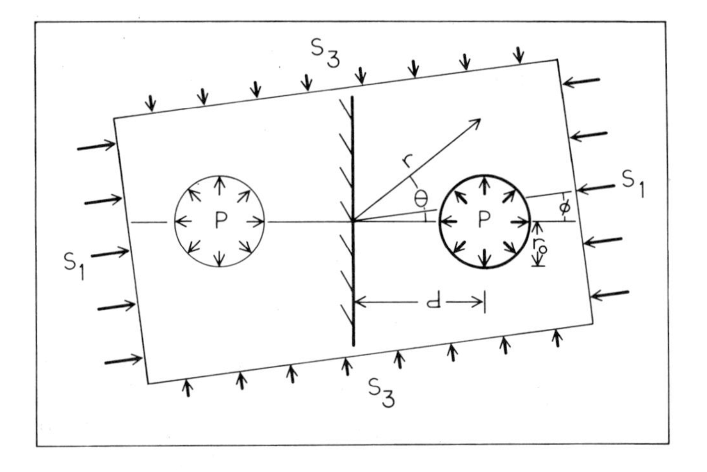

Figure 2
Boundary conditions and parameters used in our model. Two circular holes separated a distance of $2d$, both pressurized with a positive pressure P. The regional stress system, S_1; S_3, is rotated by ϕ from being perpendicular to the rigid boundary. The polar coordinate system is r; θ.

The far field stress is homogeneous throughout the two dimensional sheet and thus alters the desired boundary conditions at the source and along the rigid boundary. We will show below that an appropriate stress trajectory pattern is obtained when $(S_1 - S_3)/P \leq 0.05$. The principal stress difference introduced at the source by the far field stresses is less than 5% of P. We will show below that the boundary conditions along the rigid boundary for distances less than d from the origin are not significantly altered by this superposition. We conclude that this solution is a satisfactory approximation of the solution for the desired boundary conditions.

Differentiating the far field stress function, Φ_f, to obtain the stresses and applying some trigonometric identities we find:

$$(\sigma_{rr})_f = S_3 + (S_1 - S_3)\cos^2(\theta - \phi) \quad (1a)$$

$$(\sigma_{\theta\theta})_f = S_3 + (S_1 - S_3)\sin^2(\theta - \phi) \quad (1b)$$

$$(\sigma_{r\theta})_f = -(S_1 - S_3)\sin(\theta - \phi)\cos(\theta - \phi) \quad (1c)$$

The stresses due to the source image-system are given by:

$$(\sigma_{rr})_{si} = -(\sigma_{\theta\theta})_{si} = Pr_0^2[\xi^6 - \xi^4 d^2 \sin^2\theta - (\xi\, dr)^2 \cos^2\theta - d^4 r^2 \sin^2\theta \cos^2\theta]/[\xi^4 - d^2 r^2 \cos^2\theta]^2 \quad (2a)$$

$$(\sigma_{r\theta})_{si} = -Pr^2[d^2 \sin\theta \cos\theta(\xi^4 - 2\xi^2 r^2 + d^2 r^2 \cos^2\theta)]/[\xi^4 - d^2 r^2 \cos^2\theta]^2 \quad (2b)$$

Here $2\xi^2 = r^2 + d^2$. Note that the normal stresses are equal but opposite in sign. The total stress field is found by adding (1) and (2). The angle ω made by the maximum principal stress trajectory with the normal to the rigid boundary is given by

$$\omega = \theta + \tfrac{1}{2}\tan^{-1}[2\sigma_{r\theta}/(\sigma_{\theta\theta} - \sigma_{rr})], \quad \sigma_{rr} > \sigma_{\theta\theta} \quad (3a)$$

$$\omega = \theta + \tfrac{1}{2}\pi + \tfrac{1}{2}\tan^{-1}[2\sigma_{r\theta}/(\sigma_{\theta\theta} - \sigma_{rr})], \quad \sigma_{rr} < \sigma_{\theta\theta} \quad (3b)$$

The local principal stress difference is given by:

$$\sigma_1 - \sigma_3 = 2\{[(\sigma_{\theta\theta} - \sigma_{rr})/2]^2 + \sigma_{r\theta}^2\}^{1/2} \quad (4)$$

The equations developed above are formidable, and it is not obvious what effect changing the magnitude, $(S_1 - S_3)/P$, or the orientation, ϕ, of the far field stresses will have on the stress trajectory pattern. This problem can be simplified by noting that each stress trajectory pattern has two unique isotropic stress points where $\sigma_{rr} = \sigma_{\theta\theta}$ and $\sigma_{r\theta} = 0$. It is not difficult to visualize locations of the isotropic points if a pattern of trajectories is known. We construct a map from which values of $(S_1 - S_3)/P$ and ϕ can be found for particular locations of the isotropic points. This provides the capability of using a dike pattern to predict relative magnitude and orientation of the regional stresses.

Figure 3 shows the stress trajectories produced for selected values of $(S_1 - S_3)/P$ and ϕ. Distances are normalized by d and the stress difference is normalized by $P(r_0/d)^2$. The locations of isotropic points relative to stress trajectory patterns are illustrated in this figure. The curvature of the trajectories is greatest near these points and diminishes in every direction as one moves away from them. Trajectories tend to diverge from the isotropic points, whereas they converge radially towards the center of the source. Far from the source the trajectories are parallel to the maximum far field stress (Fig. 3b, c, d). If there is no far field stress the trajectories far from the source are radial to the point midway between the source and image (Fig. 3a). As $(S_1 - S_3)/P$ increases relative to $(r_0/d)^2$, the isotropic points move closer to the

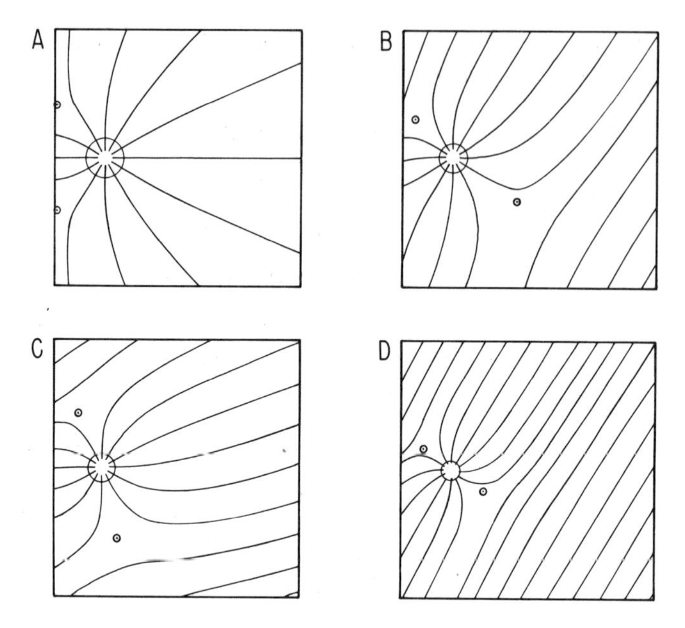

Figure 3
Stress trajectory patterns: The isotropic points for each case are shown, and a circle, centered on the intrusion with a radius of one third the distance to the nearest isotropic point, demonstrates the maximum hole size permitted by the postulates in the model. (A) No regional stress. (B) Normalized regional stress difference of 1.0; $\phi = 60°$. (C) Normalized regional stress difference of 1.0; $\phi = 20°$. (D) Normalized regional stress difference of 4.0; $\phi = 60°$.

source (Fig. 3b, d). As ϕ increases and the trajectories rotate in a counter-clockwise fashion, the isotropic points also rotate (Fig. 3b, c). The trajectories are nearly normal to the rigid boundary within a distance, d, of the origin in both Figure 3a, c. Thus addition of this regional stress does not significantly alter the stress trajectories at the rigid boundary.

Given the locations of isotropic points, Fig. 4 may be used to determine the far field stress parameters. First consider the case of a single pressurized circular hole with homogeneous far field stresses. Superposition of Φ_s and Φ_f results in

$$(\sigma_{rr})_{sf} = P(r_0/r)^2 + (S_1 - S_3)\cos^2(\theta - \phi) \tag{5a}$$

$$(\sigma_{\theta\theta})_{sf} = -P(r_0/r)^2 + (S_1 - S_3)\sin^2(\theta - \phi) \tag{5b}$$

$$(\sigma_{r\theta})_{sf} = -(S_1 - S_3)\sin(\theta - \phi)\cos(\theta - \phi) \tag{5c}$$

neglecting a hydrostatic stress which has no effect on the isotropic points. There will be two isotropic points. These lie on a line through the center of the hole in the direction of the minimum principal far field stress and at a distance, r_i, from the center. The normalized distance is

$$r_i/r_0 = \pm[2P/(S_1 - S_3)]^{1/2} \tag{6}$$

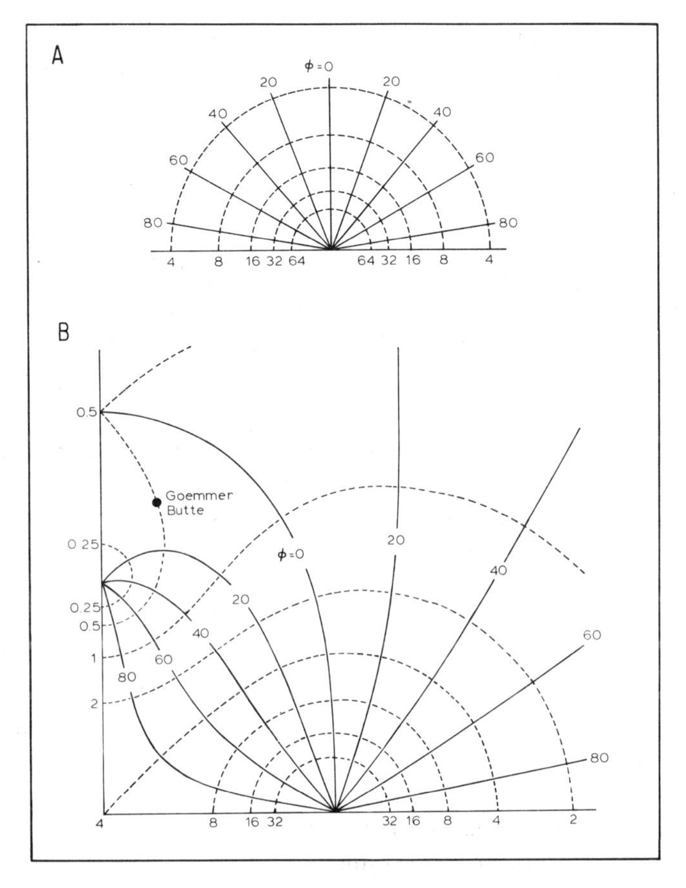

Figure 4
The isotropic point maps. Circumferential numbers refer to the angle made by the regional stress with an east–west line. Numbers along the horizontal axis refer to values for the normalized regional stress. (A) Map for no rigid boundary. (B) Map for our model.

Each possible combination of $(S_1 - S_3)/Pr_0^2$ and ϕ can be associated with a point in Fig. 4a. Lines of constant ϕ are radial and lines of constant $(S_1 - S_3)/Pr_0^2$ are circular. Given r_i/r_0 the relative stresses are determined.

In the case of an image with equal pressure located a distance, $2d$, from the source, the picture is more complicated. In Fig. 4b the hole radius, r_0, is normalized by d and the far field stress difference is normalized by Pr_0^2/d^2. Curves of constant ϕ and constant $(S_1 - S_3)d^2/Pr_0^2$ are shown. Given the locations of the isotropic points as well as r_0 and d the relative stresses are determined.

The following restrictions are put on the use of Fig. 4 in order to approximate the desired boundary conditions at the source. For the source-image system the radius of the source must be less than about $0.4d$ to avoid alteration by the image. There is a limit to the source radius beyond which the trajectories at the boundary are significantly non-radial. We choose a limit such that r_0 is less than one third the distance to the closer isotropic point. A consequence of these restrictions is that $(S_1 - S_3)/P$ must be less than 0.2.

4. *Interpretation of the Dike Pattern*

Because of detailed maps and some relative and absolute age dates now available (JOHNSON, 1968; SMITH, 1975, 1976) a selection of dikes (Fig. 1) can be made. Smith dated several radial dikes at 20 to 25 million years. We eliminate the older, sub-parallel, east-trending dikes and divide the radial dikes according to Johnson's petrologic identification.

Those of syenite and syenodiorite composition are chosen to compare with the model for several reasons. They are numerous and clearly represented in all quadrants around the peak. Thus we will have sufficient data to compare with the model and can determine the correlation on all sides of the peak. These two petrologic types are considered by Johnson, on the basis of dike intersections, to have been intruded sequentially. Therefore selecting dikes of these two closely associated compositions increases the likelihood that we are dealing with a time span during which the regional and local stresses did not vary substantially. That this is not strictly true is evidenced by syenodiorite dikes among the subparallel set. Those dikes not used in the analysis are shown in Fig. 1a in which the subparallel syenodiorite dikes appear as the thicker lines.

We now establish the orientation of the rigid boundary. Odé observed that the dike pattern on Knopf's map was roughly symmetrical on both sides of an axis trending N 75 E, which is nearly perpendicular to the long range trend of the mountain front and the Paleozoic–Mesozoic contact. His use of a line oriented N 15 W as a rigid boundary followed from this observation. On Fig. 1b this symmetry is not apparent west of West Peak where the dikes tend to be symmetrical about an E–W axis. Most of the dikes west of West Peak on Fig. 1b are absent from Knopf's map and thus were not included in Odé's analysis. Figures 3a, b, c indicate that for relatively small values of $(S_1 - S_3)/P$ or ϕ, stress trajectories between the source and rigid boundary are nearly symmetrical about an axis perpendicular to the boundary and running through the source. The orientations of these trajectories are primarily controlled by the source and boundary rather than the far-field stress. This suggests that the rigid boundary should be oriented N–S.

In addition to the pattern produced by the orientation of dikes on the map, there is the pattern produced by the density of these dikes. On Knopf's map these two patterns are both roughly symmetrical along N 75 E. On Figure 1b the pattern produced by the areal density of dikes near West Peak is nearly symmetrical along an E–W axis. This is apparent when the dike areal density is contoured. (Fig. 5a.) The orientation of the rigid boundary of the model could be responsible for this east–west symmetry.

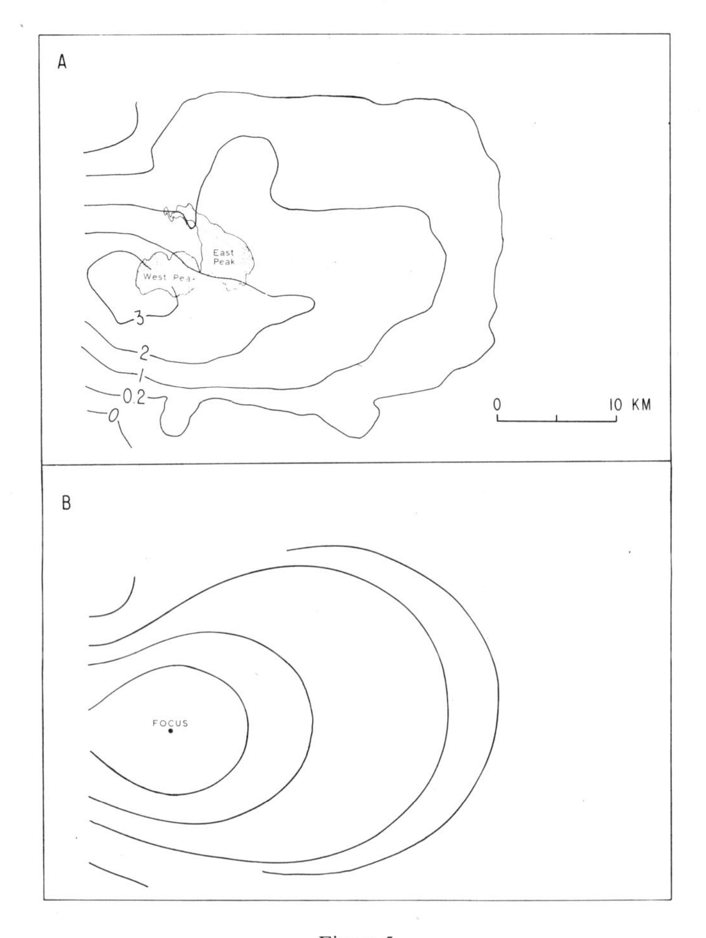

Figure 5
Relationship between dike areal density and local principal stress difference. (A) Areal dike density (number of dikes per 1.27 km^2 averaged over 62 km^2). (B) Contours of one half the local principal stress difference calculated for $\phi = 8°$; $d^2(S_1 - S_3)/Pr_0^2 = 0.5$.

We suggest that the local principal stress difference is related to dike areal density. A region having great principal stress difference could accommodate more dikes than a region having less principal stress difference. Figure 5b is a contour map of the local principal stress differences for the model determined using equation (4). We show below that values of the ratio $(S_1 - S_3)d^2/Pr_0^2$ of about 0.5 produce patterns similar to Fig. 1b. With ratios this small, very little effect is seen in the symmetry of the principal stress difference plot near the source, regardless of the values chosen for ϕ. Thus the dike areal density near West Peak should be relatively insensitive to the orientation of the regional stress and should be a good indicator of the orientation of the rigid boundary. The east–west symmetry near West Peak apparent in Fig. 5a suggests that the rigid boundary should be oriented N–S.

After locating one of the isotropic points we can use Fig. 4 to determine regional stress parameters. The principal stress difference is zero at an isotropic point; hence, an isotropic point should occur where the dike areal density is least. Figures 1b and 5a indicate probable locations near the mountain front northwest and southwest of West Spanish Peak. In the northern region there is a small syenodiorite intrusion, Goemmer Butte, with steep contacts and nearly circular plan. An intrusion with circular plan is not unexpected in a region having a small principal stress difference. We arbitrarily choose Goemmer Butte as one isotropic point.

We now determine the distance, *d*, between the source and a north–south rigid boundary. Several structures with this orientation lie west of the western termination of the dikes. These include the White Mountain intrusion, the strike of the upturned Mesozoic strata, and the contact between these strata and the intensely folded and faulted Paleozoic rocks. Any of these structures might be responsible for an abrupt increase in the elastic moduli in the horizontal plane. Such an increase would produce mechanical effects similar to the zero displacement boundary condition. A soft bed or bedding planes with low shear strength within the upturned Mesozoic strata could

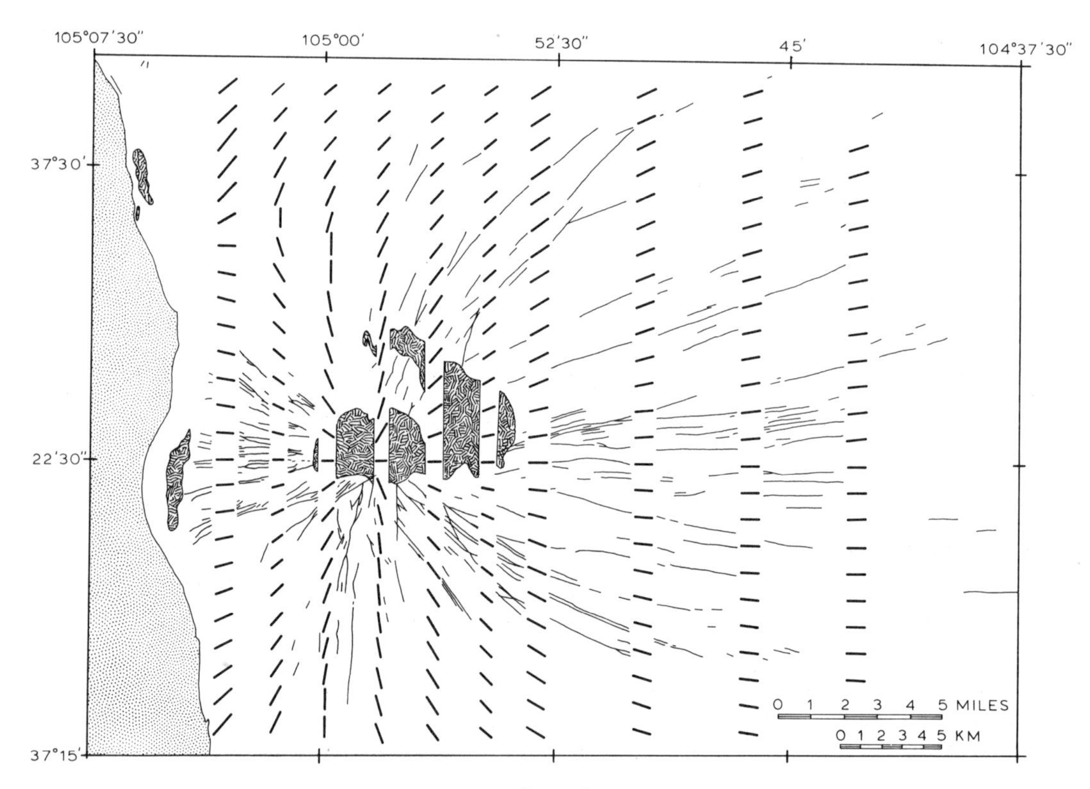

Figure 6

Map of calculated stress trajectories (short, thick lines) and the dikes used in our analysis (thin lines). Trajectories and dikes are presented in alternating strips across the map to facilitate correlation.

reduce the transmitted shear stress. This would produce mechanical effects similar to the zero shear stress boundary condition. ERDOGAN and BIRICIKOGLU (1973) investigated the propagation of a pressurized crack approaching a boundary with a material of greater modulus. The crack tip stress intensity falls abruptly toward zero near the boundary suggesting that crack propagation might cease just before the boundary. For this reason we locate the rigid boundary along the western termination of the dikes. The distance, d, from the rigid boundary to the focal point of the dikes (i.e., the center of the West Peak intrusion) is about 8 km.

The origin of polar coordinates $(r:\theta)$ is at the intersection of the rigid boundary and an east–west line through the focal point. In polar coordinates the location of the isotropic point (Goemmer Butte) is $r = 11$ km; $\theta = 79°$. From these data and Fig. 4b we find that $\phi = 8°$ and $d^2(S_1 - S_3)/Pr_0^2 = 0.5$. Because the rigid boundary is north–south this value of ϕ yields an orientation of the maximum principal regional stress of N82E. We now have values for all parameters necessary to compile a stress trajectory map. The maximum principal stress trajectories and the dikes used in the analysis are shown together in Figure 6. The correlation is excellent.

5. *Discussion*

In the preceding analysis we employed three basic assumptions: (1) linear elasticity is an appropriate theory for conditions in the Spanish Peaks region during dike emplacement; (2) a pressurized, circular hole, a rigid, planar boundary and uniform regional stresses adequately describe the boundary conditions; (3) dikes were emplaced along a unique set of trajectories of maximum principal stress. Geological and mechanical arguments were used to position the rigid boundary along a N–S line about 8 km west of the focus of the radial dike system. Goemmer Butte was selected as a point of isotropic stress. With this information we used the model to deduce a maximum regional principal stress direction of N82E and a value of 0.5 for the ratio $(S_1 - S_3)\, d^2/Pr_0^2$. This stress state was active in the Spanish Peaks region between 20 and 25 million years ago.

By estimating the source radius, r_0, the ratio of the regional stress difference to the pressure in the source can be calculated. A radius of 2.6 km circumscribes the exposed igneous complex near West Peak as well as the two focal areas identified by SMITH (1975). This seems like a reasonable upper limit for the radius, and results in an upper limit for the regional stress difference of $(S_1 - S_3) = 0.05P$. The radius of Smith's smaller focal area is about 0.5 km which yields $(S_1 - S_3) = 0.002P$. We cannot determine precisely the radius at the time of intrusion of syenite and syenodiorite dikes from geological observations, because considerable enlargement of the conduit may have occurred during subsequent magma movement, and the surface radius may be significantly different from that at depth.

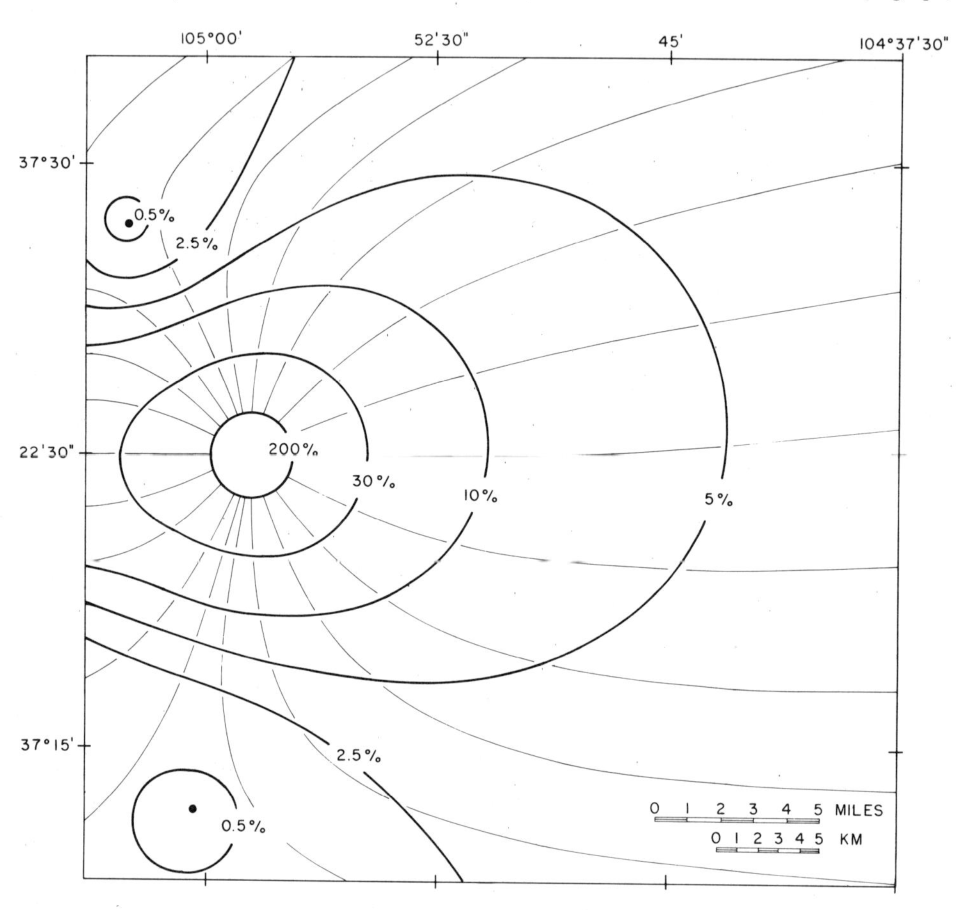

Figure 7
Local principal differences and trajectories of the maximum principal stress. The bold lines contour the stress difference as a percentage of P for the case $r_0 = 2$ km, $\phi = 8°$, and $(S_1 - S_3)\, d^2/Pr_0 = 0.5$. Thinner lines are the stress trajectories and the two solid dots are points of isotropic stress.

For the purpose of further discussion we arbitrarily choose $r_0 = 2$ km and $(S_1 - S_3) = 0.03P$. Figure 7 illustrates the proposed stress state in the horizontal plane covering a region of about 1600 km^2 around the Spanish Peaks. Maximum principal stress trajectories and contours of the local principal stress difference as a percentage of the magma driving pressure are shown. The stress field is very heterogeneous over the region. Trajectories trend in all possible compass directions and the principal stress difference varies by almost two orders of magnitude. It is clear that a few isolated measurements of stress would not adequately characterize the complete stress field. Geological discontinuities along the Sangre de Cristo mountain front and around the West Peak intrusive complex apparently coincide with mechanical discontinuities which significantly altered an otherwise uniform regional stress field. On the other hand, geological structures such as joints and lithologic layering within the Raton Basin apparently did not influence the stress field strongly at the scale we are considering. Local differences between the trajectory and dike patterns may be attributable to these structures.

To determine the regional principal stress difference we must estimate the magma driving pressure, P. Arguments based on the density contrast between magma and host rock (JOHNSON and POLLARD, 1973) suggest an upper limit of about 100 MN/m^2. A regional stress difference of this magnitude results in a local principal stress difference of 200 MN/m^2 at the boundary of the stock. The ultimate strengths of elastic sedimentary rocks at confining pressures less than 150 MN/m^2 and at high temperatures (HANDIN, 1966) are generally less than 200 MN/m^2, so this appears to be a good upper limit. Using $P \leq 100$ MN/m^2 and $r_0 \leq 2.6$ km we find $(S_1 - S_3) <$ 5 MN/m^2. For comparison, BREDEHOEFT *et al.* (1976) report an average value of 2.1 MN/m^2 for the principal stress difference in the horizontal plane in the Piceance Basin, Colorado. They made 26 measurements to depths of 0.5 km in seven different holes using hydraulic fracturing techniques. In light of this data, our predicted upper limit would seem to be quite reasonable for the Raton Basin at the time of dike intrusion.

Acknowledgements

We thank Ross Johnson of U.S.G.S. Denver, Arvid Johnson of Stanford University, and Barry Raleigh and Bill Holman of U.S.G.S. Menlo Park for critically reading the manuscript and offering useful suggestions for its improvement. Richard Smith of Climax Molybdenum Co. generously shared his data and ideas concerning the dikes southeast of West Peak.

REFERENCES

ANDERSON, E. M. (1938), *The dynamics of sheet intrusion*, Proc. Roy Soc. Eding. *58*, no. 3, 242–251.

ANDERSON, E. M., *The Dynamics of Faulting and Dyke Formation with Applications to Britain* (Hafner, New York 1972), 206p.

BIRCH, F., *Compressibility, Elastic Constants* in *Handbook of Physical Constants* (ed. S. P. Clark), The Geol. Soc. of America (New York, 1966), pp. 97–174.

BRACE, W. F. and BOMBOLAKIS, E. G. (1963), *A note on brittle crack growth in compression*, J. Geophys. Res. *68*, 3709–3713.

BREDEHOEFT, F. D., WOLFF, R. G., KEYS, W. S. and SHUTER, E. (1976), *Hydraulic fracturing to determine the regional in situ stress field, Piceance Basin, Colorado*, Geol. Soc. America Bull., *87*, 250–258.

ENDLICH, F. M. (1878), *Report on the Geology of the White River District*, Tenth Annual Report of the U.S. Geological and Geographical Survey of the Territories, 63–131.

ERDOGAN, F. and SIH, G. C. (1963), *On the crack extension in plates under plane loading and transverse shear*, J. Basic Engineering, *85*, 519–527.

ERDOGAN, F. and BIRICIKOGLU, V. (1973), *Two bonded half planes with a crack going through the interface*, Int. J. Engineering Sci., *11*, 745–766.

HANDIN, J., *Strength and ductility*, in Handbook of Physical Constants (ed. S. P. Clark) (The Geol. Soc. of America, New York 1966), 223–290.

HILLS, R. C. (1901), *Description of the Spanish Peaks quadrangle, Colorado*, U.S. Geological Survey Geol. Atlas Folio 71, 70.

HILLS, E. S., *Elements of structural geology* (Wiley, New York 1963), 483 pp.

HOEK, E. and BIENIAWSKI, Z. T. (1965), *Brittle fracture propagation in rock under compression*, Int. J. Fracture Mech. *1*, 137–155.

JOHNSON, A. M., *Physical processes in geology* (Freeman and Cooper, San Francisco 1970), 577 pp.

JOHNSON, A. and POLLARD, D. D. (1973), *Mechanics of growth of some laccolithic intrusions in the Henry Mountains, Utah, 1*, Tectonophysics, *18*, 261–309.

JOHNSON, R. B. (1960), *Brief description of the igneous bodies of the Raton Mesa region, south-central Colorado*, in *Guide to the Geology of Colorado*, Geol. Soc. America, Rocky Mtn. Assoc. Geologists, and Colorado Sci. Soc., 117–120.

JOHNSON, R. B. (1961), *Patterns and origin of radial dike swarms associated with West Spanish Peak and Dike Mountain, south-central Colorado*, Geol. Soc. America Bull. *72*, 579–590.

JOHNSON, R. B. (1968), *Geology of the igneous rocks of the Spanish Peaks region, Colorado*, Geol. Soc. America Bull. *47*, 1727–1784.

KNOPF, A. (1936), *Igneous geology of the Spanish Peaks region, Colorado*, Geol. Soc. America Bull. *47*, 1727–1784.

MUSKHELISHVILI, N. I., *Some Basic Problems of the Mathematical Theory of Elasticity* (Noordhoff, Leyden 1975), 732 pp.

ODÉ, H. (1957), *Mechanical analysis of the dike pattern of the Spanish Peaks area, Colorado*, Geol. Soc. America Bull. *68*, 567–576.

POLLARD, D. D., MULLER, O. H. and DOCKSTADER, D. R. (1975), *The form and growth of fingered sheet intrusions*, Geol. Soc. America Bull. *86*, 351–363.

POLLARD, D. D. (1973), *Derivation and evaluation of a mechanical model for sheet intrusion*, Tectonophysics, *19*, 233–269.

SIH, G. C. (1974), *Strain-energy-density factor applied to mixed mode crack problems*, Int. J. Fracture, *10*, 305–321.

SMITH, R. P. (1973), *Age and emplacement structures of the Spanish Peaks dikes, south-central Colorado*, Geol. Soc. America. Abs. with Programs (Cordilleran Sec.), *5*, 513.

SMITH, R. P. (1975), *Structure and petrology of Spanish Peaks dikes, south-central Colorado*, Ph.D. Thesis, University of Colorado, Boulder, 191 pp.

SMITH, R. P. (1976), *Compositional evolution and tectonic relations of Spanish Peaks igneous rocks, south-central Colorado*, Geol. Soc. America, Abs. with Programs (Rocky Mountain Sec.) *8*, 632.

STEVENS, B. (1911), *The laws of intrusion*, American Inst. of Mining Eng. Bull. *49*, 1–23.

TIMOSHENKO, S. and GOODIER, J. N., *Theory of Elasticity* (McGraw-Hill, New York 1951), 506 pp.

(Received 4th October 1976)

Pageoph, Vol. 115 (1977), Birkhäuser Verlag, Basel

Volcanoes as Possible Indicators of Tectonic Stress Orientation—Aleutians and Alaska[1])

By KAZUAKI NAKAMURA[2]), KLAUS H. JACOB[3]) and JOHN N. DAVIES[3])

Abstract – A new method for obtaining from volcanic surface features the orientations of the principal tectonic stresses is applied to Aleutian and Alaskan volcanoes. The underlying concept for this method is that flank eruptions for polygenetic volcanoes can be regarded as the result of a large-scale natural magma-fracturing experiment. The method essentially relies on the recognition of the preferred orientation of radial and parallel dike swarms, primarily using the distribution of monogenetic craters including flank volcanoes. Since dikes tend to propagate in a direction normal to the minimum principal stress (T-axis), the method primarily yields the direction of the maximum horizontal compression (MHC) of regional origin. The direction of the MHC may correspond to either the maximum (P-axis) or intermediate (B-axis) principal stress.

The direction of MHC obtained at 20 volcanoes in the Aleutian arc coincides well with the direction of convergence between the Pacific and North American plates. This result provides evidence that in the island arc the inferred direction of MHC is parallel to the maximum principal tectonic stress. In the back-arc region, general E–W trends of MHC are obtained from seven volcanic fields on islands on the Bering Sea shelf and the mainland coast of Alaska. These volcanic fields consist mostly of clusters of monogenetic volcanoes of alkali basalt. In the back-arc region, the trends of MHC may correspond to an E–W intermediate, a vertical maximum, and a N–S minimum principal stress.

Implications for the tectonics of island arcs and back-arc regions are: (1) volcanic belts of some island arcs, including the Aleutian arc, are under compressional deviatoric stress in the direction of plate convergence. It is improbable that such arcs would split along the volcanic axis to form actively spreading marginal basins. (2) This compressional stress at the arc, probably generated by underthrusting, appears to be transmitted across the entire arc structure, but is apparently replaced within several hundred kilometers by a stress system characterized by horizontal extension (tensional deviatoric stress) in the back-arc region. (3) The volcanoes associated with these two stress systems differ in type (polygenetic vs. monogenetic) and in the chemistry of their magmas (andesitic vs. basaltic). These differences and the regional differences in orientation of the principal tectonic stresses suggest that the back-arc stress system has its own source at considerable depth beneath the crust.

Key words: Stress field, regional; Volcanic dike pattern; Tectonics of Aleutians and Alaska; Flank crater distribution.

Introduction

The orientation of tectonic stress[4]) which has existed in recent times in the upper crust of volcanic regions can be revealed by certain volcanic surface features which

---

[1]) Lamont-Doherty Geological Observatory Contribution No. 2503.

[2]) On leave from Earthquake Research Institute, Univ. of Tokyo, Tokyo 113, Japan. Present address: Geophysics Dept., Stanford Univ., Stanford, California 94305, USA.

[3]) Lamont-Doherty Geological Observatory of Columbia University, Palisades, New York 10964, USA.

[4]) For brevity, we refer to the directions of the principal stresses of tectonic origin by the phrase 'orientation of tectonic stress'; recognizing implicitly that stress is a tensor.

indicate the trend of zones of concentration of dikes underground (NAKAMURA 1969 and 1977). The preferred orientation of dikes indicates the direction of the maximum horizontal compression (MHC).

The present paper applies the technique of finding the direction of MHC to Aleutian and Alaskan volcanoes, presents the stress orientations obtained for this region and discusses some mechanical problems related to the tectonics of island arcs and back-arc regions. The basic data for this study were derived from maps of volcanic features and active faults.

Method

Only a brief account of the method for obtaining the orientation of tectonic stress from volcanic features is given here. The method is described and discussed in detail by NAKAMURA (1977). The underlying concept is that the volcanoes may be regarded as large-scale natural magma-fracturing experiments (YODER 1976, p. 179–180), repeated over a long period of time. Essentially, the method relies on determining the trend of dike swarms which tend to develop normal to the minimum compression or parallel to the maximum horizontal compression of regional origin.

For polygenetic volcanoes (i.e., composite volcanoes[5]) and shield volcanoes of Hawaiian type), the most significant and easily recognizable surface expression of the average trend of the dikes is the direction of the long axis of the distribution of flank craters. Because flank eruptions are understood to emanate from radial dikes underground, the strikes of radial dikes are approximately inferred by lines connecting the summit or main crater with individual flank craters. Then the long axis of their distribution gives the average strike along which dikes best develop (Fig. 1). Post-caldera cones are dealt with similarly.

For volcanic fields consisting of clusters of monogenetic volcanoes, the method is rather straightforward. The average strike of eruptive fissures and alignment of monogenetic craters are taken as the surface expression of the dike trends. In this case, magma is understood to be supplied largely from below, since there is no principal polygenetic vent from which magma offshoots radially.

Some other features may be used as an auxiliary means for determining the direction of MHC: e.g., bending of radial eruptive fissures, horizontal elongation of the edifice of polygenetic volcanoes, and parallel normal faults within the volcanic edifices or fields. The bending of fissures may be caused by the change of governing stresses from magmatic pressure near the vent to regional tectonic stresses away from the central craters. Elongation of the main volcanic edifice can result from accumulation of volcanic material from flank craters, the distribution of which is similarly elongated. Normal faults may be the surface manifestation of dikes that did not reach the surface.

---

[5]) 'Composite volcano' is used for stratovolcano.

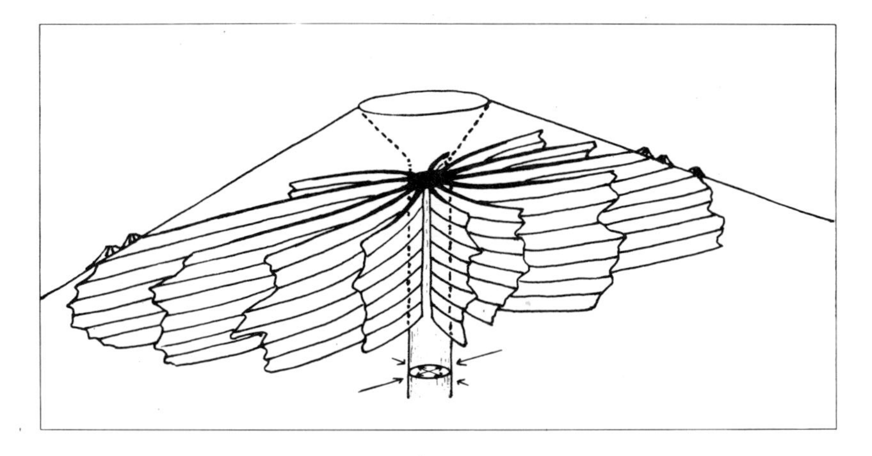

Figure 1
Diagram showing the radial dikes and flank volcanoes of a polygenetic volcano under a differential horizontal stress. The zone of flank volcanoes corresponds to the trend of radial dike concentration, and hence to the trend of the maximum compression of the horizontal component of the ambient stress (NAKAMURA, 1977).

The average strike of these features indicates the direction of either the maximum or the intermediate principal stress of regional origin in the vicinity of the volcano. Therefore, we further have to determine: (1) whether this average strike represents the maximum or the intermediate principal stress, and (2) whether the regional stress is of tectonic origin or the effect of some other source, the most important of which would be the gravitational stresses due to topography of the volcano.

To discriminate between the maximum and intermediate principal stresses within a volcanic belt, the angular relation between the trend of the volcanic belt and the preferred orientation of radial dikes can be used.

In the typical case of an island arc, in which the maximum principal stress is horizontal, the trend of the volcanic belt makes a high angle with the strike of the concentration of radial dikes at individual composite volcanoes. In the other case, for a region of extensional tectonics in which the maximum principal stress is vertical, the dike trends tend to be aligned with the axis of the volcanic belt.

The sense of relative motion across active faults distributed in the same general area can be used as well. The chemistry of magmas and types of volcanoes also may be used as auxiliary means to discriminate between the two stress systems.

If the regional stress in and around polygenetic volcanoes is caused by gravity sliding, then the zones of flank craters may be curved rather than linear; following the contour of the regional slope responsible for the downhill tensional stress. Also, if the strikes of the slopes vary, the strikes of the zones of craters may be variable among adjacent polygenetic volcanoes. Therefore, if the zones of flank craters are linear at individual volcanoes and if the strikes of the zones are common or change only gradually, tectonic stresses may most probably be responsible for the regional

Table 1

Region[1]	No.[1]	Volcano[1]	Number of vents[2]	Length (km)[3]	Width (km)[4]	Elongation[5]	Fault[6]	Azimuth (from north)[7]	Quality[8]	Lat. (°N)	Long. (°W)[9]	A, B, or C[1]	Type of Volcano[1]	Main rock types[1]
Western Alaska and Bering Sea Islands	87	Imuruk Lake	78	57 45	39 41		++ +	45° ± 15° 40°E ± 20°	G P	65.433 ~65.767	162.717 ~164.000	B? and C	scoria cones, small shield volc.	oliv. basalt, andesite
	88	Kookooligit Mtns.	200+	33	6	++		80°W ± 5°	E	63.567 ~63.650	170.092 ~170.672	A or B	sh. volc., sc. cones	oliv. basalt
	89	St. Michael Island	9+	10	2.9	+		60°W ± 10°	G	63.433 ~63.541	162.000 ~162.317	B and C	lava cones, maars?	oliv. basalt
	91	Ingakslugwat Hills	25+	30	30			90°E ± 20°	P	61.300 ~61.583	163.767 ~165.150	C	sc. cones	oliv. basalt
	93	Nunivak Island	99	50	22	+		80°W ± 10°	G	59.750 ~60.438	165.625 ~167.438	B and C	sh. volc.?, sc. cones	basalt
	96	St. Paul Island	45+	12 8	8 10		+ +	55°W ± 10° 50°E ± 20°	G G	57.108 ~57.250	170.092 ~170.417	B? or C	sc. cones, maars	basalt
	99	St. George Island	16	13	7.1		++	70°E ± 20°	G	56.540 ~56.610	169.460 ~169.782	C	sc. cones	basalt, andesite?
Alaska Peninsula and Aleutian Islands and Range	100	Buldir	(1)	(5)	(0.2)	+	++	60°W ± 10°	P	52.317	−175.767	B	Stratovolcano	basalt, andesite
	101	Kiska	(4)	(5.6)	3.8			50°E ± 25°	VP	52.100	−177.600	A	Stratovolcano	andesite
	102	Segula	ca10	6	2.1	+		40°W ± 10°	G	52.016	−178.133	A or B	Stratovolcano	basalt, andesite
	104	Little Sitkin	(1)	(3.5)	(1.1)	+	+	65°W ± 15°	P	51.950	−178.533	A	Stratovolcano	bas., and., dac., rhyodac.
	107	Gareloi	(13)	(5)	(1.7)	+		45°W ± 5°	G	51.800	178.800	A	Stratovolcano	basalt, andesite
	112	Moffett	7	8	4.6			50°W ± 10°	G	51.933	176.750	C	Stratovolcano	basalt, andesite
	113	Adagdak	(1)	(4.5)	0.5		++	60°W ± 10°	P	51.983	176.600	C	Stratovolcano	basalt, andesite
	114	Great Sitkin	6	8.5	2.4			35°W ± 10°	G	52.067	176.117	A	Stratovolcano	basalt, andesite
	131	Vsevidof	(6)	(8)	11			65°W ± 20°	P	53.133	168.700	A	Stratovolcano	basaltic and., and., rhyodac.
	132	Recheshnoi	8	18	16		+?	70°W ± 10°	P	53.150	168.550	B?	Stratovolcano	and., bas., rhyol.
	133	Okmok	55	40	14	+	+	60°E ± 10°	VP	53.417	168.050	A	shield like Stratovolcano	basalt., and., rhyol.
	135	Bogoslof	8	2.2	0.2	+		25°W ± 5°	P	53.933	168.033	A	Stratovolcano	basalt, andesite
	136	Makushin	23	24	27		+	60°W ± 15°	VP	53.867	166.933	A	Stratovolcano	basalt, andesite
	138	Akutan	(1)	(6)	0			60°W ± 15°	VP	54.133	166.000	A	Stratovolcano	basalt, andesite
	154	Dana	4	6.5	2.5	(+)		40°W ± 10°	G	55.617	161.183	B	Stratovolcano	basalt?, andesite?
	165	Veniaminof	ca50	60	(14)	+		35°W ± 5°	E	56.167	159.383	A	Stratovolcano	basalt, andesite
	157	Purple (Black) Peak	5	10	(3.5)	(+)		55°W ± 15°	G	56.555	158.785	B	Stratovolcano	basalt, andesite
	158	Antakchak	(6)	(6)	5			15°W ± 25°	P	56.883	158.167	A	Stratovolcano	dac., and., basalt
	174	Iliamna	7	10	5	+		25°W ± 10°	G	60.033	153.100	A	Stratovolcano	andesite
	178	Spurr	(2)	(3.5)	1.6	+		20°W ± 20°	VP	61.317	152.133	A	Stratovolcano	basalt, andesite?
East and Southeast Alaska	186	Edgecumbe	20	22	9	+		40°E ± 10°	E	57.012	135.767	A, B, or C	Stratovolcano	andesite, basalt

Table 1

Probable azimuths of concentration of radial and parallel dikes of Aleutian and Alaskan volcanoes. Data sources are given in the text. Description of columns:

(1) Type of volcano or volcanic field, after SMITH and SOULE (1973).
- (A) Active volcanoes having historic records of eruptions or active solfataras,
- (B) Volcanoes having no recorded eruptions nor active solfataras, but suspected to have been in eruption during historic time or past 2,000 years,
- (C) Post-Miocene volcanoes other than A or B.

(2) Number of flank and post-caldera cones other than the summit or main crater for volcanoes 100–186, and probable monogenetic vents for volcanoes 97–99. Parenthesis indicates that they are distributed only on one side of the main crater.

(3) Longest distance between craters measured in the direction of the given azimuth.

(4) Width of the zone of monogenetic craters as defined by 95% of them. Parenthesis indicate that their distribution is limited to one side of the main crater.

(5) Crosses show that the contour lines are elongated in the general direction of the azimuth.

(6) Presence of normal faults in the general direction of the azimuth is indicated by crosses.

(7) Azimuth of the probable preferred orientation of dikes as obtained principally from the distribution of craters and supplementally from elongation areas and fault trends.

(8) Quality of azimuths, E: excellent, G: good, P: poor, VP: very poor.

(9) Aerial extent of volcanic fields for volcanoes 87–99, and location of the main craters for volcanoes 100–186.

stress rather than gravitationally induced stresses. The same argument can be applied to volcanic fields of monogenetic volcanoes.

A total of 103 volcanoes and volcanic fields in the Aleutians and Alaska were listed by SMITH and SOULE (1973) as post-Miocene volcanoes. They tabulated the names of volcanoes and volcanic fields along with number (from 85 to 187), geographical location, size, degree of recent activity, type of volcanoes, and their main rock type. These volcanoes were grouped geographically by them into the following three regions:

(1) Western Alaska and Bering Sea Islands (85–99)
(2) Alaska Peninsula, Aleutian Islands and Aleutian Range (100–179).
(3) Eastern and Southeastern Alaska (179–187).

The 103 volcanoes are studied mainly by means of published geologic and topographic maps and literature, most of the references for which are given by SMITH and SOULE (1973). Among 15 volcanic fields listed for western Alaska and the Bering Sea Islands, accurate descriptions of crater distributions are available for seven fields: five from geologic maps and two (88 and 89) from topographic maps. For the remaining eight volcanic fields, little information is available regarding features indicative of orientation of dikes.

For 51 of 79 volcanoes listed in the Alaska Peninsula and the Aleutian Islands and Aleutian Range, geologic maps of variable quality were available. Among the fifty-one volcanoes, twenty have features indicative of a preferred orientation of radial dikes.

In eastern and southeastern Alaska, only one volcano, Mt. Edgecumbe, among nine shows the above features. Six volcanoes in the Wrangell Mountains are deeply ice-covered. Prindle volcano is a single scoria cone. The Rivillagigedo volcanic field consists of randomly scattered basaltic products.

For the twenty-eight volcanoes and volcanic fields which showed a preferred orientation of radial dikes, information derived from the above mentioned maps was used to obtain probable azimuths of their dike concentrations. This information and the resulting azimuths are summarized in Table 1, columns 5–9. Geographical distributions of craters are shown in Fig. 2 for the individual volcanoes listed in Table 1, except the volcanoes Ingakslugwat Hills (91), Akutan (138), and Spurr (178). Each of the latter two has only one flank eruption site.

Description

The above information given in Table 1 and Fig. 2 is supplemented by a brief description of each individual volcano in the following paragraphs. The volcano numbering system is that of SMITH and SOULE (1973) and is the same as that used in Table 1. A few volcanoes not listed in Table 1 are also described, since they have possible information on the preferred orientation of dikes.

Western Alaska and Islands of the Bering Sea

87. Imuruk Lake volcanic field (HOPKINS 1963)

The Imuruk Lake volcanic field, on the Seward Peninsula, consists of numerous scoria cones and lava flows.

> 'The volcanic vents rarely lie in sharply defined alignments: however, they do tend to cluster in major NW-trending swarms and minor NE-trending swarms generally parallel to the two systems of faults observed in the Imuruk Lake area. One exceptionally well-defined alignment is expressed by 6 vents trending northeastward.' (HOPKINS 1963)

The 'major trend' described above by Hopkins lies in a N45°W direction and is over 50 km long; the 'well-defined alignment' strikes N40°E and is 6 km long. Normal faults with NW or NE trends are widely distributed in the volcanic field. The NW-trending faults are better developed (Fig. 2).

88. Kookooligit Mountains (SMITH and SOULE 1973; US Geological Survey St. Lawrence topographic map at the scale of 1:250,000).

The Kookooligit Mountains volcanic field, which occupies the north central part of St. Lawrence Island, may be regarded as a low shield volcano which is elongated in a N80°W direction (Fig. 2). Most of the more than 200 vents, largely scoria cones, are distributed within a linear crestal zone of the shield. The zone strikes N80°W.

89. St. Michael's Island volcanic field (HOARE and CONDON 1971a and 1:63,360 series of topographic maps of Alaska, St. Michael B–1, C–1, and C–2).

St. Michael's Mountain, located in the middle of St. Michael's Island in Norton Sound, is about 170 m high. The summit area, defined by the 350 ft (110 m) contour line, is 1.2 km long, 0.3 km wide and is elongated N60°W. The contours above 150 ft (46 m) show an edifice elongated in the same direction (Fig. 2). To the northwest and southeast of the mountain, there are at least nine maar-like depressions that are more or less circular in plan view and surrounded by possible crater walls a few to several tens of meters high.

Stuart Island, which is included in St. Michael's Island volcanic field by SMITH and SOULE (1973) and located several kilometers NW of St. Michael's Island, consists of Stuart Mountain to the east and the younger West Hill to the west. Stuart Mountain is elongated in a N60°W direction and has a possible flank volcano 2.8 km from the summit, on the S85°E flank. West Hill has four possible flank craters on the NW and SE flanks. They define a linear zone 5 km long which strikes N65°–70°W.

93. Nunivak Island volcanic field (HOARE *et al.* 1968).

The topographic configuration of the Nunivak Island volcanic field on the Bering Sea shelf appears to constitute a low broad shield which is roughly elongated in a

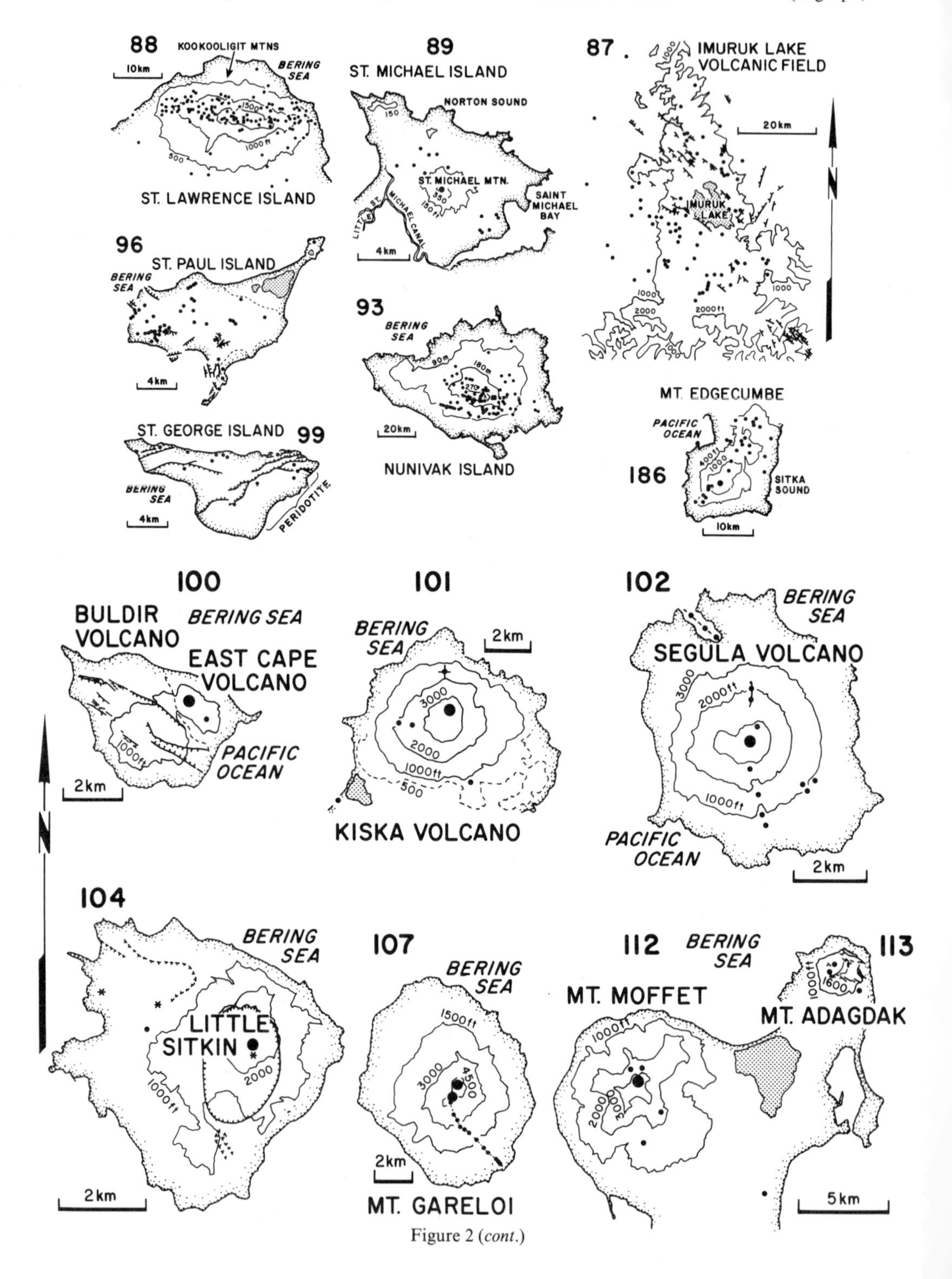

Figure 2 (*cont.*)

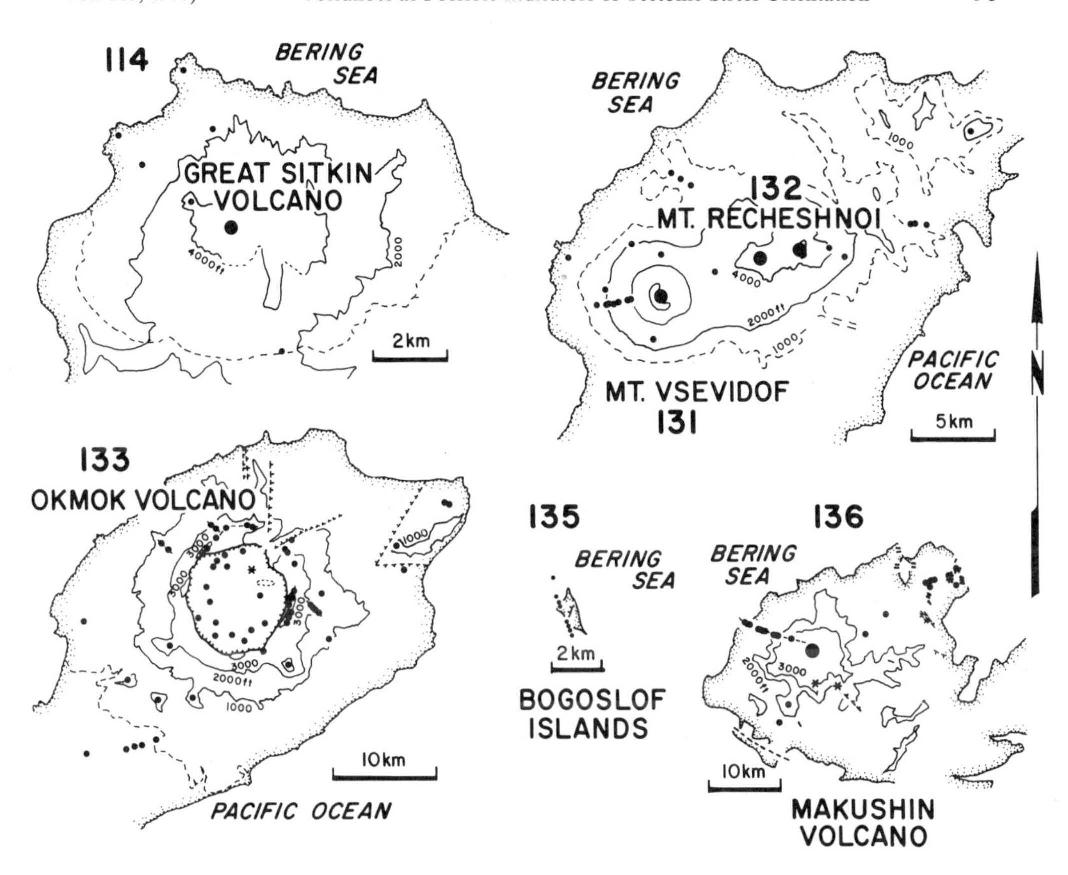

Figure 2 (*cont.*)

Distribution of craters of 25 volcanoes and volcanic fields in the Aleutian and Alaskan region as listed in Table 1 and described in the text. Larger circles: central polygenetic vents. Smaller circles: monogenetic vents. When they constitute an eruptive fissure they are tied by a line. (Broken) lines: (inferred) normal faults. Downthrown side is marked by bars. Stippled area: lake. Stars: solfataric field. Height in feet, unless otherwise indicated. Short bars for Makushin volcano (136): dikes. Thin broken lines of Kiska (101), Great Sitkin (114) and Okmok (133) volcanoes and St. Paul Island (76) are boundaries in the surface geology. Data sources are indicated in the text following the name of the volcano and the volcanic field in the chapter describing each volcano.

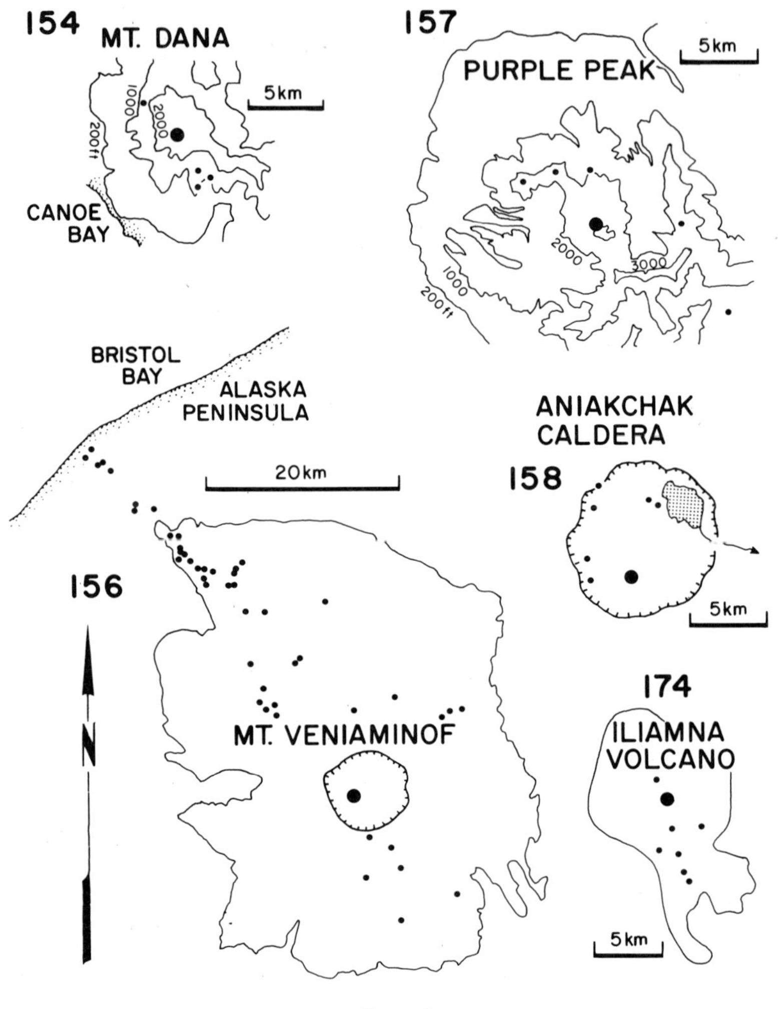

Figure 2

N75°W direction. Roberts Mountain (solid triangle in Fig. 2), 550 m in height, marks the summit of the shield. The distribution of about one hundred scoria cones on the higher part of the island is elongated N80°W. Southwestern vents about 8 km in length show alignment, striking N80°W. Most of these scoria cones are the product of very recent eruptions of alkali basalt. Volcanic activity has occurred discontinuously during the past 6.4 my. During this period, the center of activity migrated eastward on the island. The western, ventless area represents the site of earlier activity.

96. St. Paul Island (Cox *et al.* 1966).

St. Paul Island, one of the Pribilof islands on the Bering Sea shelf consists of more than 45 monogenetic volcanoes (scoria cones and tuff cones) and basaltic lava flows.

The oldest rocks are dated as 0.36 ± 0.10 my. Three NE alignments of craters are evident. Their strikes and lengths are: N55°E, 1 km; N68°E, 2 km; and N45°E, 1.5 km. A 1 km-long alignment in a N60°W direction is also observed. Normal faults are widely distributed on the island. Their strikes concentrate in ENE and NW directions (Fig. 2).

99. St. George Island (Cox *et al.* 1966).

St. George Island, one of the Pribilof Islands, is much eroded and nearly flat. Volcanic eruptions took place during 2.2–1.6 mybp on this island. About 16 vents are identified. They do not show any obvious alignment, however. Normal faults are extensively developed on this island. They strike more or less in an ENE direction. At least one vent is located on a fault. In the eastern part of the island, the faults are downthrown on the NW side. Along the SE coast, there are exposures of serpentinized peridotite which lie unconformably below the volcanic rocks (Fig. 2).

91. Ingakslugwat Hills (Coonrad 1957, Hoare and Condon 1971b).

The Ingakslugwat Hills volcanic field is one of the most recently active areas in the Yukon-Kuskokwin delta region.

> 'Numerous small spatter and breccia cones about 100 feet high occur on the NW side of the volcanic area. Some of these are alined WNW and apparently define a fracture.' (Hoare and Condon 1971b)

There is a 6.5 km-long eruption fissure striking N75°E in the NW part of the field. Another possible eruptive fissure in the middle of the field is 6.5 km long and strikes in a N70°E direction.

Alaska Peninsula, the Aleutian Islands and Aleutian Range

100. Buldir Island (Coats 1950).

Buldir Island, the westernmost Aleutian island with recent volcanic activity, consists of two composite volcanoes: Buldir volcano and the younger East Cape volcano. East Cape volcano has a dome on its southeast flank. On the dissected Buldir volcano, a series of recent normal faults strike N55°–70°W (Fig. 2). The topographically higher sides are generally downthrown.

101. Kiska volcano (Coats *et al.* 1961).

Four flank eruption sites are recognized on the geologic map of Kiska volcano (Fig. 2). One located S50°W and 8 km from the summit is completely outside the volcanic edifice. A flank eruption took place in 1962–64 on the slope due north of the summit (Coats 1964). The volcano is slightly elongated in a N70°E direction.

102. Segula volcano (Nelson, 1959).

The zone of flank craters on Segula volcano trends N30°–35°W. A cove on the north coast of the island is, in its configuration, very much like the one formed during the 1965 fissure eruption of Taal volcano, Philippines (MOORE *et al.* 1965). The cove was probably formed by phreatomagmatic explosions which occurred on a fissure trending N50°W. This fissure is not radial in direction. This may be the result of bending as is the case at Gareloi volcano (107).

104. Little Sitkin volcano (SYNDER 1959).
The only flank crater mapped on this volcano is on the western slope. There are three major fumaroles, aligned N70°W (Fig. 2). Four roughly radial normal faults are distributed on the NW and S flanks.

107. Mt. Gareloi volcano (COATS 1959).
A prominent feature of Mt. Gareloi is a radial fissure formed during its 1929–30(?) eruption. The fissure starts southward from the summit area and within 2 km it assumes a southeasterly trend (Fig. 2). There is an indication that the magma which erupted from the fissure travelled laterally, without being mixed, from the central conduit where it had been gravity-differentiated. The change in azimuth of the fissure is best explained by the increasing effect with distance away from the central conduit of the regional stress over the local magma pressure (ODÉ, 1957).

111. Kanaga volcano (COATS 1956).
Near the summit of this nearly conical volcano, there are three radial elongate depressions, one on the northwest and two on the southeast, aligned in a N65°–70°W direction over a distance of slightly more than 0.5 km (L. SNYDER, personal communication, 1976). According to him, there were major fumaroles in these depressions in 1952. Three recent flows are supposed to have erupted from a fissure 0.24 km from the center of the volcano. The orientation of the fissure is not known.

112. Mt. Moffett, Adak Island (COATS 1956).
Flank eruption sites are indicated by five basalt domes on the deeply dissected composite volcano (Fig. 2).

113. Mt. Adagdak volcano, Adak Island (COATS 1956).
At the summit of this volcano, there are two andesitic domes. In Fig. 2 these domes are marked by the same symbol as the basaltic one on the southeastern flank. The location of the former two indicates that of the main vent. The basaltic dome is S50°–60°E of the main vent. The summit area is cut by a swarm of recent normal faults trending N50°–80°W, the youngest one having been displaced possibly a few thousand years ago. The summit side of each of these faults is generally downthrown.

114. Great Sitkin volcano (SIMONS and MATHEWSON 1955).
There are six flank volcanoes on Great Sitkin. Extending several hundred meters

due south of the south-flank eruption site (Fig. 2) is a fumarolic field containing hot springs, mud pots, and fumaroles. It is not known whether the heat source of the fumaroles is related to the Great Sitkin volcano or to the surrounding Sand Bay volcanics which unconformably underlie the Great Sitkin volcanics.

131. Mt. Vsevidof, Umnak Island (BYERS 1959).

The volcano Mt. Vsevidof lies on the western flank of Mt. Recheshnoi. Five flank craters and a slightly curved eruptive fissure are on the western half of the volcano (Fig. 2). The eruptive fissure trends S75°W at the upper and EW at the western, lower end. This 2.3 km long fissure appears to be too short and too close to the central vent and its radial stress field to show distinct influence of the regional stress field. Presumably the effect of this stress field would be a straight WNW trend for the lower end of the fissure. The youngest two craters are possibly historic and located due north and south of the summit.

132. Mt. Recheshnoi, Umnak Island (BYERS 1959).

Two summit craters are suggested from the shape of Mt. Recheshnoi volcano. Flank volcanoes include three andesitic cinder cones, three rhyolite domes about 9 km due east of the eastern summit and aligned in a N80°W direction, an elongate mafic basalt vent complex which trends N60°W, and an andesitic lava dome on the northeast flank (Fig. 2).

In the northeastern sector of the volcano there is a nearly linear, N50°W-trending, topographic depression which is observable from the 1,000 ft. contour lines. We suspect that this depression is a graben bounded by normal faults.

133. Okmok volcano, Umnak Island (BYERS 1959).

Okmok volcano is a basaltic shield volcano covered with recent andesitic pyroclastic flows. The summit has collapsed, forming a caldera about 9 km in diameter. In Fig. 2, about 55 eruption sites are indicated, 15 inside the caldera and 40 on the outer slope, which include six topographically prominent pre-caldera flank cones. One of the cones located on the SE flank, Tulik (134), is counted as an independent volcano by SMITH and SOULE (1973). The overall trend of the distribution of eruption sites is in a N60°E direction. Eruption sites in the caldera are distributed in a roughly circular zone along the periphery of the caldera floor. There is a field of thermal springs elongate in a E–W direction on the eastern caldera floor. Outside the caldera on the upper northern and eastern slopes, there are two roughly concentric ridges. They are marked by eruption sites of basalt and rhyolite, and may be remnants of earlier ring fissures. Three possible eruption fissures of short length are found on the eastern and western slopes trending in a NW direction. Several faults, possibly normal faults, are inferred in the NE sector.

135. Bogoslof volcano (BYERS 1959).

Bogoslof Island, located northward and off of the major Aleutian arc ridge, is one of the 'disappearing islands', which represent the subaerial parts of a mostly submarine composite volcano with a basal diameter of over 20 km and relief of about 300 m. Several lava domes and spines of basalts of slightly alkalic composition were formed repeatedly during the last two centuries. The eruption sites fall on a nearly straight line. In Fig. 2, the outline of the island in 1935 is shown.

136. Makushin volcano, Unalaska Island (DREWES *et al.* 1961).

Makushin volcano, which occupies the northwestern part of Unalaska Island, consists stratigraphically of the Makushin volcanics and Eider Point basalt. The latter lies unconformably on the deeply dissected surface of the Makushin volcanics and older rocks. Volcanic rocks of possible flank cones of Makushin volcano are apparently mapped as the Eider Point basalt. In Fig. 2, they are mapped as flank cones. It is doubtful, however, whether all the Eider Point basalt can be regarded as products of flank vents of Makushin volcano.

Vents of the Eider Point basalt are distributed in a ENE, NW and SSW sectors of Makushin volcano. In the northwestern sector there is an eruption fissure defined by linearly aligned craters, 9 km in length, trending in a N75°W direction. The fissure slightly bends at the coastal area assuming a N60°W direction. Dikes related to Makushin volcanics are also exposed. They are most abundant in the northeastern areas where numerous vents of the Eider Point basalt are located. The trend of the dikes is variable, but E–W and NW directions prevail.

Normal faults are found trending in general WNW to NW directions. Lineaments observed on the aerial photographs have dominant NW and WNW trends, especially in the northern parts of Unalaska Island.

138. Akutan volcano (BYERS *et al.* 1949).

Only one flank eruption site, which possibly formed in 1852, is known and is 5.8 km from the summit on the northwest coast of the volcanic island.

149–153. Pavlof and adjacent volcanoes, Alaska Peninsula (KENNEDY *et al.* 1955, BURK 1965).

Mt. Emmons, Mt. Hague, Double Crater, Pavlof volcano and Pavlof Sister are composite volcanoes which form a nearly continuous ridge about 20 km long on the Alaska Peninsula. Fourteen scoria cones are distributed on the ridge. It is difficult, however, to estimate with confidence any trend of the distribution of flank eruption sites on individual composite volcanoes.

154. Mt. Dana, Alaska Peninsula (BURK 1965).

Four domes are mapped on the NW and SE flanks of Mt. Dana volcano (Fig. 2). The summit dome appears to be either elongate in a N40°W direction or composed of two domes aligned in that direction.

156. Mt. Veniaminof, Alaska Peninsula (BURK 1965).

Mt. Veniaminof is dotted with some 50 flank volcanoes, mostly scoria cones. They are concentrated in a long linear zone striking N35°W, extending through the summit (Fig. 2).

157. Purple (Black) Peak, Alaska Peninsula (BURK 1965).

Five vents are mapped on this deeply dissected volcano (Fig. 2). The volcanic edifice is slightly elongated in a N40°W direction.

158. Aniakchak volcano, Alaska Peninsula (BURK 1965).

Six post-caldera cones are distributed in a sector NW of Vent Mountain, possibly the principal crater (Fig. 2). No craters are mapped outside the caldera.

174. Iliamna volcano, Cook Inlet region (GRANTZ *et al.* 1963).

Based on topographic considerations, seven subsidiary craters including a possible 4 km long eruptive fissure, are inferred from a geologic map showing the distribution of lava flows of Iliamna volcano (Fig. 2). The possible fissure strikes S20°E. The volcanic edifice is elongate in a S30E direction.

178. Spurr volcano, Cook Inlet region (JUHLE *et al.* 1955).

Mt. Spurr last erupted in 1953 from a subsidiary crater, 3.3 km south of the highest peak. If we regard the peak as a part of the eastern caldera rim (R. L. SMITH, personal communication, 1976), then the eruption site is located to the southeast of the center of the volcano which is covered by snow and ice.

Eastern and southeastern Alaska

816. Edgecumbe volcano (LONEY *et al.* 1975).

Scoria cones on the northeastern slope might be flank volcanoes of a possibly independent composite volcano, 'Crater ridge', located just northeast of Edgecumbe volcano. In Table 1, they are included among the flank volcanoes of Edgecumbe. Even if they are excluded, the azimuth given in the Table remains the same (Fig. 2).

Results

Probable azimuth of dike concentration

For 28 volcanoes and volcanic fields discussed above, probable azimuths of zones of concentration of flank craters, which are thought to represent the trends of dikes are summarized in column 9 of Table 1 and shown in Fig. 3. These azimuths are derived from the data presented in Fig. 2 and in the previous chapter. For most volcanoes the azimuth is obtained directly from the distribution of craters. This is the case for volcanoes 88, 89, 91, 112, 114, 135, 138, 154, 156, 157, 174, 178, and 186.

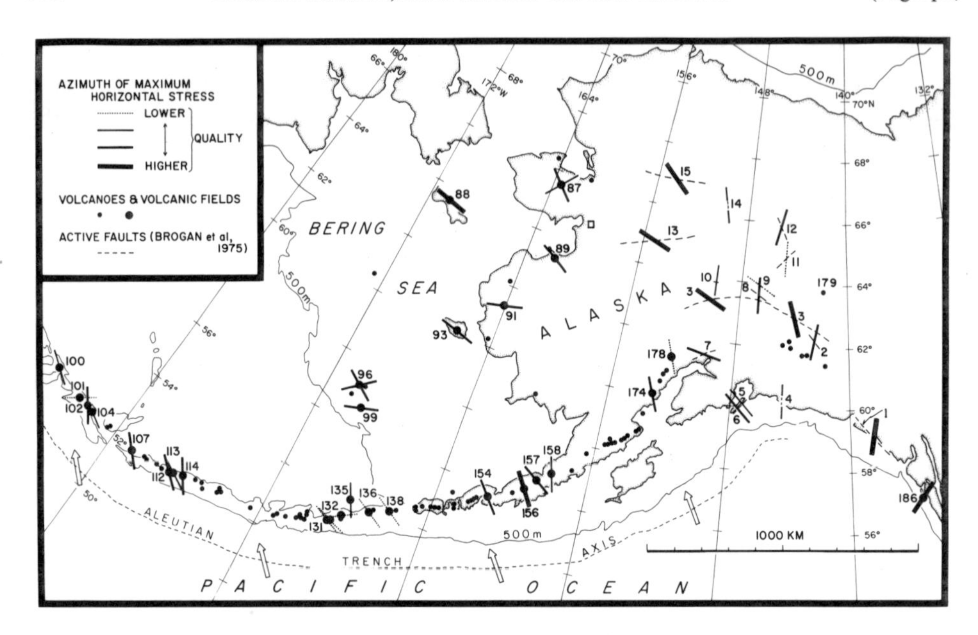

Figure 3
Azimuth of the maximum horizontal compression due to the contemporary stress system deduced from post-Miocene volcanoes (SMITH and SOULE 1973) and active faults (BROGAN *et al.*, 1975). Volcanoes (solid circles) for which the azimuth is indicated are marked by larger circles. Azimuth is given four different grades of quality. Numbers attached to volcanoes and faults are the same as in Tables 1 and 2. Open arrows: direction of motion of Pacific plate relative to North American plate after MINISTER *et al.* (1974). Square: epicenter of the earthquake shown in Fig. 4. For details see text.

When flank craters are distributed on one side of the main polygenetic vent, the azimuth of the zone of crater distribution is less well defined. Volcanoes 131 and 158 are examples.

When bending of a radial fissure is observed, the final direction which is assumed at the portion of the fissure most distant from the summit is taken as the most significant direction. Volcanoes 102, 107, 131, and 136 are examples.

In the western Alaska–Bering Sea shelf region the probable azimuths of zones of dike concentration have a dominant east–westerly trend. In the Aleutian Islands–Alaska Peninsula–Aleutian Range region these azimuths have a dominant northwesterly trend, with two exceptions: Kiska (101) and Okmok (133). Within each of these regions (especially the second one) the individual azimuths derived for a given volcano or volcanic field are strikingly consistent with the dominant regional trend.

Evaluation of quality of the azimuth

The quality or reliability of the obtained azimuth as representative of the average trend of zones of dike concentration is different for different volcanoes. For example, the N35°W direction at Mt. Veniaminof leaves little doubt about accuracy and temporal stability, because the zone of flank eruption is very well defined by a large number of monogenetic vents. In contrast, the N20°W direction at Mt. Spurr is rather uncertain and does not offer any evidence of temporal stability. Therefore, we classify the azimuths into four quality ranks: excellent (E), good (G), poor (P) and very poor (VP).

High ranks are assigned when: (1) the azimuth is well defined by the crater distribution; that is, when the number of craters is large, the zone of crater distribution is long, the ratio between 'length' and 'width' is large and the azimuth of the zone parallels individual eruptive fissures, (2) flank craters are distributed on both sides of the main crater, (3) elongation of the edifice of polygenetic volcanoes is observed and/or subparallel normal faults are distributed in the volcanoes and volcanic fields trending in a direction similar to that of crater distribution.

In practice, the highest quality (E) is given to volcanoes with more than 20 craters in a zone with 'length' longer than 20 km and with the 'length'/'width' ratio larger than 2, and when at least one of the subsidiary features ('elongation' and 'faults') is observed, and the azimuth is defined within $\pm 10°$. The lowest quality, VP, is assigned to volcanoes whose zones of distribution of flank craters are poorly defined or when: (1) only one flank crater is known and no other features are observed (Spurr, 178 and Akutan, 158), (2) the azimuth changes drastically when a single crater is removed (Kiska, 101), (3) the trend of eruptive fissures makes a high angle with the trend of the zone defined by the overall distribution of craters (Okmok, 133 and Makushin, 136).

Stress orientation obtained from active faults

For obtaining the average orientation of tectonic stress during recent geologic times, active faults have a significance similar to the present method that uses volcanic features. Active faults in Alaska studied by Brogan *et al.* (1975) are used both to check the result obtained by the present study of volcanoes and to cover a wider area so that regional discussions may be more meaningful. Data described by Brogan *et al.* (1975) are used almost exclusively, because they mapped the distribution of active faults in a wide enough area with uniform standards.

What is obtained from the strike and sense of movement of faults is primarily the direction of principal horizontal shortening (PHS; Lensen 1961 and Lensen *et al.* 1971). The direction of PHS is perpendicular to the fault strike for a pure reverse fault, about 45° to the strike for a pure strike-slip fault and parallel to the strike for a pure normal fault. The direction of PHS is assumed here to be a good approximation for that of the maximum horizontal compression (MHC) as discussed, e.g. by Lensen *et al.* (1971) and Pavoni (1971).

Azimuths of PHS obtained from 15 active faults are tabulated in Table 2 and mapped in Fig. 3. Brogan *et al.* (1975) described 22 faults. Seven of them are discarded in Table 2, however, because they are either unclear in the sense of motion

Table 2

Azimuth of maximum horizontal shortening associated with active faults in Alaska

Name[1]	Strike (N)[1]	Sense[1-2]	Length (km)[1]	Azimuth (N)[3]	Quality[4]
1. Fairweather	38°W	D	200	7°E	E
2. Totschunda	35°W	D	65	10°E	G
3. Denali	E–W	D	2,000 <	70° – 15°W	E
4. Ragged Mountain	N	N	24	0°	P
5. Hanning Bay	NE	R	6	45°W	P
6. Patton Bay	NE	R	62	45°W	G
7. Castle Mountain	55°E	D	45	80°W	G
8. McGinnis Glacier	NW	D	50	0°	G
9. Donnelly Dome	55° ~ 85°W	N	8	55° ~ 85°W	VP
10. Healy Creek	E	R	18	0°	P
11. Shaw Creek	45°E	S?	150	0°	VP
12. Tintina	25° ~ 30°E	D	100	15° ~ 20°E	G
13. Kaltag	65°E	D	400	70°W	E
14. Dall Mountain	NS ~ 23°W	N	30	NS ~ 23°W	P
15. Kobuk-Alatna Hills	EW	D	270	45°W	E

[1] After BROGAN *et al.*, 1975.
[2] D: dextral strike-slip, S: sinistral strike-slip, N: normal, R: reverse.
[3] Azimuth of the maximum horizontal shortening given by strike and dominant sense of movement of the fault.
[4] Same as in Table 1. For details, see text, p. 17.

(six faults) or less than 5 km in length (one fault). Because the azimuths are deduced solely from the dominant sense of motion, they may bear errors of up to a few tens of degrees.

The quality of the obtained azimuth is evaluated and classified as excellent, good, poor or very poor, keeping in mind the quality given to those azimuths obtained from volcanoes. Excellent quality is given when the length of a fault system is longer than 200 km and the sense of motion is well established. Very poor quality is given to a fault system either when less than 10 km in length or the sense of displacement is highly uncertain.

Most of the active faults mapped by BROGAN *et al.* (1975) and shown in Fig. 3 are in interior Alaska. Therefore, most of the azimuths of PHS determined from faults are not directly comparable to the azimuths of MHC determined from volcanoes. However, for the three faults closest to the Bering Sea shelf – western Alaska region (western Denali, Kaltag, and Kobuk-Alatna Hills) the azimuths of PHC (all of the highest quality) are in the sector WNW to NW, in good agreement with the azimuths of MHC determined from volcanoes for this region. In south central Alaska, near the eastern terminus of the Aleutian arc, the azimuths of PHS are in the sector WNW to N, in only poor agreement with the two azimuths of MHC (174 and 178) determined for the Aleutian Range. In southeastern Alaska the north-northeasterly azimuth of PHS for the Fairweather fault (highest quality) is in good agreement with the northeasterly azimuth of MHC for Mt. Edgecumbe (186).

Discussion

Western Alaska and Bering Sea (back-arc) region

Probable azimuths of dike concentrations for volcanoes (hereafter simply referred to as azimuths) are generally oriented east-westerly (E–W ± 50°) in this back-arc region that extends 1,000 km in N–S and 500 km in E–W directions. In addition to this general feature, the azimuths systematically change trend from northwesterly in the northern part to southwesterly in the southern part of the region (Fig. 4). This systematic change in trend becomes better defined when the quality of

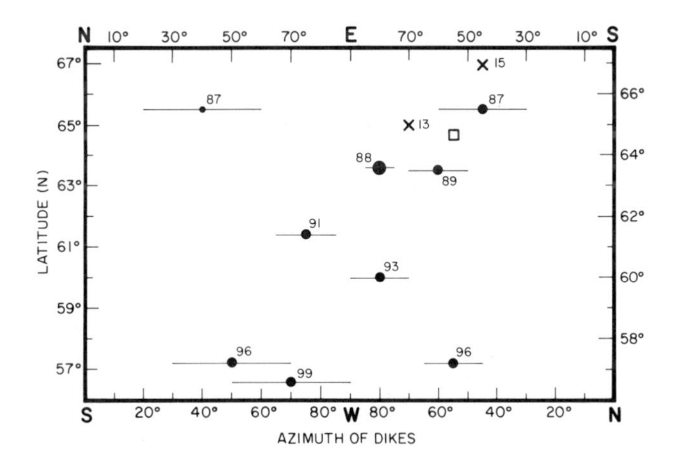

Figure 4

Change of the probable azimuths of dike concentrations with latitude for volcanoes of the western Alaska–Bering Sea region. Volcanoes are shown by solid circles. Size of the circle corresponds to the quality of the obtained azimuth. Attached numbers refer to those used in Table 1. Note that two possible azimuths are plotted for volcanoes 87 and 96. Crosses are those obtained by active faults whose data are given in Table 2. Square shows the orientation of the intermediate-stress axis of an earthquake (location in Fig. 3) with focal mechanism solution of normal-fault type (SYKES and SBAR 1974).

the azimuth determination is considered; e.g., the northeasterly azimuth of the Imuruk Lake field (87) is more poorly determined than the northwesterly one (Fig. 2).

The systematic geographical distribution of volcano-derived azimuths and their conformity with the azimuths obtained from active faults (Figs. 3 and 4) strongly suggest that they represent the orientation of the maximum horizontal compressive stress of tectonic origin in the entire region.

Another notable feature of this region is that two different azimuths are obtained sometimes at single volcanic fields (87 and 96). This may indicate that the direction of the maximum horizontal compression changed or rather fluctuated in orientation with lapse of time. This might be the case if the magnitudes of the two horizontal principal stresses are similar. The horizontal stress is likely, though not conclusively, to be either the minimum or intermediate principal stress. Then the resultant tectonic style would be extensional. This is also supported by the presence of an active normal fault (14) and of a normal-faulting type earthquake (Figs. 3 and 4; SYKES and SBAR 1974) in the same general area. The following volcanic features may also support this conclusion:

(1) Volcanoes in this back-arc region do not define a narrow volcanic belt.
(2) Their magmas are characteristically olivine-basalt or alkali basaltic.
(3) They consist of clusters of monogenetic volcanoes and are classified as volcanic fields by SMITH and SOULE (1973). Kookooligit Mountains (88), St. Michael's Island (89) and Nunivak Island (93) might be classified as central volcanoes forming low shields, however (Fig. 2).

Moreover, there is a possibility that the same stress field, with the maximum stress in the vertical direction, prevails in the entire back-arc region north of the Aleutian–Wrangell volcanic belt. This is suggested by a rather uniform petrographic province of alkalic affinity which may stretch from the western Alaska volcanic fields to Prindle volcano, 179 (FOSTER *et al.* 1966) in the east, some 200 km northeast of Wrangell Mountains.

Alaska Peninsula and the Aleutian Islands and Range

Ninety percent of the probable azimuths of dike concentration of volcanoes in this narrow volcanic belt trend N45°W $\pm$ 30° which is normal to oblique to the strike of the volcanic belt (Table 1, Fig. 3). This fact strongly suggests that the maximum horizontal compressive stress, which determined the azimuths, is tectonic in origin and that the azimuths are parallel to the maximum principal stress (NAKAMURA 1977). This conclusion is supported by comparison of the azimuths with the directions of relative motion between the Pacific and North American plates (Fig. 5). Ninety percent of the volcano-derived azimuths are within 20° of the expected sloping line that gives the azimuths of the relative plate motion at individual volcanoes derived from computations by JACOB *et al.* (1977).

The scatter of individual azimuths around the value expected from the relative plate motion may be the result of local mechanical heterogeneities, rather than deviations in the regional stress field. This is suggested by the better fit of the azimuths of higher quality to the expected value (Fig. 5). Consider the azimuths at two volcanoes, Kiska (101) and Okmok (133), which strongly deviate from the expected orientation; the azimuth for Kiska volcano (101) is dependent primarily on the location of a flank crater on the extreme southwest (Fig. 2). Excluding this point, this still poorly determined azimuth becomes northwesterly. Since this particular

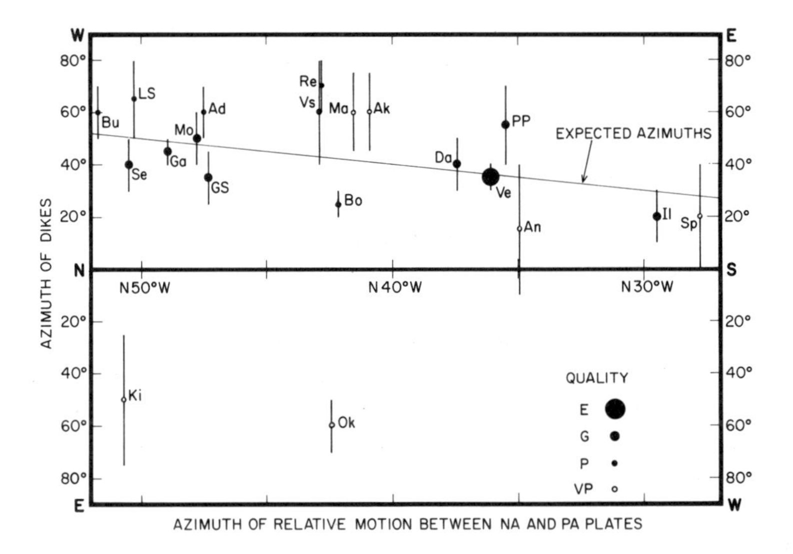

Figure 5
Comparison between the probable azimuths of dike concentration of volcanoes in Aleutian arc (Table 1) and the direction of relative motion between the Pacific and North American plates at individual volcanoes (JACOB *et al.* 1977). The diagonal solid line indicates the locus along which the two directions (slip vectors and dike distributions) coincide. Quality grades (Table 1) of the azimuths are shown by the different circle sizes; highest quality has the largest circle, 'poor' and 'very poor' are not differentiated.

vent is located totally outside the volcanic edifice, there might be a problem in assigning it as a flank vent of Kiska volcano. For this reason, the azimuth, N50°E, for Kiska volcano is considered rather questionable.

The N60°E azimuth for Okmok volcano (133) is ranked as very poor in quality because the NW trend of three eruption fissures is nearly normal to the long axis of the distribution of flank craters. In addition to this feature, Okmok volcano has some other features that are exceptional for volcanoes in the Aleutian arc. Okmok is essentially a basaltic shield volcano, while all the other volcanoes in the Aleutian arc are composite volcanoes (SMITH and SOULE 1973). The development of circular instead of radial structures is remarkable on this volcano. Post-caldera cones, aligned along the periphery of the caldera floor (Fig. 2) appear to indicate ring fissures. Outside the caldera, the northern arcuate ridge and the eastern arcuate ridge also represent parts of ring fissures. The scatter in trends of roughly radial normal faults is so large that some make a right angle with each other (Fig. 2). All these features are more or less suggestive that the magnitude of two horizontal principal stress axes are rather close and that the maximum principal stress is vertical, at least near the summit region. Alternatively, the stress field in the vicinity of Okmok volcano may have been quite variable in time. The data are insufficient to establish this explanation, however.

Comparing the azimuths obtained from the easternmost Aleutian volcanoes (174, 178) with the one obtained from the Castle Mountain fault (7), we find poor agreement (Fig. 3). However, the direction of PHS indicated by the Castle Mountain fault (7) may actually strike in a more northwesterly direction if the dip-slip component given in Table II is taken into account. This fault has a dip-slip component with the north side up. Such dip-slip displacement exists also on neighboring thrust faults (5, 6 and 10). Therefore, assuming that the dip-slip component is of a reverse-faulting type, the given azimuth of N80°W should rotate somewhat clockwise. Furthermore, according to DETTERMAN *et al.* (1974), the displacement of the Castle Mountain fault 'has involved predominantly dip-slip reverse movement in the Late Tertiary (mainly post-Oligocene)' and 'the most recent breaks along the fault most probably originate through roughly 2.3 m of dip-slip movement'. On the other hand, the azimuth (N20°W) for Spurr volcano is of very poor quality (see description). The azimuth for Iliamna volcano could be oriented westerly, if we consider a possible down-slope bend towards east of the inferred fissure on the southeast flank (Fig. 2).

Eastern Alaska region

Only one azimuth has been obtained in this region. Edgecumbe volcano (186), however, may well represent the direction of the maximum horizontal tectonic stress of the surrounding region. The azimuth, N40°E, has high quality and coincides very well with the direction of the principal horizontal shortening N35°–40°E, indicated by a nearby, possibly active, dextral strike-slip fault, the Chatham Strait fault, that runs in a N10°W direction for more than 400 km (LONEY *et al.* 1975).

Arc and back-arc region

From the results thus far discussed and shown in Fig. 3 the overall pattern of the azimuths of maximum horizontal compression in the Aleutian and Alaskan regions can be seen to have the following features: (1) Along the consuming plate boundary, the azimuths are nearly parallel to the directions of relative motion between the North American and Pacific plates. In this case it is reasonable to infer that the azimuth of MHC corresponds to the direction of the maximum principal stress. (2) Along the dextral strike-slip or transform segments of the plate boundaries, the azimuths are obliquely rotated in a clockwise sense from the strike of the boundary. (3) The stress trajectories implied by the pattern of these azimuths spread inland in a curved fan shape with the axis of symmetry running roughly in a N25°–30W direction from the eastern end of the Aleutian trend (Fig. 3). These trajectories curve westward into the Bering Sea shelf back-arc region where they become almost perpendicular to the azimuths of MHC for the central Aleutian arc. The overall pattern is crudely similar to the result of PAVONI (1971). In his global map of 'principal horizontal pressure as derived from recent and late Cenozoic tectonics', four trajectory lines gradually approach a west-northwesterly trend in the Bering Sea from an almost north-south trend in eastern Alaska.

Considering the time during which the volcanoes and faults have been active, the above overall stress pattern may have been maintained for at least tens to hundreds of thousands years and probably longer.

Most features obtained by the present study appear to be explained, at least qualitatively, by the relative plate motion alone. Even the more east-westerly trend in northwest Alaska may be explained by a simple elastic plate model with driving forces of ridge push, trench pull and drag force (RICHARDSON *et al.* 1976). The nearly right-angle relation between azimuths of the central Aleutians and the immediate back-arc region may be difficult, however, to explain by such a simple model. Another explanation of the observed pattern might be that the stress system in the back-arc region has its own origin compatible with an upward movement beneath the back-arc lithosphere. This suggestion is supported by the difference in geographical distribution and in chemistry of associated magmas between volcanoes of the arc and back-arc regions, respectively.

Provided the origins of the tectonic systems are different for the arc and back-arc regions, an interesting implication emerges which demands that the compressive stress at the arc, probably generated by the convergence of plates, is transmitted towards the back-arc region for only about 500 km where it is gradually replaced by a tensional stress system of different origin. Such a situation should in some cases favor opening of marginal seas (KARIG 1974). On the other hand, the observed pattern implies also that if back-arc spreading takes place at all, it should occur not within the island arc volcanic belt, but within a subparallel linear region a few hundred kilometers behind the arc. Within arcs where the strike of the maximum horizontal compression is normal or oblique to the arc axis, e.g., as in the Aleutians (this study), Japan and central Chile (NAKAMURA 1977), splitting of the arc along its axis is mechanically impossible.

In terms of the three-dimensional principal stress system, the obtained azimuths of MHC correspond most probably to the direction of the maximum principal stress in the overthrusting plate near the island arc, while these azimuths probably correspond to the direction of the intermediate principal stress in the back-arc region. In the Aleutian arc case, the boundary between compressional and tensional stress regions passes somewhere south of the Pribilof Islands (96 and 99), north of the Healy Creek fault (10), and somewhere between the Wrangell Mountains and Prindle volcano, north of the Denali fault system (3).

Conclusion

Application of the method that uses volcanic features as indicators of the average, long-term orientation of tectonic stress, as proposed by NAKAMURA (1977), is demonstrated in the Aleutian–Alaska region by mapping the obtained azimuths of maximum horizontal compression. The internal consistency of this mapping and its successful correlation with the expected stress field in the Aleutian arc reconfirms our basic premise that flank eruptions result from the formation of radial dikes, the average orientation of which is controlled by tectonic stress. This confirmation also encourages us to extend the method to radial and parallel dikes of the geologic past. Extension to the past would imply that if combined with accurate dating of intrusive rocks one could reconstruct similar maps for trajectories of tectonic stress orientation for various geologic epochs.

The long-term stress pattern can be checked by, or provide an evaluation of, short-term or transient stress indicators such as seismic focal mechanism solutions and *in situ* stress or strain measurements.

The strike and sense of motion on major faults also provide long-term information on the orientation of tectonic stress. The mapped pattern of the azimuths of the maximum horizontal compression obtained from volcanoes and faults is internally consistent and a reasonable interpretation is possible in terms of relative motion of elastic plates of the Pacific and North America. In the back-arc area, however, it is necessary to assume an independent stress source which causes tensional deviatoric stress in the overlying lithosphere.

Acknowledgement

Useful information was kindly provided by R. R. Coats, J. M. Hoare and R. L. Smith on the Aleutian and Alaskan volcanoes and by G. Plafker on active faults of Alaska. The manuscript was critically reviewed by Lynn R. Sykes and Paul Richards. This work was carried out mostly while Nakamura was recipient of the Senior Postdoctoral Fellowship at Lamont-Doherty Geological Observatory and partly during his following visit to the Geophysics Department, Stanford University. Research was supported by the Earth Science Section, National Science Foundation Grants DES 75–03640 and HES 75–17623–A01, and partly by Energy Research and Development Administration Contract EY 76–S–02–3134.

References

Brogan, G. E., Cluff, L. S., Korringa, M. K. and Slemmons, D. B. (1975), *Active faults of Alaska*, Tectonophysics *29*, 73–85.

Burk, C. A. (1965), *Geology of the Alaska Peninsula, Part 2: geological map*, Geol. Soc. Amer. Mem. *99*.

Byers, F. M., Jr. (1959), *Geology of Umnak and Bogoslof Islands Aleutian Islands, Alaska*, US Geol. Surv. Bull. *1028–L:* 267–369.

Byers, F. M. and Barth, T. F. W. (1949), *Volcanic activity on Akun and Akutan Islands*. Proc. 7th Pacific Sci. Congr. New Zealand *2*, 382–397.

COATS, R. R. (1953), *Geology of Buldir Island, Aleutian Islands, Alaska*, US Geol. Surv. Bull. *989–A*, 8–9.

COATS, R. R. (1956), *Geology of northern Adak Island, Alaska*, US Geol. Surv. Bull. *1028–C*, 47–67.

COATS, R. R. (1959), *Geologic reconnaissance of Gareloi Island, Aleutian Islands, Alaska,* US Geol. Surv. Bull. *1028–J*, 249–256.

COATS, R. R. (1964), *Aleutian islands and Alaska*, Bull. Volc. Eruptions *4*, 4–5.

COATS, R. R., NELSON, W. H., LEWIS, R. Q. and POWERS, H. A. (1961), *Geologic reconnaissance of Kiska Island, Aleutian Islands, Alaska,* US Geol. Surv. Bull *1028–R*, 563–581.

COONRAD, W. L. (1957), *Geologic reconnaissance in the Yukon–Kuskokwin Delta region, Alaska.* US Geol. Surv. Misc. Geol. Investigations Map, I–223, 1:500,000.

COX, A., HOPKINS, D. M. and DALRYMPLE, G. B. (1966), *Geomagnetic polarity epochs: Pribilof Islands, Alaska*, Geol. Soc. Amer. Bull. *77*, 883–910.

DETTERMAN, R. L., PLAFKER, G., HUDSON, T., TYSDAL, R. G. and PAVONI, N. (1974), *Surface geology and Holocene breaks along the Sustina segment of the Castle Mountain fault, Alaska*, US Geol. Surv. Misc. Field Studies, Map MF 618.

DREWES, H., FRASER, G. D., SNYDER, G. L. and BARNET, H. F., JR. (1961), *Geology of Unalaska Island and adjacent insular shelf, Aleutian Islands, Alaska,* US Geol. Surv. Bull. *1028–S*, 583–676.

FOSTER, H. L., FORBES, R. B. and RAGEN, D. M. (1966), *Granulite and peridotite inclusions from Prindle volcano, Yukon–Tanana upland, Alaska,* US Geol. Surv. Prof. Paper *550–B*, 115–119.

GRANTZ, A., ZIETZ, I. and ANDREASEN, G. E. (1963). *An aero-magnetic reconnaissance of the Cook Inlet area, Alaska*, US Geol. Surv. Prof. Paper *316–G*, 117–134.

HOARE, J. M. and CONDON, W. H. (1971a), *Geologic map of the St. Michael quadrangle, Alaska*, US Geol. Surv. Misc. Geol. Investigations Map I–682.

HOARE, J. M. and CONDON, W. H. (1971b), *Geologic map of the Marshall quadrangle, western Alaska,* US Geol. Surv. Misc. Geol. Investigations Map I–668.

HOARE, J. M., COX, A., DALRYMPLE, G. B. and CONDON, W. H. (1968), *Geology, paleomagnetism and potassium-argon ages of volcanic rocks, Nunivak Island, Alaska*, Geol. Soc. Amer. Mem. *116*, 377–413.

HOPKINS, D. M. (1963), *Geology of the Imuruk Lake area Seward Peninsula, Alaska*, US Geol. Surv. Bull. *1141–C*, 1–101.

JACOB, K. H., NAKAMURA, K. and DAVIES, J. N. (1977), *Trench-volcano gap along the Alaska–Aleutian arc: Facts, and speculations on the role of terrigenous sediments for subduction.* AGU Monogr., Ewing Symposium on Island Arcs and Back-arc Regions., in press.

JUHLE, W. and COULTER, H. (1955), *The Mt. Spurr eruption, July 9, 1953*, Trans. Amer. Geophys. Union *36*, 199–202.

KARIG, D. E. (1974), *Evolution of arc systems in the western Pacific*, Ann. Rev. Earth Planet. Sci. *2*, 51–75.

KENNEDY, G. C. and WALDRON, H. H. (1955), *Geology of Pavlof volcano and vicinity, Alaska*, US Geol. Surv. Bull. *1028–A*, 1–19.

LENSEN, G. J. (1961), *Principal horizontal stress directions as an aid to the study of crustal deformation*, Publ. Dom. Obs. Ottawa *49*, 389–397.

LENSEN, G. J. and OTWAY, P. M. (1971), *Earthshift and post-earthshift deformation associated with the May 1968 Inangahua earthquake, New Zealand*, Roy. Soc. New Zealand Bull. *9*, 107–116.

LONEY, R. A., BREW, D. A., MUFFLER, L. J. P. and POMEROY, J. S. (1975), *Reconnaissance geology of Chichagof, Baranof, and Kruzof Islands, southern Alaska*, US Geol. Surv. Prof. Paper *792*, 1–105.

MOORE, J. G., NAKAMURA, K. and Alcaraz, A. (1966), *The 1965 eruption of Taal volcano*, Science *151*, 955–960.

NAKAMURA, K. (1969), *Arrangement of parasitic cones as a possible key to regional stress field*, Bull. Volc. Soc. Japan *14*, 8–20 (in Japanese with English abstract).

NAKAMURA, K. (1977), *Volcanoes as possible indicators of tectonic stress orientation: Principle and proposal*, J. Volc. Geoth. Res. *2*, in press.

NELSON, W. H. (1959), *Geology of Segula, Davidof and Khvostof Islands, Alaska*, US Geol. Surv. Bull. *1028–K*, 257–266.

PAVONI, N. (1971), *Recent and late Cenozoic movements of the earth's crust*, Roy. Soc. New Zealand Bull. *9*, 7–17.

RICHARDSON, R. M., SOLOMON, S. C. and SLEEP, N. H. (1976), *Intraplate stress as an indicator of plate tectonic driving force*, J. Geophys. Res. *81*, 1847–1856.

SIMONS, F. S. and MATHEWSON, D. E. (1955), *Geology of Great Sitkin Island, Alaska*, US Geol. Surv. Bull. *1028–B*, 21–43.

SMITH, R. L. and SOULE, C. E., *Western Alaska and Bering Sea Islands; Alaska Peninsula and the Aleutian Islands and Range* in *Data Sheets of the Post-Miocene volcanoes of the world with index maps* (IAVCEI, Rome 1973).

SNYDER, G. L. (1959), *Geology of Little Sitkin Island, Alaska*, US Geol. Surv. Bull. *1028–H*, 169–210.

SYKES, L. R. and SBAR, M. L., *Focal mechanism solutions of intra-plate earthquakes and stresses in the lithosphere* in *Geodynamics of Iceland and the North Atlantic Area*, (ed. Kjartansson), 207–224.

YODER, H. S., *Generation of Basaltic Magma* (National Academy of Sciences, Washington, D.C. 1976), 265 pp.

(Received 14th February 1977)

Pageoph, Vol. 115, 1977), Birkhäuser Verlag, Basel

Stress and Shear Fracture (Fault) Patterns Resulting from a Suite of Complicated Boundary Conditions with Applications to the Wind River Mountains[1])

By GARY COUPLES[2])

Summary – The Airy stress function is used, via the Principle of Superposition and the series summation concept, to obtain stress states in a static, self-gravitating elastic beam subjected to boundary stresses. The boundary conditions investigated are more complicated than those previously published and include cases with sawtooth-, step-, and sinusoidally-shaped lower-boundary loads, with and without additional tectonic end loads. Potential shear fracture (fault) patterns derived from the calculated stress fields indicate co-existing (simultaneous) regions of lateral shortening and extension. Application of one of the cases to the study of the structural geometry of the Wind River Mountains of Wyoming yields a good 'fit' and forms a possible explanation for the observed rotations and zones of shortening and extension.
Key words: Stress field; Fracture patterns; Stress due to loading.

1. Introduction

Analytical solutions of the type presented by HAFNER (1951) have proven useful as aids in understanding crustal deformations such as the uplifts of Precambrian basement blocks found in the Rocky Mountain foreland (STEARNS, 1971; 1975). However, models of this type have seen only limited application, possibly because solutions for only a limited range of boundary conditions were available in the literature. New solutions were needed to provide a larger suite of loading conditions to choose from when selecting a model. The present study was undertaken to generate solutions (i.e., internal stress fields and associated features, such as fractures) for a suite of new boundary conditions described in the results.

Why were new solutions desired? Final geometries or shapes of the uplifts in the foreland led me to speculate that the 'geometries' of the initial loading conditions may have had forms similar to those of the resulting structures. Assymetric uplifts, like the Wind River Mountains, and plateau-like uplifts, such as the Beartooth Mountains, suggested that there was reason to investigate lower-boundary normal stresses with the 'shape' of sawtooth or step functions. The proximity of some of the uplifts to the

[1]) Paper derived from MA Thesis submitted to Rice University.

[2]) Department of Geology, Rice University, Houston, Texas, 77001, present address: Department of Geology, Texas A&M University, College Station, Texas, 77843, USA.

Idaho–Wyoming thrust belt also suggested that horizontally-directed normal stress might be an important part of the imposed boundary conditions.

Therefore, plane strain solutions (i.e., stress fields and derived features) were prepared for combinations of end loads and/or sawtooth-, step-, or sinusoidally-shaped normal stresses applied to the lower boundary of an elastic beam. Each stress system calculated in this study included components related to the standard state stress field) due to the effect of gravity). Potential shear fracture trajectories were drawn at $\pm 30°$ to the lines of most-compressive principal stress (i.e., σ_1 trajectories[3]). The usage of the solutions is illustrated by applying the potential shear fracture (fault) trajectories resulting from one of the cases studied to the explanation of the Laramide structural development of the Wind River Mountains of Wyoming.

2. *Method*

Problems in structural geology have traditionally required reduction to some simplified concept. The general approach, taken by HAFNER (1951) and followed here, is to investigate the mechanics of a rectangular beam of arbitrary thickness – i.e., a two-dimensional problem. By appropriate arguments and assumptions, and by proper selection of material properties, etc., the theoretical elastic response of the two-dimensional beam can be extended to represent the equivalent distortions of an appropriate segment of the earth's crust. In the case considered here, the 'appropriate' segment of the crust is taken to be the upper 15 km of continental basement (see discussion in STEARNS (1971) for the definition of basement). The many problems associated with that extension are discussed further below.

The rectangular beam under consideration in this study is depicted in Fig. 1a, along with conventions for spatial coordinates. Figure 1b illustrates the sign convention for stress used throughout this paper.

Two-dimensional problems in the deformation of continua are either of the plane strain or plane stress type. Because plane stress rarely arises in natural geologic situations, plane strain is assumed, and the Airy stress function is used to solve for the stresses in the beam.

HAFNER (1951) outlined the pertinent theoretical aspects and the basic approach of using the Airy stress function in geologic deformations. Further details can be found in most texts on the theory of elasticity (e.g., JAEGER, 1969; JAEGER and COOK, 1969; TIMOSHENKO and GOODIER, 1970). The mathematical analysis requires the selection of the Airy stress function, $\Phi(x, z)$, such that

$$\nabla^2\{\nabla^2\Phi(x, z)\} = \nabla^4\Phi(x, z) = 0. \tag{1}$$

[3]) Compressive stresses are considered negative. Thus, σ_1 is the algebraically-least principal stress.

There are infinitely many functions, Φ, which satisfy equation (1), but the 'trick' of the analysis is to choose Φ which yields the desired boundary stresses, determined by evaluating

$$\sigma_{xx} = \frac{\partial^2\Phi(x, z)}{\partial z^2}$$
$$\sigma_{zz} = \frac{\partial^2\Phi(x, z)}{\partial x^2}$$
$$\tau_{xz} = \frac{\partial^2\Phi(x, z)}{\partial x\, \partial z} \tag{2}$$

at the boundary values of X and Z.

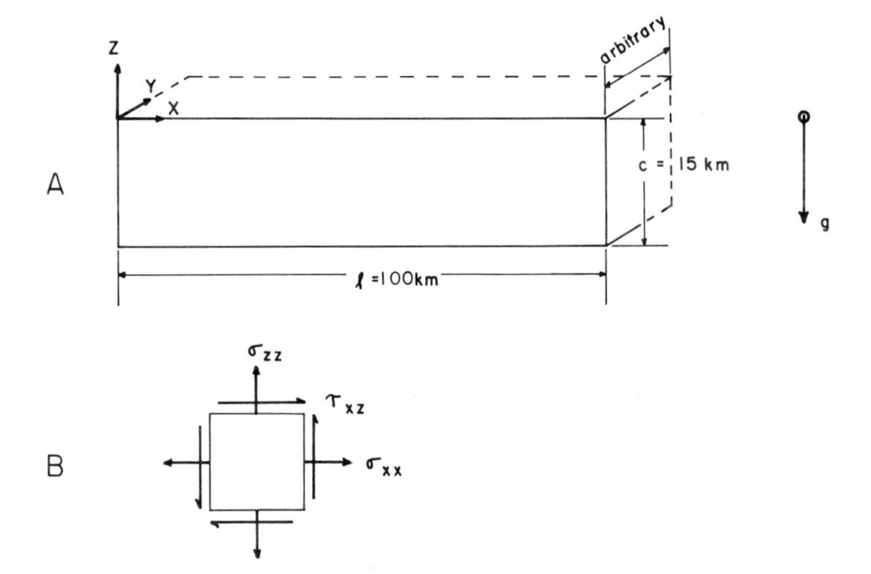

Figure 1
(A) Geometry and dimensions of rectangular beam studied in this paper. Coordinate system depicted with origin in upper-left corner of beam. Arrow at right labeled g indicates gravitational attraction acting parallel to Z. (B) Components of two-dimensional stress field acting in the X-Z plane. Shear and normal stresses are considered positive as shown.

Many different forms of Airy stress function may satisfy equation (1), but of primary interest to this study are the polynomial and hyperbolic-trigonometric (sinusoidal) forms. Specifically, let us consider the choices

$$\Phi(x, z) = D\cdot x\cdot z^2 \tag{3}$$

$$\Phi(x, z) = \sin \alpha x\cdot\{C_1 \cosh \alpha z + C_2 \sinh \alpha z + C_3 z \cosh \alpha z + C_4 z \sinh \alpha z\}. \tag{4}$$

Equation (3) describes a supplementary stress system[4]) consisting of a constant load applied to the end of the beam, with static equilibrium being maintained by shear stresses acting on the base and end surfaces. Figure 6 of HAFNER (1951) depicts the result of superposing a 'standard state' stress system (see below) and the supplementary stress system described by equation (3). Similarly, the stress field due to equation (4) superposed with the 'standard state' results in stress and fracture trajectories as illustrated in Plates 1A, 1B, 1C, and 1D of HAFNER (1951).

The terminology 'standard state' was introduced by ANDERSON (1942, p. 137) who used it as meaning the hydrostatic state, or an isotropic state where

$$\sigma_{\text{vert}} = \sigma_{\text{horiz}} = \rho g z \tag{5}$$

where ρ = density and
g = acceleration due to gravity.

By 'standard state', I mean that state of stress in a body (in this case, a portion of the crust) in the absence of imposed loads. The 'elastic' stresses in a laterally-constrained, self-gravitating body are given by

$$\sigma_{\text{vert}} = \rho g z$$

$$\sigma_{horiz} = \frac{v}{1 - v} \cdot \sigma_{vert}. \tag{6}$$

In such a stress system, $\sigma_{\text{horiz}} < \sigma_{\text{vert}}$ unless $v = 0.5$, where v is Poisson's ratio. HAFNER (1951) was aware of this theoretical state of stress (equations (6) of this paper) and pointed out that the hydrostatic state (equations (5)) is achieved in the body only by the application of additional horizontal stress (unless $v = 0.5$). Nevertheless, he chose to use the hydrostatic state for his standard state, opting for a definition consistent with that used by ANDERSON (1942).

Unfortunately, there is no generally accepted state of stress for the 'natural' standard state. We must choose to use either a theoretical state (equations (6)) or an assumed state (equations (5)). Heim's rule – that the natural state of stress tends to become hydrostatic over long periods of time (JAEGER, 1969, p. 172) – does not resolve the problem, particularly for basement rocks. This is because Heim's conclusion was based on the assumption that the crustal rocks would flow, thereby effectively removing the lateral constraints and allowing a hydrostatic state of stress to be achieved.

Reference to data published by HANDIN (1966) and by BORG and HANDIN (1966) indicates that 'granitic' rocks are able to undergo only a limited amount of permanent deformation before failing when subjected to the maximum temperatures (up to 500°C) and confining pressures (about 5 kb) that would be experienced by rocks at the bottom of a 15 km thick portion of continental crust. It should be noted that the differential

[4]) A supplementary stress system is one which is the result of a set of imposed boundary loads. It is distinct from a stress system related to the effects of gravity; gravity does not influence the supplementary stress system.

stresses necessary to deform the specimens in those experiments was on the order of 5–10+ kb. Equations (6) predict about 3 kb maximum differential stress in the standard state stress field. I interpret this information to mean that 'granitic' rocks do not flow as assumed in Heim's rule, and that the natural standard state in crustal rocks is not necessarily the hydrostatic state.

Inasmuch as the *in situ* state of stress in the crust prior to Laramide deformation is unknown, I have chosen the stress system given by equations (6) as more likely than the hydrostatic state. Determination of a value to be used for Poisson's ratio was made by reference to JAEGER (1969, p. 58), from which I selected $v = 0.25$ as being typical for 'granitic' rocks.

One additional consideration was 'built into' the standard state as used in this study. I planned to apply the results to the Rocky Mountain foreland, where, prior to Laramide deformation, the basement was overlain by a sedimentary veneer some 2.5 to 4.5 km thick. This depth of burial was simulated by adding a uniform 0.5 kb compressive stress acting in the z direction. Assuming a constant $\rho = 2.75$ for the basement rocks and a constant $g = 9.8\ \text{m/sec}^2$ leads to the standard state used in this study:

$$\begin{aligned} \sigma_{zz} &= \rho g z - 0.5\ \text{kb} \\ \sigma_{xx} &= \frac{v}{1-v}\cdot\sigma_{zz} = \frac{\sigma_{zz}}{3} \\ \tau_{xz} &= 0. \end{aligned} \tag{7}$$

In the examples referred to above (Fig. 6 and Plates 1A–D of HAFNER), the Principle of Superposition was employed. For linearly-elastic materials, this principle is: if ψ_1 and ψ_2 are valid static stress fields defined over the same physical region, a linear addition of the two fields (i.e., summing corresponding components of the stress tensor) results in another valid, statically-admissible stress field, ψ_3, whose boundary conditions are the sum of the boundary conditions which would have caused ψ_1 and ψ_2 individually.

All of the solutions reported in this paper contain the standard state stress system described above (equations (7)). The stress fields due to one or more of a variety of boundary loads are superposed onto this standard state, resulting in stress fields related to the boundary conditions more complicated than those previously published. The basic development of the equations necessary to calculate the interior stresses caused by the end load boundary condition is found in HAFNER (1951), as is the development of the expressions giving values for the stress components in a beam subjected to a sinusoidally-shaped lower-boundary supplementary normal stress.

The components of the supplementary stress system due to an end load are determined by applying equations (2) to equation (3):

$$\begin{aligned} \sigma_{xx} &= 2\cdot D\cdot x \\ \sigma_{zz} &= 0 \\ \tau_{xz} &= -2\cdot D\cdot z. \end{aligned} \tag{8}$$

Substitution of $(l - x)$ for x in equation (3) allows the end load to be applied to the opposite end of the beam. Notice that the supplementary stress system described by equations (8) contributes nothing to the vertical normal stress component, σ_{zz}.

The equation for the lower-boundary (i.e., $z = -c$) supplementary normal stress in the case of a simple, sinusoidally-varying load has the form:

$$\sigma_{zz}(x, -c) = A \cdot \sin \alpha x. \tag{9}$$

However, this study employs the Fourier series approach discussed in TIMOSHENKO and GOODIER (1970, p. 53–61) to obtain lower-boundary supplementary normal stresses that vary laterally as sawtooth- and step-shaped functions. Briefly summarized, this approach proceeds as follows.

Consider p different cases of sinusoidal supplementary normal stress, each with its own amplitude, A_n, and wave number, α_n. For each of them we have

$$\sigma_{zz_n}(x, -c) = A_n \cdot \sin \alpha_n x. \tag{10}$$

Using the Principle of Superposition, let us sum all the different stress fields (i.e., $\psi_1 + \psi_2 + \cdots + \psi_p$), each with boundary stresses as in equation (10). Simultaneously, we sum their boundary stresses

$$\underset{\text{total suppl.}}{\sigma_{zz}} (x, -c) = \sum_{n=1}^{p} A_n \cdot \sin \alpha_n x, \tag{11}$$

where A_n is a function of Z, being constant at a given depth. Thus, the σ_{zz} component of the stress tensor has the form:

$$\underset{\text{total suppl.}}{\sigma_{zz}} (x, z) = \sum_{n=1}^{p} A_n(z) \cdot \sin \alpha_n x. \tag{12}$$

Equation (11) forms a Fourier sine series approximation for $\sigma_{zz}(x, -c)$ of some shape. The process described above (equations (9)–(12)) can be repeated for the case where $\cos \alpha x$ is substituted for $\sin \alpha x$ in equation (4) so that there are series in $\cos \alpha_n x$ and in $\sin \alpha_n x$ available for approximation. This allows $\sigma_{zz}(x, -c)$ to take any shape, or, *any* $\sigma_{zz}(x, -c)$ may be approximated, as long as the shape function is 'balanced' so that it requires no constant term in its approximation.

The other stress components (σ_{xx}, τ_{xz}) can be analysed in similar fashion, generating expressions for their boundary and interior values. Other Airy stress functions with different forms could be used to generate a suite of 'base elements' for all the stress components. Thus, any statically-admissible set of supplementary boundary loads could be duplicated (approximated) by a sufficiently complicated scheme of superposition using the 'base' boundary conditions, with a consequent ability to determine the interior stress-state of the body. Therefore, any statically-loaded beam which meets the formal mathematical-mechanical assumptions of the analysis *can* be modeled for interior stress-states by using a suite of Airy stress functions and the Principle of Superposition, although this approach may not always prove to be practical.

Those formal assumptions can be stated in an hierarchial sequence. The body is (1) a *continuum* in (2) *static equilibrium*. The body (3) behaves in a *linear elastic* fashion and is *homogeneous* and *isotropic*. The geometry of the body is that of (4) a *rectangular beam* of arbitrary thickness. Elastic distortions of the beam occur under a condition of (5) *plane strain*. Assumptions made in extending the results to explain natural deformations do not affect the theoretical development and are treated in the discussion below.

The discussion above has been general, aimed only at aiding in understanding the concept of series approximation. The equations below (see also equations (16)–(23) of HAFNER (1951)) are sufficient for calculating the total stress field due to those lower-boundary supplementary normal stresses that can be approximated by sine series. The stress components of each term in the series are given by

$$\begin{aligned} \sigma_{xx_n}(x, z) &= \sin \alpha_n x \cdot (-K_1 \cdot F_1 + K_2 \cdot F_2) \\ \sigma_{zz_n}(x, z) &= \sin \alpha_n x \cdot (-K_1 \cdot F_3 - K_2 \cdot F_4) \\ \tau_{xz_n}(x, z) &= \cos \alpha_n x \cdot (K_1 \cdot F_4 - K_2 \cdot F_1), \end{aligned} \tag{13}$$

with

$$\begin{aligned} F_1(z) &= \sinh \alpha_n z + \alpha_n z \cosh \alpha_n z \\ F_2(z) &= 2 \cdot \cosh \alpha_n z + \alpha_n z \sinh \alpha_n z \\ F_3(z) &= \sinh \alpha_n z - \alpha_n z \cosh \alpha_n z \\ F_4(z) &= \alpha_n z \sinh \alpha_n z, \end{aligned} \tag{14}$$

and

$$\begin{aligned} K_1 &= \frac{A_n \alpha_n c \cosh \alpha_n c + A_n \sinh \alpha_n c}{\sinh^2 \alpha_n c - \alpha_n^2 c^2} \\ K_2 &= \frac{A_n \alpha_n c \sinh \alpha_n c}{\sinh^2 \alpha_n c - \alpha_n^2 c^2}. \end{aligned} \tag{15}$$

The constant c is the depth of the beam, and l is its length (refer to Fig. 1a). α_n is determined by

$$\alpha_n = \frac{n\pi}{l}. \tag{16}$$

The A_n terms of the series (in equations (15)) allow for the necessary variation to calculate stress systems for various shapes of boundary stress. For a simple sawtooth wave (see Fig. 2a) we find

$$f(x) = \frac{2w}{\pi} \cdot \sum_{n=1}^{\infty} \frac{(-1)^{n+1}}{n} \cdot \sin \frac{n\pi x}{l}, \tag{17}$$

and for a step function (Fig. 2b)

$$f(x) = \frac{4w}{\pi} \sum_{n=1,3,5,\ldots}^{\infty} \frac{1}{n} \cdot \sin \frac{n\pi x}{l}. \tag{18}$$

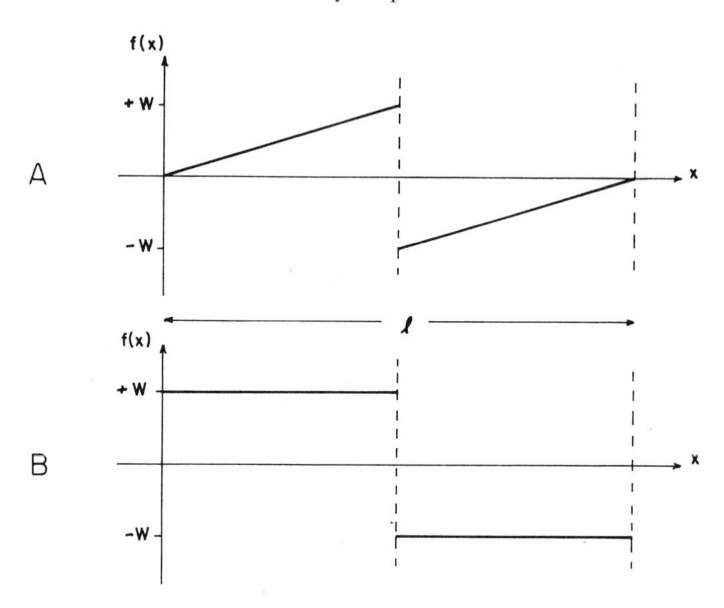

Figure 2
(A) Sawtooth function $f(x)$ with period l and amplitude $\pm W$. (B) Step function $f(x)$ with period l and amplitude $\pm W$.

Using these relationships, the following expressions were derived for the sawtooth load

$$A_n = \frac{2 \cdot A \cdot (-1)^{n+1}}{n \cdot \pi}, \tag{19}$$

and for the step load

$$A_n = \frac{4 \cdot A}{n \cdot \pi} \quad (n = 1, 3, 5, \ldots). \tag{20}$$

For the sinusoid there is one term

$$A_1 = A. \tag{21}$$

In these equations, A is the amplitude of the wave being approximated. In the case of the step and the sawtooth, it is measured at the discontinuity – when $A = 1.0$ kb, the σ_{zz} value jumps from 1.0 kb below standard state to 1.0 kb above standard state at the discontinuity point.

In order to obtain smooth shapes for the lower-boundary normal stresses, I made preliminary calculations which showed that a 15 term series gave the smoothest sawtooth before the appearance of a Gibb's spike.[5]) A 10 term series gave an acceptably smooth step function with minimal spike development.

---

[5]) A Gibb's spike is a numerical phenomenon that is observed when approximating a function that has a discontinuity point. It becomes manifest as more terms are added to the series approximation.

In this study, the state of stress is calculated at grid points distributed across an X-Z plane of the beam. The grid is rectangular, with grid points spaced at 2.0 km horizontally and 1.5 km vertically. Thus, there are 51 grid points in the X-direction and 11 in the Z-direction for a total of 561. The standard state stress field (equations (7)) is calculated at each grid point, and equations (8) (with possible variations) and/or equations (13)–(15) are added to the respective stress components at each grid point.

From the stress components at each point,

$$\begin{bmatrix} \sigma_{xx} & \tau_{xz} \\ \tau_{xz} & \sigma_{zz} \end{bmatrix}$$

the principal stresses and their orientations are calculated. Possible shear fracture trajectories (geologically representative of fault trajectories) are determined at a characteristic angle, θ, relative to the direction of the most-compressive principal stress, σ_1. For the purposes of this study, I choose as representative a value of φ (angle of internal friction) of 30° (HANDIN, 1966) resulting in $\theta = 45° - \varphi/2 = 30°$. Diagrams are drawn to depict principal stress trajectories and potential shear fracture trajectories in the beam; this is a standard means of presentation. In this study, I employ a Calcomp 1136 plotter to draw tangent lines to each of the principal stresses and each of the fracture trajectories at each grid point. Families of curves are drawn by hand on the machine output using the tangent-line information. The curves are drawn with a spacing designed to aid visual clarity, but the spacing has no exact physical significance. This discussion of technique is necessary here only to point out the possibility for subjective bias and human error in the drafting process.

Appendix A briefly discusses the organization of the computations involved in the study.[6]) Also included there is a description of the steps involved in plotting the results.

3. *Results*

Table 1 summarizes the suite of boundary conditions investigated in this study. For each case calculated, principal stress trajectories and potential shear fracture trajectories have been drawn. The location of each set of stress trajectory/shear fracture trajectory diagrams is indicated in Table 1.

In Figs. 3 through 11, the boundary stresses are illustrated on the stress trajectory diagram in each figure. All boundary normal stresses are compressive, with their magnitudes depicted by scaled arrows directed *toward* the beam. The normal stresses on the base consist of a standard state component and a supplementary component. The sawtooth or other shape can be visualized as having been added to a constant. Thus the positive and negative variation of the supplementary stress becomes a variation of magnitude, but not of sign, when superposed with the standard state.

---

[6]) Listings of the programs used, with documentation, can be obtained at cost from the author.

Table 1
Summary of imposed boundary Conditions)*

Shape of lower-boundary normal stress	End loads†)		Location of stress and fracture trajectory diagrams
Sawtooth	*Left*	*Right*	
0.4 kb	0.0 kb	0.0 kb	Fig. 3
1.0	0.0	0.0	Fig. 4
0.4	0.0	1.0	Fig. 8
0.4	0.0	3.0	Fig. 9
1.0	3.0	0.0	Fig. 10
1.0	2.0‡)	1.0	Fig. 11
Step	*Left*	*Right*	
0.4 kb	0.0 kb	0.0 kb	Fig. 5
1.0	0.0	0.0	Fig. 6
Sinusoid	*Left*	*Right*	
1.0 kb	0.0 kb	1.0 kb	Fig. 7

*) All imposed boundary conditions in addition to the standard state as described in text.
†) End loads necessitate basal shear; 0.45 kb for 3.0 kb end load and 0.15 kb for 1.0 kb end load.
‡) This end load is the net result of 1.0 kb end load on left with 0.15 kb shear on base added to 1.0 kb uniform horizontal compression with no associated shear.

Shear stresses acting on the boundaries of the beam are portrayed by half-arrows to indicate *sense of shear*. The half-arrows are drawn to the same scale as the normal stress arrows. No shear stresses are present on the top of any of the beams. Shear stresses are always present on the ends of the beams – hence these planes are not principal planes. Shear stresses are present along the bottom of the beam only when there is an applied end load (Note: the absence of shear stresses on the base of the beam is a function of the loading system selected – i.e., the absence is a designed absence. Other loading systems could, and possibly should, be calculated to investigate the effects of variable shear stresses distributed on the base of the beam).

The shear fracture trajectory diagram in each figure is, in one sense, easily explained. The two families of lines (for right-lateral and left-lateral senses of shear) are nothing more than orientations drawn at $\pm 30°$ to the σ_1 trajectory. On a phenomenological basis, these lines do not represent the prediction of specific shear fractures, either in location or orientation. They represent a field of trajectories (orientations) for the fractures that may possibly form. As described here, no knowledge is implied about the *location* of any fracture that forms as a result of the imposed loads.

This disclaimer is not to be taken in reverse: that the shear fractures which actually do form will not have orientations like the drawn trajectories. Arguments can be advanced (see Discussion) which elevate the trajectories as drawn into a status of 'near-prediction'. For the purpose of describing the diagrams containing the results of this study, let us assume that the shear fracture trajectories represent possible orientations of the shear fractures that actually would form as a result of the imposed loads.

Figures 3 and 4 form a pair. Both of these cases have a sawtooth-shaped lower-boundary supplementary normal stress. Both have the same standard state. The difference in loading conditions is in the magnitude of the sawtooth. Figures 3 and 4 are examples of ± 0.4 kb and ± 1.0 kb differential loads, respectively. Notice that the stress fields induced by the loads are quite similar in terms of the orientation of the principal stresses. The main difference is that the increased differential load in Fig. 4 causes a singular point (where principal stresses undergo a 90° rotation – i.e., an isotropic point, or a point of zero differential stress) to 'move up' into the beam on the right side. The fracture trajectories likewise necessarily exhibit great similarity. 'Shallow' fractures (near the top of the beam) imply lateral shortening (in the X-direction) on the left side and lateral extension on the right side of both beams. 'Deep' fractures indicate lateral expansion on the left side in both cases, but opposite senses of gross displacement on the right side. A very important feature to note is that a single mechanical system (i.e., the loaded beam) indicates or requires simultaneous lateral shortening and lateral expansion.

The other results (Figs. 5–11) can be described in a similar fashion. Figures 5 and 6 form another pair. This time the supplementary normal stress on the lower boundary has the shape of a step. Differential loads are ± 0.4 kb and ± 1.0 kb, respectively. Similar comments are in order with respect to singular points and regions of shortening and extension. Notice that the main difference between the pairs (Figs. 3 and 4 and

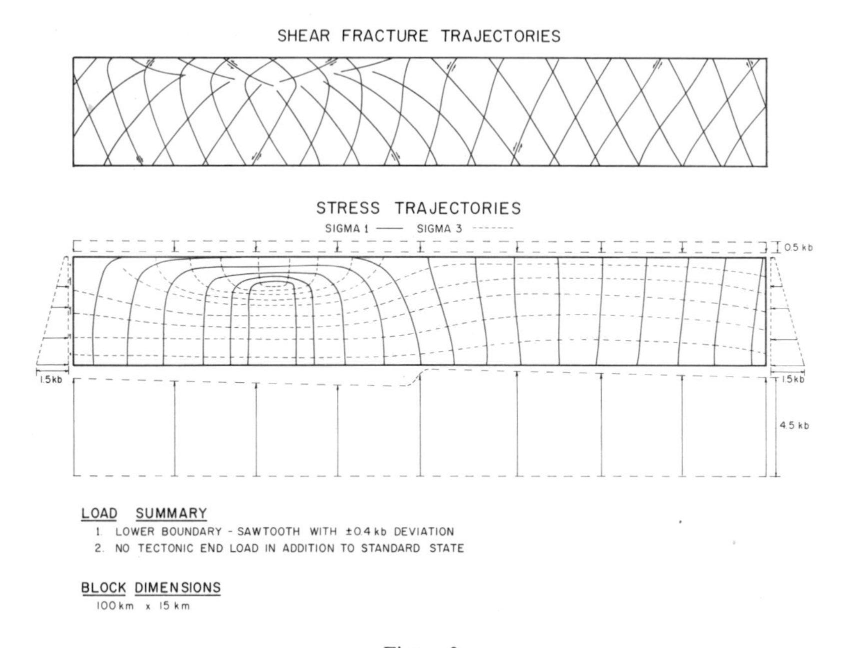

Figure 3

SHEAR FRACTURE TRAJECTORIES

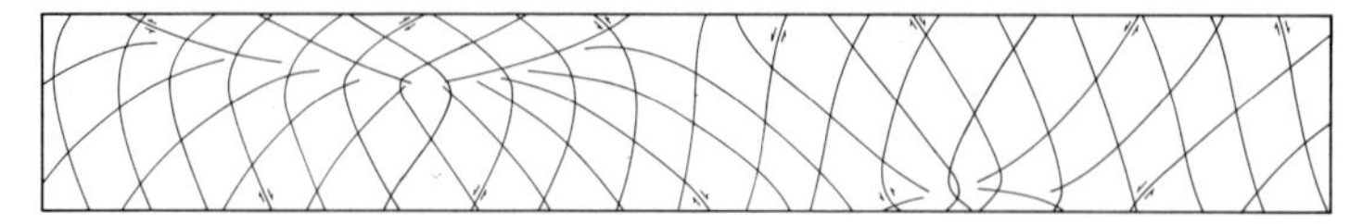

STRESS TRAJECTORIES

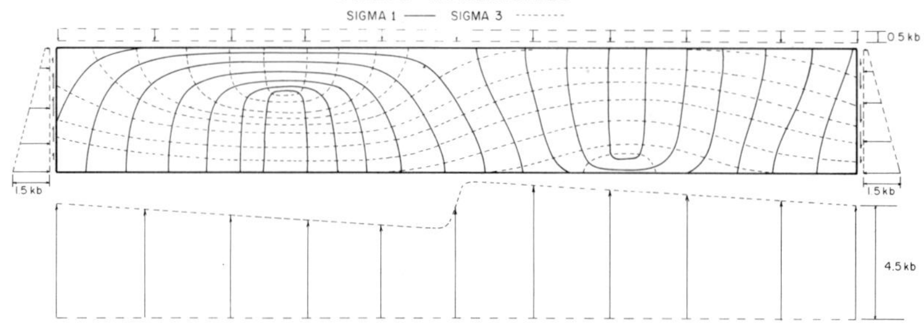

LOAD SUMMARY

1. LOWER BOUNDARY – SAWTOOTH WITH ± 1.0 kb DEVIATION
2. NO TECTONIC END LOAD IN ADDITION TO STANDARD STATE

BLOCK DIMENSIONS

100 km. x 15 km.

Figure 4

SHEAR FRACTURE TRAJECTORIES

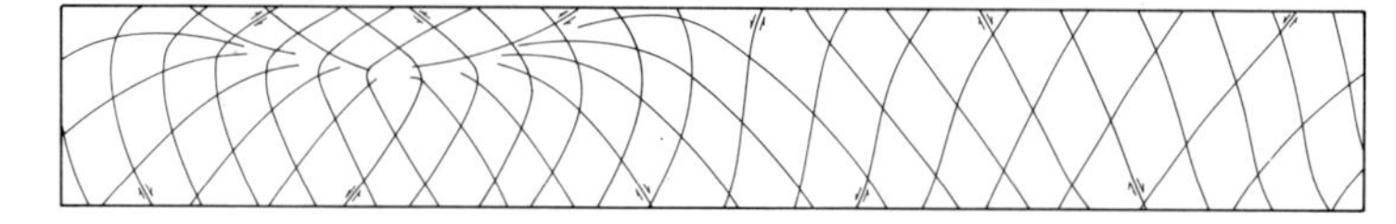

STRESS TRAJECTORIES

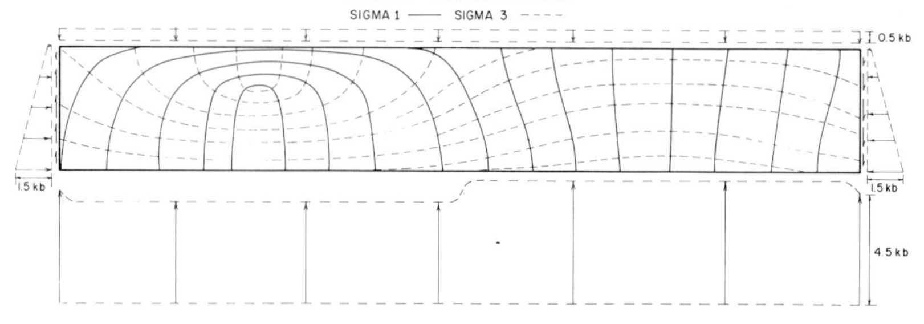

LOAD SUMMARY

1. LOWER BOUNDARY – STEP WITH ± 0.4 kb DEVIATION
2. NO TECTONIC END LOAD IN ADDITION TO STANDARD STATE

BLOCK DIMENSIONS

100 km. x 15 km.

Figure 5

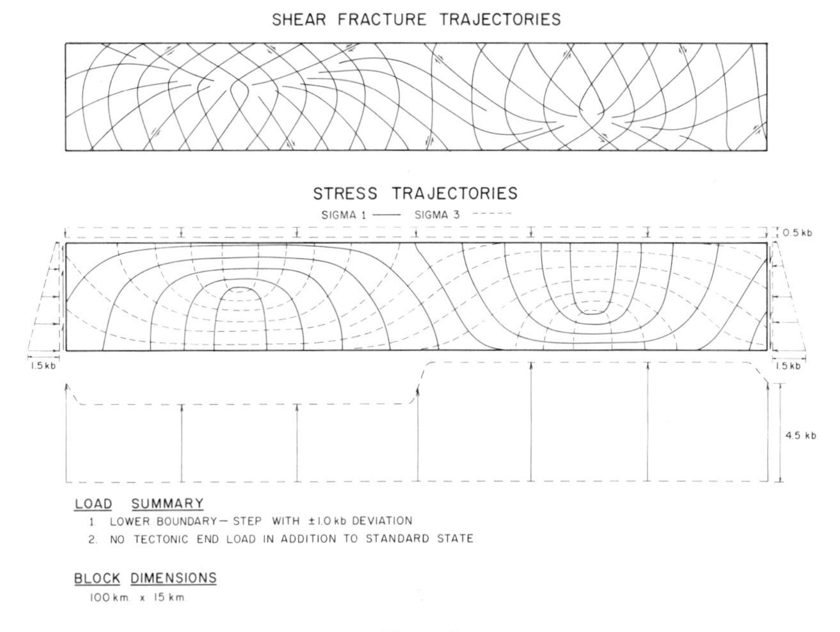

Figure 6

Figs. 5 and 6) is a smoother curvature of trajectories in the case of a step-shaped loading condition. This difference is most apparent when comparing the stress trajectories of Figs. 4 and 6.

The four cases discussed thus far have had no imposed end loads. Because of this, they are periodic in the X-direction. One end of each diagram matches the other. However, the addition of end loads removes that symmetry. Figure 7 has a sinusoidal lower-boundary supplementary normal stress with a difference of ± 1.0 kb; it also has an end load of 1.0 kb imposed on the right. Notice that the normal stress magnitudes drawn on the end of the beam are separated into: (1) a part due to the imposed load, and (2) a part related to the standard state. All of the diagrams which follow are consistent in this way. In Fig. 7, the relative 'smoothness' of the loading stress is reflected in the smoother curvature of the trajectories; there are certainly fewer 'shoulders' (or regions of high curvature) present along the stress trajectory lines than in either Fig. 4 or Fig. 6.

Figures 8 and 9 form another pair. Here, the difference is the magnitude of the end load, imposed on the right. In Fig. 8, the end load is 1.0 kb with 0.15 kb shear stress on the base of the beam. The end load is 3.0 kb in Fig. 9, and the shear stress on the base is 0.45 kb. The appearance of Fig. 8 is very similar to Fig. 3; the differences in the trajectories are close to the right end of the beam where the end load has its greatest effect. (Recall from equations (8) that σ_{xx} decreases linearly away from the end load.) However, Fig. 9 is not similar to Fig. 3 at all. The end load has caused the region of shortening on the left side to increase in area (the singular point has been 'moved'

SHEAR FRACTURE TRAJECTORIES

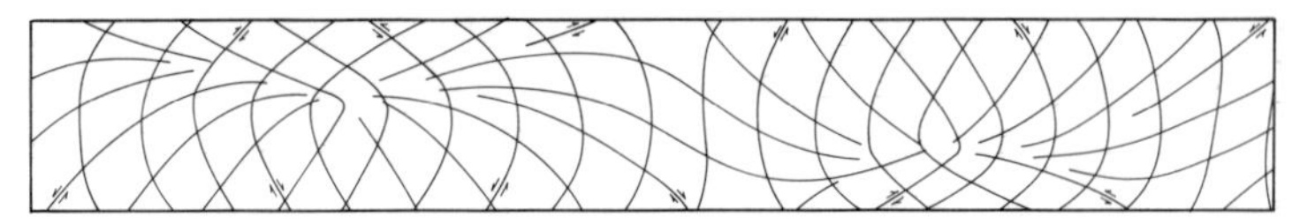

STRESS TRAJECTORIES

SIGMA 1 —— SIGMA 3 - - - - -

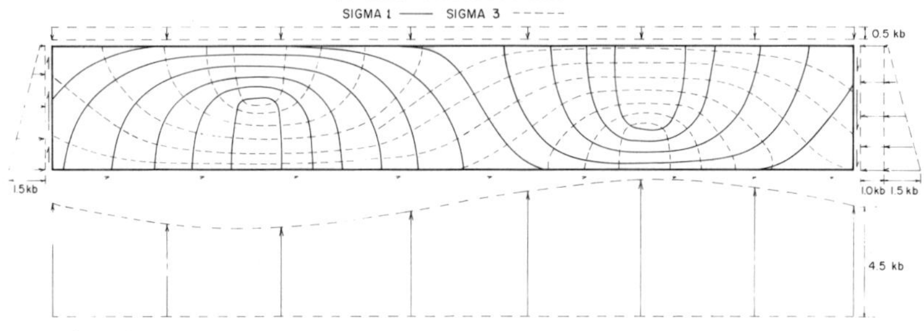

LOAD SUMMARY

1. LOWER BOUNDARY – SINUSOID WITH ±1.0 kb DEVIATION
2. 1.0 kb TECTONIC END LOAD IN ADDITION TO STANDARD STATE, WITH 0.15 kb SHEAR ALONG BASE

BLOCK DIMENSIONS

100 km x 15 km.

Figure 7

SHEAR FRACTURE TRAJECTORIES

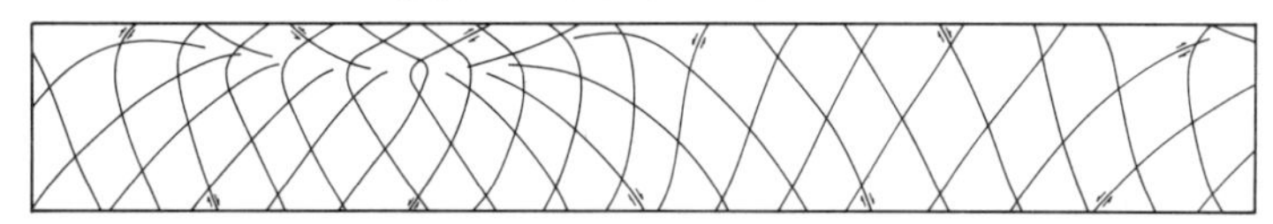

STRESS TRAJECTORIES

SIGMA 1 —— SIGMA 3 - - - - -

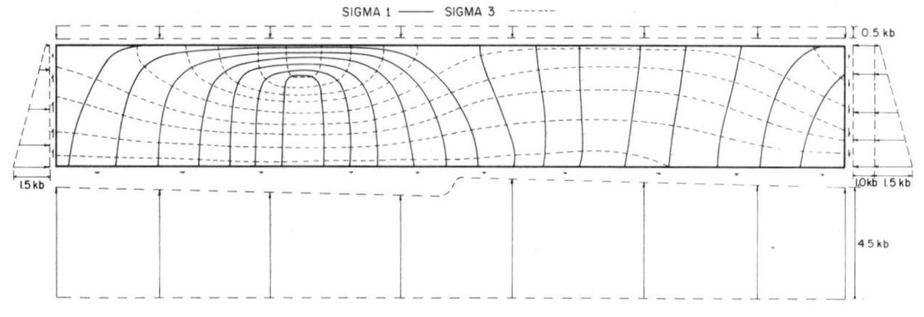

LOAD SUMMARY

1. LOWER BOUNDARY – SAWTOOTH WITH ±0.4 kb DEVIATION
2. 1.0 kb TECTONIC END LOAD IN ADDITION TO STANDARD STATE, WITH 0.15 kb SHEAR ALONG BASE

BLOCK DIMENSIONS

100 km. x 15 km.

Figure 8

down). The right side of the beam is totally different; in Fig. 9 it is a broad region containing fractures that would lead to lateral shortening. The top of the beam indicates all shortening, except for a small extension region in right-center. Note an increase in the number of singular points as the complexity of the loading system increases.

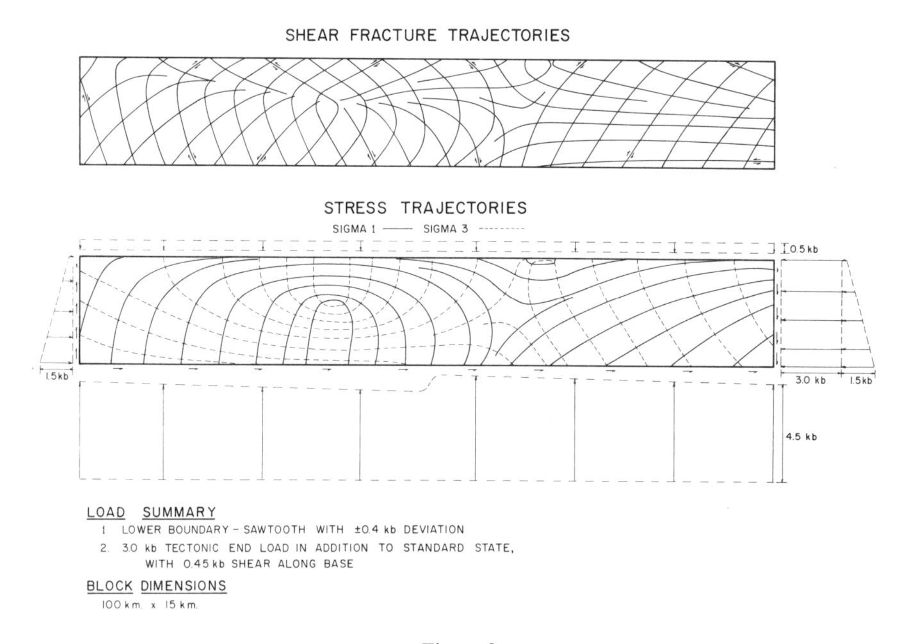

Figure 9

Figure 10 is included to demonstrate the effects of a left end load on a ± 1.0 kb sawtooth. Notice that the upper-left region of lateral shortening is much increased in size relative to Fig. 4. Also note the singular point developed in the lower-left corner. Figure 11 is another diagram illustrating the effects of end loads on the trajectories. The loading combination of 1.0 kb left end load, 1.0 kb uniform horizontal compression, and ± 1.0 kb sawtooth leads to trajectories that are quite similar to those in Fig. 10. The lower-left singular point of Fig. 10 is not observed in Fig. 11, but a singular point is fully developed on the right side of the beam.

It is significant to note that, by-and-large, the imposition of horizontal loads does not greatly alter the potential fracture trajectories, particularly in the region immediately above the 'jump' in the lower-boundary normal stress. Figure 9 is the only case investigated in which the potential fault trajectories in a substantial portion of the beam show primarily horizontal motions. However, the magnitudes of the loads depicted in Fig. 9 are 'disproportionate' when compared to load ratios in the other cases investigated.

SHEAR FRACTURE TRAJECTORIES

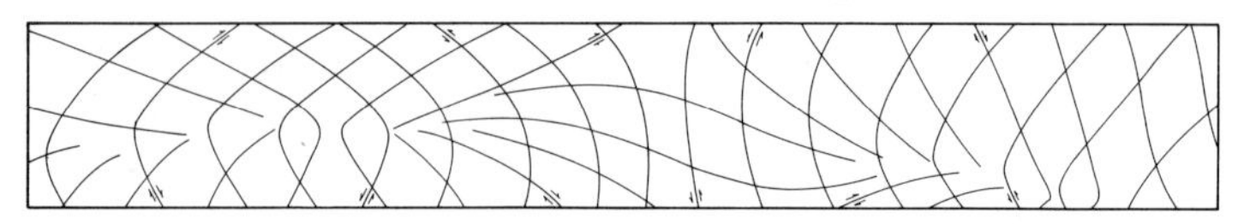

STRESS TRAJECTORIES

SIGMA 1 —— SIGMA 3 -------

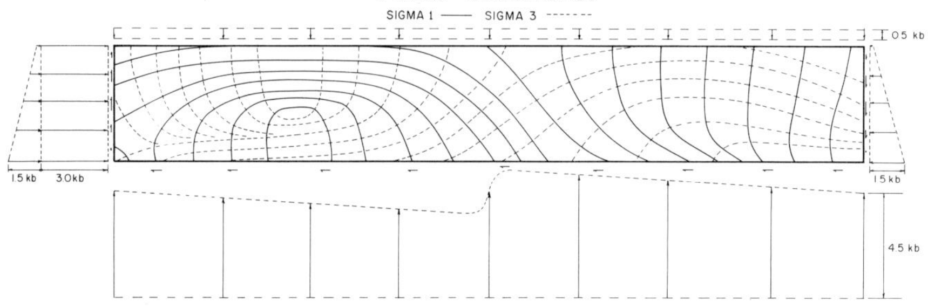

LOAD SUMMARY

1. LOWER BOUNDARY – SAWTOOTH WITH ± 1.0 kb DEVIATION
2. 3.0 kb TECTONIC END LOAD IN ADDITION TO STANDARD STATE, WITH 0.15 kb SHEAR ALONG BASE

BLOCK DIMENSIONS

100 km. × 15 km.

Figure 10

SHEAR FRACTURE TRAJECTORIES

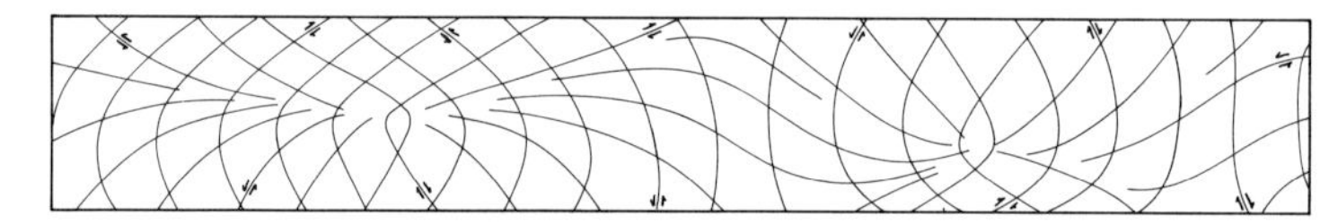

STRESS TRAJECTORIES

SIGMA 1 —— SIGMA 3 - - - -

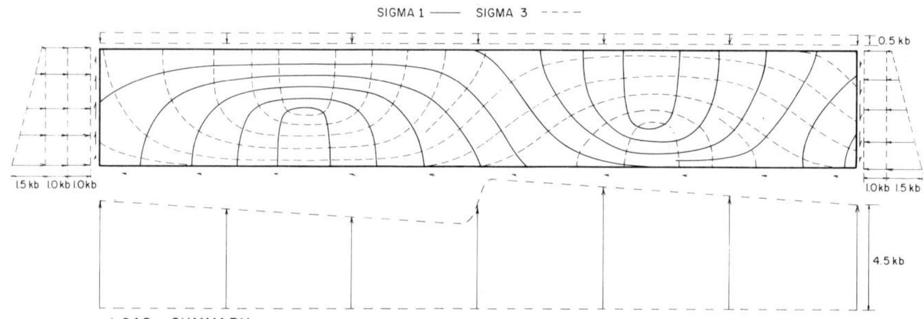

LOAD SUMMARY

1. LOWER BOUNDARY – SAWTOOTH WITH ± 1.0 kb DEVIATION
2. 1.0 kb TECTONIC END LOAD IN ADDITION TO STANDARD STATE, WITH 0.15 kb SHEAR ALONG BASE
3. 1.0 kb UNIFORM HORIZONTAL COMPRESSION

BLOCK DIMENSIONS

100 km. x 15 km.

Figure 11

4. Discussion

The impetus for generating this larger suite of stress and shear fracture trajectories resulting from the solution of more boundary value problems arose from a longer-term study of Rocky Mountain foreland deformation; however, the solutions are certainly 'independent' of the Rockies and can be used in studying similar deformations elsewhere.

One of the more important aspects involved in solving boundary value problems is the selection of a solution technique. For static problems, analytic methods are easiest to implement. When specifying boundary conditions in terms of stresses, the Airy stress function is used. Mixed boundary conditions (both stresses and displacements specified on the boundaries) can be handled with equivalent mathematics (SANFORD, 1959; HOWARD, 1966), but I chose to specify loads in terms of stresses (actually, the loads are composed of traction vectors). Numerical methods can also be employed, but their use amounts to 'overkill' for static problems, as well as being somewhat more expensive[7]) (J.-CL. DEBREMAECKER, pers. commun., 1976).

A feature of the stress and fracture-trajectory diagrams worth noting can be seen by comparing Figs. 4 and 5. Although the loading conditions in these two cases differ only in the magnitude of the sawtooth, the fracture trajectories are not coincident. HAFNER (1951) attempted to present the results of such a sequence of increasing loading conditions by drawing only one diagram and indicating regions of 'stability' for each load increment. By inference, each loading system in the sequence resulted in identical fracture trajectories. The differences in trajectories evidenced by Figs. 4 and 5 rule out that data-presentation shortcut used by HAFNER.

In general terms, the fracture trajectories assume rather unusual patterns in the regions around singular points. The trajectories tend to splay around or out of these areas, developing a vague resemblance to feathers. That these patterns seem unreasonable is not surprising. The singular points are locations of zero differential stress, and the immediate surroundings have low stress differences. Thus, it is unlikely that fractures would initiate in these regions. HAFNER (1951) did not attempt to depict fracture trajectories in these regions of low stress difference, but they are shown here: 1) for completeness, and 2) because the problem of fracture propagation, especially into or through such regions, is very poorly understood. Drawing these trajectories may provide information that may aid in advancing that understanding.

The formal assumptions listed above (p. 119) are quite restrictive. *Sensu stricto* they limit the analysis to loading conditions below a 'failure load'. In other words, the formation of the first crack strictly violates the assumption of *continuum*, at least in the immediate vicinity of the crack. It is possible to circumvent this problem by defining

[7]) At the current rate ($360.00 per hour) for computer time on an IBM 370/155, each case costs only $5.00 to compute using the Airy stress function. Plotting costs average about $12.00 (based on a rate of $36.00 per wall-hour) when plotted on a Calcomp 1136.

continuum to include cracks – or, that in bulk, the material behaves as a *continuum*. Then, the question affecting this study becomes: does the presence or formation of a fracture alter the stress field such that the calculated fracture trajectories are invalid? Or, restated, do the actually-formed fractures propagate along the trajectories as initially calculated, or along some other path?

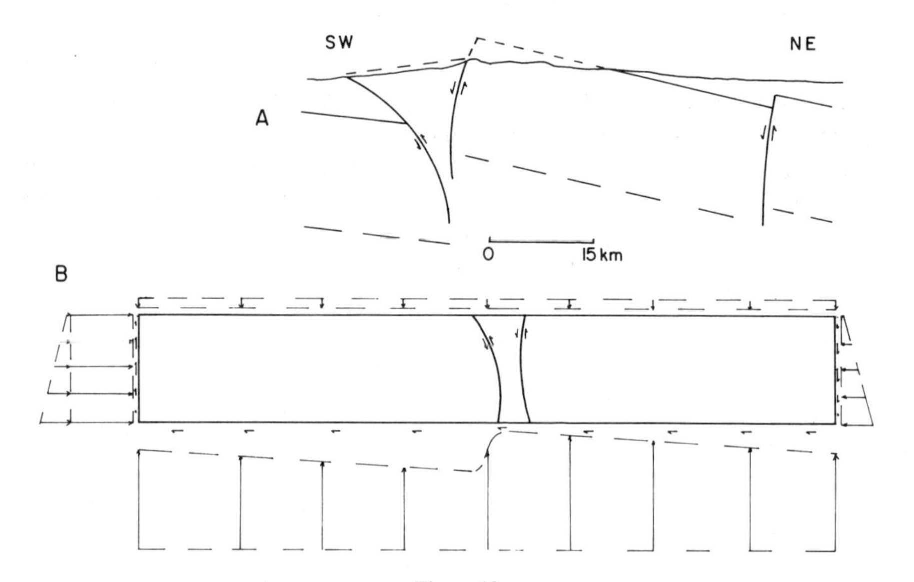

Figure 12
(A) Hypothetical cross-section of the southern Wind River Mountains passing near the town of Lander. Dashed line is drawn to facilitate visualizing the upper 15 km of basement. Scale is same in both A and B. (B) Simplified version of Figure with only two of the potential fault trajectories shown.

The means of answering that question calls upon empiricism – the question cannot be answered on a theoretical basis at this time. The approach appeals to the degree of ordering found among natural fractures. STEARNS (1968; 1972) has presented field evidence pertinent to this point. For the Bonita fault near Tucumcari, New Mexico, STEARNS (1972) has shown that there is a high degree of ordering for shear fractures both parallel to the normal fault and conjugate to it. He has also shown that fractures associated with folded rocks exhibit remarkable degrees of ordering (STEARNS, 1968).

It is very difficult to conceive of these fractures (those in ordered sets) having *all* formed simultaneously; they must have developed sequentially. Thus, for real rocks, the ordered orientations of fractures indicate that the formation of one crack, and even the formation of many fractures, does not appreciably alter the orientation of the causative stresses, otherwise, the ordering would not be obtained. As a corollary, it seems reasonable to accept that fracture orientations predicted from theoretical solutions such as those presented here are not invalidated by the limiting assumptions encountered in the formal analysis.

The ultimate test of results such as those presented here will not come from theoretical considerations. To be sure, advances in the understanding of theoretical mechanics and modeling techniques will be incorporated by future workers and will lead to more sophisticated solutions. But the test of the results must again come from empiricism – i.e., from the usefulness of the solutions. If theoretical solutions aid in the understanding of geological problems, if they help explain a possible origin for a previously enigmatic structure, then the solutions attain value and a sense of validity. For systems as complex as those being discussed here, the solutions will never be 'correct'; they can only become 'more correct'.

STEARNS (1975, Figs. 10–13, p. 156–157) used one of HAFNER's (1951) solutions to explain a possible mode of origin for the structures observed in a cross-section drawn across the entire northern Big Horn Basin in Wyoming. In this region, a 'rigid' basement was overlain by a 'passive' sedimentary veneer prior to Laramide deformation. The observed (and interpreted) final motions of the rotated and differentially uplifted and downdropped basement blocks can be explained as resulting from the motion of rigid blocks along curved faults. The locations, curvatures, and senses of shear of the faults is remarkably well matched (explained) by HAFNER's solution.

However, the structural style in the more southwesterly portions of Wyoming is not as well characterized by high-angle faults as is the northern region studied by STEARNS (1971; 1975). In this southern part of the state, there are many more low-angle faults and basement 'overhangs'. This geometry has variously been termed 'upthrust'. Attempts to explain this geometry via the solutions of HAFNER (1951), SANFORD (1959), or HOWARD (1966) have seen only limited success. The curving reverse fault, and, in some cases, even the normal-fault zone, have been either predicted or modeled by analog. But the *combination* of proper rotation of blocks, and proper locations and geometries of faults has not previously been explained.

Figure 12a depicts a typical 'upthrust' structure – the Wind River Mountains of Wyoming. The cross-section as shown is very schematic. Surface and near-surface attitudes and geometries are honored as far as they are known. The configuration for the faults shown at depth is purely speculative, and merely represents my concept of one possible geometry. Note that the correct rotations are produced with no 'room' problems, as is the shortening and extension.

Figure 12b is a simplified version of Fig. 11 with only two faults drawn. Suppose that these two faults form in the beam and that movement occurs along those faults with the indicated sense of shear. The resulting geometry would be very similar to the geometry of the 'real' world shown directly above in Fig. 12a. The inexact 'fit' resulting from this process speaks to the need for additional solutions, and for allowances for finite deformations and post-initiation changes in shape. A forthcoming paper (COUPLES, in prep.) will discuss the implications of such allowances, as well as treat other examples involving the usage of the new solutions presented here. That paper will also discuss some of the aspects that must be considered in applying theoretical solutions to the study of geologic deformations.

5. *Conclusions*

An inexpensive and relatively simple technique has been described which allows for the calculation of stress-states in an elastic beam subjected to a complicated set of boundary loads. Procedures for presenting the stress field information and for deriving the associated potential shear fracture pattern have also been outlined.

The conclusions of this study may be summarized as follows:

1. Loads with similar geometry but of different magnitudes result in non-identical stress trajectories, and, consequently, in different potential shear fracture patterns.
2. Potential fault patterns are not greatly altered even by imposing large lateral loads.
3. Arguments derived from the ordering observed in natural fractures indicate that the unsolved problem of fracture initiation and propagation may not be important in limiting the usefulness of static solutions such as those presented here.
4. Previously enigmatic structures like the Wind River Mountains can be better explained by loading systems with geometry similar to that of the resulting mountains – assymetric. Regardless of whether or not the solutions used here 'exactly' explain the structure, the hypothetical geometry suggested by the solutions 'explains' a greater portion of the observed natural relationships than did previous attempts.

6. *Acknowledgements*

I would like to thank the members of my advisory committee at Rice University, J.-Cl. DeBremaecker and H. G. Ave Lallemant, and my chairman, Clark Burchfiel, for advice and support during my work on this project, and for assistance in improving the manuscript. Chuck Wielchowsky, Dave Weinberg, Dave Parrish, and John Handin suggested many improvements in the text; their contributions are greatly appreciated.

I especially wish to thank my fellow graduate students, both at Rice and Texas A&M, for their many stimulating discussions. The students and staff of the Center for Tectonophysics, Texas A&M University, were helpful in clarifying certain ideas in my mind. Dave Weinberg deserves special credit for helping me develop my ideas about the 'upthrust' problem. Dave Stearns was instrumental in inspiring me to undertake this project. His guidance and concern are deeply appreciated and hereby gratefully acknowledged.

I was supported during this project by a National Science Foundation Graduate Fellowship. Funding for computer time was supplied from Rice University funds.

Appendix A

SERRA (1973) presented a computer program for the calculation and plotting of stress trajectories and associated fracture trajectories based on HAFNER (1951). His program provided a very helpful guide in organizing my calculations, but certain changes were necessary. SERRA's program calculated the total stress components at each grid point and immediately plotted a trajectory symbol. In the method described in this paper, the stress components are determined as series sums. This necessitates storing the values in arrays in the memory of the computer. After the total stress field was calculated, the arrays were carried into a separate plotting routine. Changes also were required in this step. SERRA plotted both principal stress-trajectory tangents at the same time. I found it much easier to plot each one separately, draw in the trajectory lines, and then superimpose them into one diagram. The benefits of following this procedure become even more apparent when trying to draw the shear fracture trajectories which are non-orthogonal. One other modification was made in SERRA's procedure. I plotted symbols from left to right on one pass and from right to left on the return. This was done to minimize costly free-travel of the plotter.

REFERENCES

ANDERSON, E. M., *The dynamics of faulting* (Oliver and Boyd, London, 1942), 183 p.

BORG, I., and HANDIN, J. (1966), *Experimental deformation of crystalline rocks*, Tectonophysics *3*, 249–368.

COUPLES, G. (in prep.), *Analysis of certain mechanical parameters – applications to Laramide basement faulting in the Rocky Mountain Foreland*, in *Laramide Folding Associated with Basement Block Faulting in the Rocky Mountain Region* (ed. Vince Matthews), Geol. Soc. Am. Mem.

HAFNER, W. (1951), *Stress distributions and faulting*, Bull. Geol. Soc. Am. *62*, 373–398.

HANDIN, J. (1966), *Strength and ductility*, in *Handbook of Physical Constants* (ed. S. P. Clark), Geol. Soc. Am. Mem. *97*, 223–289.

HOWARD, J. H. (1966), *Structural development of the Williams Range Thrust, Colorado*, Bull. Geol. Soc. Am. *77*, 1247–1264.

JAEGER, J. C., *Elasticity, Fracture and Flow: with Engineering and Geological Applications* (Methuen and Co., Ltd., London, 1969), 268 p.

JAEGER, J. C. and COOK, N. G. W., *Fundamentals of Rock Mechanics* (Chapman and Hall, London, 1969), 515 p.

SANFORD, A. R. (1959), *Analytical and experimental study of simple geologic structures*, Bull. Geol. Soc. Am. *70*, 19–52.

SERRA, S. (1973), *A computer program for the calculation and plotting of stress distribution and faulting*, Math. Geol. *5*, 397–407.

STEARNS, D. W., *Certain aspects of fractures in naturally deformed rocks*, in *Rock Mechanics Seminar*, (ed. R. E. Riecker) (Air Force Cambridge Research Laboratories, Bedford, Mass., 1968), 97–116 pp.

STEARNS, D. W. (1971), *Mechanisms of drape folding in the Wyoming Province*, Twenty-third Ann. Field Conf., Wyo. Geol. Assn. Guidebook, 125–143.

STEARNS, D. W. (1972), *Structural interpretation of fractures associated with the Bonita Fault*, Twenty-third Ann. Field Conf., N. Mex. Geol. Soc. Guidebook, 161–164.

STEARNS, D. W. (1975), *Laramide basement deformation in the Bighorn Basin – the controlling factor for structures in the layered rocks*, Twenty-seventh Ann. Field Conf., Wyo. Geol. Soc. Guidebook, 149–158.

TIMOSHENKO, S. P. and GOODIER, J. N., *Theory of Elasticity* (McGraw-Hill, New York, 1970), 566 p.

(Received 28th October 1976)

Pageoph, Vol. 115 (1977), Birkhäuser Verlag, Basel

Preliminary Stress Measurements in Central California Using the Hydraulic Fracturing Technique

By Mark D. Zoback,[1]) John H. Healy[1]) and John C. Roller[1])

Abstract – Use of the hydraulic fracturing technique for determining *in situ* stress is reviewed, and stress measurements in wells near the towns of Livermore, San Ardo, and Menlo Park, California are described in detail. In the Livermore well, four measurements at depths between 110 and 155 m indicate that the least principal compressive stress is horizontal and increases from 1.62 to 2.66 MPa. The apparent direction of maximum compression is N 70° E ($\pm 40°$). At the San Ardo site the least principal stress is that due to the overburden weight. At depths of 240.2 and 270.7 m the minimum and maximum horizontal stresses are estimated to be 11.4 and 22.5 MPa, and 12.0 (± 1.1) and 15.8 (± 3.3) MPa, respectively. From an impression of the fracture at 240.2 m, the direction of maximum compression appears to be about N 15° E. The rock in the Menlo Park well is too highly fractured to yield a reliable measurement of the horizontal stresses. The data indicate, however, that the least principal stress is vertical (due to the overburden weight) to a depth of 250 m.

Key words: Stress hydrofracture; Tectonics of California.

1. Introduction

The importance of making *in situ* stress measurements has long been recognized by earth scientists and engineers. The prediction and control of crustal earthquakes, the safe design of underground facilities, the efficient exploitation of the earth's resources, and the proper interpretation of a variety of geologic phenomena all require knowledge of the *in situ* stress field. Although a number of techniques have been devised for determining the magnitude and direction on *in situ* stresses, we have chosen to develop a stress measurement system employing the 'hydrofrac' method. Based on the capabilities of our truck-mounted drill rig, we routinely operate to depths around 300 m, but our basic plan has been to develop an independent and flexible system capable of operating in almost any well that might become available.

This report presents the first of our measurements from wells we drilled near San Ardo, Livermore, and Menlo Park, California. The work at San Ardo is part of a profile of stress measurements made perpendicular to the San Andreas fault. A report on the results of this study is being prepared by J. Bredehoeft, R. Wolff and W. S. Keys of the USGS Water Resources Division. Because of the widespread interest in the potential usefulness of the hydraulic fracturing technique, we describe the data in detail and discuss hydrofrac operations at some length.

---

[1]) Office of Earthquake Studies, U.S. Geological Survey, Menlo Park, California, USA.

Basic theory of hydraulic fracturing

The hydraulic fracturing technique involves pressurizing a well or borehole until a tensile fracture is induced in the rock. If we presume that one of the principal stresses is vertical, the principal horizontal stresses can be determined if a vertical hydraulic fracture is induced in a vertical well-bore. According to the theory of hydraulic fracturing (HUBBERT and WILLIS 1957, KEHLE 1964), a vertically oriented hydraulic fracture will form parallel to the direction of the maximum horizontal compression when the borehole pressure (in an impermeable rock with interstitial pore pressure P_0) reaches the breakdown pressure P_b. P_b is given by the equation

$$P_b = T + 3S_3 - S_1 - P_0 \tag{1}$$

where S_1 and S_2 are the greatest and least principal horizontal matrix stresses (compression is positive) and T is the tensile strength of the rock.

Five distinct measurements are theoretically necessary to determine the *in situ* field; (1) the normally existing pore pressure in the rock, P_0, (2) the pressure at fracture initiation or breakdown, P_b, (3) the pressure necessary merely to keep the fracture open (this pressure, called the instantaneous shut-in pressure (ISIP) is presumed to be equal to the least compressive stress, S_3), (4) the tensile strength of the unfactured rock (usually determined by laboratory measurements on recovered core samples), and (5) the orientation of the induced fracture. Knowing the densities and thicknesses of the overlying rock units, the vertical principal stress S_2 is determined as that due only to the overburden weight.

Most techniques designed to measure *in situ* stress are of the strain release or overcoring type. The disadvantages of these methods are that they must be made fairly close to exposed surfaces (where one might expect the active stress field to be perturbed) and that it is necessary to compute stress from measured strain. These disadvantages do not apply to the hydraulic fracturing technique. However, in hydraulic fracturing we do make three critical assumptions. First, we assume that the rock behave elastically and that we generate the theoretically anticipated fracture and do not open a preexisting fracture or weakness. Second, we rely on knowledge of the rock's tensile strength in order to compute the maximum compressive stress. Third, since we are limited to measuring the orientation of the fracture at the wellbore, we assume that the fracture propagates with the same orientation. As shall be discussed, all of these assumptions present problems to investigators. Even when a 'theoretical' fracture is formed, the use of P_b and the tensile strength of the rock for computing S_1 from (1) can be unreliable (see ZOBACK *et al.* 1976). Nevertheless, the successful stress measurement of RALEIGH *et al.* (1972) and HAIMSON (1973) at the Rangely oil field has given renewed confidence to us and other investigators. *In situ* stress measurements using the hydrofrac technique have also been reported in Iceland (HAIMSON and VOIGHT 1976e) and near the Hohenzollern-graben structure in southwest Germany (RUMMEL and JUNG 1975). HAIMSON (1976c) offers a review of hydrofrac stress measurements in the continental United States.

Field operations

In order to fracture a well or borehole hydraulically, the part of the hole to be fractured (subjected to high pressure) must be isolated. This is normally achieved by using an inflatable straddle packer. As fluid is pumped through drill steel or high-pressure tubing and into the packer from the surface, the inflatable units expand and form a high-pressure seal above and below the straddled interval. A valve between the packers is then opened (either manually from the surface, or by automatically exceeding a pre-set pressure) which allows the fracturing fluid to contact the borehole. If the packers are not set in a relatively smooth borehole, fluid can leak around the packers and thus nullify the experiment. Normally, the pressure of the fluid in the borehole is recorded at the surface but self-contained downhole instruments are also used for pressure measurements.

Once a hydraulic fracture has been formed, it is necessary to measure its orientation accurately in order to determine the direction of the maximum compressive stress. This can be accomplished in two ways. The first technique involves recording an impression of the fracture by inflating a rubber or plastic impression sleeve against the fractured borehole wall. By using a downhole compass and orienting device, one can then determine the orientation of fracture impressions. However, because well bores are not straight, smooth, and uniform, the impression sleeves can be marred by travelling in and out of the borehole, and identification of the fractures can be difficult in some cases. The other technique for determining fracture orientations is to use a device such as an ultrasonic borehole televiewer and thus 'visually' attempt to detect the fractures in the well bore (see BREDEHOEFT *et al.* 1976).

It is important to establish field procedures that prevent getting 'stuck' in the hole. Uncased wells are not ideal geometrical and mechanical entities, and the oil-field logic that has led to the design of commercially available packers has been one of saving the well and not the instruments. Thus packers are made from relatively soft, millable steels, and many critical and expensive parts are not made for repeated use.

The first step in our procedure is to gain as much information as possible. For example, caliper logs are extremely useful for detecting irregularities in the well diameter that could prevent the packers from setting properly. Retrieving as much core as possible can indicate the degree to which the rock is prefractured or jointed. Wells drilled into poorly consolidated sediments may require circulation to remove debris and improve the well condition before fracturing is attempted.

Experience has proved the importance of careful packer maintenance. After each trip with a packer, it is completely broken down and cleaned. All of the O-rings and seals are replaced as necessary, and critical connectors are carefully inspected. When reassembled, the packers are commonly surface tested by inflation in a steel casing before being used downhole.

It has also been our experience that hydraulic fracturing at moderate depths can be done through well-maintained and repeatedly inspected drill steel and that special high pressure tubing is not required. We have constructed a special valve that is mounted between the straddle packer and the drill steel which allows us to pressurize the drill steel as it is lowered into the well without setting the straddle packer. By continuously testing the drill steel we can isolate and remove potentially leaky pipe.

We have used a Failing Co. 'Holemaster'[2]) drill rig in the experiments reported here. The inflatable straddle packer used in these experiments was composed of a pair of Lynes Production-Injection Packers, which isolated 2.1 m of the borehole. At the Livermore site a Petrodyne 6B airdriven double-acting pump was used to pressurize the well-bore. This pump could deliver 8.89 liters/minute at a maximum pressure of 13.8 MPa. A Sperry-Sun downhole compass and single-shot camera were used to orient impression sleeves and hence the fractures. From the surface this compass is lowered through the drill steel into the center of a 6.4 m section of non-magnetic steel mounted just above the packers. At the San Ardo and Menlo Park sites, a diesel-driven radial piston pump (pumping at a rate of 6.35 liters/minute) and a Kuster single-shot camera and downhole compass were used. At all sites downhole pressures were measured with a retrievable Kuster KPG gauge, and uphole pressures were measured with a Tyco Instruments pressure transducer.

2. *Experimental sites*

The San Ardo site (Fig. 1) is located in Peach Tree Valley (near Pancho Rico Creek) in the southern Salinas Valley area. The site is approximately 18 km NE of San Ardo and 3 km SW of the San Andreas fault. In this region the creep rate on the San Andreas fault is very high and seismic activity low (WESSON *et al.* 1973). The well was drilled to a depth of 290 m in relatively undisturbed Tertiary sediments and was fractured at depths centered at 116.4, 130.1, 167.0, 215.8, 240.2 and 286.6 m. The lower Pliocene Pancho Rico Formation, which crops out at the surface, is chiefly a well-sorted, fine-grained and noncalcareous, and massive sandstone that ranges from friable to well-cemented. This formation conformally overlies the upper Miocene Santa Margarita Formation, which is medium-to-coarse-grained calcareous sandstone ranging also from friable to well-cemented. The contact between these two formations is commonly difficult to determine. Pre-Tertiary granitic basement rocks lie at depths of 300–450 m. Although recent uplift is indicated by many landslides and prominent erosional scarps, in the region between the Los Lobos thrust fault (22 km to the southwest) and the San Andreas fault, very few faults are mapped in the flat-lying sediments (DURHAM 1974).

The Livermore drill site (Fig. 1) is located in the northeastern section of Livermore Valley, which trends east–west through the Diablo Range of west central California.

[2]) Use of trade names is for descriptive purposes only and does not constitute endorsement of these products by the U.S. Geological Survey.

Below 65 m of unconsolidated Quaternary alluvium, the Livermore gravels (semi-consolidated Pliocene and Pleistocene gravel, sand, and clay) extend for several thousand feet. The well was hydraulically fractured at depths of 110.3, 119.5, 142.3, 148.4, and 154.4 m. The upper four fractures were in semi-consolidated sandy gravel. The deepest fracture was in somewhat clayey sandy shale.

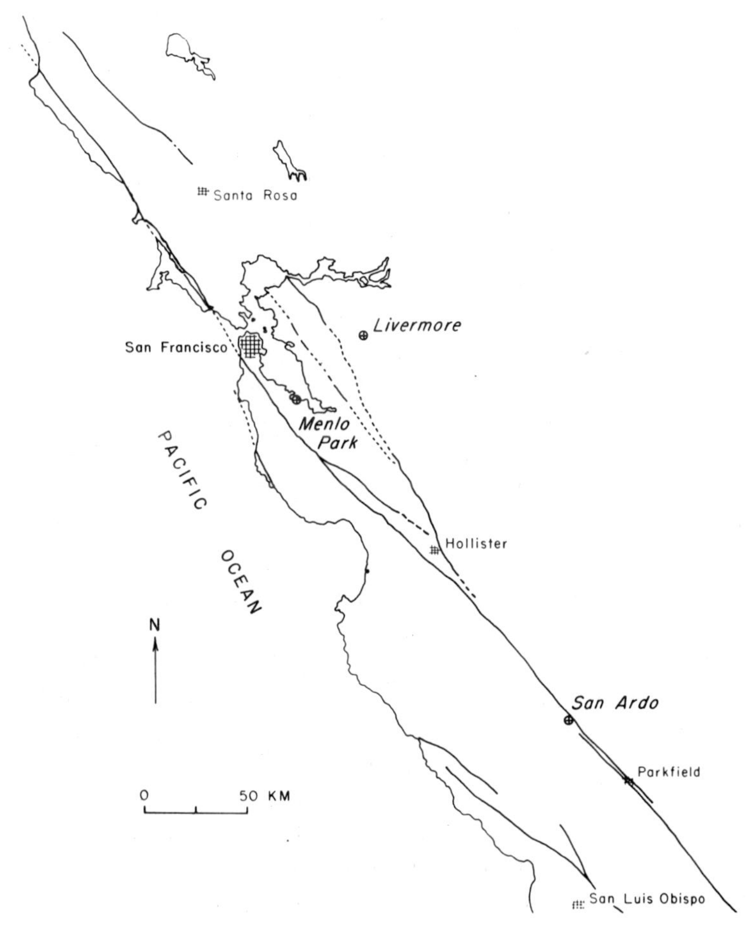

Figure 1
Locations of the San Ardo, Menlo Park, and Livermore hydrofrac sites in central California.

The Menlo Park drill site (Fig. 1) is located east of town only 50 m from the edge of San Francisco Bay. The well was drilled through 183.0 m of Holocene mud and then cased. The well was then extended to a depth of 249.0 m through graywacke of the Franciscan formation. Recovered core and caliper logs indicate that between 183–214 m the graywacke was too fractured (and the hole too irregular) to attempt fracturing. However, the well was successfully fractured at depths of 223.9, 230.0, 242.2, and 246.8 m in competent but highly jointed rock.

3. Hydrofrac data

San Ardo

At San Ardo, the well was fractured by USGS Water Resources Division personnel at depths of 116.4, 130.1, and 167.0 m. The technique they used was to pressurize the borehole with water until breakdown was observed. Pumping then continued as leucite and walnut shell were injected to prop open the fractures for subsequent detection with a borehole televiewer. When shut-in, the borehole pressures dropped rapidly to 2.67, 2.78, and 3.87 MPa at 116.4, 130.1, and 167.0 m respectively (J. D. BREDEHOEFT and R. G. WOLFF oral communication). These pressures closely correspond to the approximate weight of the overburden material for a mean bulk density of 2.3 gm/cm^3.

Laboratory work has shown that correctly oriented hydraulic fractures can be generated in prefractured rock if a high-viscosity fluid is used as the fracturing fluid (see ZOBACK *et al.* 1976). Thus at depths of 215.8, 240.2, 270.9, and 283.8 m, we used drilling mud as the frac fluid and the wells were shut-in immediately after breakdown was observed (Fig. 2). On subsequent pumping cycles, successively more pumping was allowed before shutting in the well.

At 215.8 m (Fig. 2a) failure of the valve that diverted fluid pressure from the packers into the packed-off interval caused the lower packer to burst at 16.7 MPa. However, from the nature of the record and the consistency of the shut-in pressures (4.69 MPa), a fracture was evidently formed with the release of high pressure into the wellbore beneath the upper packer. As at the shallower depths, the shut-in pressure is approximately equal to the weight of the overburden (assuming the fracture formed at about 216 m). An interesting characteristic of the second, third, and fourth pressuring cycles is that a very sharp breakdown was observed at a pressure about 0.7 MPa lower than the pumping pressure (see inset in Fig. 2a).

At 240.2 m (Fig. 2b) a different type of pressure-time record was observed. Breakdown occurred at 25.24 MPa, and the initial shut-in pressure was 11.38 MPa. As at 270.7 m (Figure 2c) the pressure drop preceding breakdown was caused by momentarily shutting off the pump. In subsequent pressurizing cycles breakdown occurred at a pressure approaching 9.31 MPa. On cycle 2, the well was shut-in when a slight pressure drop was observed at 12.90 MPa, and the shut-in pressure decayed very little. On cycles 3–6 the fractures seemed to open suddenly, and the shut-in decay was initially quite rapid but then became distinctly slower. This characteristic shut-in behavior leads us to believe that on cycle 2 we did not reopen the fracture and pressure drop instigating shut-in was perhaps a minor pump malfunction. In these data, then, two shut-in pressures are apparent: one that corresponds to the abrupt change in decay rate (or ISIP), and the long-time or asymptotic shut-in pressure (ASIP). The ASIP, at 5.13 MPa, is about the expected overburden weight.

At 270.7 m (Fig. 2c) the quality of the shut-in pressures was marred by a leak through a pipe joint (which led to development of the pipe checking valve mentioned above). The breakdown pressure on the first cycle was 29.0 MPa and on subsequent cycles approached 17.5 MPa. Because of the leaky pipe, the shut-in records are not as distinctive as those at 240.2 m. It does not appear, however, that a change in decay rate occurs at about 12.0 (± 1.1) MPa.

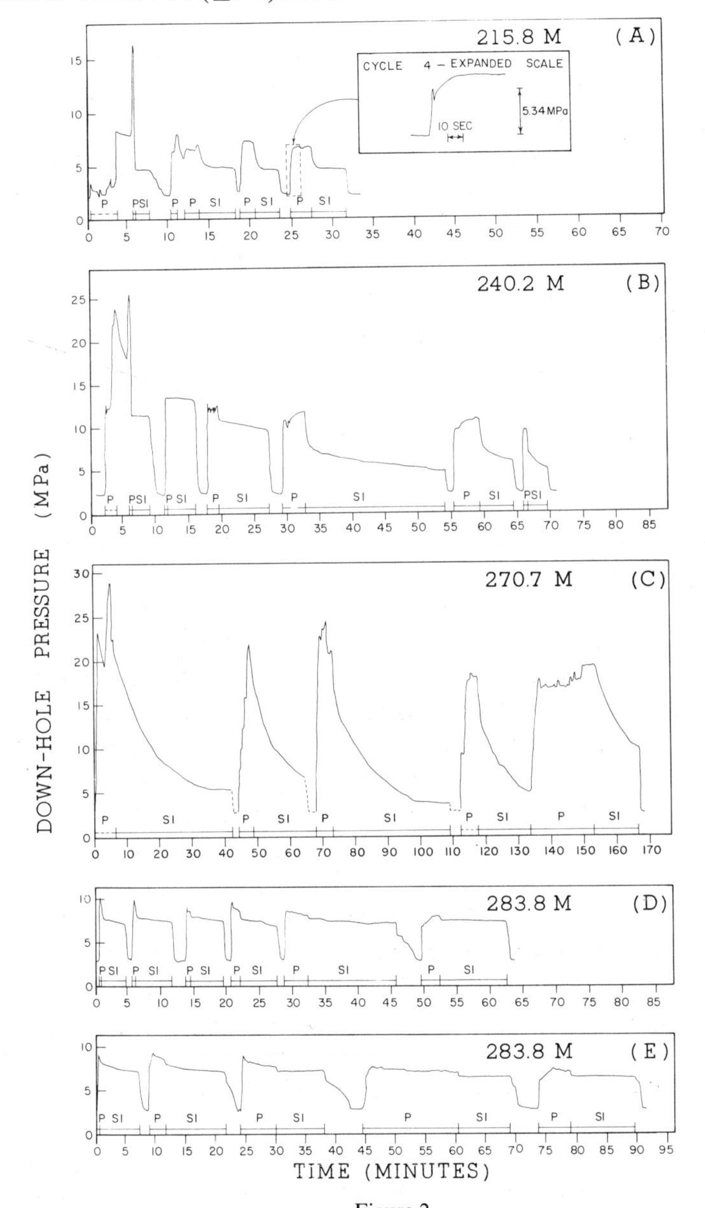

Figure 2

Pressure-time records from the San Ardo site. The depth of each measurement is shown. These data are digitized and replotted up-hole measurements corrected for the hydrostatic head.

At 286.6 m (Figs. 2d, e) the well was fractured after a 3.7-cm i.d. pipe was cemented in from the surface for subsequent heat flow measurements in the well. The initial breakdown pressure was 10.07 MPa (Fig. 2d), and since the breakdown pressures were nearly the same on the following cycles, we apparently had opened a preexisting fracture in the rock. The shut-in pressures in Fig. 2d are all about 7.6 MPa. Two weeks after the initial fracturing, the well was repressurized (Fig. 2e). The initial shut-in pressure is higher than in the previous measurements (8.2 MPa). It is not clear whether this implies that a different fracture opened, that there was a significant change in the *in situ* stress field, or that there is undetected error or uncertainty in the measurements. After sufficient pumping the shut-in pressures decreased to a value of 6.52 MPa, again a value closely corresponding to the weight of the overburden.

In an attempt to measure the orientation of the fracture at 240.2 m, a semi-cured rubber wrap was epoxied to a packer element and expanded against the borehole wall. Although a well-delineated fracture impression was not apparent, two small vertical impressions were observed at an azimuth of about N 15° E. If these features were caused by the fracture, this direction is presumed to be that of the maximum compressive stress.

Livermore

A modification of the conventional hydrofrac technique was employed at the Livermore site and consisted of pressurizing the borehole by expanding a flexible plastic membrane (commercially available shrink tubing). This technique has the advantages of recording a fracture impression at the time it is generated and preventing fluid intrusion into preexisting cracks. However, the membrane can also cause unpredictably high breakdown pressures (ZOBACK *et al.* 1976).

At 110.3 and 119.5 m (Figs. 3a, b) the breakdown pressures are quite low ($\sim$7 MPa) and on subsequent pumping cycles the fractures appear to open slowly at very low pressures. At 110.3 m a fairly distinct shut-in pressure of 1.62 MPa is indicated on cycles 1 and 2. At 119.5 m cycles 1–3 also indicate a fairly distinct shut-in pressure (1.53 MPa).

At 142.3 m (Fig. 3c), the irregular nature of the pressure build-up resulted from intermittent pumping to allow the plastic sleeve to expand against the borehole. The breakdown pressure was again quite low (7 MPa), but on subsequent pumping cycles, the fractures appear to open suddenly as in the San Ardo records. At this depth the shut-in pressure was 2.24 MPa, and the final 'breakdown' pressure was 3.11 MPs.

At 148.4 m the initial breakdown pressure was substantially higher than at shallower depths (15.4 MPa), and the shut-in pressures again showed an initially rapid then slow decay rate changes (ISIP) w-s 4.12 MPa and the long-time shut-in pressure (ASIP) seemed to approach 2.8 MPa, except on cycle 1.

At 154.5 m (Fig. 3e) the sharp pressure drops preceding breakdown resulted from adjustments of the pump. The breakdown pressure (12.6 MPa) was initially high as at

148.4 m and approached 4.35 MPa on the last cycles. The shut-in pressure consistently appeared to be 2.66 MPa.

The vertical and inclined indentations on the impression sleeve from the Livermore well (Fig. 4) illustrate limitations of this technique. It is hard to distinguish with certainty marks made by passage in and out of the borehole from those associated with hydraulic fractures. Nonetheless, superposition of the data tends to indicate vertical fractures roughly oriented at N 70° E (±40°).

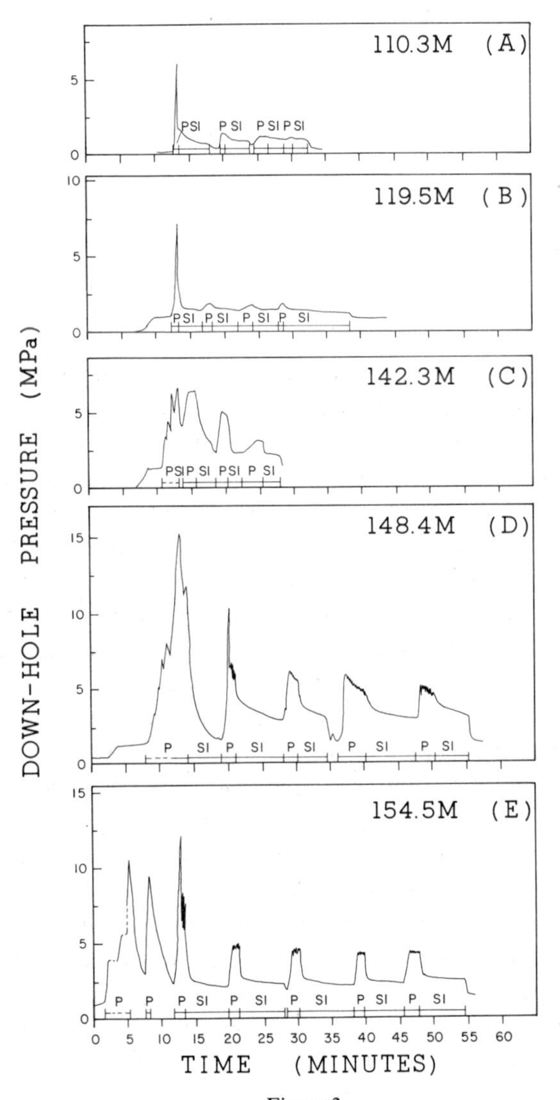

Figure 3
Replotted downhole pressure-time records from the Livermore site.

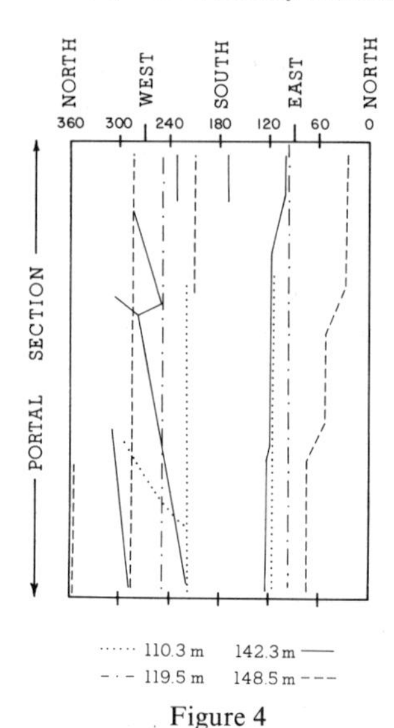

Figure 4
Superposition of impression sleeve indentations as a function of azimuth (from Livermore site).

Menlo Park

At Menlo Park fracturing could be done only within the lowest 35 m of the well. At 223.9 m (Fig. 5a) the initial breakdown pressure was quite low (10.38 MPa), and on the third pressurizing cycle we observed that the upper packer was leaking fluid up the well. This observation negates the apparent shut-in pressures on the first and second cycles since they are the same as the leak pressure in cycle 3. This observation, we feel, is good justification for conducting multiple fracturing cycles.

At 230.0 m (Fig. 5b) the initial breakdown pressure was again quite low (8.97 MPa) and did not change between the first and second cycles, thus indicating that a pre-existing fracture was being opened. As pumping increased on subsequent cycles, the breakdown pressure decreased only slightly. Fairly uniform shut-in pressure decays are observed on the records, and the only characteristic shut-in pressure seems to be a sudden decay acceleration apparent at 6.21 MPa on cycles 3 and 4.

At 236.1 m (Fig. 5c) the initial breakdown pressure was 14.8 MPa, and from the small decrease between the first and second cycles, it again appears that a pre-existing fracture was opened. As for other records discussed above, a noticeable slowing of the decay rates is observed at 5.86 MPa on several cycles.

At 242.2 m (Fig. 5d) we again appeared to open a preexisting fracture (at 15.1 MPa) which showed a very interesting shut-in behavior. On cycles 1 and 2, before substantial fluid was pumped into the fracture, the shut-in pressure decayed toward hydrostatic pressure (prompting the test cycle after cycle 2). However, after the fracture is extended on cycle 3, the shut-in indicated a very distinct decrease in slope at 5.69 MPa and less severe subsequent decay. This behavior is quite similar to that of the shut-in decays at 148.5 m at Livermore (Fig. 3d).

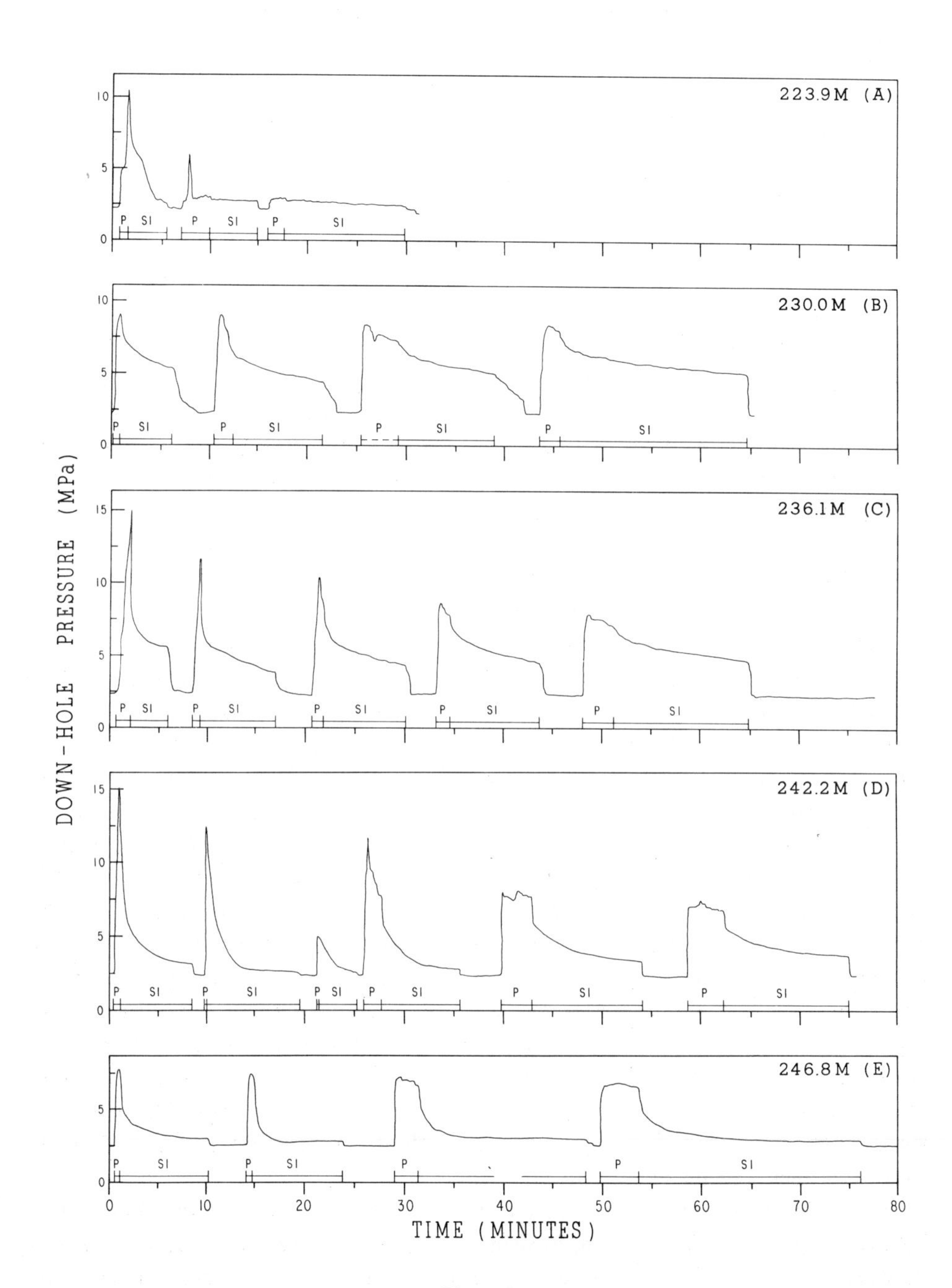

Figure 5
Pressure-time records from Menlo Park. Data are replotted up-hole measurements corrected for the hydrostatic head.

At 246.8 m, the deepest fracture was produced by setting a single packer and fracturing the interval between 244.6 m and the bottom of the well (249.0 m). The pressure-time record from this depth (Fig. 5e) is much like the others; the breakdown behavior indicates that a preexisting fracture was opened, and although fairly difficult to detect, a decrease in the slope of the shut-in decay curves occurred at 4.97 MPa.

4. *Discussion and interpretation*

Shut-in pressures from the three hydrofrac sites are presented in Fig. 6. Three types of characteristic shut-in pressures were identified on the pressure-time records; an instantaneous shut-in pressure (ISIP) or sharp decrease in the shut-in decay rate (circles), a long-time or asymptomatic shut-in pressure (ASIP), (dots), and an acceleration in the shut-in decay rate (open square). Clearly, we must first explain these shut-in behaviors if we are to interpret these data in terms of the *in situ* stresses.

At 240.2 m at San Ardo (Fig. 2b), the ASIP indicates that the least compressive stress is that due to the overburden (as the ISIP's at the other depths do) and the ISIP is somewhat higher (as is that at 270.7 m). To investigate the cause of these high ISIP's, let us first consider the initiation of hydraulic fractures in a region where the least principal stress is vertical.

In his theoretical analysis of hydraulic fracture formation, KEHLE (1964) attempted to show that if the least principal stress was vertical, a horizontal fracture could be formed at the inside edges of the packer elements owing to an induced vertical stress. However, A. LACHENBRUCH (oral commun.), has indicated that there may be a serious shortcoming in Kehle's analysis caused by the model he used for a straddle packer configuration. Kehle assumed the packers to be rigid solids in uniform contact with the wellbore. From this model the induced vertical stress responsible for horizontal fracture initiation arose from the discontinuous application of radial pressure at the packer's edge. However, since actual packers are inflated and *deformable*, the pressure in the packers is equal to the pressure between them, and the only discontinuity in radial pressure is at the *outer* edges of the packer (where fluid could not contact any fracture that might form). Thus, we conclude that with inflatable packers not only is Kehle's analysis not applicable, but also the only way for horizontal fractures to form seems to be for fluid to seep into preexisting flaws such as bedding planes, foliations, or fractures and then cause these features to open. HAIMSON and FAIRHURST's (1970) inability to generate 'horizontal' fractures in true triaxial laboratory experiments seems to support this hypothesis.

Returning now to discussion of the shut-in pressures at San Ardo (Fig. 6), we believe that at 240.2 and 270.7 m, by using a viscous fracturing fluid, we were able to generate vertical fractures at the well bore that we assume are perpendicular to the least principal horizontal stress. Since, when a well is shut-in, the rate of pressure drop in the borehole is merely a measure of the rate at which fluid bleeds out into the fracture, the characteristic fast and slow shut-in decays are what would be expected if, (1) flow were initially through an open vertical fracture (which closes at the pressure indicated by the change in decay rate and then slowly continued through this 'closed' fracture into horizontal planes, or (2) through a vertical fracture which 'rolls over' into a horizontal plane. This, then, explains why the ASIP is equal to the weight of the overburden and leads us to estimate the least principal horizontal stress at 240.2 as 11.38 MPa. This is consistent with the estimate from 270.7 m (12 ± 1.1) and the poorly constrained estimate from 287 m (>8.2 MPa). It is interesting to speculate

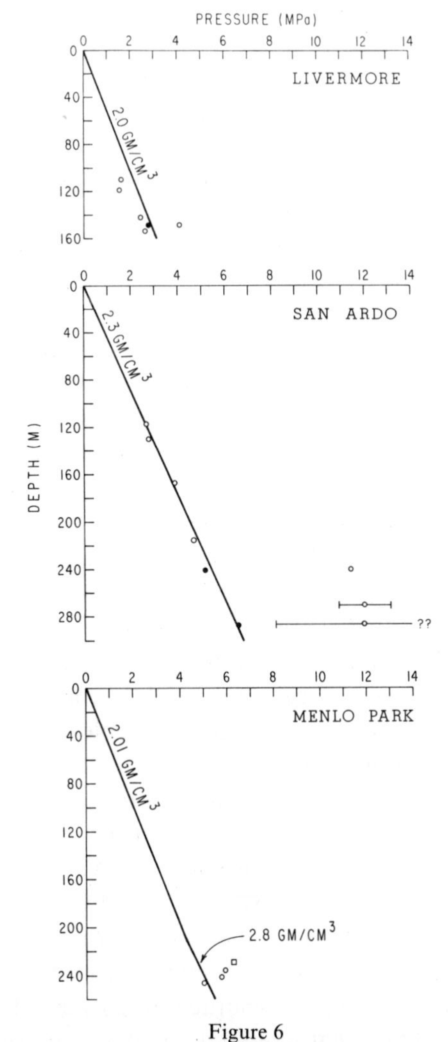

Figure 6
Shut-in pressures from Livermore, San Ardo, and Menlo Park, California, and reference stress gradients. Circles are instantaneous shut-in pressures, the dots are long-time shut-in pressures, and the open square is a sudden decay acceleration during shut-in.

that the drop in the ISIP's that accompanied increased pumping at 240.2 and 287 m may result from the rotation of the initially opened fracture toward a horizontal plane. We offer a third possible explanation of why two shut-in pressures are observed. HAIMSON (1976d) discusses observing two shut-in pressures in areas where the least stress is vertical and also suggests that the fractures turn from a vertical plane to a horizontal one as they propagate.

At Livermore interpretation of the shut-in pressures is fairly straightforward. As indicated by the ISIP's (and the ASIP at 148.5 m), the least principal stress is less than the probable minimum overburden pressure. That corresponding to a bulk density of 2.0 gm/cm^3 (implying about 38% porosity) is shown in Fig. 6 as a lower limit for reference. We thus anticipate that these shut-in pressures are measures of the least principal horizontal stress. As evidenced by the spurious ISIP and impression sleeve record, a preexisting fracture was probably opened at 148.5 m. This is further indicated by the first cycle shut-in decay which, like the 242.2 m Menlo Park record (Fig. 5d), behaves as if we were initially opening a naturally propped preexisting fracture. Disregarding the impression sleeve record from 148.5 m, the data in Fig. 4 still indicates that the direction of maximum compression at Livermore is about N 70° E maximum stress orientation.

Data from the Menlo Park site (Fig. 6) are somewhat less useful than those from San Ardo and Livermore. The Holocene mud extending to a depth of 183 m has a mean bulk density of 2.01 gm/cm^3. If we extend this density through the highly broken graywacke that extends to 214 m and use the measured density of 2.6 gm/cm^3 below this depth, we arrive at an anticipated overburden stress of 5.1 MPa at 246.8 m, quite close to the measured shut-in pressure of 4.97 MPa. Thus, at this site as at San Ardo, the least principal stress appears to be that due to the overburden down to the maximum well depth. The higher shut-in pressures at the other depths apparently correspond to the preexisting fractures that were opened at unknown orientation.

Also of interest in the Menlo Park data is the nature of the shut-in decay record at 230.0 m. The sudden acceleration in decay rate is very similar to a record reported by AAMODT (1975). He interprets the acceleration in decay rate to represent the closing of one of several fractures that intersect the borehole. A similar explanation is possible for the Menlo Park well as the fracture normal to the higher stress component closes, it can act as a temporary source of fluid and thus the rate of pressure drop in the borehole is lessened. It is feasible, however, that the initially fast and then slow shut-in decays mentioned above also correspond to two fractures intersecting the borehole. If, as one fracture closes, its loss as a pressure sink overwhelms its temporary role as a pressure source, then a noticeable decrease in the shut-in decay rate might be observed. Thus, the fast and then slow decay character described above (where the ASIPs correspond to the overburden weight) could also be caused by a vertical and horizontal fracture that intersects the borehole. HAIMSON (1976a) also observed indications of both horizontal and vertical fractures on impression sleeves, and he too reported two shut-in pressures.

From equation (1) we could now estimate the greatest principal stress from knowing the least principal stress, pore pressure (assumed to be hydrostatic), breakdown pressure, and tensile strength of the rock. However, because it is uncertain that P_b is a reliable indication of fracture initiation (see ZOBACK *et al.* 1976), we choose another technique for estimating S_1 that has also been used by BREDEHOEFT *et al.* (1976). If we assume that the breakdown pressure on secondary pumping cycles corresponds to the sudden opening of the fractures (as is certainly suggested in Fig. 2a), then equation (1) can be used for estimating S_1 with $T = 0$. In accordance with our interpretation of the shut-in pressures, however, this analysis is possible only for selected data from Livermore and San Ardo.

At Livermore, the computed value for S_1 (Table 1) is clearly incorrect since it is lower than S_3. A near-hydrostatic value for P_0 is expected since the well is in such permeable material, and while errors in S_3 most seriously affect the computation of S_1, we believe the likely source of error at Livermore to be misinterpretation of P_b when $T = 0$. At Livermore, this pressure appears not to be the pressure at which the fracture opens, but rather some higher pressure at which the fractures are sufficiently open to accept fluid faster than the pump can supply it. In future operations, to make this method of analysis useful we must detect the pressure at which the fracture first opens.

If the P_b values used at San Ardo do correspond to the crack-opening pressure, we must consider the reasonableness of the high values of S_1 that the analysis indicates. For example, at 240.2 m, the effective stresses appear to be $\sigma_v = 2.72$ MPa, $\sigma_{H\min} =$

Table 1
Well summaries

	Depth meters	ISIP (MPa)	ASIP (MPa)	S_v (MPa)	$S_{H\min}$ (MPa)	$S_{H\max}$ (MPa)
San Ardo	116.4	2.67		2.67		
	130.1	2.78		2.87		
	167.0	3.87		3.87		
	215.8	4.69		4.69		
	240.2	11.38	5.13	5.13	11.4	22.5
	270.9	12.0 (±1.1)			12.0 (±1.1)	15.8 (±3.3)
	287.0	7.6, 8.2		6.52	8.2	
Livermore	110.3	1.62			1.62	
	119.5	1.53			1.53	
	142.3	2.24			2.24	
	148.5	4.12	2.8		2.8	
	154.5	2.66			2.66	2.11(?)
Menlo Park	230.0	6.21 (see text)				
	236.1	5.86				
	242.2	5.69				
	246.8	4.97		4.97		

9·02 MPa, and $\sigma_{H\max} = 20.14$ MPa. Considering a simple Coulomb failure law (of the form $\sigma_1 - C_0 + A\sigma_3$), however, we see that the estimate for $\sigma_{H\max}$ is quite reasonable for unfaulted material assuming even conservative values of 20–30 MPa and 3–5 for C_0 and A, respectively. Two other criteria seem to indicate the reliability of this $\sigma_{H\max}$ estimate. First, the P_b value used apparently represents a minimum value (as it should). Second, while the P_b value is less than the shut-in pressure (as would be expected in tectonic areas), the pumping pressure builds up to the approximate value of the least principal stress.

4. *Summary and conclusions*

We believe that our preliminary results in using the hydrofrac technique to make in-site stress measurements are quite encouraging. We can apparently generate hydraulic fractures in the less than ideal material commonly found near major faults and can interpret the results with respect to the principal compressive stresses in the region. Our experience at the Menlo Park site, however, taught us that site selection remains an important criterion controlling the success of the technique. We feel that our approach to hydraulic fracturing is the correct one. Only with a flexible and totally independent system such as ours will sufficient data become available to determine the ultimate usefulness of stress measurements at relatively shallow depths.

Experience has shown that even at depths where the least principal stress is vertical, proper experimental techniques and correct interpretation of the data can yield the tectonically relevant horizontal stresses. The necessity for making enough shallow measurements to determine the rate at which the tectonic stresses increase with depth is apparent as well as the need for selected deep tests. At San Ardo, although we cannot estimate the rate at which the tectonic stresses increase with depth, we conclude that the least tectonic stress increases at a rate less than the overburden stress because of the nearby strike-slip faulting that evidently occurs at depth. At Livermore, the least principal tectonic stress apparently increases at a rate greater than the overburden stress. The reason for this is not clear (perhaps the explanation is increased strength with depth, and the depth to which this trend continues must be determined.

Refinement of field techniques remains an on-going process. We must implement a more reliable (and efficient) method for determining the orientations of the hydrofracs, and incorporating a borehole televiewer into our system may well be required. We are also investigating several techniques for more reliable determination of the pressure at which fractures reopen on secondary pumping cycles.

In conclusion, the reliable and routine measurement of *in situ* stress remains an important and challenging task. With time, however, we are confident that stress measurements made with the hydrofrac technique will be relatively routine and will make an important contribution to tectonic interpretation and seismic risk analysis.

Acknowledgments

The authors wish to thank Bob Jones, Dennis Peifer, and Ralph Lamson for their aid in this investigation. The televiewer survey of the San Ardo well was provided by W. S. Keys, and the well was fractured and the shut-in pressures determined at 116.4, 130.1, and 167.0 m by Roger Wolff and John Bredehoeft, all of the USGS Water Resources Division. We thank Dave Pollard and Roger Stewart for suggestions which improved this manuscript.

References

AAMODT, R. L. (1974), *An experimental measurement of in situ stress in granite by hydraulic fracturing*, Los Alamos Scientific Laboratory Report 5605-MS.

BREDEHOEFT, J. D., WOLFF, R. G., KEYS, W. S. and SHUTER, E. (1976), *Hydraulic fracturing to determine the regional in situ field, Picenace Basin, Colorado*, Geol. Soc. Amer. Bull. *87*, 250–258.

DURHAM, D. L. (1974), *Groundwater geology of the southern Salinas valley area, California*, U.S. Geological Survey Prof. Paper *819*, 111 pp.

FORD, R. S., *Groundwater geology of Livermore valley – a satellite urban area* in *Urban Environmental Geology in the San Francisco Bay Region* (ed. E. A. Daneby) (San. Fran. Section Assoc. of Engl. Geol., 1969), 162 pp.

HAIMSON, B. C. and FAIRHURST, C., *In situ stress determination at great depth by means of hydraulic fracturing* in *Rock Mech. Theory and Practice, Proc. of 11th Symp. of Rock Mech. Univ. of Ca., Berkeley, CA., June 1969* (ed. W. H. Somerton) (Soc. Mining Eng. AIME, New York, 1970).

HAIMSON, B. C. and STAHL, E., *Hydraulic fracturing and the extraction of minerals through wells* in *3rd Symp. on Salt* (Northern Ohio Geol. Soc. Cleveland, Ohio 1972) 421–432.

HAIMSON, B. C., *Earthquake related stresses at Rangely, Colorado* in *New Horizons in Rock Mechanics, Proc. 14th Symp. on Rock Mechanics* (eds. Hardy and Stefanko) (ASCE, New York 1973) 689–708.

HAIMSON, B. C., LACOMB, J., GREEN, S. J. and JONES, H. H., *Deep stress measurements in tuff at the Nevada test site* in *Proc. 3rd Int. Congr. Rock Mech., Denver, Colorado*, Vol. 2 (National Academy of Sciences, Washington, D.C. 1974) 557–562.

HAIMSON, B. C. (1976a), Hydrofracturing stress measurements in the Blue Ridge Belt of South Carolina, *EOS*, *57*, 289.

HAIMSON, B. C. (1976b), *Crustal stress measurements through an ultra deep well in the Michigan Basin*, EOS, *57*, 326.

HAIMSON, B. C. (1976c), *Crustal stress in the continental United States as derived from hydrofracturing tests* in *Symposium on the Nature and Physical Properties of the Earth's Crust*, in press.

HAIMSON, B. C. (1976d), *The hydrofracturing stress measuring technique – method and recent field results* in *U.S., Proc. Int. Soc. for Rock Mechanics Symp. on Investigation of Stress in Rock* (*Sydney, Australia*) 23–30.

HAIMSON, B. C., VOIGHT, B. (1976e), *Stress measurements in Iceland,* EOS Trans. AGU, 1007.

HERD, D. G. (1975), *The Las Positas fault: An active, northeast-trending, left-lateral fault in eastern Alameda County, California.*

HUBBERT, M. K. and WILLIS, D. G. (1957), *Mechanics of hydraulic fracturing*, Pet. Trans. AIME *210*, 153–168.

HUEY, A. S. (1948), *Geology of the Tesla quadrangle, California*, Div. Mines and Geology Bull. *140*.

KEHLE, R. O. (1964), *The determination of tectonic stresses through analysis of hydraulic well fracturing*, J. Geophys. Res. *69*, 259–273.

RALEIGH, C. B., HEALY, J. H. and BREDEHOEFT, J. D., *Faulting and crustal stress at Ranely, Colorado* in *Flow and Fracture of Rocks* (ed. H. C. Heard, I. Y. Borg, N. C. Carter and C. B. Raleigh) (Amer. Geophys. Union, Washington, D.C. 1972).

RUMMEL, F. and JUNG, R. (1975), *Hydraulic fracturing stress measurements near the Hohenzollern-graben structure, SW Germany*, Pure appl. Geophys. *113*, 321–330.

WESSON, R. L., BURFORD, R. O. and ELLSWORTH, W. L., *Relationship between seismicity, fault creep, and crustal loading along the central San Andreas fault* in *Proc. Conf. on Tectonic Prob. of the San Andreas Fault System* (eds. R. L. Kovach and A. Nur) (Stanford University Press, Stanford, CA. 1973).

ZOBACK, M. D., RUMMEL, F., JUNG, R., ALHEID, H. J. and RALEIGH, C. B. (1976), *Rate controlled hydraulic fracturing experiments in intact and pre-fractured rock*, Int. J. Rock Mech. Mining Sci., in press.

(Received 1st November 1976)

Pageoph, Vol. 115 (1977), Birkhäuser Verlag, Basel

Crustal Stress in Iceland

By Bezalel C. Haimson[1]) and Barry Voight[2])

Abstract – Hydrofracturing stress measurements have been carried out to about 0.4 km in two boreholes in Quaternary volcanic rocks in Reykjavik, Iceland, on the flank of the Reykjanes–Langjökull continuation of the Mid-Atlantic Ridge. The measurements indicate a dominant orientation of $\sigma_{H\,\max}$ approximately perpendicular to the axial rift zone, in contrast to earthquake focal mechanism solutions from within the axial rift zone of the Reykjanes Peninsula. In one hole (H32) a depth-dependent change in stress orientation is indicated, with σ_1 horizontal above a depth of about 0.25 km, and vertical below it; however the orientation of $\sigma_{H\,\max}$ remains unchanged. The data thus suggest reconciliation of an apparent conflict between the dominantly compressive indications of shallow overcoring stress measurements and dominant extension as required by focal mechanism solutions. The measured stresses are supported by the more reliable of overcoring measurements from southeast Iceland, and by recent focal mechanism solutions for the intraplate Borgarfjördur area. A fundamental change in crustal stresses appears therefore to occur as a function of distance from the axis of the axial rift zone. The data seem reasonably explicable in terms of a combination of thermoelastic mechanisms associated with accretion and cooling of spreading lithosphere plates. Stresses directly associated with the driving mechanisms of plate tectonics apparently do not dominate the observed stress pattern.

Key words: Stress hydrofracture; Stress at plate boundaries; State of stress; Tectonics of Iceland.

1. Introduction

Some years ago it was suggested that the spreading mechanism of the Mid-Atlantic Ridge could produce compressive stresses within the North American plate – even in zones of former extension – if, e.g., that plate were moving more slowly than the convecting mantle (Voight 1969, p. 960–961). Available data were suggestive but did not provide a reliable test; later, the more extensive evidence collected by Sbar and Sykes (1973) provided strong support for the concept. It seemed likely that stress measurements could provide one of the best clues to the fundamental question of driving mechanisms of plate tectonics (Voight *et al.* 1969, p. 662–667, 671–674, Voight 1970, p. 127, Sbar and Sykes 1973, p. 1861, Sykes and Sbar 1974, Ranalli and Chandler 1975, Solomon *et al.* 1975, Richardson *et al.* 1976), given the assumption that the plate tectonic hypothesis was correct.

It had further seemed likely that crustal extension should prevail in the general region of Mid-Ocean Ridges, although the overall dimensions of this zone were not at all well established; this view was in general supported by the rift structure of

[1]) Dept. of Mineral Engineering, University of Wisconsin, Madison, Wisconsin 53706, USA.
[2]) Dept. of Geology & Geophysics, Penn. State University, University Park, Pennsylvania 16802, USA.

Iceland, and by earthquake focal mechanism solutions at Mid-Ocean Ridge locations reported by SYKES (1967). But static rock stress measurements made at about the same time by HAST (1969) raised a new problem, for Hast's Icelandic measurements indicated relatively high values of compressional stress. Because Hast's measurements were conducted at very shallow depth, it was suggested that 'stress history' or thermoelastic perturbations might have been dominant, rather than deepseated tectonic processes (VOIGHT 1969, p. 959; cf. VOIGHT and ST. PIERRE 1974). These surface data were thus not considered as necessarily reliable indications of crustal stress conditions at depth. Nevertheless, their anomalous nature was noted as 'worthy of recognition and hopefully, careful evaluation in the future' (VOIGHT 1970, p. 129; cf. WARD 1971, p. 2992, PÁLMASON and SAEMUNDSSON 1974, p. 43, SYKES and SBAR 1974, p. 221).

During the intervening period the hydrofracturing technique was being developed as a procedure for stress measurements (HAIMSON and FAIRHURST 1970, HAIMSON 1974). Elementary elastic relationships exist between recorded hydraulic pressures and *in situ* rock stresses and between fracture direction and principal stress orientation. Laboratory results confirmed these relationships, even for anisotropic rock (HAIMSON and AVASTHI 1975), and in the last five years a number of successful field measurements were carried out between very near the surface down to 5 km (HAIMSON 1973, 1974, 1975a, 1975b, 1976a, 1976b, 1976c, ZOBACK *et al.* 1977, RUMMEL and JUNG 1975).

The possibility of a joint investigation of Icelandic rock stresses by the hydrofracturing technique arose during our collaboration at the 1974 Ghost Ranch Workshop (SHOEMAKER (ed.) 1975, p. 23–26). With the encouragement and the assistance of Gudmundur Pálmason and his colleagues at the National Energy Authority in Iceland, project objectives were more specifically defined and the first set of hydrofracturing measurements were completed in the summer of 1976 under an NSF grant. The basic data and interpretations associated with these measurements are presented in this paper.

2. *Hydrofracturing stress measurements*

General location

Seven successful hydrofracturing experiments were carried out in two boreholes in Quaternary igneous rocks in Reykjavik, Iceland (lat. 64° 10′N, long. 21° 52′W), on the flank of the Reykjanes-Langjökull continuation of the Mid-Atlantic Ridge. The holes are located about 20–25 km northwest of the axis of the active zone of rifting and volcanism (Figs. 1 and 2). These holes are relatively small diameter observation wells drilled 15–20 years ago (Table 1) near the periphery of the Laugarnes hydrothermal system.

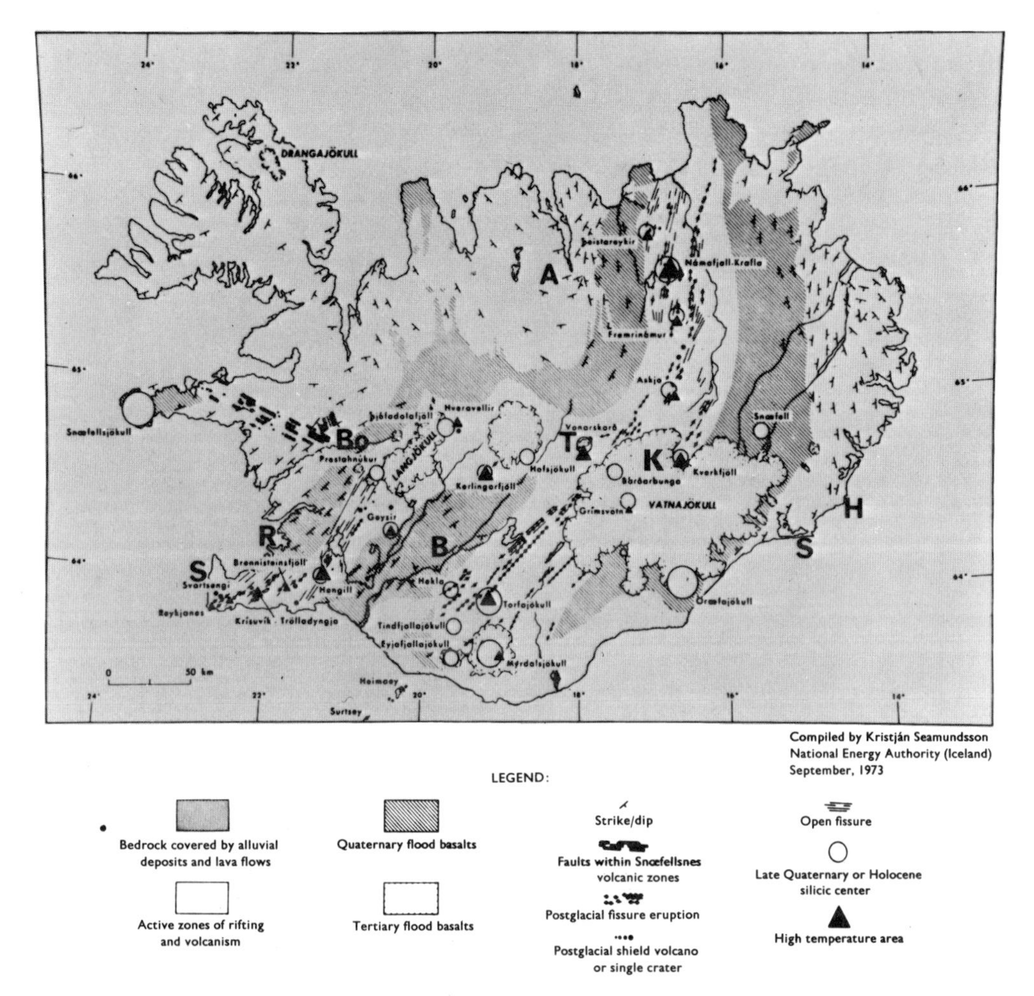

Figure 1
Icelandic stress measurement sites, and localities mentioned in text. Geologic base map after K. Saemundsson. Reykjavik hydrofracturing site, R. overcoring sites: Sandgerdi, on Reykjanes Peninsula, S; Burfell, B; Akureyri, A; Hvalnes, H; Stokksnes, S. Borgarfjödur earthquake location, Bo. Hypothetical plume centers at Tungnafellsjökull, T, and Kverkfjöll, K.

Geologic setting

The relationship of Iceland to Mid-Atlantic Ridge has been recently and ably reviewed (PÁLMASON and SAEMUNDSON 1974, KRISTJANSSON (ed.) 1974). The subject is of much interest and has been relatively well explored, for if ocean-floor spreading is taking place in the manner portrayed by plate tectonics, then Iceland must be splitting apart. The active zone of the Reykjanes Ridge segment (strike N40°E) of the Mid-Atlantic Ridge comes ashore at the tip of the Reykjanes Peninsula; the zone crosses Iceland in a relatively complicated manner, with two branches in southern

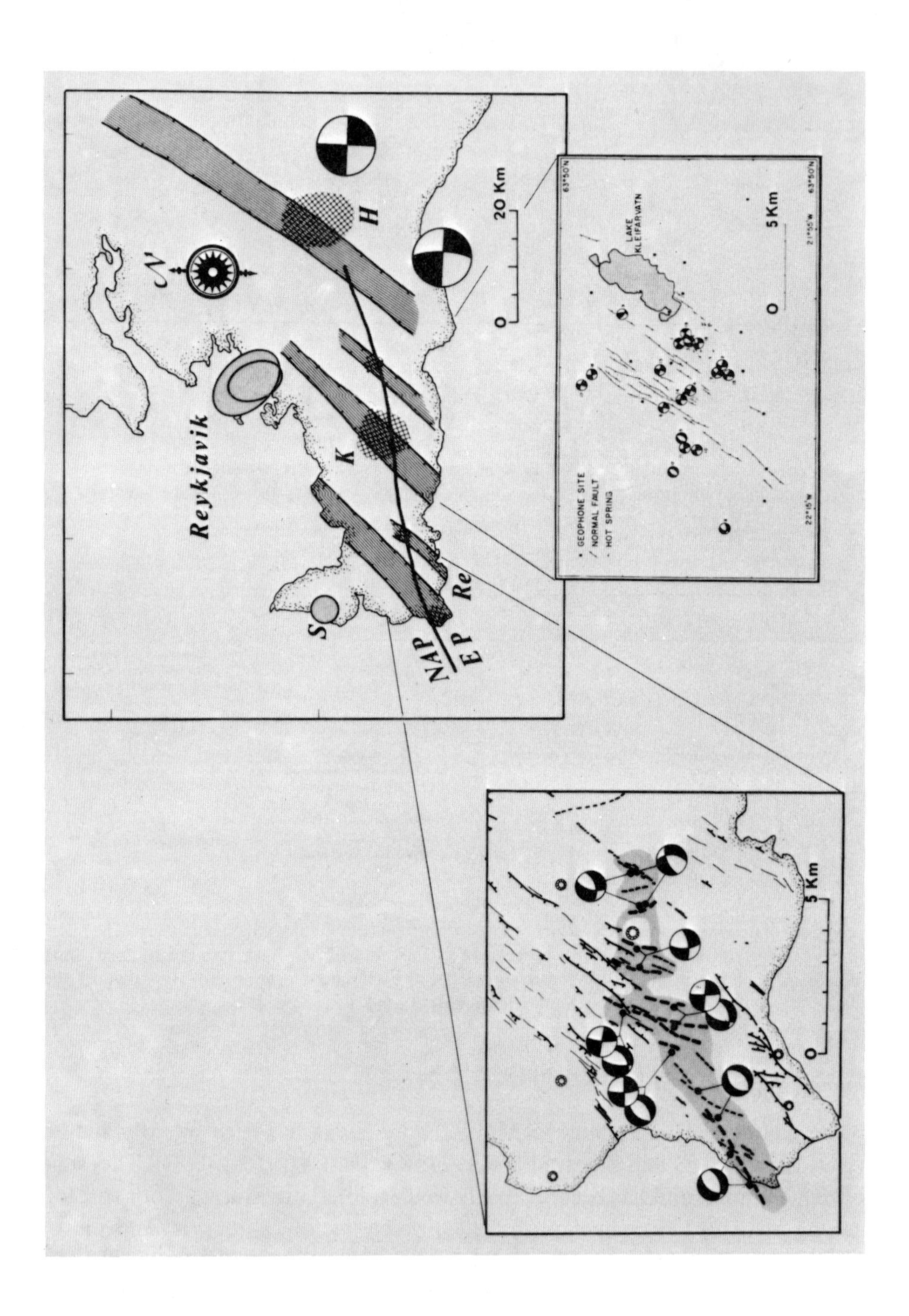
Reykjavik
K
H
S
NAP
EP
Re
20 Km
0
LAKE KLEIFARVATN
GEOPHONE SITE
NORMAL FAULT
HOT SPRING
5 Km
0
5 Km
0

Table 1
Boreholes used for hydrofracturing experiments

Hole Number	Year Completed	Elevation*) (m)	Depth (m)	Diameter (mm)	Inflatable Packer Diameter (mm)	Casing Depth (m)	Bottom Hole Temp. (°C)
H18	1956	8.42	697	75	65	152	119
H32	1961	33.27	606	101	90	32	75

*) Above mean sea level.

Iceland, and finally connects along an oblique offshore zone with the Kolbeinsey (Iceland-Jan Mayen) Ridge.

The axial zones of rifting and volcanism contain a great variety of volcanic and instrusive forms, cut by numerous faults and open fissures running mainly NE–SW in southern Iceland, with a more northerly trend in northern Iceland. On the Reykjanes Peninsula the general trend of the axial rift zone is about N70°E (Figs. 1 and 2; cf. KJARTANSSON 1960). Individual eruptive and open fissures and faults, however, strike about N35–45°E and are thus en-echelon to the general trend (JÓNSSON 1967, KJARTANSSON 1960, NAKAMURA 1970). The nearest normal faults to the borehole sites (about 6 km to the southeast) strike N45°E. The general trend of the axial rift zone then undergoes a profound change to about N35°E, parallel to the trend of individual fissures and faults. This trend is maintained from south of Hengill, 30 km east of Reykjavik, to Langjökull (Figs. 1 and 3). Measured rate of subsidence in the axial rift zone is estimated at about 4–6 mm/yr maximum, decreasing toward the boundaries of a 30 km zone on each side of the axis. Near Reykjavik the rate of subsidence has been about 1.0–1.5 mm/yr (TRYGGVASON 1974a, p. 243, 261). Precise

Figure 2
Tectonic map of the Reykjanes Peninsula, Iceland, forming part of the transition zone between the Reykjanes Ridge to the southwest and the south-Iceland transform fault to the east (after KLEIN *et al.* 1977). Boundary between North American (NAP) and Eurasian (EP) Plates, based on linear seismic zone of KLEIN *et al.* (1973), denoted by heavy line. Shaded areas are zones of concentrated extensional faults and open fissures, arranged in a left-lateral en-echelon pattern. Hachured zones are geothermal areas, which occur at intersections of fault-fissure swarms and linear seismic zone; major high temperature areas at Hengill (H), Krisuvik (K), and Reykjanes (Re). Borehole stress measurement sites at Reykjavik and Sandgerdi (S). Ellipses for stress conditions at about 0.3 km at H18, H32. Two earthquake (mag 5–6) focal plane solutions indicated (upper hemisphere) as discussed by WARD (1971). Detailed insets of tectonics and focal mechanism solutions for Krisuvik (KLEIN *et al.* 1973) and Reykjanes (KLEIN *et al.* 1977); geology after J. Jónsson. Faults denoted by barbed lines, fissures by thin lines; shield volcanos, double circles; crater row, circle chains. Seismic lineations are from trends of epicenters during individual subswarms of September 1972 swarm; dense concentration, heavy dashed lines; diffuse concentrations, light dashed lines. Stippled pattern indicates seismic zone for 1972 swarm. Focal mechanism solutions at Krisuvik represent actual events; those for Reykjanes represent typical patterns. Overlapping focal spheres indicating overlapping tectonic regimes. Upper hemisphere projections; compressional quadrants (i.e., containing axis of minimum compression) black.

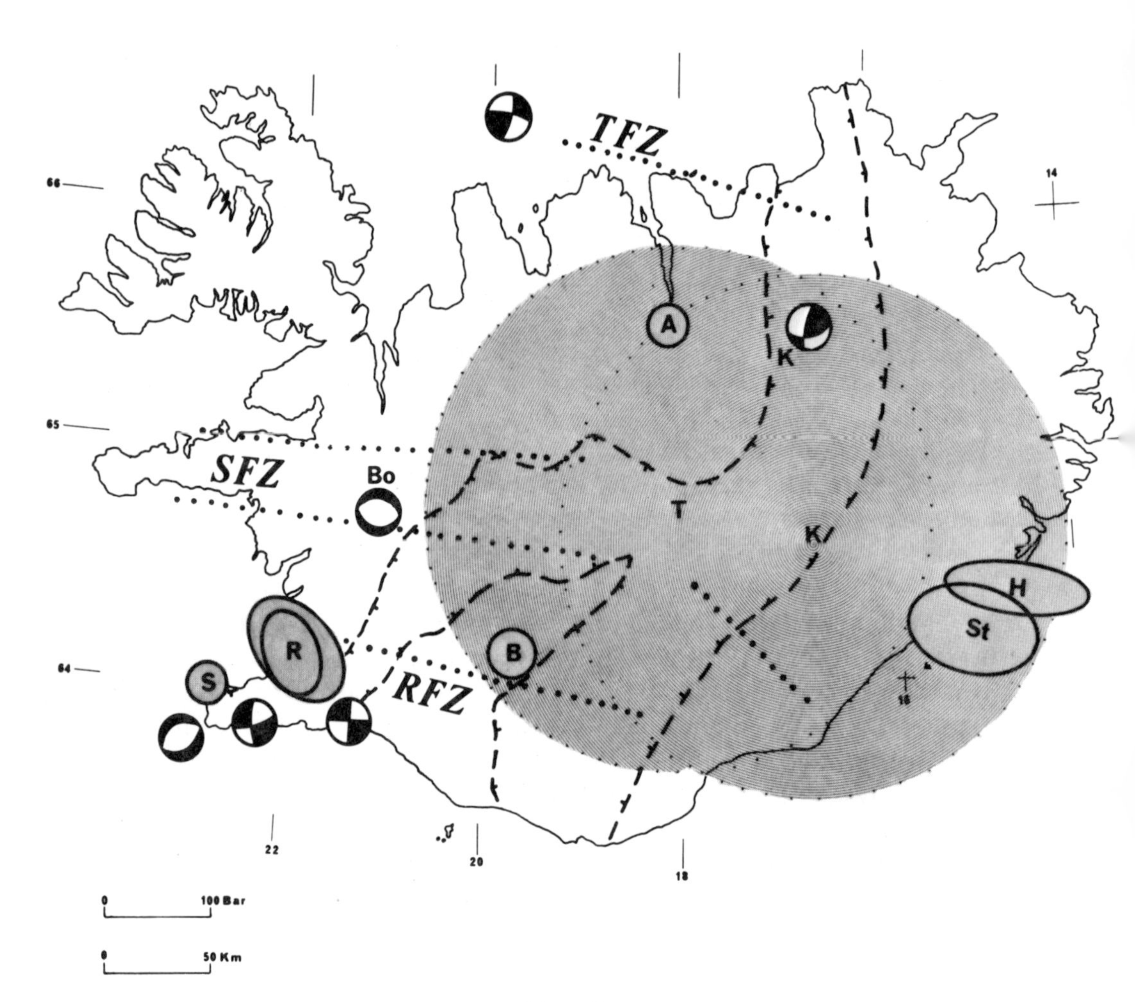

Figure 3
Tectonic map of Iceland indicating observed stress conditions. Axial rift zone boundaries denoted by dashed heavy lines with barb. Tjörnes (TFZ), Snaefellsnes (SFZ), and Reykjanes (RFZ) Fracture zone boundaries indicated by dotted lines. Another possible fracture zone boundary passes through Oraefajökull near the southeast coast (cf. Fig. 1). Hypothetical plume centers at Kverkfjöll (K) and Tungnafellsjökull (T) indicated. Hypothetical plume radii taken as 140 km. Focal mechanism solutions for Reykjanes Peninsula indicate typical microearthquake patterns for Reykjanes and Krisuvik areas (KLEIN *et al.* 1973, 1977), and a magnitude 5 earthquake discussed by WARD (1971). Solutions also shown for Borgarfjördur area Bo; EINARSSON *et al.* 1977), Krafla (K; WARD *et al.* 1969), and Skagafjördur, off the north coast (STEFANSSON 1966, SYKES 1967). Upper hemisphere projections except for Skagafjördur. Overcoring stress measurements at Sandgerdi (S), Akureyri (A), Burfell (B), Hvalnes (H), Stokksnes (St); circle or ellipse diameter indicates average horizontal compressive stress. Ellipses are presented only for Hvalnes and Stokksnes, as discussed in text. Hydrofracturing stress elipses shown for Reykjavik (R), for about 0.3–0.4 km depth. H32 is the smaller ellipse.

distance measurements suggest a combination of left-lateral and extensional movement on the Reykjanes Peninsula (BRANDER *et al.* 1976).

The Reykjanes Peninsula and an adjacent zone due east of the abrupt bend in the axial rift zone, below the latitude of Reykjavik, contains numerous located earthquake epicentres (TRYGGVASON 1973, BJÖRNSSON and EINARSSON 1974). The overall seismic zone, perhaps 100 km wide,[3]) has been interpreted as a complex fracture zone (viz. the Reykjanes Fracture Zone) which exhibits transform motion between the Reykjanes Ridge and the eastern axial rifting zone of Iceland (Fig. 3, WARD *et al.* 1969, WARD 1971, p. 2999). Motion is not simple and on the Reykjanes Peninsula the plate boundary apparently has both transform fault and spreading ridge characteristics (KLEIN *et al.* 1973, 1977; BJÖRNSSON and EINARSSON 1974, BRANDER *et al.* 1976). An equivalent seismic zone off the north coast (Tjörnes Fracture Zone) has been similarly interpreted in reference to transform motion between the eastern axial rifting zone and the Kolbeinsey Ridge (SYKES 1967, WARD 1971, p. 3001); the concept has found substantial support in field investigations (SAEMUNDSSON 1974). A third fracture zone has been suggested at 64°40′N for the Snaefellsness Vatnajökull trend (SIGURDSSON 1970, SCHÄFER, 1972). The Reykjavik boreholes investigated in this study are located near the imprecisely-defined northern boundary of the Reykjanes Fracture Zone.

The axial rift and volcanic zones are flanked symmetrically by, primarily, Quaternary volcanics which in turn are bordered by volcanics of Tertiary (16 m.y. maximum) age (Fig. 1). The strata in general dip toward the active zone (Fig. 1); this is interpreted as suggesting continuous volcanism and crustal spreading during the last 7 m.y. (FRIDLEIFSSON 1973, SAEMUNDSSON and NOLL 1975; see however, EINARSSON 1968). Boreholes in Reykjavik are in Quaternary rocks ranging in age from about 2.8–1.8 m.y. (TÓMASSON *et al.* 1975). There is evidence here for ten glaciations in the volcanic succession, which is characterized by sequences of subaerial lava flows intercalated, at intervals corresponding to glaciation, by volcanic hyaloclastites and moraines. During this span of time two central volcanoes were locally active, the Stardalur (the younger) and Kjalarnes central volcanoes. The rate of volcanic production was higher near the central volcanoes, with the result that exceptionally thick hyaloclastite[4]) units accumulated nearby. The volcanoes were further characterized by an abundance of local sheet and dike intrusives. These volcanoes are now deeply eroded (erosion in the Reykjavik area amounts to an estimated 1.1–1.4 km (Saemundsson, oral communication)), and their core regions are noted as the sites of

---

[3]) Epicenters of large earthquakes in South Iceland seem more narrowly restricted, i.e. to a zone perhaps 20 km wide (TRYGGVASON 1973, BJÖRNSSON and EINARSSON 1974, p. 230).

[4]) Subglacial volcanics tend to accumulate under the ice near the eruptive orifice. Eruptive fissures 2–10 km long are the most common volcano form in Iceland and during major subglacial fissure eruptions a series of parallel hyaloclastite (volcaniclastic) ridges 1–5 km broad and several hundred metres thick typically formed along the active zones. In subsequent subaerial eruptions lava flows banked up against the hyaloclastite ridges and eventually buried them (TÓMASSON *et al.* 1975, FRIDLEIFFSON 1975). Thicker hyaloclastite accumulations occur near central eruptions.

positive gravity anomalies (Fig. 4; EINARSSON 1954; TÓMASSON *et al.* 1975). The Reykjavik geothermal fields are located on the southern margin of the Kjalarnes central volcano; these low temperature fields are superimposed upon extinct, eroded, high temperature (> 200° at depth < 1 km) fields (TÓMASSON *et al.* 1975).

Thermal gradients to the west of the volcanic zone in southwest Iceland increase regularly from about 70°C/km in Tertiary rocks 100 km west to about 165°C/km in early Quaternary rocks 20 km west (PÁLMASON 1973, 1974; TÓMASSON *et al.* 1975). Variable thermal gradients, including trend reversals, are observed in the Quaternary

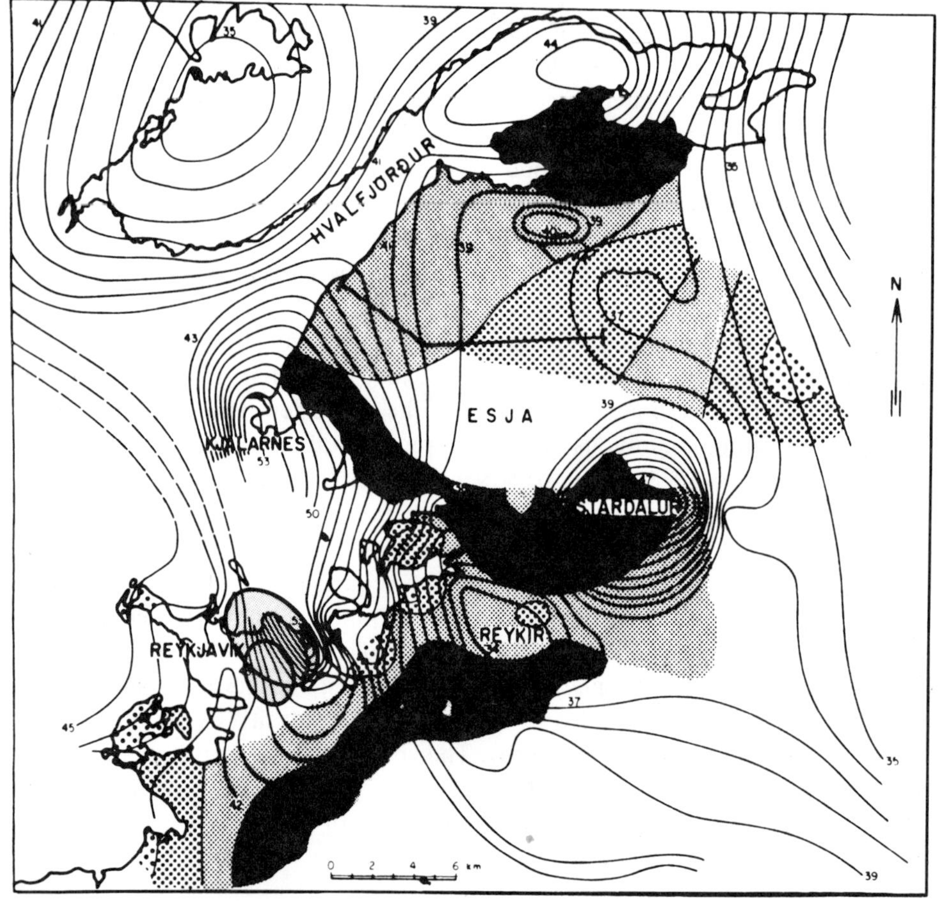

Figure 4
Bouger anomaly and electrical resistivity map of the Reykjavik province. Resistivity data for 0.9 km depth. (After TÓMASSON *et al.* 1975). Note stress ellipse orientations relative to the Kjalarnes gravity anomaly trend.

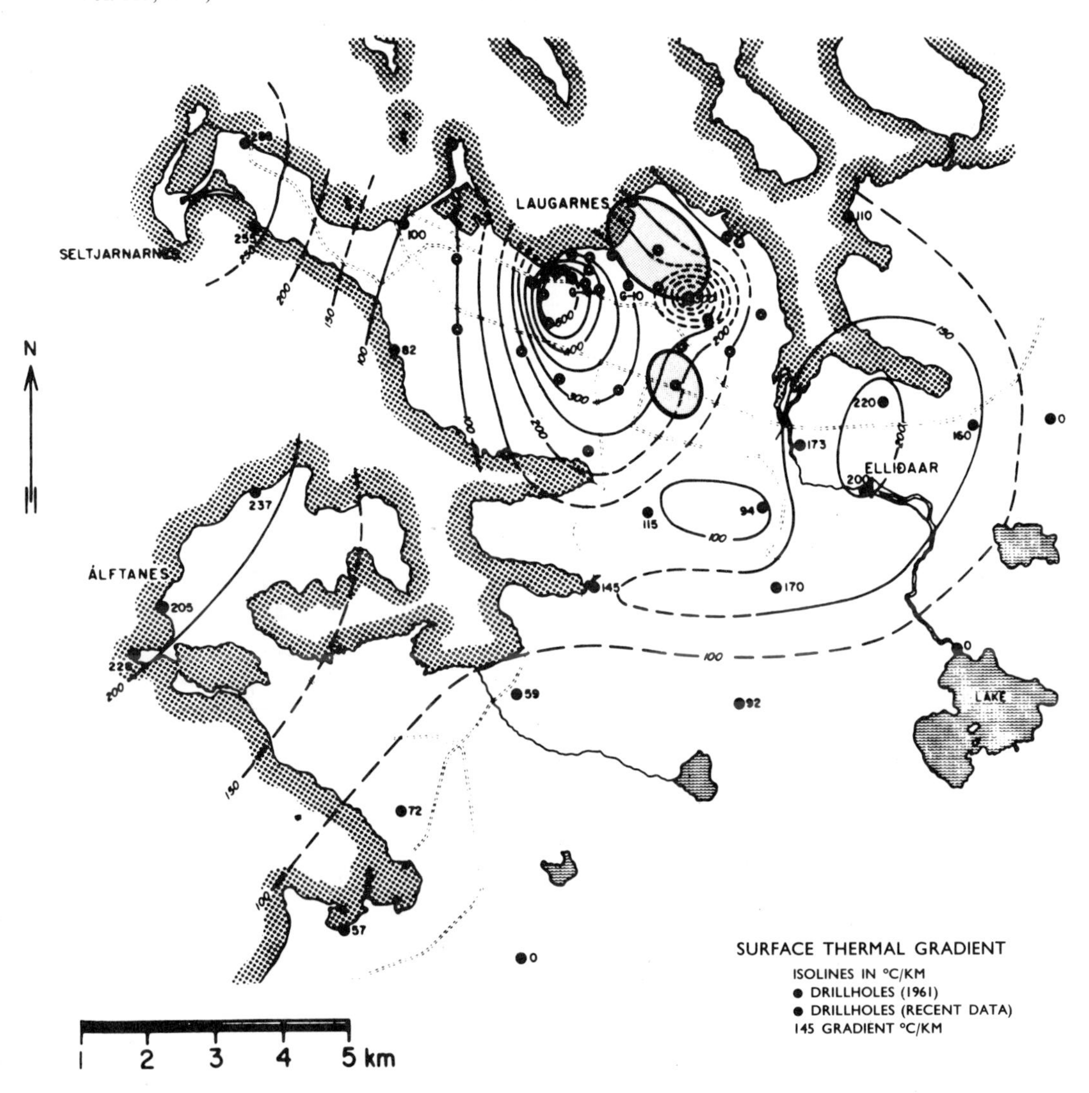

Figure 5
Thermal gradient map of the Reykjavik area, showing anomalies associated with Laugarnes, Seltjarnarnes, Ellidaár, and Áftanes fields (after PÁLMASON 1975). Isolines in °C/km. Stress ellipses for H18 (the larger) and H32, for about 0.3 km depth.

strata due to increased water circulation (Fig. 5). In the Laugarnes area the surface gradient varies rather substantially, with the range for most holes about 100–400°C/km (PÁLMASON 1975, TÓMASSON *et al.* 1975). There is little transport of water from depth in rocks between the thermal areas. Surface gradients of 0°C/km occur 5–10 km or so southeast of the Reykjavik thermal fields, due to cold ground water penetrating young volcanics to depths as great as 0.75 km.

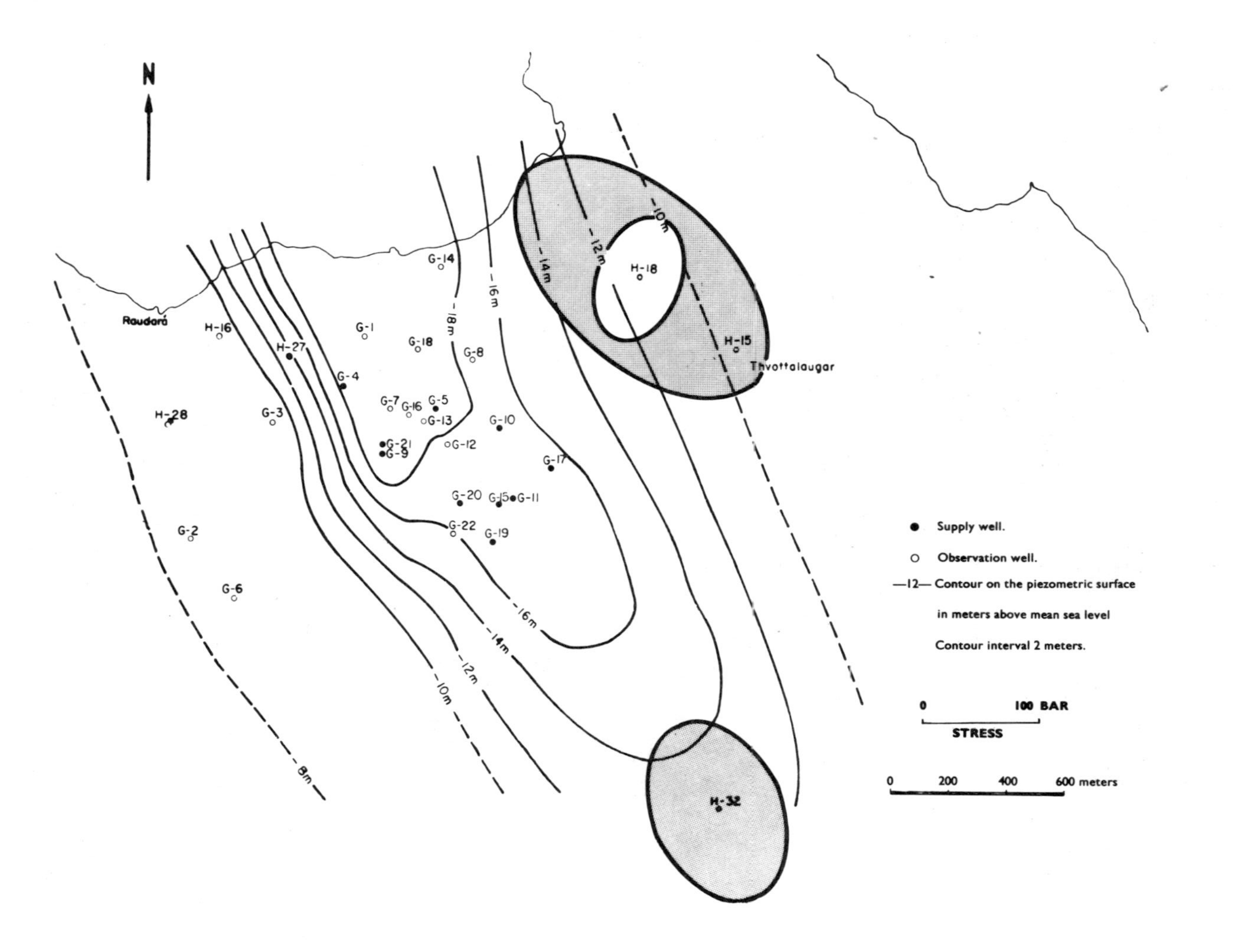
N
Raudará
Thvottalaugar
H-15
H-16
H-18
H-27
H-28
H-32
G-1
G-2
G-3
G-4
G-5
G-6
G-7
G-8
G-9
G-10
G-11
G-12
G-13
G-14
G-15
G-16
G-17
G-18
G-19
G-20
G-21
G-22
8m
10m
12m
14m
16m
18m
Supply well.
Observation well.
—12— Contour on the piezometric surface
in meters above mean sea level
Contour interval 2 meters.
0
100 BAR
STRESS
0
200
400
600 meters

Laugarnes borehole site

The Laugarnes system is one of three apparently separate hydrothermal systems within 6 km of the center of Reykjavik (Fig. 6). Boreholes were drilled in the area in the period 1928–1963. Details of exploitation and hydrology are given by THORSTEINSSON and ELIASSON (1970) and TÓMASSON *et al.* (1975); holes are of the open type with shallow near-surface cemented casing. Earlier wells were drilled with calyx rigs using steel shot or diamond bits and are designated by the prefix H (Table 1); wells completed by the rotary method have the prefix G. Bottom hole temperatures for a series of holes considered for this project varied from 8–119°C, for hole depths of 286–691 m.

The Laugarnes area is covered by a coarse grained olivine basalt flow, up to 30 m thick, resting on interglacial sediments up to 60 m thick (Fig. 7; THORSTEINSSON and ELIASSON 1970). Underneath, basalt flows alternate with hyaloclastites to at least 2.2 km (depth of the deepest borehole); down to 1250 m the intercalations of hyaloclastites and sediments are relatively thick and many of the flow series exhibit pillow structure. The deeper strata appear to dip 3–12°SE, based on identification of key beds in wells (Fig. 7). Local intrusions (dolerite) are associated with the Kjalarnes central volcano. Two stress measurement experiments were conducted in intrusive dolerite (in borehole H18). The remaining measurements were made in fractured olivine or tholeiite basalt in various stages of alteration. Limited amount of rock core was available for material property investigations (Table 2).

Description of experiments

Two series of hydrofracturing stress measurements were conducted, the first in borehole H32 located at the corner of Miklabraut and Greinsavegur streets, in the eastern part of Reykjavik, and the second in hole H18 situated at the corner of Reykjavegur and Sundlaugarvegur, in the northeastern part of Reykjavik (Fig. 5). The two holes are about 1.8 km apart. Testing procedures and equipment were identical in both holes except that hole H32 required a larger size of inflatable packers than H18 (Table 1).

The lowering and raising of the necessary equipment in the hole was carried out using NQ (70 mm outside diameter) drilling pipe, and a National Energy Authority owned Mayhew truck-mounted drilling rig capable of lifting 30 tons. Our testing depth was limited by the total length of available NQ pipe (400 m). Each hydrofracturing test required at least two 'trips' to the planned level in the hole, one for pressurization and fracturing and a second for borehole packer impression of the

Figure 6
Piezometric surface elevations for the Laugarnes system. Data for 15 November 1967 (after THORSTEINSSON and ELIASSON 1970). Stress ellipses for H18, at 180 m (the smaller ellipse) and 290 m, and for H32 at 350 m (orientation of $\sigma_{H\max}$ is N20–25°W for three H32 measurements). Scale for ellipse radius indicated. Cross section through H32 given in Fig. 7.

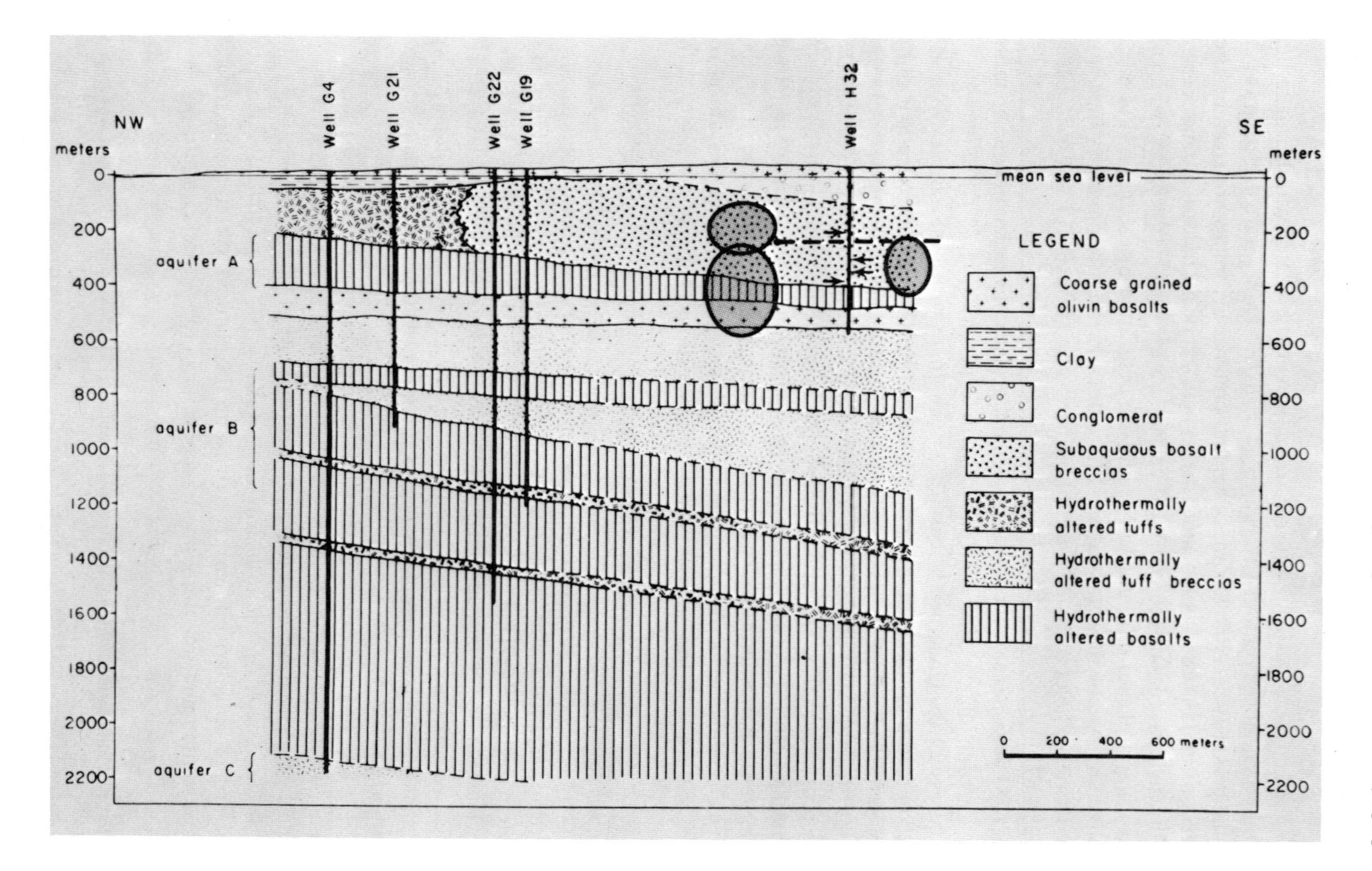
NW
SE
meters
meters
Well G4
Well G21
Well G22
Well G19
Well H32
mean sea level
0
200
400
600
800
1000
1200
1400
1600
1800
2000
2200
aquifer A
aquifer B
aquifer C
LEGEND
Coarse grained olivin basalts
Clay
Conglomerat
Subaquaous basalt breccias
Hydrothermally altered tuffs
Hydrothermally altered tuff breccias
Hydrothermally altered basalts
0 200 400 600 meters

Table 2
Physical properties of rock tested)*

Hole	Depth (m)	Density (g/cc)	T†) (bars)	Temp. (°C)	C_0‡) (bars)	E§) (bars × 10^5)
H18	180	2.92	102	46	100	4.3
	180	2.93	117			
	222	2.94		53	1340	3.5
	222	2.94				
	306		107	65		
	490		21	98		
	634	2.85	134	115	1575	4.4
	634	2.85				
H32	201	2.55	83		600	5.0
	201	2.53			440	1.7
	201	2.58				
	254		35			
	351	2.98	91		4140	7.3
	351	2.96			2570	8.2
	351	2.96				
	372		97			
	457	2.63	76		530	2.2
	457	2.50				

*) Based on laboratory testing of available core samples, except for temperature in H18 which was logged in the borehole in 1976.
†) T – Laboratory hydrofracturing tensile strength (See Haimson, 1973).
‡) C_0 – Uniaxial compressive strength.
§) E – Young's modulus in compression.

tested zone. In some cases a pretest impression was also taken. For the pressurization stage of the test a 'straddle packer' installation was used which included two inflatable rubber seals straddled by the hydrofracturing interval. In the larger diameter hole (H32) the seals were 1.0 m long and the interval was 1.7 m. In H18 the seals were 0.8 m long and the interval was 1.5 m. Attached to the bottom of the straddle packer was an Amerada-type self-contained pressure recorder which continuously plotted down-hole pressure vs. time. The straddle packer assemblage was lowered into position using the drilling pipe. The latter also served as a conduit for the water pumped from the surface to inflate the packers and then pressurize the straddled interval. Packer inflation pressure was kept at about 100 bars. Pressurization was generated on surface by using two high pressure air-activated pumps set in parallel to yield a flow of 0.5 lit/sec at 150 bars. The air was supplied to the pumps by a

Figure 7
Generalized cross section through Laugarnes area on northwest-southwest trend. For well locations see Fig. 5 (after Thorsteinsson and Eliasson 1970). Arrows on H32 indicate stress measurement locations. Stress ellipses approximately indicate stress conditions as a function of depth in northwest-southwest plane. Note change in stress regime denoted by dashed line. Above line, maximum compression stress is horizontal, indicated by long axis of ellipse. Below line, maximum compression is vertical.

15 m^3/min air compressor. In addition to the downhole pressure recorder, a pressure gauge, a pressure transducer and a flowmeter were installed in the hydraulic line on the surface. Oscillographic recorders continuously plotted pressure and flow variations with time.

Testing procedure followed that described by HAIMSON (1974). Packers were first inflated to the preset pressure so that they adhered to the borehole wall and sealed-off the straddled interval. The valve to the interval was then opened and pressurization of the isolated borehole segment began. Hydrofracture was indicated by a sudden drop in pressure followed by stabilization at some lower pressure value.

Pressurizations were repeated several times in order to obtain and verify the magnitudes of the hydrofracturing tensile strength and the shut-in (level-off) pressure. In all tests the pressurization rate was kept at approximately 100 bars/min.

The packer impression stage of the test was carried out using an inflatable packer covered with a soft rubber sleeve. The impression packer (1.2 m long) was lowered to the center of the hydrofractured interval, inflated to some 100 bars pressure and kept pressurized for about 45 min. During this time an orienting tool was lowered into a nonmagnetic string of pipe at the top of the impression packer and a photograph was taken of a compass face showing not only the direction of magnetic North but also that of a scribe marked earlier on the packer. The orientation of the impression was then used to determine the azimuth of the hydrofracture. The recorded pressures and hydrofracture orientation were then used to calculate the magnitudes and directions of the horizontal principal stresses.

Results

Four successful hydrofracturings in as many attempts were obtained in hole H32. In hole H18, however, seven attempts were needed to obtain three successful tests due to the prefractured nature of the rock. The depths were selected based on the condition of available core remnants (only representative samples have been preserved), the lithology as logged by the National Energy Authority, and a driller's log of the extracted core. Packer impressions revealed in each case a number of inclined joints. In successful tests these joints appeared to be sufficiently tight or healed so as not to interfere with pressurization. In such cases a vertical hydrofracture was induced. To verify that hydrofractures were actually obtained two packer impressions were taken in tests H32–3 and H18–2, one before and one after pressurization. In each test both impressions showed preexisting inclined cracks in the rock but only the second impression showed also the existence of a vertical fracture traversing the length of the interval.

The breakdown and shut-in pressures recorded are given in Table 3. P_c and P_s are the downhole breakdown and shut-in pressures, respectively; both are average values based on downhole and surface recordings. T is the hydrofracturing tensile strength determined from the difference between the breakdown pressure and the

Table 3
Hydrofracturing pressures recorded in Iceland tests

Hole-Test No.	Depth (m)	P_H*) (bars)	P_0**) (bars)	P_c^1†) (bars)	P_c^2††) (bars)	P_c^3‡) (bars)	P_c‡‡) (bars)	T§) (bars)	P_s^1†) (bars)	P_s^2††) (bars)	P_s^3‡) (bars)	P_s‡‡) (bars)
H32–1	203	20	15	117	137	133	135	90	19	39	43	41
H32–4	285	28	23	69	97	91	94	34	21	49	50	49
H32–2	350	34	29	110	144	150	147	83	28	62	54	58
H32–3	375	37	32	172	209	203	206	134	21	58	58	58
H18–2	180	18	13	69	87	—	87	41	21	39	—	39
H18–7	290	28	23	110	138	134	136	69	45	73	—	73
H18–6	324	32	27	148	180	176	178	83	48	80	81	80

*) P_H – head pressure to the depth tested.
**) P_0 – pore pressure at the depth tested.
†) P_c^1, P_s^1 are the measured *surface* breakdown and shut-in pressures respectively.
††) $P_c^2 = P_c^1 + P_H$ and $P_s^2 = P_s^1 + P_H$.
‡) P_c^3, P_s^3 are the measured *downhole* breakdown and shut-in pressures respectively.
‡‡) P_c, P_s are the downhole breakdown and shut-in pressures as averaged between P_c^2 and P_c^3, and P_s^2 and P_s^3, respectively.
§) T is the hydrofracturing tensile strength determined in each test from the difference between the breakdown pressure and the peak pressure obtained during repressurization.

peak pressure obtained during the second pressurization (provided the flow rate was kept approximately the same). Hydrofracturing tensile strength values were also obtained in the laboratory (Table 2) using a method described by HAIMSON (1973). Average laboratory values of T were used previously in calculating the maximum horizontal compressive stress (HAIMSON and VOIGHT 1976). However, in the present paper the field values of T were used because we did not always have core samples from the corresponding intervals that were tested, the rock mass was heterogeneous, and the two boreholes were so old that mechanical changes could have occured in the *in situ* rock which would not necessarily be reflected in the laboratory samples.

The following relationships were used in calculating the principal horizontal stresses (HAIMSON 1974, 1976a):

$$\sigma_{H\min} = P_s \tag{1}$$

$$\sigma_{H\max} = T + 3\sigma_{H\min} - P_c - P_0 \tag{2}$$

where $\sigma_{H\min}$ and $\sigma_{H\max}$ are the least and largest horizontal compressive stresses, respectively, and P_0 is the formation pore pressure at the depth tested.

Equation (2) is a simplified version of equation (6) in HAIMSON (1976a) in which stresses due to borehole–fluid penetration into the rock are also taken into account. However, as Fig. 1 in HAIMSON (1976a) shows, the difference between the two relationships is very small for values of P_c such as those encountered in our tests. The reason for our use of equation (2) was that we were not able to determine the poro-elastic parameter of the rock due to lack of sufficient core samples.

The vertical stress, σ_v, was calculated from the overburden load at the depth tested (d);

$$\sigma_v = \gamma d \tag{3}$$

where γ is the vertical pressure gradient due to weight of overburden. The latter was estimated from laboratory measurements of rock density (Table 2).

The calculated stresses are given in Table 4 and plotted with respect to depth in Fig. 8. The data suggests that each borehole site represents a distinct stress population[5]). Linear regressions yield the following:
for H32:

$$\begin{aligned} \sigma_v &= 0.27\, d \\ \sigma_{H\,\min} &= 20 + 0.11\, d \\ \sigma_{H\,\max} &= 47 + 0.07\, d \end{aligned} \tag{4}$$

and for H18:

$$\begin{aligned} \sigma_v &= 0.29\, d \\ \sigma_{H\,\min} &= -\ 13 + 0.29\, d \\ \sigma_{H\,\max} &= -\ 23 + 0.47\, d, \end{aligned} \tag{5}$$

where stresses (σ) are in bars and depths (d) are in meters. The quality of fit achieved by linear regressions is excellent for $\sigma_{H\,\min}$ in both instances (cf. Fig. 8); the fit is good with respect to $\sigma_{H\,\max}$.

The results for H32 are very consistent. Both $\sigma_{H\,\min}$ and $\sigma_{H\,\max}$ increase steadily with depth, at about the same rates which are substantially less than the rate of increase of overburden pressure. The relative orientation of $\sigma_{H\,\max}$ and $\sigma_{H\,\min}$ remain virtually constant with depth (Table 4), i.e., $\sigma_{H\,\max}$ direction is N23° (± 3)W; $\sigma_{H\,\max}/\sigma_{H\,\min}$ ratios are in the range 1.2–1.5.

On the other hand, because of the difference in stress gradient, a reorientation of principal stresses can be expected as a function of depth. Within the range of data a transition occurs at about 250 m; above this depth the vertical stress is the (assumed) intermediate principal stress; below it, the vertical stress is the greatest principal stress. For sufficiently high shear stresses, such stress orientations imply strike-slip and normal fault regimes, respectively; the maximum observed shear stresses are only 21 bar, however, although the regressions indicate that shear stresses increase with depth.

---

[5]) Linear regressions based on *all* data give $\sigma_{H\,\min} = 17.5 + 0.14\, d$ and $\sigma_{H\,\max} = 42 + 0.14\, d$. The quality of fit achieved for $\sigma_{H\,\min}$ is marginally satisfactory; however $\sigma_{H\,\max}$ values are widely scattered, and the linear trend for $\sigma_{H\,\max}$ appears dominated by the H32 data. The regression for $\sigma_{H\,\max}$ based on a lumping of all data is therefore regarded as generally misleading; the data for each borehole is more adequately considered as belonging to separate populations.

Table 4
Principal stress magnitudes and directions based on hydrofracturing testing in Iceland

Hole-Test No.	Depth (m)	σ_V*) (bars)	$\sigma_{H\,\min}$†) (bars)	$\sigma_{H\,\max}$‡) (bars)	$\sigma_{H\,\text{ave}}$§) (bars)	$\sigma_{H\,\max}$ (direction)
H32–1	203	54	41	63	52	N25°W
H32–4	285	76	49	64	57	N25°W
H32–2	350	93	58	81	70	—
H32–3	375	100	58	70	64	N20°W
H18–2	180	51	39	58	49	N20°E
H18–7	290	83	73	129	101	N45°W
H18–6	324	92	80	118	99	—

*) Vertical stress.
†) Least horizontal stress.
‡) Largest horizontal stress.
§) Average horizontal stress ($= (\sigma_{H\max} + \sigma_{H\min})/2$).

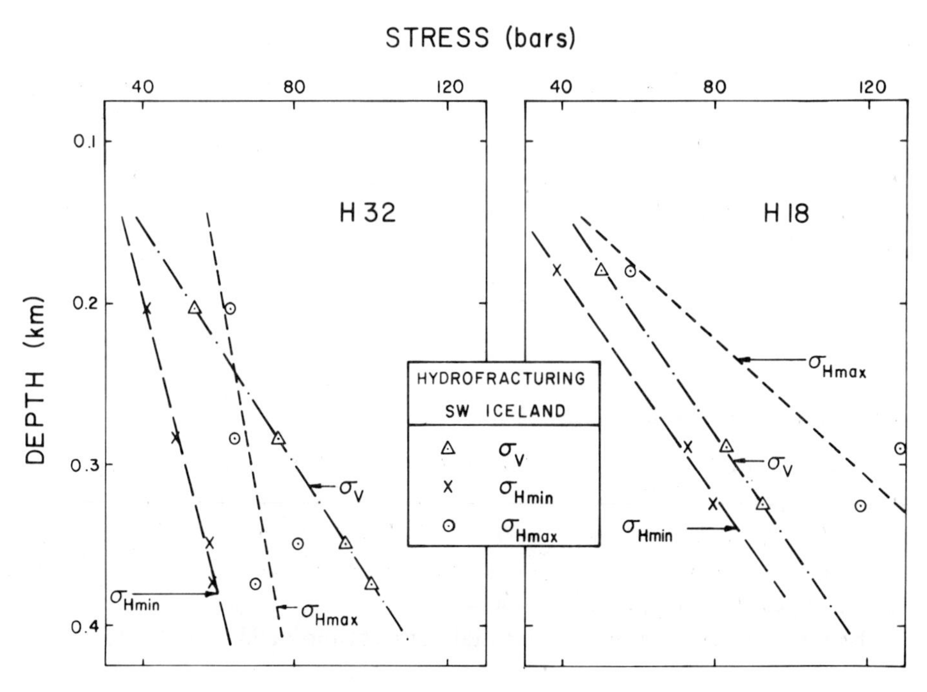

Figure 8
Measured stresses at H18 and H32 and linear regressions of stress depth relationship.

It is uncertain how far extrapolations can be carried out beyond the range of observations. Another principal stress transition is suggested, however, at 100 m; above this depth the vertical stress is the least principal stress (corresponding to a thrust fault regime), with near-surface horizontal compression estimated in the approximate range 20–40 bar. The data also suggest the possibility of another transition on the order of 1 km depth, with σ_2 and σ_3 (both horizontal) exchanging orientations. More measurements would be necessary to test this hypothesis, however; the extrapolation from existing data is simply too great to encourage confidence.

The results for H18 are also relatively consistent. $\sigma_{H\,min}$ and σ_v seem to increase steadily with depth at about the same rate. $\sigma_{H\,max}$ also increases with depth, but perhaps at a substantial rate. Three data points are insufficient to permit much confidence in the illustrated stress gradient, but a relatively high magnitude of horizontal compression seems well defined in dolerite at about 0.3 km. A change in orientation of $\sigma_{H\,max}$ with depth seems indicated by a total of two measurements; the indicated shear stress for the shallower measurement is low, ca. 10 bar, so that perhaps small stress variations could account for principal stress orientation changes. Extrapolation of $\sigma_{H\,max}$ trends are clearly unsafe; however, at face value, the possibility of near surface horizontal *tension* (or equivalent fissuring) is suggested for $d < 60$ m, whereas an increase of $\sigma_{H\,max}$ compression is suggested for $d > 325$ m. Throughout the range of observations the vertical stress is taken as the intermediate principal stress, thus corresponding to a potential strike-slip fault regime. Regression suggests the possibility of vertical σ_1 (normal fault regime) for $d < 150$ m.

Thus the data from each site displays internal consistency, although different states of stress are suggestive for boreholes H32 and H18. On the whole the data indicate a dominant NNW or NW orientation for $\sigma_{H\,max}$; at H18 the stress state favors strike-slip faulting (for $d > 150$ m), whereas this regime seems restricted (say 100 m $< d <$ 250 m) at H32. A normal fault regime seems dominant at H32[6]) (for $d > 250$ m).

3. *Interpretation*

Orientation of principal stresses and local structure

As summarized in Table 4, five determinations of the direction of $\sigma_{H\,max}$ were made at Reykjavik. The data are illustrated in Fig. 6 on the local scale of the well field, and in Figs. 2–5 on regional scales. The NNW trend of hole H32 was reproducible to a surprising degree, whereas H18 provided both NW and NNE trends, indicating some local perturbation of the stress field.

The direction of maximum horizontal compression at H32 points toward the

[6]) Among other reasons, the suggestion of a change in predicted fault orientation with depth makes attractive the consideration of deeper measurements at this site.

center of the piezometric depression associated with the Laugarnes well cluster, raising the possibility of a causative influence of groundwater exploitation on ground distortion and hence on stress orientation. However, THORSTEINSSON and ELIASSON (1970, p. 1201; cf. LOHMAN 1961) computed the thickness reduction of the Laugarnes aquifer due to piezometric pressure reduction; even under the most liberal assumptions, reduction in thickness due to elastic deformation would not exceed 1 cm. Relevelling in 1961 and 1969 of benchmarks established in 1948–50 at Laugarnes revealed *no* relative movement of bench marks in excess of the limits of accuracy of the levels (i.e., 1.4–3.9 mm/km; THORSTEINSSON and ELIASSON 1970, p. 1202), thus indicating the absence of appreciable elastic or plastic deformation of the aquifers and aquicludes. These data therefore suggest that the observed stresses are *not* an effect of local well field ground distortion.

The NNW axial alignment of the piezometric depression bears no obvious relation to nearby surface deformational structures, characterized by NE–striking step faults. The more likely explanation is that the elongate piezometric depression reflects anisotropic permeability, probably involving high porosity hyaloclastite channels separated by low porosity lavas and intrusives; the orientation of such channels is undoubtedly influenced by the NNW axial alignment of the Kjalarnes central volcano, as suggested by geophysical resistivity and gravity surveys (Fig. 4; EINARSSON 1954, TÓMASSON *et al.* 1975).

The fact that $\sigma_{H\,\max}$ directions at H32 (and also H18, *if* approximated by the acute angle bisector or by the deeper measurement) are parallel to the piezometric depression axis therefore could simply be a reflection of the influence of the Kjalarnes central volcano on both stress and hydraulic conductivity trends. A structural influence of the volcano on stress distributions could be indirect (elastic anisotropy tends to cause limited realignment of regional $\sigma_{H\,\max}$ toward NNW direction of maximum stiffness), direct (post-emplacement distortion of flows and intrusives by ground distortion in vicinity of volcano), null (volcano is aligned parallel to regional compression, hence, no reorientation need occur), or lacking entirely. Available data do not permit complete evaluation or discrimination between these hypotheses. However, stress data indicates that lateral compression ($\sigma_1 = \sigma_{H\,\max}$) is important, particularly for H18, whereas distortion about a bulging volcano should be primarily extensional in nature. The two holes seem to be located in a similar structural position on the volcano flank (Fig. 4), so that the difference in stress state does not seem readily explained by direct structural influences. Still, stress reorientation due to complex multilayer flexuring might account for observations. The data do not seem to offer strong arguments either in support of or in opposition to the hypothesis of local stress reorientation due to elastic anisotropy. Neither can the null hypothesis be rejected. The required circumstances do not seem necessarily fortuitous because of the symmetry of the observed stress field with the axial rift zone; indeed, a similarly orientated stress field might have been responsible for the alignment of the volcano in the first place.

Mapped surface thermal gradients (PÁLMASON 1975, Fig. 1) display a local pattern similar in many details to those of the gravity surveys; these data seem to also reflect the axial trend of the Kjalarnes volcano, and thus permit interpretations similar to those given above. In addition, however, the possibility of a thermoelastic effect must be raised. It indeed seems likely that a lithospheric plate containing inclusions of abnormally hot rock masses will display an inhomogeneous stress pattern due to the thermoelastic effect (cf. VOIGHT and ST. PIERRE 1974). These thermoelastic stresses must be superimposed upon the regional stress field, and their total effect would depend upon the relative magnitudes of each component.

As a highly idealized (plane stress) example we consider here a cylindrical hot inclusion of radius a and uniform temperature T_0 in an otherwise cold, unstressed plate. The solution is:

$$\sigma_r = \sigma_\theta = +\tfrac{1}{2}E\alpha T_0, \qquad \tau_{r\theta} = 0, \quad r < a,$$

$$\sigma_r = -\sigma_\theta = +\tfrac{1}{2}E\alpha T_0(a^2/r^2), \qquad \tau_{r\theta} = 0, \quad r > a.$$

As a general property of such stress states there is assumed a stress discontinuity across the inclusion surface. Letting $E = 5 \times 10^5$ bar (cf. Table 2), $\alpha = 8 \times 10^{-6}\,°C^{-1}$, and $T_0 = 100°C$, we have a uniform value of $\sigma_r = \sigma_\theta = 200$ bar (compression), in the inclusion, and variable σ_r (compression) and σ_θ (tension) in the plates. The latter diminish as $(1/r)^2$; at $r = a$, $\sigma_r = -\sigma_\theta = 200$ bar, at $r = 2a$, $\sigma_r = -\sigma_\theta = 50$ bar, and so forth. The mathematical effect is the same, incidentally, if we start with a uniformly hot plate and nonuniformly cool it (the inclusion is cooled to a lesser degree); the latter description more adequately fits the thermal history of Icelandic rocks as discussed previously. The result, which must be interpreted in terms of more or less continuous rather than discontinuous stress fields, is that high values of compression might be expected within thermal anomalies, with radial compression and circumferential tension (diminishing rapidly with radial distance) expected beyond them. Stresses due to overburden pressure and initial conditions must be superimposed upon these predicted thermoelastic changes so that absolute tension might not ordinarily be expected; in addition, actual stress changes should be expected to be less than theoretical estimates because of rock strength limitations.

Our measurement sites may be considered to lie on the flank of the thermal anomaly (Fig. 5). The boundary of the anomaly is not sharp, as in the inclusion model, but involves a relatively broad zone with variation in three dimensions. The observed lateral stresses do not indicate overwhelming compression with respect to the vertical (overburden) stress or to the mean stress; with respect to the simple model, interpretation must therefore be couched as $r > a$. Four of five measurements suggest (approximately) radial compression and circumferential extension with respect to the central zone of the anomaly (Fig. 5). The remaining measurement (N45°W, H18) could conceivably be related to the eastern bulge in the anomally located immediately to the southeast of H18. The stress values seem of the right order of magnitude, although the lack of *strong* radial compression does not seem favorable to the hypothesis; still, detailed evaluation does not seem possible, lacking further information on initial stresses (prior to imposition of thermal effects) and details of the thermal gradient history expressed as functions of space and time. On balance, the possibility of some thermoelastic stress influence seems supported by available evidence; in any event the hypothesis of a local thermoelastic effect cannot be safely rejected.

It should also be noted that the greatest magnitudes of measured stress ($\sigma_{H\max} = \sigma_1$) occurred in intrusive dolerite of H18. All other measurements were made in basalt. The possibility must therefore be entertained that the formational history of the involved rocks may have influenced the present stress fields.

To conclude this section, we note that several hypotheses concerning local geologic relationships, not necessarily mutually exclusive, are available for interpretation of the Reykjavik stress data. Some should be eliminated, however, by future experiments designed to test their specific attributes.

Comparison with overcoring measurements

Stress measurements by the overcoring technique were performed by Nils Hast in 1967–8; some details have been reported by HAST (1969, p. 174–5, 193–196, 205–206, 1973, p. 413–416). The numerical data (previously unpublished) are summarized in Table 5. Measurements were made in shallow vertical boreholes, at three points only a few meters from the shoreline. The deepest measurements (about 50 m) were made at Burfell, a hydroelectric site in the interior of Iceland (10 km west of the notorious volcano Hekla).

Table 5
Overcoring stress measurements in Iceland)*

Locality	Rock Type	Depth (m)	σ_V (bars)	$\sigma_{H\min}$ (bars)	$\sigma_{H\max}$ (bars)	$\sigma_{H\text{ave}}$ (bars)	$\sigma_{H\max}$ (direction)
Burfell†)	basalt	65	18	17	70	44	N52°E
Akurery (a) ‡)§)	basalt	3	1	−2	45	22	N4°W
(b)	basalt	10	3	20	53	36	N1°W
(c)	basalt	13	4	20	51	35	N1°E
Sandgerdi§)	basalt	4	1	28	44	36	N75°E
Stokksnes	gabbro	30	8	79	135	107	N73°W
Hvalnes¶)	granophire	29	8	45	135	90	N79°W

*) Measurements by N. HAST (written communication with B. VOIGHT).

†) Measurement point from tunnel about 60 m from steep face in faulted mountain block, 100 m above valley floor; topographic effect; adjacent nearly-vertical fault systems at about N65°E and N17°E (TÓMASSON, 1967).

‡) Probable topographic effect.

§) Shallow measurements within effect of surface temperature fluctuations.

¶) Estimated erosion here about 1.8 km (G. P. L. WALKER, oral communication with B. VOIGHT).

The most important aspects of these measurements is the indicated prevalence of compressive stresses, with average magnitudes greater than overburden pressure at the depth of measurement. Hast also attempted to use the inferred principal stress orientations at each site for geotectonic interpretation, but we believe that few of his measurements could provide reliable indications of crustal stress directions at depth. Perhaps none do. Some are quite shallow (see Table 5) and are undoubtedly affected by near-surface thermoelastic effects, which are known to be quite severe (HEIDECKER 1968, HOOKER and DUVALL 1971, VOIGHT and ST. PIERRE 1974). The measurements near Akureyri and Burfell are strongly affected by topographic features; whereas the measured stress directions at these locations are no doubt reasonably accurate, they reflect local topography at least as much as deep seated geotectonic processes and thus must be rejected for regional analysis. At Burfell, for example, the measurement site is located about 60 m from a steep NE trending mountain face, about 100 m above the valley floor. It is therefore not surprising that the NW component of stress seems relatively small (Table 5).

Equally important, our data show that prevalent near-surface compression (e.g., as indicated by the overcoring measurements) is *not* a necessary indication of similar conditions at depth. As demonstrated by the H32 regressions of Fig. 8, a deep-seated zone of lateral extension (normal fault regime) can be overlain by a zone of lateral compression (strike-slip or thrust fault regimes). Indeed, the surface principal stresses predicted by extrapolation of the H32 data are 18 and 43 bar, both compressive; these values are very nearly the same as those measured by overcoring at Sandgerdi (Fig. 2, Table 5) in an equivalent tectonic position. The similarity of these data is undoubtably fortuitous; however it is absolutely clear that the near-surface data by themselves do not provide a reliable indication of deeper stress states.

The only measurements by Hast which conceivably could provide reliable indications of regional stress orientation are those at Hvalnes and Stokksnes in eastern Iceland. The two sites are perhaps sufficiently deep (30 m) to avoid surface temperature change effects; they are about 26 km apart, but recorded $\sigma_{H\,\max}$ orientations are almost identical, viz. N76° (±3)W. Now the Hvalnes site is located near the base of a 600 m high, NW trending ridge (Hvalnesfjall) and is therefore probably within a zone of topographic stress concentration; the agreement of stress measurements at the two sites could therefore be fortuitous. The data are so remarkably similar, however, that the possibility also seems reasonable that they could reflect a common deep-seated environment.

Stress ellipse data for the eastern Iceland sites are therefore included in Fig. 3, for comparison with data from our hydrofracturing experiments and with regional geologic features. Locations for other overcoring sites are also shown, with average compressive stress magnitudes indicated by circle diameter; measured stress orientations at these sites may be accurate but are not considered reliable for regional analysis, as discussed above, and thus are not shown. None of the overcoring sites appear to be associated with thermal anomalies.

The stress orientation data from hydrofracturing and overcoring experiments as given on Fig. 3 each display a similar disposition in relation to adjacent axial rift valleys. The Reykjavik measurements indicate maximum horizontal compression (i.e., $\sigma_{H\max}$, not necessarily σ_1) roughly normal (highly oblique) to the trend of the Reykjanes–Langjökull axial rift zone, whereas the Hvalnes–Stokksnes data indicate a similar trend in reference to the eastern axial rift zone. Although data at neither location is truly susceptible to unequivocal interpretation, the circumstantial evidence seems strong enough to consider analysis in regional terms.

Comparison with focal mechanism solutions

Few major earthquakes in Iceland have been recorded to the extent that a reliable focal mechanism could be obtained. STEFANSSON (1966) and SYKES (1967, p. 346–347) reported on a magnitude 7 earthquake north of Iceland (66.29°N, 19.78°E); the axis of minimum compression was nearly horizontal at N62°E, with the axis of maximum compression also horizontal with NW trend (Fig. 3). The alignment of one of the nodal planes with epicenters in the Tjörnes seismic zone was taken by Sykes an an indication of a major transform fault between the eastern axial rift of Iceland and the Kolbeinsey Ridge (cf. SAEMUNDSSON, 1974).

WARD (1971, p. 2996–2998) discussed two focal mechanism solutions for South Iceland (Fig. 2). Both indicated strike-slip movement and a nearly horizontal NW-trending axis of minimum compression. These solutions also were interpreted as supporting the hypothesis of a WNW or E–W striking transform fault near latitude 64°N, which had been proposed by WARD *et al.* (1969) on the basis of historic epicenters and a more narrowly defined microearthquake zone.

The stress distribution noted by Ward was in general supported by focal mechanism solutions for about 470 small earthquakes on the Reykjanes Peninsula.[7]) The common feature of most of these solutions is the nearly horizontal NW-trend of the axis of minimum compression, a direction approximately perpendicular to the NE strike of individual near-surface fissures and faults (KLEIN *et al.* 1973, Figs. 6, 12; 1977, JÓNSSON 1967). Intermediate and maximum compression axes vary in direction, apparently indicating slip components of normal (σ_1 vertical), strike-slip (σ_1 horizontal), or oblique (σ_1 inclined) faulting. Solutions of various types sometimes appear mixed, occurring in close juxtaposition to one another, and on the whole deformation cannot be simply described as small ridge segments sharply bounded by small transform faults. Some systematic trends are evident; thus, the western part of the Reykjanes area seismic zone is characterized by normal faulting on conjugate NE-striking planes, oblique faulting is dominant in the eastern part, and the central part contains a domain of strike slip faulting overlying (at depth <3 km) a domain of normal faulting (Fig. 2; KLEIN *et al.* 1977).

---

[7]) 433 of these earthquakes were associated with the Reykjanes swarm of September 1972 (KLEIN *et al.* 1977).

The center of rotation for relative motion of North American and Eurasian plates proposed by PITMAN and TALWANI (1972) requires that small circles about the pole strike N80°W in southern Iceland. The minimum compression axes derived from focal mechanisms therefore do not parallel the small circles of rotation. Neither are they perpendicular to the axial rift zone on the peninsula which defines the mid-Atlantic plate boundary, as located by surface geology (N70°E) or by micro-earthquakes (N70°–90°E; Fig. 2); they are more or less perpendicular to the average NE trend of surface fractures and to the axis of the Reykjanes Ridge south of Iceland.

The zone of hypocenters reveal a narrow zone of brittle deformation on the Reykjanes Peninsula about 2–5 km wide, 40 km long, and 5 km or so deep (KLEIN *et al.* 1973, 1977). About 90% of the located earthquakes had depths between 1.5–6 km (cf. WARD and BJÖRNSSON 1971, KLEIN *et al.* 1973, 1977). Shallower deformation seems aseismic, perhaps reflecting tensile deformation (cf. SCHOLZ *et al.* 1969). The deepest earthquakes had hypocenters near 6–7 km at Reykjanes, 8–9 km at Krisuvik (KLEIN *et al.* 1973, p. 5088), and 13 km near Hveragerdi (WARD and BJÖRNSSON 1971, p. 3974). Maximum hypocentral depth therefore decreases toward the oceanic ridge, where maximum depths are likely to be about 5 km. The boundary between layer 3 (defined by P-wave velocity of 6.5 km/sec, PÁLMASON 1971) and layer 4 (P-wave velocity of 7.2 km/sec, upper mantle; see PÁLMASON and SAEMUNDSSON 1974, p. 40) occurs at about 8–9 km, and thus most hypocenters are located above or within the upper part of layer 3[8]). The decrease in frequency of earthquake occurrence below 6 km probably reflects the increased ductility of the rocks as a function of temperature and mean pressure.

A focal mechanism solution (based on 14 events) was reported for the Krafla area in the eastern axial rift zone, the only consistent source of microearthquakes found in northern Iceland (WARD *et al.* 1969, p. 673–4; cf. WARD and BJÖRNSSON 1971, p. 3972). The suggested nodal planes both imply oblique slip with a significant thrust component. The maximum compression axis plunges from the NW, whereas the axis of minimum compression plunges from the NE.

More recently a (1974) earthquake sequence involving both swarm and foreshock-mainshock-aftershock activity was observed in Borgarfjördur, about 80 km north of Reykjavik (EINARSSON *et al.* 1977). The main epicentral zone as defined by 563 hypocenters is about 4 km wide by 25 km long, trending nearly east-west; a secondary branch extends about 9 km northwest of the main zone. Most focal depths ranged from 0 to 8 km, and none occurred below the inferred crust-mantle boundary at 10 km.

About 18 'reliable' focal mechanism solutions were determined (11 are composite solutions), all of which indicate normal faulting. The maximum compression (σ_1) axes cluster about the vertical axis. The average azimuth of the axis of minimum

[8]) The deepest located hypocenters in Iceland occurred at 20–30 km depth in association with the Heimaey eruption (BJÖRNSSON and EINARSSON 1974, p. 238), about 8–18 km below the upper mantle boundary at this location.

compression (σ_3) is about north-south approximately perpendicular to observed surface fault breaks in the spicentral zone (Fig. 3); the inference is that the Borgarfjördur area is under horizontal N–S (relative)[9]) tension (EINARSSON *et al.* 1977).

It is significant that the majority of microeathquakes recorded in Iceland are associated with major geothermal areas (WARD *et al.* 1969, WARD and BJÖRNSSON 1971). The specific causative mechanism of the microearthquakes is therefore open to question. Microearthquakes could be due to fundamental tectonic movements; it also seems possible that they could reflect geothermal pore-pressure perturbations unrelated to direct tectonic action. In the latter instance the geothermal effect would be the 'trigger' and subsequent fault motions would correspond to existing rock stresses, whatever their cumulative cause. Data are few, but no correspondence of microearthquake activity to geothermal well operations has yet been noted (WARD and BJÖRNSSON 1971, p. 3977); this may reflect the shallow depth of pumped wells or the fluid pressure decrease associated with pumping. Indeed, observed horizontal ground movements at the Reykjanes geothermal area is known to be significant (as great as 16 mm in 4 years), and apparently is related to discharging wells (BRANDER *et al.* 1976, Fig. 7, p. 204). WARD and BJÖRNSSON (1971) note that it is not clear whether or not high fluid pressures exist in Icelandic geothermal areas. Nonetheless the increase in pore pressure required for faulting may not be substantial if the state of stress is at a nearly critical level with respect to mass rock strength. The data are consistent with the hypothesis that water in geothermal areas leads in some way to weakening of the crustal rocks. Finally, it has been noted that seismicity in earthquake swarms migrates laterally at velocities appropriate to fluid diffusion processes (KLEIN *et al.* 1977).

Data on magnitudes are not generally available from the earthquake studies, so that it is not possible to compare in detail the state of stress related to earthquakes with borehole stress measurements. Stress orientations are interpreted, however, and thus it is known that where σ_1 is vertical, as seems commonly the case along the Reykjanes Peninsula and at Borgarfjördur (Fig. 3), the average value of horizontal stress is less than the average value of overburden pressure. This is in apparent contrast with results of overcoring and the H18 hydrofracturing measurements surveys, but we have demonstrated from hydrofracturing data that near-surface lateral compression is not *necessarily* indicative of abnormally high lateral stresses at depth (see data for H32, Fig. 8). Systematic changes in principal stress orientation with depth (in both cases predominantly involving a change in σ_1 from the horizontal to the vertical)

[9]) The ground effects seem associated primarily with a single event, the main shock of June 12. The focal mechanism data do not cluster about the mean. Instead, two populations seem to be involved, with associated average least-compression axes oriented at N25°E, and N25°W. These stress axes seem to correspond to observed fault sets with WNW trend (typical of an echelon tensional (fault, volcanic) features of the Snaefellsnes Peninsula) and NE trend (EINARSSON *et al.* 1977, Fig. 8), respectively. An observed NNW-trending fault set seems unrepresented by local mechanism axes. The data thus suggest local variations of $\pm 24°$ or more about mean regional stress orientations.

have been documented at Reykjavik from hydrofracturing data, and at the central sector of the Reykjanes seismic zone from focal mechanism solutions.[10])

Perhaps more important, static (hydrofracture, overcoring) measurements just beyond the boundary of the axial rift zone, and the focal mechanism solutions of the Borgarfjördur earthquakes, suggest a direction of $\sigma_{H\max}$ roughly normal to the axial rift zone, whereas microeathquake data from the Reykjanes Peninsula consistently suggest an extension component normal to the rift zone at the axial location ($\sigma_{H\min}$ is perpendicular to individual fissures en-echelon to the axial zone itself, so that a shear stress component also exists parallel and perpendicular to the axial zone). Although the role of geothermal water in microearthquake generation is uncertain, it is likely that fluid pressure variation would not affect the above conclusion, i.e., the relative *orientation* of principal stress axes in the hydrothermal areas probably remains the same irrespective of values of effective stresses. Existing data therefore suggest that a systematic increase in σ_H, normal to the axial rift zone, occurs as a function of increasing distance from the axial rift zone (Fig. 3).

Comparisons of WARD's (1971) two focal solutions with the hydrofracturing data leads to a similar anomaly. Earthquake hypocenters and hydrofracturing sites are both located within the 'Reykjanes Fracture Zone' as defined by Ward; whereas Ward's data is consistent with a NE compression axis and left lateral slip, some Reykjavik data suggest NW compression[11]) and the possibility of right lateral slip (Fig. 2). It should however be noted that in theory the transform fault segment does not extend beyond the axial rift valley, and that beyond this point the fracture zone might not exist as a zone of slip, or, alternatively, could even behave as a transcurrent (rather than transform) fault zone.[12]) The hydrofracturing and focal mechanism data are not therefore in conflict, although the drastic change in principal stress axis obviously requires explanation. It is possible that the orientation of $\sigma_{H\max}$ is controlled by the *gross* orientation of the axial rift zone, and the so-called Reykjanes Fracture Zone exerts little or no mechanical influence on the stress state in the Reykjavik area.

---

[10]) Similar changes have been observed near Lake Jocassee, South Carolina where hydrofracturing measurements (HAIMSON 1976c) and focal mechanism solutions (TALWANI 1977) show the vertical stress to vary from least principal stress near surface, to intermediate stress at about 1 km, to largest stress at greater depths.

[11]) The data do not necessarily suggest that slip is occurring; this is merely the orientation of $\sigma_{H\max}$ (not necessarily σ_1). In H32, e.g., the stress state suggests a dominance of potential normal faulting.

[12]) In this regard it might be observed that numerous northern Iceland earthquakes occur along the Tjörnes Fracture Zone *west* of its connection with the Kolbeinsey Ridge. The 28 March 1963 event, in which right-lateral slip was indicated by SYKES' (1967) focal mechanism solution, is one of these. Its location 40 km *west* of the Kolbeinsey Ridge suggests that it may be located on a transcurrent fault segment of the fracture zone, rather than on an existing transform fault segment; if so, the employment of its focal mechanism solution as seismological evidence for transform faulting is erroneous. The question should be re-examined. The western part of the zone of seismicity could be related to transcurrent motion along a preexisting transform fault associated with the now extinct Langjökull–Skagi axial rift zone discussed by SAEMUNDSSON (1974, p. 500).

We further note that the Reyjavik stress data ($\sigma_{H\,max}$) are more or less consistent with the right lateral movements in a transcurrent zone postulated by SIGURDSSON (1970) for the Snaefellsnes region in western Iceland (Fig. 3); however, the Borgarfjördur data suggest that $\sigma_{H\,max}$ in that zone is orientated E–W, and the entire body of data seem in better agreement with the model of $\sigma_{H\,max}$ orientation as influenced by the gross orientation of the axial rift zone. The $\sigma_{H\,max}$ axis at Reykjavik is also roughly perpendicular to an 'anticline' at Borgarnes, about 40 km distant (SAEMUNDSSON 1967) and a similar relationship apparently exists between measured stress and a prominent zone in Tertiary basalts bordering the eastern Iceland axial rift zone (WALKER 1964, 1974, SAEMUNDSSON 1974). These observations are noted in order to complete the record as regards the geometry of structure and stress state, and no direct causative relationship is thereby implied. For example, the flexure zones (in some cases anticlines) have been most generally interpreted as a reflection of sagging within the axial rift zone (WALKER 1974, p. 185–186, SAEMUNDSSON 1974, p. 500–502, PÁLMASON and SAEMUNDSSON 1974, p. 43), rather than of compressive buckling (EINARSSON 1968). The geometric relationship of extinct flexure zone and (existing) stress state is probably due to indirect causes, i.e., both seem to be geometrically related to the orientation of the axial rift zone.

The solution for the Krafla microearthquakes also requires explanation. WARD *et al.* (1969, p. 678) noted that the focal mechanism for Krafla gives the intermediate stress along the axis of the axial rift zone, which is thus 'consistent with SYKES' (1967) results'. They fail to emphasize, however, a maximum compression axis highly oblique to the axial rift zone and a significant thrust component (Fig. 3), neither of which fits SYKES' (1967) model of normal faulting with the inferred minimum compression axis approximately perpendicular to the ridge. More data will be required to resolve the issue; the existing solution is a composite one, and its reliability may be questionable. We note however, that (1) the Krafla data are not incompatible with the results of static stress measurements to date, and that (2) differences between stress state and microearthquake behavior as exhibited by the Reykjanes Peninsula and eastern Iceland axial rift zones could reflect basic differences in stress, thermal and growth history (see, e.g., SIGURDSSON, 1970, WARD, 1971, SCHÄFER, 1972, SAEMUNDSSON 1974).

Interpretation in terms of plate tectonic models

Stress measurements outside the rift zones to date suggest relatively strong component of crustal compression ($\sigma_{H\,max} = \sigma_2$ or σ_1) roughly normal to the axial rift zones (Fig. 3). Considered by itself, such data might be taken as an indication that the lithospheric plates are being forced apart. However, at least in southwest Iceland, focal mechanism solutions suggest the contrary view, i.e., that crustal extension is occurring across the axial rift zone ($\sigma_{H\,min} = \sigma_3$). Both sets of data seem to be correct indications of the crustal stress environment at different physical locations; the problem is to reconcile them.

Now it is well known that parts of the lithosphere can be subjected to lateral extension at the same time that adjacent parts are subjected to lateral compression. This feature is, for example, predicted by mathematical models of lithosphere subjected to convective drag (see, e.g., VOIGHT *et al.* 1969, Fig. 3). However, such models ordinarily require a substantial distance separating the regions of lateral compression and extension; in Iceland this distance seems abnormally small.

Indeed, the basic problem of stress interpretation appears to be at present irreconcilable if posed in terms of a boundary value problem involving an initially unstressed plate subjected solely to arbitrary stress boundary conditions. We therefore believe that the interpretive solution is to be found in the contrary view – that the spreading lithospheric plates are stressed in a complex manner, with the stresses at any location a function of the loading and thermal history endured by the rocks at that location, as well as of existing boundary conditions. Viewed in this way, extension across the Reykjanes Peninsula axial rift zone directly reflects active processes of ocean floor spreading, whereas the orientation of $\sigma_{H\,\max}$ normal to the axial rift zones, as observed elsewhere in Iceland, represents the addition of stress history effects which may or may not reflect spreading processes. Significant lateral stress components, induced by one or a combination of processes, have somehow influenced the rock mass beyond the axial rift zones.

Some of these mechanisms have already been referred to in the section on Principal Stress Orientation and Local Structure. Thermoelastic effects near local thermal anomalies, for example, can cause crustal compression over and adjacent to the anomaly. However, inasmuch as the orientation of $\sigma_{H\,\max}$ seems potentially regional in scope and not merely associated with local thermal anomalies,[13]) some fartherreaching explanation seems required. A strong possibility involves thermoelastic effects which exist on a regional scale in association with the accretion and motions predicted by plate tectonics theory. According to this view, progressive cooling of the lithospheric plates as a function of spreading from the axial zone of accretion results in thermal contraction in a direction parallel to the trend of the axial zone (TURCOTTE 1974, COLLETTE 1974), and basal accretion and subsequent cooling of the newly accreted, hot material, results in contraction perpendicular to the axial zone (SYKES and SBAR 1974). These two mechanisms both favor alignment of principal stresses with axial rift zones or associated isochrons. The basal accretion and cooling mechanism mainly induces a systematic increase in the σ_H component normal to the rift zone as a function of distance from the rift zone whereas axial accretion and cooling tends to cause regional alignment of $\sigma_{H\,\min}$ parallel to the axial rift zone. The existing stress data seem to correspond well with the predicted patterns associated with accretion-cooling models.

According to this view, the Snaefellsnes Peninsula is a topographic manifestation of thermal contraction fissuring and graben formation, with the volcanism a secondary

[13]) Further tests will be necessary for documentation.

effect reflecting pressure release beneath rifted lithosphere. Certainly the state of stress associated with the Borgarfjördur earthquakes is supportive, although complications are to be expected at this latitude owing to shifts in the plate boundary (see, e.g., SAEMUNDSSON 1974). Similarly, the H32 data at Reykjavik suggest the possible interpretation of the Reykjanes Fracture Zone as a thermal contraction structure.

Furthermore, for ridge segments separated at intervals of 100 km or so by fracture zones, thermal contraction stresses can be partially relieved by bending (TURCOTTE 1974). In so doing, the upper part of the lithosphere goes into compression, whereas lower parts remain in (relative) tension. Therefore, lithosphere 'skin' compression is indeed predicted beyond the axial rift zone, as observed by the overcoring and hydrofracturing measurements; the transition from near-surface thrust or strike slip stress regimes to deeper normal fault regimes is also predicted, as documented by our Reykjavik measurements. Stress components generated in response to ocean floor spreading driving mechanisms would involve the entire plate and would thus be superimposed upon this stress system. Whether these stress components could be generally recognizable or not would depend upon their relative magnitude in comparison with the local stresses; in Iceland there is room for doubt in this regard.

SYKES and SBAR (1974, p. 222) suggest that the 'radial orientation' of maximum compressive stress in Iceland, as determined from Hast's overcoring measurements, agrees with the stress distribution suggested by Morgan (cf. VOGT 1974) for a mantle plume. This view seems in error, for plume type convection does not seem well supported by the stress data for the following reasons:

(1) Most overcoring measurements are affected by topographic features or near-surface processes; local stress directions thus obtained must be rejected for regional analysis, as discussed previously in greater detail. There is a suggestion of a regional trend involving compression roughly normal to the axial rift zones, although more data are required for substantiation. In any case it cannot be claimed that existing data form a radial trend (Fig. 3).

(2) Near surface stresses in flanks of the axial rift zones but within the crestal zone of the plume as defined by $R = 140$ km are dominantly compressive (Fig. 3), rather than tensile in both radial and tangential directions as predicted by plume theory (Tryggvason 1974b). To be sure, the values of 'tension' given by Tryggvason do not include lateral stress components associated with overburden pressure; thus the 'predicted' tension must be suitably modified (see, e.g., HUBBERT 1951, p. 163), and accordingly absolute tension is unlikely beyond some critical depth. The critical point is therefore not merely that lateral compression exists, but rather that the average lateral compressive stress at shallow depths is greater than can be accounted for by lateral components due to overburden superimposed upon tensile components. Indeed, the average horizontal stress is equal to or greater than full overburden pressure at most shallow measurement sites (Tables 4 and 5).

(3) In addition, radial flow from plumes must induce tangential tension throughout the entire radial flow region; again, these predicted values must be superimposed on lateral stress components due to overburden. The orientation of stresses at Borgarfjördur are consistent with this view (EINARSSON *et al.* 1977), but the measurements at Reykjavik show the tangential stresses to be highly compressive, often in excess of full overburden pressure (Fig. 3; Table 3).

These arguments appear to be basically valid irrespective of whether the center of the plume is taken to be Tungnafellsjökull or Kverkfjöll (TRYGGVASON 1974b, p. 252, SIGVALDASON *et al.* 1974, SIGVALDASON and STEINTHÓRSSON 1974), or whether the model is modified to account for extensional deformation or intrusions for radii $< R$. We cannot conclude from these arguments that no plume exists; but we do suggest that an Icelandic plume does not seem to be effective in dominating the crustal stress pattern as thus far observed. Much the same can be said for driving of the lithosphere plates by conveyor-belt mantle convection. Observed stress orientations are compatible, inasmuch as stresses are at least in a rough way aligned parallel to inferred directions of plate motion. But the large magnitudes of near-surface compression are thus left unexplained and require some other mechanism. The transport mechanisms of plate tectonics are apparently not overriding other mechanisms in influencing the crustal stress pattern.

Forcible wedging of lithospheric plates (LLIBOUTRY 1969, KOIDE and BHATTACHARJI 1975) at the spreading axis does not seem supported by data in Iceland. The microearthquake and surface displacement data for the Reykjanes Peninsula indicate extension virtually throughout the full thickness of lithosphere.[14]) The average lateral stress within the axial rift zone is probably compressive, but its value is certainly less than the weight of overburden. Lithosphere thickness here is probably only on the order of 10 km, and the resulting lateral force cannot be very significant. It is insufficient as a global tectonics driving mechanism.

Finally, PÁLMASON (1973) has introduced a quantitative kinematic model of crustal accretion at an axial zone, based on concepts of plate tectonics, and tested it by comparison with general patterns involving Icelandic data on regional dips of lava beds, heat flow, and dyke distribution as a function of depth (cf. BODVARSSON and WALKER 1964, CANN 1974, SCHÄFER 1975). The model consists of a two-dimensional rift zone column, bounded vertically by rigid lithospheric plates which move apart with a specified drift velocity. Although forces moving the plates were not specified (they were merely assumed to originate in the underlying mantle), the horizontal strain is calculated. This strain is extensional (and therefore was equated

[14]) The upper mantle boundary occurs here at 8–9 km and may represent the lithosphere boundary, because the upper mantle is probably in a state of partial fusion (PÁLMASON and SAEMUNDSSON 1974, p. 40). Earthquakes occur with less frequency in the range 6–9 km, probably because of rock weakness and ductility induced by high temperature and hydrolytic effects (500–1000°C; PÁLMASON and SAEMUNDSSON 1974, p. 37, VOIGHT 1974b, p. 1403); even so these also seem to reflect extension (e.g., KLEIN *et al.* 1973, Fig. 5). Extension is thus the dominant lithospheric deformational mode at the axial zone.

to the intensity of dyke intrusion). The nature of the strain field can be deduced by analysis of lava isochrons and trajectories (see, e.g., PÁLMASON 1973, Fig. 4b, Table 1). Associated stresses would naturally be a function of imposed boundary conditions; although Pálmason's model is purely kinematic and therefore is independent of explicit consideration of stress, the involved strains within the axial zone suggest that the corresponding lateral stresses at any point would ordinarily be less than overburden pressure. This view of extension is supported by the microearthquake data. Beyond the axial zone Pálmason's model assumes the lithosphere to behave in a different (i.e., rigid) manner. Our Reykjavik measurements demonstrate the likelihood of a specific difference in behavior. Thus although stresses as such are not explicitly considered in Pálmason's kinematic model, the model appears to be compatible with observations of the state of stress if allowance is made for thermoelastic effects particularly beyond the axial rift zone.

Possible role of glacio-isostatic plate flexure

The true post-glacial uplift of nearshore areas in Iceland during the past 11000 yr is about 80–160 m, completed by about 9000 yr B.P. (TRYGGVASON 1974a, p. 243, cf. EINARSSON 1966) and uplift in the interior was perhaps as great as 350 m (EINARSSON 1966, p. 164). Lateral stress changes associated with such glacial-induced deflections must have been substantial (see, e.g., BROTCHIE *et al.* 1961; BROTCHIE and SILVESTER 1969), resulting in radial and tangential (flexural) compression within the hinge line, at the surface of the subglacial lithosphere. With the unflexing associated with post-glacial uplift, all glacially-induced flexural stresses could ideally disappear; however, perhaps they may not have completely vanished, for the following reasons:

(1) At the same time that compression occurred at the top of the lithosphere, extension occurred at its base. This stress field could have encouraged formation of brittle fracturing and igneous intrusions; many of these intrusions extended fully to the surface, resulting in sub-glacial volcanism. Zones of extension were thus 'filled', by intrusions or dilatational volume changes, and the associated strains were possibly not fully relieved when the causative glacial load was removed. Compressional strains could thus have been locked into the lithosphere. Because the crust is weaker along the axial rift zones, the zone of intrusions and fracturing would on the whole have been preferentially aligned with this trend. The associated remanant $\sigma_{H\,\max}$ direction should therefore be about perpendicular to the axial rift zone, just as seems indicated by much of available data.

(2) During the 3 m.y. interval in which glacial depression of the lithosphere was active, accretion of mantle to the base of the lithosphere might have been significant. The newly accreted portion of lithosphere could have prevented full release of flexural strains.

This hypothesis perhaps can account at least in part for the magnitudes and orientations of the measured stresses. Continuing spreading-related deformation in the axial rift zone would be superimposed upon a pre-stressed plate; the axial zone itself would thus be largely relieved of glacially-induced residual stress, with the acting stress field dominantly reflecting normal faulting and thinning of the axial rift zone. Whether beyond the axial zone the rocks could still be affected by stress components reflecting the glacial history of the rock mass is however uncertain, particularly inasmuch as glacial flexing and isostatic unflexing occurred numerous times during the Pleistocene; the effectiveness of the mechanisms noted above may have been thereby reduced. To the extent that plate tectonics mechanisms are acceptable to the reader, thermoelastic stress-producing mechanisms can hardly be denied, whereas the appropriate role to be assigned to glacio-isostatics is still uncertain.

On views contrary to plate tectonics models

EINARSSON (1968) has been the chief opponent to crustal drift in Iceland. As pointed out by PÁLMASON and SAEMUNDSSON (1974, p. 43) many of his arguments are no doubt valid in that one should not extrapolate to Iceland the relatively simple models of crustal spreading commonly assumed (not necessarily correctly) for submarine parts of the Mid-Oceanic Ridge. But it also appears that many of his arguments are not incompatible with spreading if the possibilities of shifting of active zones, oblique spreading, and complex transform faults are considered (cf. PÁLMASON and SAEMUNDSSON 1974). Einarsson viewed the main elements of the 'zig-zag' Mid-Atlantic Ridge as conjugate shear planes. He suggested these shears were generated by an E–W principal compression (N–S least compression) in southern Iceland, and by a NE–SW principal compression in northeastern Iceland (EINARSSON 1968, p. 7567); these predictions are not very well substantiated by existing stress measurements (Fig. 3). EINARSSON (1968, p. 7572) himself recognized some aspects of this discrepancy, and suggested that some mechanisms might be depth-dependent, involving a tectonic layer possibly 700 km thick; the latter suggestion has not found support.

In the view of HAST (1969, 1973) the ocean floor behaves as a rigid plate stressed by horizontal and shear forces; these forces are presumed to arise at the contact between ocean floor and continental crust. The Mid-Atlantic Ridge and associated fracture zones are assumed to represent orthogonal shear zones. Hast considered his Iceland data as crucial to this theory. He noted that his measurements of high compression seemed inconsistent with ocean floor spreading, as previously discussed; he argued further that measured principal shear directions were parallel to the active volcanic zones and ridge segments. There are some problems in this interpretation; e.g.,

(1) The Akureyri measurement does not occur at the point at which the Mid-Atlantic Ridge intersects the northern shore, as HAST (1973, p. 416) assumed; maximum compression at this locality is aligned approximately parallel to

the nearest axial rift zone (northern part of the eastern rift zone), not at 45° to it.

(2) Similarly, the Burfell measurement was compared with distantly-located alignments of the rift zones, rather than with closer zones with attitudes not in correspondence to the shear hypothesis (cf. HAST 1973, Figs. 4, 5).

(3) The interpretation of stresses tending to 'shear the whole island right across parallel to the direction of the Mid-Atlantic Ridge and along volcanic areas approximately following it' is not sustained by the data. Indeed the sense of shear seems reversed for north and south Iceland.

Furthermore, the direction of principal stress orientation suggested by Hast for southwest Iceland is not sustained by our Reykjavik measurements. Indeed, we believe that the directions indicated for this and several other points must be rejected for regional tectonic analysis, as discussed previously. Hast does not raise the problem of earthquake focal mechanism solutions.

KEITH (1972) has argued in favor of ocean-floor convergence and describes the Mid-Ocean Ridge as compressional and isostatically-supported by the buoyancy of a thickened wedge of light material formed by intermixture of crust and mantle. He follows EINARSSON (1967) in interpreting Icelandic flexures as reflecting compression. Our H18 Reykjavik measurements could perhaps be cited in support of this interpretation, but the H32 data and Borgarfjördur focal solutions do not support it, particularly in view of the thin lithosphere in Iceland. Keith conceded that earthquake focal mechanism solutions from the axial rift and fracture zones support ocean floor spreading rather than convergence. He has questioned the interpretation of focal mechanism solutions within axial rift zones in general, inasmuch as these are not supportive of his hypothesis.

4. *Conclusions*

Although the small scale effects of local structural and thermal heterogeneity have yet to be thoroughly investigated, the hydrofracturing stress measurements at Reykjavik suggest a dominant regional orientation of $\sigma_{H\max}$ approximately perpendicular to the axial rift zone (Fig. 3). This orientation is furthermore supported by shallow overcoring measurements in southeast Iceland, and by recent earthquake focal mechanism solutions for the Borgarfjördur intraplate area. The hydrofracturing measurements indicate a depth dependence of stress environment, with associated changes in principal stress orientation. Conditions of high lateral stress are represented in hole H18, probably analogous to the large compressive stresses recorded in shallow overcoring experiments. Stress conditions analogous to regimes of normal faulting are also represented (H32); this condition, and its orientation, are both consistent with the Borgarfjördur data. Perhaps similar data could be obtained from deeper measurements at H18.[15]) The flanks of the axial rift zone thus appear mainly characterized by trajectories of $\sigma_{H\,\min}$ parallel to isochrons. The associated principal stresses may however vary as a function of spatial position, relative to latitude, longitude, and depth, and large lateral compression may exist, particularly at shallow crustal levels.

This state of stress is fundamentally different from that in the axial rift zones themselves. In the rift zones, $\sigma_{H\,\min}$ is consistently aligned perpendicular to individual rift zone fissures and faults; $\sigma_{H\,\min}$ may thus be oblique to the gross orientation of the rift zone itself, where (as on the Reykjanes Peninsula) the zone seems best described as an obliquely-spreading ridge or a leaky transform fault.

The existing data seem consistent with the hypothesis that thermal stress components are introduced as a function of spreading, due to both axial zone lithospheric accretion and cooling, and basal lithospheric accretion and cooling. The notion that fracture zones and transform faults are initiated by thermal contraction seems strengthened. The data do not offer support to the plume hypothesis as an important driving mechanism in plate tectonics. Forcible separation of lithospheric plates at the axial zone itself seems contraindicated, inasmuch as axial zone extension prevails throughout the extent of lithosphere, the base of which cannot be much below the crust mantle boundary at about 10 km depth. As thus interpreted, thermoelastic stresses are induced as a function of spreading from the axial zone, and could become dominant stress components in lithosphere at and beyond the rift zone flank. If so, it may be insufficient to attribute observed intraplate stresses only to plate tectonic forces presently acting on the edges and bases of lithospheric plates, as attempted in recent model studies (SOLOMON *et al.* 1975, RICHARDSON *et al.* 1976). RICHARDSON *et al.* (1976, p. 1848) recognized a need for some caution in this regard, but concluded that 'there are grounds for believing that the effects of such additional stress-producing mechanisms either are minor or can be minimized by scrutiny of the stress observations chosen to compare against the predictions of force models'. This view is rather optimistic, if our assessment of the Icelandic data is correct.

Acknowledgements

We are extremely grateful to the Icelandic National Energy Authority and in particular Dr. Pálmason, who helped organize the research program and contributed

[15]) We emphasize that further studies in Iceland are clearly necessary in order to test and expand upon the interpretations presented here. Existing data do not permit a complete assessment of possible smaller-scale influences on the stress field, such as local thermoelastic or structural effects; the possible regional role of glacio-isostatics remains poorly understood. Further hydrofracturing studies in selected holes at different locations should help to provide answers to some of these specific questions. Our study was restricted to 0.4 km simply by available drilling equipment; future studies should make more complete use of the much greater depth available in existing boreholes and perhaps employ new boreholes drilled specifically for geophysical tests. Such future studies could provide invaluable information on lithospheric stress variation as a function of surface position and depth. The implications of the Icelandic data are, as we and others have attempted to suggest, far-reaching.

to it in many ways. During our stay in Iceland we worked most closely with Mr. Thorsteinn Thorsteinsson, whose generous assistance proved invaluable. Thanks are also due to the Icelandic National Research Council for permitting us to carry out research in Iceland. Discussions or field excursions with many individuals contributed to our understanding of Icelandic problems; the following deserve special mention: Jens and Haukur Tómasson, Sigurdur Saemundsson, Jon Jónsson, Ingvar Fridleiffson, Svein Jakobsson, Sigurdur Thorarinsson, George Walker, MacKenzie Keith, Wolf Jacoby, Nils Hast, and Charles Sammis.

Special thanks are due to University of Wisconsin graduate students J. Avasthi and L. Cheung who assisted with the field testing.

This research was supported by the Division of Earth Sciences of the National Science Foundation, NSF Grant EAR76–03821. Preliminary Studies were supported in 1974 by the Faculty Research Fund of the Pennsylvania State University and by the University of Wisconsin.

References

Björnsson, S. and Einarsson, P. (1974), *Seismicity of Iceland* in *Geodynamics of Iceland and the North Atlantic Area* (ed. L. Kristjansson) (D. Reidel Publishing Company, Dordrecht-Holland/Boston, U.S.A., 1974).

Bodvarsson, G. and Walker, G. P. L. (1964), *Crustal drift in Iceland*, Geophys. J. Roy. Astron. Soc. *8*, 285–300.

Brander, J., Mason, R. G. and Calvert, R. W. (1976), *Precise distance measurements in Iceland*, Tectonophysics *31*, 193–206.

Brotchie, J. F., Penzien, J. and Popov, E. P. (1961), *Analysis of stress concentrations in thin rotational shells of linear strain hardening material*, Res. Dept. 100–111, Inst. Eng. Res., Univ. Calif., Berkeley.

Brotchie, J. F. and Silvester, R. (1969), *On crustal flexure*, J. Geophys. Res. *74*, 5240–5252.

Cann, J. (1974), *A model for oceanic crustal structure developed*, Geophys. J. Roy. Astron. Soc. *39*, 187–196.

Collette, B. J. (1974), *Thermal contraction joints in a spreading seafloor as origin of fracture zones*, Nature *251*, 299–300.

Einarsson, Tr. (1954), *A survey of gravity in Iceland, Rit 30, Reykjavik*, Soc. Sci. Island. 22 p.

Einarsson, Tr. (1966), *Late and post-glacial rise in Iceland and sub-crustal viscosity*, Jökull *16*, 157–166.

Einarsson, Tr. (1967), *The Icelandic fracture system and the inferred causal stress field* in *Iceland and Mid-Ocean Ridges, Rit. 38, Reykjavik* (ed. S. Björnsson) Soc. Sci. Island. 209 p.

Einarsson, Tr. (1968), *Submarine ridges as an effect of stress fields*, J. Geophys. Res. *73*, 7561–7576.

Einarsson, P., Klein, F. W. and Björnsson, S. (1977), *The Borgarfjördur Earthquakes of 1974 in West Iceland*, Bull. Seism. Soc. of Am., Vol. 67, 187–208.

Fridleifsson, I. B. (1973), *Petrology and structure of the esja quaternary volcanic region, Southwest Iceland*, D. Phil. Thesis, Oxford, 208 p.

Fridleifsson, I. B. (1975), *Lithology and structure of geothermal reservoir rocks in Iceland*, Proc. 2nd U.N. Symp. Development and Use of Geothermal Resources, 21 p.

Haimson, B. C., *Earthquake related stress at Rangely, Colorado* in *New Horizons in Rock Mechanics*, (Proc. Fourteenth Symp. on Rock Mechanics) (eds. H. R. Hardy and R. Stefanko) (Am. Soc. of Civil Engineers, N.Y., 1973), p. 689–708.

Haimson, B. C. (1974), *A simple method for estimated in-situ stresses at great depths* in *Field Testing and Instrumentation of Rock*, Am. Soc. for Testing and Materials Special Tech. Publ. *554*, 156–182.

Haimson, B. C. (1975a), *The state of stress in the earth's crust*, Rev. Geophys. Space Phys. *13*, 350–352.

Haimson, B. C. (1975b), *Deep in situ stress measurements by hydrofracturing*, Tectonophysics *29*: 1–4, 41–47.

Haimson, B. C. (1976a), *The hydrofracturing stress measuring technique – methods and recent field results in the U.S.*, Proceedings Intl. Symposium on Investigation of Stress in Rock, Institute of Engineering, Sydney, p. 23–30.

HAIMSON, B. C. (1976b), *Crustal stress in the continental United States as derived from hydrofracturing tests*, Proc. Symp. on Physical Properties of the Earth's Crust, Vail, Colorado, in press.

HAIMSON, B. C. (1976c), *Preexcavation deep-hole stress measurements for design of underground chambers – case histories*, Proc. – 1976 Rapid Excavation and Tunnelling Conference (eds. R. J. Robbins and R. J. Conlon), Soc. of Mining Engineers of AIME, New York, p. 699–721.

HAIMSON, B. C. and AVASTHI, J. M. (1975), *Stress measurements in anisotropic rock by hydraulic fracturing* in *Applications of Rock Mechanics*, Proc. 15th Symp. on Rock Mechanics (ed. E. R. Hoskins Jr.), Am. Soc. of Civil Engineers, N.Y., p. 135–156.

HAIMSON, B. and FAIRHURST, C. (1970), *In-situ stress determination at great depth by means of hydraulic fracturing* in *Rock Mechanics—Theory and Practice*, (Proc. Eleventh Symp. on Rock Mechanics), (ed. W. H. Somerton), American Inst. of Mining Engineers, N.Y., p. 559–584.

HAIMSON, B. C., LACOMB, J., GREEN, S. J. and JONES, A. H. (1974), *Deep stress measurements in tuff at the Nevada test site*, 3rd Int'l. Congress in Rock Mechanics, Denver, Proceedings, Vol. II, p. 557–562.

HAIMSON, B. C. and VOIGHT, B. (1976), *Stress measurements in Iceland* (abstract), EOS Trans. AGU, *57*, 1007.

HAST, N. (1969), *The state of stress in the upper part of the earth's crust*, Tectonophysics *8*, 169–211.

HAST, N. (1973), *Global measurements of absolute stress*, Phil. Trans. Roy. Soc. London *274*, 409–419.

HEIDECKER, E. (1968), *Rock pressures induced by weathering and physicochemical processes*, Proc. Austral, Inst. Min. Met. *226*, 43–45.

HOOKER, V. E. and DUVALL, W. I. (1971), *In situ rock temperature. Stress investigations in rock quarries*, U.S. Bur. Mines Dept. Inv. 7589, 12p.

HUBBERT, M. K. (1951), *Mechanical basis for certain familiar geologic structures*, Geol. Soc. Am. Bull. *62*, 355–372.

JÓNSSON, J. (1967), *The rift zone and the Reykjanes Peninsula* in *Iceland and Mid-Ocean Ridges, Rit 38, Reykjavik* (ed. S. Björnsson) Soc. Sci. Island., 209 p.

KEITH, M. L. (1972), *Ocean floor convergence: a contrary view of global tectonics*, J. Geol. *80*, 249–276.

KJARTANSSON, G. (1960), *Geological map of Iceland, sheet 1, Reykjavik*, Mus. Natur. Hist.

KLEIN, F. W., EINARSSON, P. and WYSS, M. (1973), *Microearthquakes on the Mid-Atlantic plate boundary on the Reykjanes Peninsula in Iceland*, J. Geophys. Res. *78*, 5084–5099.

KLEIN, F. W., EINARSSON, P. and WYSS, M. (1977), *The Reykjanes Peninsula, Iceland, earthquake swarm of September 1972 and its tectonic significance*, Journal of Geophysical Research, vol. 82, 865–888.

KOIDE, H. and BHATTACHARJI, S. (1975), *Mechanistic interpretation of rift valleys*, Science *189*, 791–793.

KRISTJANNSSON, L., *Geodynamics of Iceland and the North Atlantic Area* (D. Reidel, Dordrecht, 1974), 324 pp.

LLIBOUTRY, L. (1969), *Sea-floor spreading, continental drift and lithosphere sinking with an asthenosphere at melting point*, J. Geophys. Res. *74*, 6525–6540.

LOHMAN, S. W. (1961), *Compression of elastic artesian aquifers*, U.S. Geol. Survey Prof. Paper 424–B, p. 47.

NAKAMURA, K. (1970), *En echelon features of Icelandic ground fissures*, Acta Natur. Island, *II*(8), 15 p.

PÁLMASON, G. (1971), *Crustal structure of Iceland from explosion seismology, Rit 40, Reykjavik*, Soc. Sci. Island, 18 pp.

PÁLMASON, G. (1973), *Kinetics and heat flow in volcanic rift zone with application to Iceland*, Geophys. J. Roy. Astron. Soc. *33*, 451–481.

PÁLMASON, G. *Heat flow and hydrothermal activity in Iceland* in *Geodynamics of Iceland and the North Atlantic Area* (ed. L. Kristjannson) Proc. NATO Advanced Study Inst. in Reykjavik, 1974, (D. Reidel, Dordrecht, 1974), p. 297–306.

PÁLMASON, G. (1975), *Geophysical methods in geothermal exploration*, Proc. 2nd U.N. Symp. Development and Use of Geothermal Resources, 37 p.

PÁLMASON, G. and SAEMUNDSSON, K. (1974), *Iceland in relation to the Mid Atlantic Ridge*, Ann. Rev. Earth Plan. Sci. *2*, 25–50.

PITMAN, W. C. and TALWANI, M. (1972), *Sea-floor spreading in the North Atlantic*, Geol. Soc. Am. Bull. *83*, 619–646.

RANALLI, G. and CHANDLER, T. E. (1975), *The stress field in the upper crust as determined from in-situ measurements*, Geol. Rundschau *64*, 653–674.

RICHARDSON, R. M., SOLOMON, S. C. and SLEEP, N. H. (1976), *Intraplate stress as an indicator of plate tectonic driving forces*, J. Geophys. Res. *81*, 1847–1856.

RUMMEL, F. and JUNG, R. (1975), *Hydraulic fracturing stress measurements near the Hohenzollern-Graben-structure, SW Germany*, Pure appl. Geophys. *113*, 321–330.

SAEMUNDSSON, K. (1967), *An outline of the structure of SW Iceland, Iceland and Mid-Ocean Ridges*, Soc. Sci. Islandica publ. 38 (ed. S. Björnsson), p. 151.

SAEMUNDSSON, K. (1974), *Evolution of the axial rifting zone in northern Iceland and the Tjörnes fracture zone*, Bull. Geol. Soc. Am. *85*, 495.

SAEMUNDSSON, K. and NOLL, H. (1975), *K: Ar ages of rocks from Husafell in western Iceland and the development of the Husafell central volcano*, Jökull *24*.

SBAR, M. L. and SYKES, L. R. (1973), *Contemporary compressive stress and seismicity in eastern North America: an example of intra-plate tectonics*, Geol. Soc. Am. Bull. *84*, 1861–1882.

SCHÄFER, K. (1972), *Transform faults in Iceland*, Geol. Rundschau *61*, p. 942.

SCHÄFER, K. (1975), *Horizontal and vertical crustal movements in Iceland*, Tectonophysics *29*, 223–231.

SCHOLZ, C. H., WYSS, M. and SMITH, S. W. (1969), *Seismic and aseismic slip on the San Andreas fault*, J. Geophys. Res. *74*, 2049.

SHOEMAKER, E. M. (ed), *Continental drilling, Rep. 1974 Workshop on Continental Drilling at Ghost Ranch, N.M.* (Carnegie Inst., Washington, 1975), 56 pp.

SIGURDSSON, H. (1970), *Structural origin and plate tectonics of the Snaefellsnes volcanic zone, western Iceland*, Earth Planet. Sci. Letters *10*, 129.

SIGVALDASON, G. E. and STEINTHÓRSSON, S., *Chemistry of tholeiitic basalts from Iceland and their relation to the Kverkfjöll hot spot* in *Geodynamics of Iceland and North Atlantic Area* (ed. L. Kristjansson) (Reidel, Dordrecht–Holland, 1974), p. 155–164.

SIGVALDASON, G. E., STEINTHÓRSSON, S., OSKARSSON, N. and IMSLAND, P. (1974), *Compositional variation in recent Icelandic tholeiites and the Kverkfjöll hot spot*, Nature *251*, 579–582.

SOLOMON, S. C., SLEEP, N. H. and RICHARDSON, R. M. (1975), *On the forces driving plate tectonics: inferences from absolute plate velocities and intraplate stress*, Geophys. J. Roy. Astron. Soc. *42*, 769–801.

STEFANSSON, R. (1966), *Methods of focal mechanism studies with application on two Atlantic earthquakes*, Tectonophysics *3*, 210–243.

SYKES, L. R. (1967), *Mechanism of earthquakes and nature of faulting on the mid-oceanic ridges*, J. Geophys. Res. *72*, 2131.

SYKES, L. R. and SBAR, M. L. (1974), *Focal mechanism solutions of intraplate earthquakes and stresses in lithosphere* in *Geodynamics of Iceland and North Atlantic Area* (ed. L. Kristjansson) (Reidel, Dordrecht–Holland, 1974), p. 207–227.

TALWANI, P. (1977), *Stress distribution near Lake Jocassee, South Carolina*, Pure appl. Geophys. this volume.

THORSTEINSSON, T. and ELIASSON (1970), *Geohydrology of the Laugarnes hydrothermal system in Reykjavik, Iceland*, Geothermics, Spec. Issue 2, p. 1191–1204.

TÓMASON, H. (1967), *Jardsraedirannsoknir Virkjunarstadarins vid Búrfell, Timariti Verkfraedingafelags Islands*, 3. 6. Hefti 1966 Reykjavik, Steindorsprent H.F.

TÓMASSON, J., FRIDLEIFSSON, I. B. and STEFÁNSSON, V. (1975), *A hydrologic model for the flow of thermal water in SW Iceland with a special reference to Reykir and Reykjavik thermal areas*, Proc. 2nd U.N. Symp. Development and use of Geothermal Resources, 15 p.

TRYGGVASON, E. (1973), *Seismicity, earthquake swarms and plate boundaries in the Iceland region*, Bull. Seis. Soc. Am. *63*, 1327.

TRYGGVASON, E., *Vertical crustal movement in Iceland* in *Geodynamics of Iceland and North Atlantic Area*, (ed. L. Kristjansson) (Reidel, Dordrecht–Holland, 1974a), p. 241–262.

TRYGGVASON, E., *Surface deformation in Iceland and crustal stress over a mantle plume* in *Continuum Mechanics Aspects of Geodynamics and Rock Fracture Mechanics* (ed. P. Thoft-Christensen) (Reidel, Dordrecht–Holland, 1974b) p. 245–254.

TURCOTTE, D. L. (1974), *Are transform faults thermal contraction cracks?*, J. Geophys. Res. *79*, 2573–2577.

VOGT, P. R., *The Iceland Phenomenon: imprints of a hot spot on the ocean crust and implications for flow beneath plates* in *Geodynamics of Iceland and North Atlantic Area* (ed. L. Kristjansson) (Reidel, Dordrecht–Holland, 1974), p. 105–126.

VOIGHT, B. (1969), *Evolution of North Atlantic Ocean: Relevance of rock pressure measurements in North Atlantic – Geology and Continental Drift*, Am. Assoc. Petrol. Geol. Mem. *12*, 955–962.

VOIGHT, B. (1970), *Prediction of 'in situ' stress patterns in the earth's crust*, Int'l. Soc. Rock Mec., Determination of Stresses in Rock Masses Symposium, Lisbon, 1969, 111–131.

VOIGHT, B. (1974a), *A mechanism of 'locking-in' orogenic stress*, Am. J. Sci. *274*, 662–665.

VOIGHT, B. (1974b), *Deformable-plate tectonics: ductile deformation of old and new lithosphere*: Am. Assoc. Petrol. Geol. Bull. *58*, 1403–1406.

VOIGHT, B. and ST. PIERRE, H.P.B. (1974), *Stress history and rock stress*, Proc. 3rd congress, Int. Soc. Rock Mec., Denver.

VOIGHT, B., TAYLOR, J. W. and VOIGHT, J. P. (1969), *Tectonophysical implications of rock stress measurements*, Geol. Rundschau *58*, 655–676.

WALKER, G. P. L. (1964), *Geological investigations in eastern Iceland*, Bull. Volcanol. *27*, 1–15.

WALKER, G. P. L. (1974), *The structure of eastern Iceland* in *Geodynamics of Iceland and North Atlantic Area* (ed. L. Kristjansson) (Reidel, Dordrecht–Holland, 1974), pp. 177–188.

WARD, P. L. (1971), *New interpretation of the geology of Iceland*, Geol. Soc. Am. Bull. *82*, 2991–3012.

WARD, P. L. and BJÖRNSSON, S. (1971), *Microearthquakes, swarms and the geothermal areas of Iceland*, J. Geophys. Res. *76*, 3953.

WARD, P. L., PÁLMASON, G. and DRAKE, C. (1969), *Microearthquake survey and the Mid-Atlantic Ridge in Iceland*, J. Geophys. Res. *74*, 665–684.

ZOBACK, M. D., HEALY, J. H. and ROLLER, J. C. (1977), *Preliminary stress measurements in central California using the hydraulic fracturing technique*, Pure appl. Geophys. this volume.

(Received 3rd December 1976)

Pageoph, Vol. 115 (1977), Birkhäuser Verlag, Basel

Shear and Tension Hydraulic Fractures in Low Permeability Rocks

By P. Solberg,[1]) D. Lockner[1]) and J. Byerlee[1])

Abstract – Laboratory hydrofracture experiments were performed on triaxially stressed specimens of oil shale and low-permeability granite. The results show that either shear or tension fractures could develop depending on the level of differential stress, even in specimens containing preexisting fractures. With 1 kb of confining pressure and differential stress greater than 2 kb, hydraulic fluid diffusion into the specimens reduced the effective confining pressure until failure occurred by shear fracture. Below 2 kb of differential stress, tension fractures occurred. These results suggest that hydraulic fracturing in regions of significant tectonic stress may produce shear rather than tension fractures. In this case *in situ* stress determinations based on presumed tension fractures would lead to erroneous results.

Key words: Stress hydrofracture; Hydrofracture technique.

1. *Introduction*

The use of hydraulic fracture data for *in situ* stress measurements has become increasingly popular because of their simplicity and applicability to a wide variety of situations. The hydraulic fracturing technique, which was developed in 1947 for secondary oil recovery, consists of sealing off a section of a borehole and then pressurizing it with fluid until fracture occurs. Hubbert and Willis (1957) published the first theoretical study of the relations between hydraulically induced tension fractures and surrounding *in situ* stresses. Scheidegger (1962), Kehle (1964), and Fairhurst (1964) presented further theoretical refinements. These workers recognized the effect of pore pressure in the surrounding rock formation on hydraulic fracturing, and noted that fracture fluid penetration into the rock away from the borehole could affect results. Gretener (1965) criticized the use of hydraulic fracturing data for determining *in situ* stresses, noting that the distribution of fracture fluid around the borehole wall significantly affected breakdown pressure and that the effect was virtually impossible to assess accurately. Haimson (1968) presented the first theoretical analysis of the effect of fracture fluid penetration, which predicted breakdown pressures lower than those calculated from the simple elastic model of Hubbert and Willis. Haimson (1968), Haimson and Fairhurst (1969), Haimson and Stahl (1970), and Haimson and Edl (1972) conducted laboratory hydraulic fracturing experiments

[1]) U.S. Geological Survey, Menlo Park, CA, USA.

to test the theoretical hypotheses. They conclude that all fractures produced were tension fractures and that experimental results were in accord with theoretical predictions. LAMONT and JESSEN (1963) conducted early experiments on the effects of existing fractures on the extension of hydraulic fractures. More recently ZOBACK *et al.* (1976) have found it possible to nullify the effects of existing fractures by the use of high-viscosity drilling mud but note that this may cause anomalously high breakdown pressures.

The possibility of failure by shear, rather than tension, in hydraulic fracturing experiments has not received adequate attention. PAULDING (1968) presented a theoretical discussion of the shear failure possibility in hydraulic fracturing. LOCKNER and BYERLEE (1977), in a series of laboratory experiments on the hydraulic fracturing of Weber Sandstone, found that either shear or tension fractures were produced depending on the differential stress, the rock permeability, and the viscosity and injection rate of the fracturing fluid. If shear fractures are formed during field hydraulic fracturing experiments designed to measure *in situ* stresses, serious errors could result if failure under tension is assumed. We have conducted the following additional laboratory hydraulic fracturing experiments with a low permeability granite and an oil shale to investigate this problem.

2. Experimental methods

The experimental method employed in this study is the same as that used by LOCKNER and BYERLEE (1977). All samples were machined into cylinders 2.54 cm in diameter and 6.35 cm in length. A 0.25 cm diameter hole was drilled along the axis to accept the hydraulic fracturing fluid. Each sample was subjected to a confining pressure of 1 kb and an additional axial load of from 1 to 7 kb to simulate hydrostatic loading and tectonic stresses. Fluid was then injected into the axial hole at 3.3×10^{-6} cc/s to 3.3×10^{-3} cc/s using a constant-flow-rate pump. Shell Tellus Oil No. 15 was used as the fracturing fluid for all experiments. Its viscosity ranges from 20 cp at atmospheric pressure to 90 cp at 1kb of pressure. Failure was indicated by a drop in borehole pressure.

Westerly Granite, with a permeability of 3.5×10^{-8} darcies at 1 kb confining pressure (BRACE *et al.* 1968), was used because of its physical homogeneity and relatively well known physical and mechanical properties. Eocene oil shale from the Mahogony Zone of the Green River Formation was supplied by the Rifle Oil Shale Facility of the Energy Resource and Development Agency, Rifle, Colorado. It is referred to as oil shale in this study. Oil shale was selected because it contains no measurable porosity or permeability at room temperature and pressure (J. WARD SMITH, personal communication). It is much weaker than Westerly Granite. Several oil shale specimens contained preexisting fractures that were oriented either parallel to the borehole axis or parallel to the ends of the specimen. It was decided to include these specimens in order to evaluate the effect of preexisting fractures on the hydraulic fracture mechanism.

3. Experimental results

Five Westerly Granite and nine oil shale specimens were hydraulically fractured under triaxial conditions at differential stresses between 1 and 6.9 kb. At differential stresses greater than 2 kb specimens failed by shear fracture, whereas at lower differential stress they failed in tension. This relation held even in oil shale specimens that contained preexisting fractures.

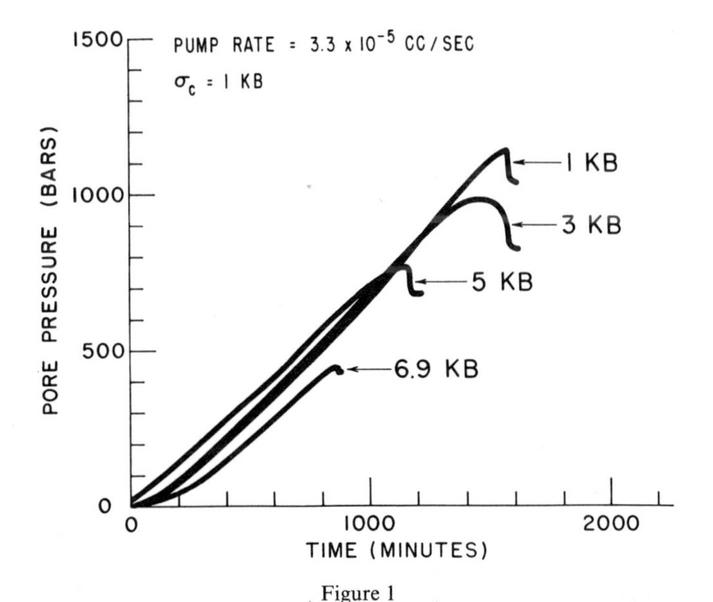

Figure 1
Breakdown curves for four experiments conducted on Westerly Granite at 1 kb confining pressure and an injection rate of 3.3×10^{-5} cc/s. Differential stress ranged from 1 to 6.9 kb.

The experiments on Westerly Granite shown in Fig. 1 were conducted at 1 kb confining pressure and a constant hydraulic fluid injection rate of 3.3×10^{-5} cc/s. The specimens loaded to 3, 5, and 6.9 kb differential stress failed by shear fracture, and increased levels of differential stress resulted in decreased breakdown pressures. The shear strength determined in these experiments is closely in accord with that determined by MOGI (1966) and BYERLEE (1967) if confining pressure is replaced with the confining pressure minus the borehole pressure. Thus, consistent with the effective stress concept, it seems that at failure the pore pressure equalled the borehole pressure in the region of fracture propagation. The Westerly samples which failed in shear were found to contain incipient fractures surrounding the borehole. HAIMSON and EDL

(1972) have shown experimentally that when the differential stress exceeds the unconfined compressive strength, localized shear fracture occurs at the borehole wall even before fluid injection. The specimen at 1 kb differential stress failed by tension fracture at a fluid pressure of 1110 bars.

The hydraulic pressure-breakdown curves for the three oil shale samples run at 3.3×10^{-5} cc/s injection rate are plotted in Fig. 2. The specimen loaded to 2.5 kb differential stress failed by shear fracture, whereas those loaded to 1 and 2 kb failed in tension. Results from specimens run at injection rates of 3.3×10^{-4} and 3.3×10^{-3} cc/s are consistent with those run at the slower injection rate. HAIMSON and FAIRHURST (1969) have shown experimentally that increased injection rates result in increased

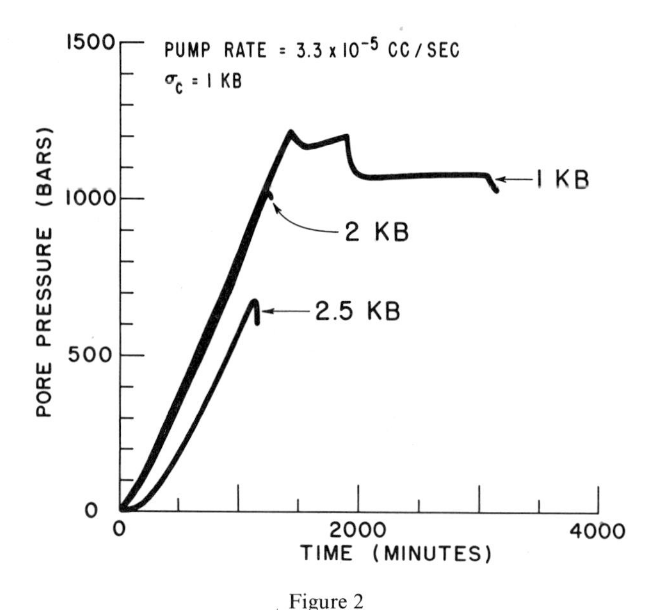

Figure 2
Breakdown curves for three experiments conducted on oil shale at 1 kb confining pressure and an injection rate of 3.3×10^{-5} cc/s. Differential stress ranged from 1 to 2.5 kb.

breakdown pressures. ZOBACK *et al.* (1976) in related work suggest that anomalously high breakdown pressures at high injection rates may be due to pressure losses at the propagating crack tip. Apparently, at the slow injection rates used in our experiments such pressure losses and anomalous breakdown pressures do not occur.

Both the oil shale specimens loaded to 2 and 2.5 kb differential stress contained preexisting fractures that intersected the borehole. The specimen run at 2 kb differential stress failed at a borehole pressure of 1020 bars along the axially oriented fracture. The crack reopened as hydraulic fluid pressure exceeded the 1 kb confining pressure. The specimen run at 2.5 kb differential stress contained a preexisting crack parallel to the sample end but failed at a borehole pressure of 680 bars by shear fracture.

The sample loaded to 1 kb differential stress failed in tension at 1217 bars. It did not contain any preexisting cracks. The specimen exhibited multiple breakdown pressures (Fig. 2) and was found to contain several small tension fractures. The opening of these may have caused transient pressure drops as the pump rate was temporarily exceeded by the ability of the fracture to accept fluid. Repetition of such a process could cause the apparent multiple breakdowns.

4. Discussion

Regardless of the fracture type (shear or tension), hydraulic fluid diffusion from the borehole into our samples apparently took place in all cases, even in seemingly impermeable oil shale. This could be due to increased permeability with increased effective differential stress. ZOBACK and BYERLEE (1975) studied the relations between permeability and differential stress for Westerly Granite and found that before failure the permeability increased by approximately a factor of 3, owing to dilatant microcrack opening. Presumably the same phenomenon occurs in our samples. Virgin oil shale may seem impermeable because pores and preexisting microcracks are filled with high-viscosity oil, but increased differential stress and resulting crack opening of even small magnitude could provide many new pathways for the hydraulic fluid and dramatically increase permeability.

Samples that failed in tension did so at significantly less than twice the confining pressure, the theoretically predicted breakdown pressure for impermeable material using Kirsh's solution given by TIMOSHENKO and GOODIER (1951) and applied by HUBBERT and WILLIS (1957). GRETENER (1965) suggested that in permeable rocks hydraulic fluid diffusion away from the borehole increases the pore pressure, thus reducing the effective confining pressure and the breakdown pressure for tensile failure. HAIMSON (1968) has shown theoretically how the breakdown pressure could decrease due to fluid penetration, and HAIMSON and EDL (1972) published experimental results verifying the theoretical expectations.

At the increased differential stresses at which shear fracture occurs samples become increasingly dilatant and hence permeable to the fracturing fluid. Fluid diffuses throughout the sample, which at some stage can no longer support the differential stress. Shear fracture occurs when the rock's shear strength is exceeded. It is well known that shear strength is a function of effective pressure (confining pressure minus pore pressure). As effective pressure increases, shear strength increases. Thus at a fixed confining pressure, as in our experiments, an increase in pore pressure (due to hydraulic fluid diffusion from the borehole) will cause a decrease in shear strength and therefore in breakdown pressure when failure occurs by shear.

The oil shale sample loaded to 2 kb differential stress failed by the reopening of a preexisting macroscopic fracture at a fluid pressure of 1020 bars. As stated previously, Kirsh's solution predicts that the 1 kb confining pressure would cause a 2 kb circumferential concentration of stress at the borehole wall, and one might expect that 2 kb fluid pressure would be necessary to reopen the crack. Once formed, however, a macroscopic fracture may be propped open by loosened grains on the fracture surface. Hydraulic fluid would need to flow into the crack only a short distance before the stress concentration would diminish and fluid pressure could overcome the normal stress, opening the crack. ZOBACK *et al.* (1976) have found that in some cases preexisting fractures become leaky at substantially lower fluid pressures than the normal stress holding the crack closed. In a similar manner the preexisting crack in the oil shale specimen loaded to 2.5 kb differential stress may have acted as an extended source of hydraulic fluid. Fluid could permeate out into the sample through the crack as well as through the borehole wall. Thus the presence of preexisting fractures tends to increase the overall permeability and favor the production of shear rather than tension fractures at sufficiently high levels of differential stress.

In regions with high tectonic stresses, such as those simulated in our experiments, the existing pore pressure may be sufficient to cause shear fracturing even before fracture fluid injection. If not, but with differential stress levels greater than the unconfined compressive strength of the rock, fracture fluid diffusion away from the borehole may raise the pore pressure sufficiently to cause shear fracturing. Field hydraulic fracturing experiments designed to measure *in situ* stresses commonly use more rapid injection rates than those of our experiments to insure the production of tension rather than shear fractures. The experimental procedure consists of injecting the fracturing fluid into the borehole until a drop in pressure indicates fracture initiation. Pumping is continued to propagate the fracture and then stopped to record the instantaneous shut-in pressure. During shut-in, the reservoir of fracture fluid trapped in the fracture slowly diffuses out into the surrounding rock. Therefore, even if a tension fracture is produced initially, fluid diffusion out into the rock during shut-in might raise the pore pressure enough to cause shear fracture in tectonically active areas.

Unrecognized shear fracturing may have occurred in hydraulic fracturing experiments in the field. AAMODT (1974) discussed *in situ* stress determinations from the hydraulic fracturing of a 758 meter drillhole in northern New Mexico for the Los Alamos Dry-Hot-Rock Geothermal Energy Program. He concluded that *two* anomalous pressure drops after shut-in were the result of closing hydraulically produced vertical and horizontal tension fractures. An equally reasonable explanation however, is that only one large tension fracture was produced and that later pressure drops were the result of shear fracturing due to fluid diffusion from the tension fracture after shut-in. Recent seismic evidence from Los Alamos (ALBRIGHT and HANOLD 1976) suggests that this may well be so. Although not a hydraulic fracturing experiment designed to measure *in situ* stresses, hydraulically induced fractures at the Rocky Mountain Arsenal disposal well were originally assumed to be tension fractures, but later seismic studies (HEALY *et al.* 1968) indicated failure in shear due to massive fluid diffusion.

5. *Conclusions*

This study suggests that even in very low permeability rocks such as granite or oil shale hydraulic fracturing may produce shear rather than tension fractures if significant tectonic stresses exist, regardless of the presence of preexisting fractures. The measured breakdown pressures and orientations of hydraulically produced tension fractures in our experiments are in accord with the results of previous workers (HAIMSON 1968, HAIMSON and FAIRHURST 1969, HAIMSON and STAHL 1970, HAIMSON and EDL 1972). New fractures are oriented in axial planes containing the borehole, as expected. If tension fractures are produced in the field, these results indicate that *in situ* stress determinations based on the principles developed by HUBBERT and WILLIS (1957) are useful. But under differential stresses greater than 2 kb all samples failed by shear rather than tension fracture. Unfortunately, the differentiation between shear and tension fractures in deep boreholes is difficult with present day technology. Misidentification of fracture type in field studies could lead to gross errors in the determination of both orientation and magnitude of *in situ* stresses.

REFERENCES

AAMODT, R. L. (1974), *An experimental measurement of in situ stress in granite by hydraulic fracturing*, L.A.S.L. in house publication LA-5605-MS, 1–4.

ALBRIGHT, J. N. and HANOLD, R. J., *Seismic mapping of hydraulic fractures made in basement rocks* in *E.R.D.A. 2nd Annual Symp. on Enhanced Oil and Gas Recovery* (Tulsa, Oklahoma 1976).

BRACE, W. F., WALSH, J. B. and FRANGOS, W. T. (1968), *Permeability of granite under high pressure*, J. Geophys. Res. *73*, 2225–2236.

BYERLEE, J. D. (1967), *Frictional characteristics of granite under high confining pressures*, J. Geophys. Res. *72*, 3639–3648.

FAIRHURST, C. (1964), *Measurement of in situ rock stresses, with particular reference to hydraulic fracturing*, Rock, Mech. and Engineering Geol. *2*, 129–147.

GRETENER, P. E. (1965), *Can the state of stress be determined from hydraulic fracturing data?* J. Geophys. Res. *70*, 6205–6212.

HAIMSON, B. C. (1968), *Hydraulic fracturing in porous and nonporous rock and its potential for determining in situ stresses at great depth*, PhD. Thesis, University of Minnesota.

HAIMSON, B. C. and FAIRHURST, C. (1969), *Hydraulic fracturing in porous-permeable materials*, AIME Petrol. Trans. 811–817.

HAIMSON, B. C. and FAIRHURST, C. (1970), *In-situ stress determination at great depth by means of hydraulic fracturing* in *Rock Mech. – Theory and Practice, Proc. of 11th Symposium on Rock Mechanics* (ed. Somerton, Soc. Mining Engineers of AIME), 559–584.

HAIMSON, B. C. and STAHL, E. J. (1970), *Hydraulic fracturing and the extraction of minerals through wells* in *3rd Symp. on Salt, Northern Ohio Geol. Soc.* Cleveland, Ohio, 421–432.

HAIMSON, B. C. and EDL, J. N., JR (1972), *Hydraulic fracturing of deep wells*, AIME, Petrol. Trans. *4061*, 1–12.

HEALY, J. H., RUBEY, W. W., GRIGGS, D. T. and RALEIGH, C. B. (1968), *The Denver earthquakes*, Science *161*, 1301–1310.

HUBBERT, M. K. and WILLIS, D. G. (1957), *Mechanics of hydraulic fracturing*, Trans. AIME *210*, 153–163.

KEHLE, R. O. (1964), *The determination of tectonic stresses through analysis of hydraulic well fracturing*, J. Geophys. Res. *69*, 259–273.

LAMONT, N. and JESSEN, F. W. (1963), *The effects of existing fractures in rocks on the extension of hydraulic fractures,* AIME Petrol. Trans. 203–209.

LOCKNER, D. and BYERLEE, J. (1977), *Hydrofracture in Weber Sandstone at high confining pressure and differential stress*, J. Geophys. Res. in press.

MOGI, K. (1966), *Some precise measurements of fracture strength of rocks under uniform compressive stress,* Intl. J. Rock Mech. Geomech. *4*, 41–55.

PAULDING, B. W. (1968), *Orientation of hydraulically induced fractures* in *Proc. 9th Symp. on Rock Mechanics*, 461–483.

SCHEIDEGGER, A. E. (1962), *Stresses in the earth's crust as determined from hydraulic fracturing data,* Geologie und Bauwesen *7*, 45–53.

TIMOSHENKO, S. and GOODIER, J. N., *Theory of Elasticity*, 3rd ed. (McGraw-Hill, New York 1951).

ZOBACK, M. D. and BYERLEE, J. D. (1975), *The effect of microcrack dilatancy on the permeability of Westerly Granite*, J. Geophys Res. *80*, 752–755.

ZOBACK, M. D., RUMMEL, F., JUNG, R. and RALEIGH, C. B. (1976), *Laboratory hydraulic fracturing experiments in intact and prefractured rock*, Intl. J. Rock. Mech. Geomech., in press.

(Received 28th October 1976)

Pageoph, Vol. 115 (1977), Birkhäuser Verlag, Basel

Stress in the Lithosphere: Inferences from Steady State Flow of Rocks

By Jean-Claude C. Mercier, Douglas A. Anderson and Neville L. Carter[1])

Abstract – Mechanical data and flow processes from steady state deformation experiments may be used to infer the state of stress in the lithosphere and asthenosphere. Extrapolations of flow equations to a representative geologic strain rate of 10^{-14}/sec. for halite, marble, quartzite, dolomite, dunite and enstatolite are now warranted because the steady state flow processes in the experiments are identical to those in rocks and because the geotherms are reasonably well established. More direct estimates are obtained from free dislocation densities, subgrain sizes and recrystallized grain sizes all of which are functions only of stress. Using the last of these techniques, we have estimated stress profiles as a function of depth from xenoliths in basalts and kimberlites, whose depths of equilibration were determined by pyroxene techniques, from four different areas of subcontinental and suboceanic upper mantle. The results are similar and indicate stress differences of about 200 to 300 bars at 40 to 50 km, decaying to a few tens of bars at depths below 100 km. These stresses are reasonable and are in accord with extrapolations of the mechanical data provided that allowance is made for a general increase in strain rate and decrease in viscosity with depth.

Key words: Stress in lithosphere; Flow of rocks; Paleopiezometer.

1. *Introduction*

Knowledge of the state of stress in the lithosphere and asthenosphere is fundamental to the physical understanding of a wide variety of important geological and geophysical phenomena which, in turn, are of both practical and scientific interest. The physical processes involved in plate motions and in their interactions that give rise to first-order lithospheric structures are all dependent on the state of stress and material properties in this part of the earth. Advances in earthquake seismology, including a basic understanding of the origin of earthquakes, depend on the knowledge of the stress state as well as the long term strength properties of rocks. Economic recovery of certain forms of geothermal energy will depend on the stress state and knowledge of this parameter, together with other structural information should lead to discoveries of new mineral and fossil fuel reserves. Methods that are currently available for determinations of the state of stress, documented in this volume, include *in situ* measurements, seismic techniques, mechanical force-balance modelling, estimates from regional geologic and geodetic observations and estimates from laboratory data. All of these methods have limitations in accuracy and in depth of

[1]) Dept. of Earth and Space Sciences, S.U.N.Y., Stony Brook, New York 11794, USA.

effectiveness in the heterogeneous lithosphere but when applied together, they should ultimately yield a reasonable approximation of the stress state.

Our approach to this topic involves application of experimental steady state mechanical data and attendant flow processes to obtain estimates of the variation of stress and equivalent viscosity with depth. Specifically, available data for a fairly wide variety of rock types can be extrapolated to the lower rates of natural deformations with some confidence provided that the deformational processes are identical and that the geotherms are reasonably well known; fortunately, these conditions are met in most instances. More direct estimates, for comparison with all the methods mentioned above, come from determinations of free dislocation densities and from subgrain sizes, both of which are dependent only on stress and may be calibrated by laboratory experiments. In addition, it appears as if recrystallized grain size is also dependent on stress and virtually independent of other physical variables. This discovery offers a particularly promising technique for stress estimates because of the inherent stability of the new grains as compared with subgrain size and dislocation densities both of which may be altered readily during the journey to the surface and, some instances, during the sample preparation process. Preliminary results suggest that the method based on grain sizes may provide an extremely powerful and accurate way of estimating stress magnitudes in both crustal and upper mantle rocks. These results, as well as others, are presented here and are applied primarily to xenoliths in basalts and kimberlites of several areas.

Unless specified otherwise, the following units are used throughout this paper: kilobars for pressures and for stresses ($\sigma_1 - \sigma_3$), micrometers (μm) for subgrain sizes and μm^{-2} for dislocation densities.

2. *Temperature and strain rate in the crust and upper mantle*

The most important physical parameter controlling flow stress in the ductile regime is the homologous temperature, the ratio of the absolute ambient temperature, T, to the melting temperature, T_m. It is, therefore, of prime importance to obtain good estimates of thermal gradients under continents and ocean basins. Temperature gradients in the shallow crust are measured directly and, for thick sedimentary basins of the Gulf Coast, range from 22°C/km (Moses, 1961) to 36°C/km (Nichols 1947). Lachenbruch (1968) has calculated maximum and minimum thermal profiles for the Sierra Nevada on the basis of surface observations, borehole data and assumptions concerning distribution of radioactive heat sources; his gradients fall between the extremes noted above. Herrin (1972) has calculated similar values on the same basis for the Canadian Shield and Basin and Range provinces.

Temperature profiles deeper in the earth are, of course, more uncertain and are based mainly on geophysical and geochemical constraints, such as those incorporated in the theoretical gradients (Fig. 1) of Clark and Ringwood (1964) and of Griggs

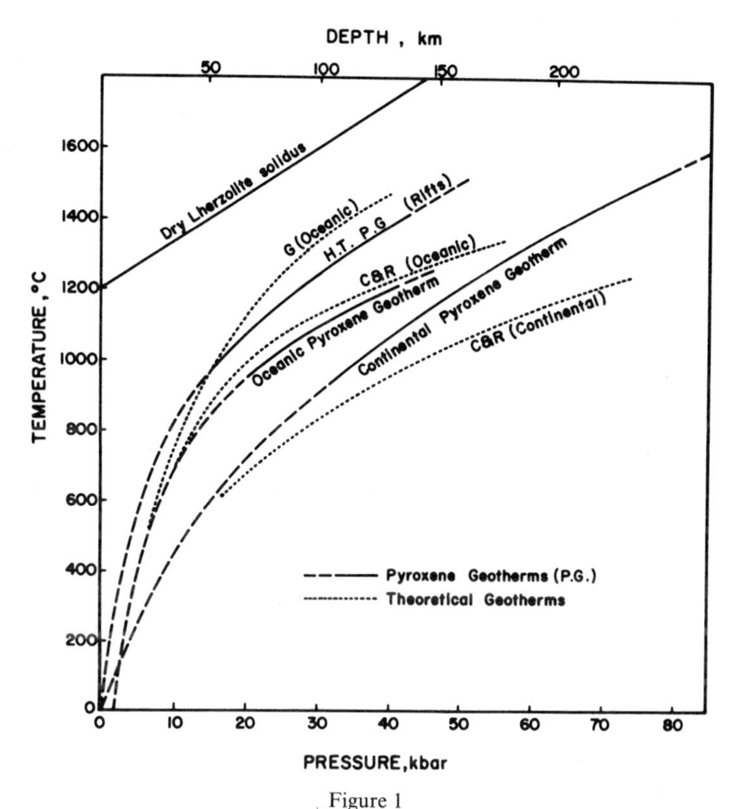

Figure 1
Pyroxene geotherms (solid curves; MERCIER and CARTER 1975) and theoretical geotherms based on various geophysical and geochemical constraints (dashed curves; CLARK and RINGWOOD 1964; GRIGGS 1972). Experimental dry lherzolite solidus after KUSHIRO (1973).

(1972). Attempts have also been made to account for thermal convection (e.g., STACEY 1969; TORRANCE and TURCOTTE 1971; MCKENZIE *et al.* 1974; HOUSTON and DEBREMAECKER 1975; FROIDEVAUX and SCHUBERT 1975) with attention focused on the thermal structure beneath spreading ridges (e.g., LANGSETH *et al.* 1966; OXBURGH and TURCOTTE 1968; SCLATER and FRANCHETEAU 1970) and in descending lithospheric slabs (e.g., MCKENZIE 1969; MINEAR and TOKSOZ 1970; GRIGGS 1972). More direct estimates of equilibration temperatures and pressures in the upper mantle have become possible through comparisons with results from experimental studies at high pressures and temperatures on certain systems. Specifically, for equilibrium assemblages of coexisting diopside, enstatite and an Al-phase (garnet or spinel) temperatures and pressures may be estimated, respectively, from the experimentally determined Di(En) solvus and from Al_2O_3 isopleths for enstatite. Using this more direct approach, and raw Al_2O_3 data, Boyd (1973) has estimated an ancient geotherm beneath Africa by analyzing garnet peridotite specimens from kimberlites and MACGREGOR (1974) estimated an oceanic geotherm from analyses of spinel lherzolites and garnet pyroxenites in alkali basalts from Hawaii and other sources. More recently, MERCIER and CARTER (1975) have used the pyroxene technique, applying several corrections for extra elements, and have derived the pyroxene geotherms shown in Fig. 1. The geotherms so derived are in excellent accord with those estimated on the basis of other geochemical and geophysical observations and will be used as reference geotherms for this paper.

Inasmuch as steady state flow processes are thermally activated rate processes, it is important also to have good estimates of natural strain rates. Such estimates have been derived from geodetic measurements of surface displacements, as those along the San Andreas fault (WHITTEN 1956), those due to rebound from ancient water and ice loads (e.g., HASKELL 1935; CRITTENDEN 1967; MCCONNELL 1968; WALCOTT 1970; CATHLES 1975) and estimated rates of shortening in orogenic regions (GILLULY 1972). The strain rates so obtained fall in the range 10^{-13} to 10^{-15}/sec. and are in agreement with those calculated assuming that plate displacements result from homogeneous shearing strain over depth intervals of from 100 to 1000 km. Accordingly, we shall adopt, as a representative geological steady-state strain rate, a value of 10^{-14}/sec. and in certain analyses given below this value will be held constant so as to allow comparisons of effects of temperature on flow stress and equivalent viscosity of different materials. This is only an average value, however, and there are preliminary indications that the strain rate increases with depth in the depth range of interest here.

3. *Steady state flow of rocks*

In this paper we shall be concerned only with the flow of rocks in the steady state (secondary creep; constant stress and strain rate) which is probably a reasonable approximation of the average flow behavior of the asthenosphere and of most of the lithosphere. We do not mean to imply that transient (primary creep) and accelerating (tertiary creep) creep are not important in the shallow crust and locally at depth but too little is known of creep in these regimes to warrant discussion. Steady state data available for common rock forming materials are given in Table 1 and Fig. 2 presents flow stresses and equivalent viscosities ($\eta = \sigma/3\dot{\varepsilon}$) as a function of temperature, calculated from these data at a strain rate of 10^{-14}/sec. The steady-state data are all best fit by an equation of the form (WEERTMAN 1968)

$$\dot{\varepsilon} = A \exp\left(- \frac{Q}{RT10^{-3}} \right) \sigma^n \tag{1}$$

where $\dot{\varepsilon}$ is the strain rate, A is a nearly temperature-insensitive material constant, Q is the creep activation energy, n is the stress exponent and R and T have their usual meaning. In the steady state, the motion of slip dislocations accounts for much of the creep strain and dynamic recovery is accomplished primarily by the climb of

Table 1

Material	A[1])	$Q\left(\frac{\text{kcal}}{\text{mole}}\right)$	n	Source
Halite	3×10^{-6}	23.5	5.5	HEARD (1972)
Marble (1 cyls.)	6×10^{-13}	62	8.3	HEARD and RALEIGH (1972)
Dolomite	1×10^{-22}	83	9.1	HEARD (1976)
Quartzite (dry)	6.7×10^{1}	64	6.5	HEARD and CARTER (1968; reduced by PARRISH *et al.* 1976)
Quartzite (wet)	7×10^{3}	55	2.6	PARRISH *et al.* (1976)
Dunite (dry)	1.8×10^{8}	100	3.0	KIRBY and RALEIGH (1973)
Dunite (dry)	4.3×10^{8}	126	3.0	POST (1976)
Dunite (dry)	5.1×10^{9}	111	3.3	CARTER and AVE'LALLEMANT (1970)[2])
Dunite (wet)	1.2×10^{3}	54	2.1	CARTER and AVE'LALLEMANT (1970)[2])
Dunite (wet)	4.3×10^{8}	94	3.0	POST (1976)
Enstatolite (dry)	2×10^{4}	70	2.4	RALEIGH *et al.* (1971)
Enstatolite (wet)	2.3×10^{3}	65	2.8	ROSS (1977)

[1]) A is expressed in units of $\text{bar}^{-n}\,\text{sec}^{-1}$ for halite, marble and dolomite, and of $\text{kbar}^{-n}\,\text{sec}^{-1}$ for the remaining materials.

[2]) Revised (Carter, 1975).

edge segments and cross-slip of screw segments, commonly combining to form subgrains or polygons. Syntectonic recrystallization also aids the softening process, permitting steady state flow, and the creep rate in this regime is probably most commonly limited by vacancy diffusion.

For the crustal rocks, flow stresses and viscosities (Fig. 2) at comparable temperatures increase in the sequence halite, marble, dry quartzite, dolomite, a sequence that might have been anticipated from field observations. Whether wet quartzite will be more or less viscous than the marble depends on the temperature, this ambiguity resulting from the high temperature dependence of viscosity of the wet quartzite. The slopes of the lines in Fig. 2A are proportional to Q/n; the shallow slopes for halite, marble, dry quartzite and dolomite, indicating a low temperature dependence of stress and viscosity, are for $Q/n < 10$ whereas the steeper slopes for the remaining materials, indicating a high temperature dependence, are for $Q/n > 20$, a relationship noted also by HEARD (1976). The mantle rocks all have a high temperature dependence of stress and viscosity, the absolute values of which for dunite depend critically on the choice of data. The dry enstatolite data are suspect because they yield stresses and viscosities lower than the wet enstatolite and all results for dry dunite, a relationship contrary to that expected from field and experimental observations. For more complete discussions of steady state flow of rocks, the reader is referred to recent reviews by HEARD (1976) and by CARTER (1976).

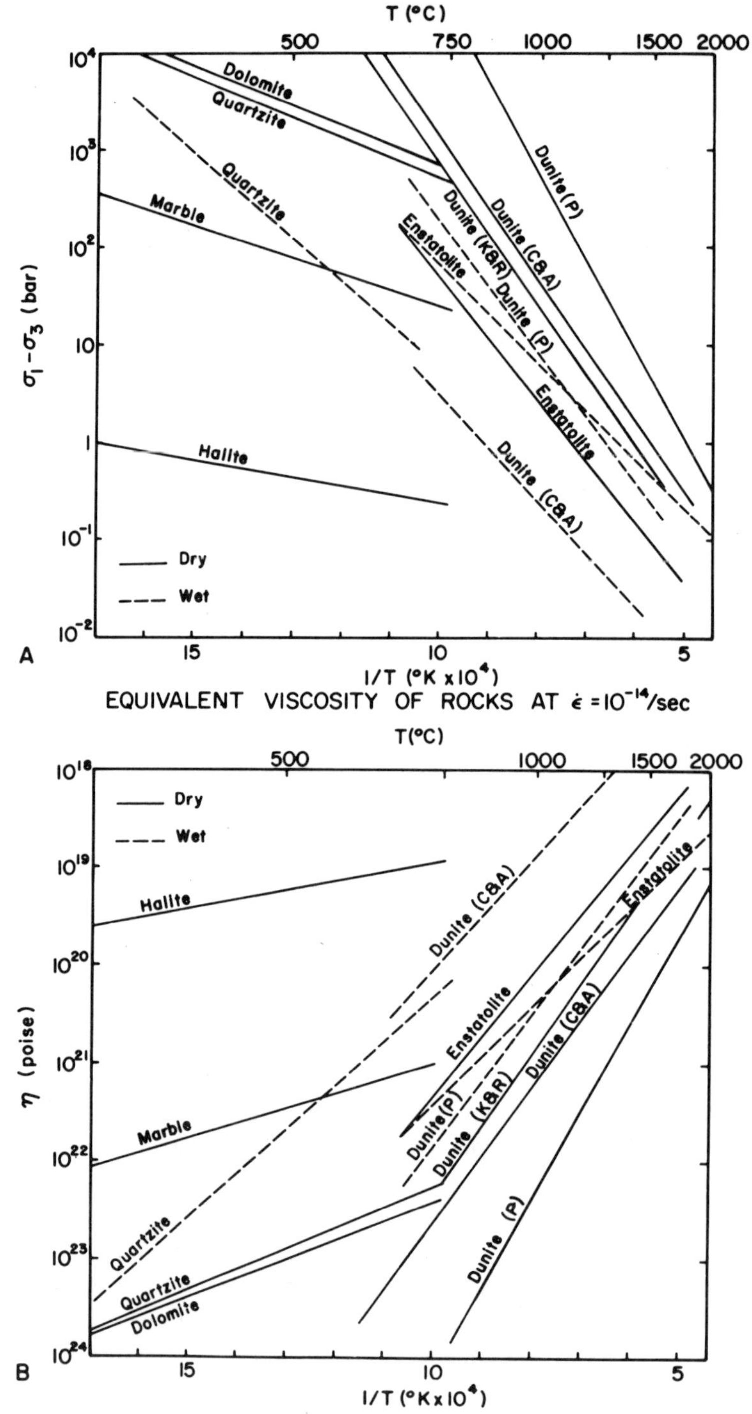

Figure 2

Steady state flow stress (A) and equivalent viscosity (B; $\eta = \sigma/3\dot{\varepsilon}$) of rocks as a function of temperature, calculated at a strain rate of 10^{-14}/sec from equation (1) and the flow parameters given in Table 1 (CARTER 1976).

4. *Stress estimates from laboratory measurements*

Estimates from the mechanical data

As was noted in the introduction, the flow processes observed for steady state experiments are commonly identical to those observed for naturally deformed materials and hence extrapolations to depths and rates of natural deformations are warranted. In Fig. 3A, such an extrapolation, to a constant strain rate of 10^{-14}/sec, has been accomplished for crustal rock types using the flow parameters of Table 1, equation (1), and the high temperature oceanic geotherm of MERCIER and CARTER (1975). The extrapolation was first made using the continental geotherm but this resulted in flow stresses at depth much too high to be reasonable as is shown, for example, by those for Yule Marble (dashed curve, Fig. 3A). We reason, therefore, that the continental geotherm, which was obtained for cratons, has a crustal thermal gradient (20°C/km near the surface) too low to account for deformation in orogenic regions. The oceanic geotherm used probably has too high a crustal gradient (36°C/km near the surface) but is believed to be more nearly representative on the basis of the mechanical data. In any instance, the flow stresses so obtained for all materials are reasonable (<3 kb) at depths below about 10 km although, for quartzite and dolomite, they increase to impossibly high values at shallower depths. This must mean that at the shallower depths: (a) the power flow law of equation (1) breaks down at stresses near a kilobar to an exponential stress dependence for which there is some experimental evidence, HEARD 1976; SHERBY and BURKE 1968; (b) that transient creep prevails in this region; (c) the strain rate decreases toward the surface; or (d) a combination of these effects.

Whereas pressure need not be taken into account for the crustal materials because experimental pressures employed are in the appropriate range, this is not true for extrapolation of steady data for mantle materials. In the absence of information on the activation volume for creep, the approximation employed by WEERTMAN (1970) is used whereby

$$\dot{\varepsilon} = Df(\sigma) = A \exp(-aT_m/T)f(\sigma) \tag{2}$$

where D is the diffusion coefficient and a is Q/RT_m. This equation assumes, with some experimental justification, that the pressure effect on D is given approximately by its effect on the melting temperature T_m. The melting temperature assumed to be most pertinent to the dunite data is that for Fo_{90} extended to depth by KIRBY and RALEIGH (1973) using KRAUT and KENNEDY'S (1966) melting relation (Fig. 3B). These values have then been combined with the continental and high temperature oceanic pyroxene geotherms of MERCIER and CARTER (1975) to yield the T/T_m curves shown in Fig. 3B and the values used in equation (2). Variations in the flow stress as a function of depth, using the oceanic geotherm, have been calculated from the empirical constants (Table 1) of CARTER and AVE'LALLEMANT (1970), KIRBY and RALEIGH (1973) and POST (1976), and are shown in Fig. 3B. In all instances, the stresses decay

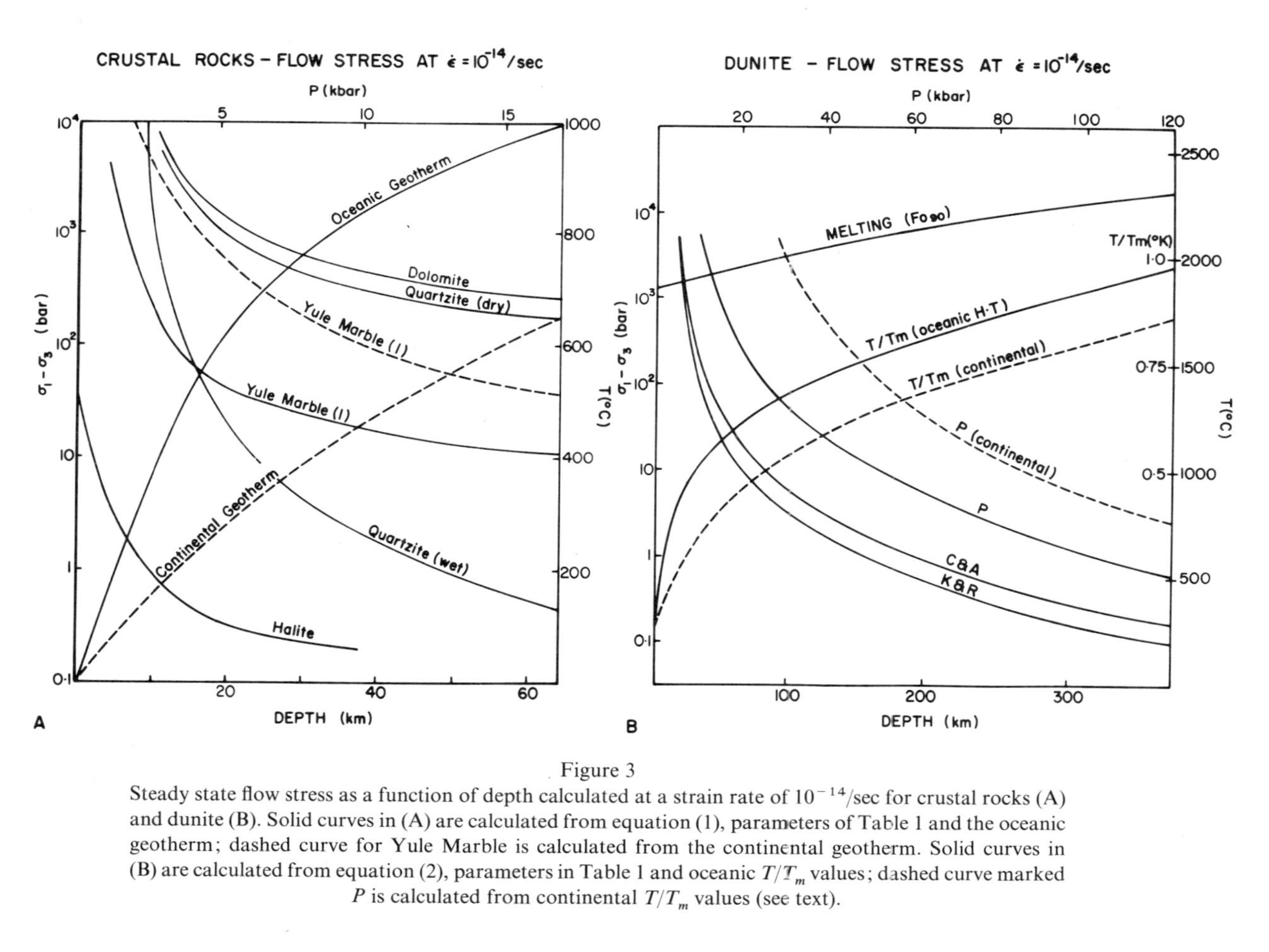

Figure 3

Steady state flow stress as a function of depth calculated at a strain rate of 10^{-14}/sec for crustal rocks (A) and dunite (B). Solid curves in (A) are calculated from equation (1), parameters of Table 1 and the oceanic geotherm; dashed curve for Yule Marble is calculated from the continental geotherm. Solid curves in (B) are calculated from equation (2), parameters in Table 1 and oceanic T/T_m values; dashed curve marked P is calculated from continental T/T_m values (see text).

rapidly in the first hundred kilometers, levelling off to a few bars at most thereafter. Flow stresses calculated for the continental geotherm, and shown for POST's (1976) data, are about an order of magnitude higher.

Estimates from dislocation densities and subgrain sizes

More direct estimates of flow stresses may be obtained by determining the free dislocation densities and subgrain sizes. For metals, it has been recognized for some time that the free dislocation density increases with stress (e.g., WEERTMAN and WEERTMAN 1970) according to the relation:

$$\rho_f = A_f(\mu b)^{-2}\sigma^2 \quad (3)$$

where A_f is a constant, μ is the shear modulus and b is the Burgers vector. GOETZE and KOHLSTEDT (1973) and GOETZE (1975) have shown that equation (3) applies to olivine as well. GOETZE's (1975) data are reproduced in Fig. 4A and indicate that $\rho_f = \sigma^2$; that is $A_f \cdot (\mu b)^{-2} = 1$. This technique generally requires TEM determinations and is thus tedious but is potentially sufficiently important to merit the effort.

The size of subgrains produced during both static and dynamic recovery is also a function of stress and is given by

$$d \propto (\mu b/\sigma)^m = A_d(\mu/\sigma)^m \quad (4)$$

where d is the average diameter measured normal to the longest dimension of subgrains and A_d and m are constants. The value of the exponent, m, is found to be near unity for most materials but may range from 0.5 (Al at 25°C; BALL 1957) to 1.2 (Fe; GLOVER and SELLARS, 1973). RALEIGH and KIRBY (1970) first made use of (4) by determining subgrain sizes produced by polygonization of olivine during creep experiments and found m to be nearly equal to 1 (Fig. 4B). Subsequently, GREEN and RADCLIFFE (1972a, 1972b) examined, using TEM, the same samples studied optically by RALEIGH and KIRBY (1970) and found the subgrain size to be much smaller (Fig. 4B). We believe that the optical technique yields too large an average subgrain size as some very low-angle boundaries are not detected whereas the TEM estimates are biased toward regions in which the sub-boundary spacing is exceptionally small. We have, therefore, employed the dislocation decoration technique of KOHLSTEDT *et al.* (1976) for which the three-dimensional dislocation distributions can be observed in thin sections made from oxidized specimens. Some examples of decorated olivine-rich specimens are given in Fig. 5 and the preliminary results of our study based on this method are given in Fig. 4B where it is seen that the data fall between the optical and TEM results. We obtain

$$d = 115/\sigma \quad (5)$$

the average subgrain size observed being about 0.6 times that observed by RALEIGH and KIRBY who found $d = 190/\sigma$.

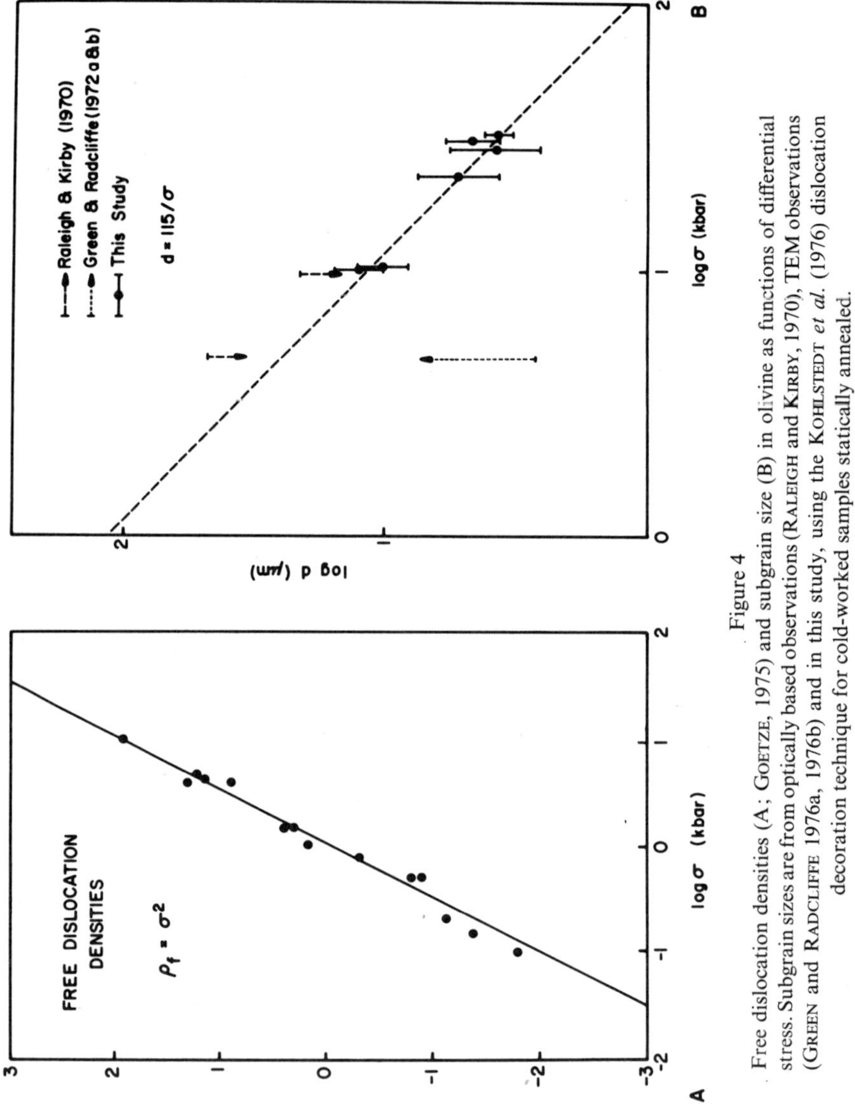

Figure 4

Free dislocation densities (A; GOETZE, 1975) and subgrain size (B) in olivine as functions of differential stress. Subgrain sizes are from optically based observations (RALEIGH and KIRBY, 1970), TEM observations (GREEN and RADCLIFFE 1976a, 1976b) and in this study, using the KOHLSTEDT *et al.* (1976) dislocation decoration technique for cold-worked samples statically annealed.

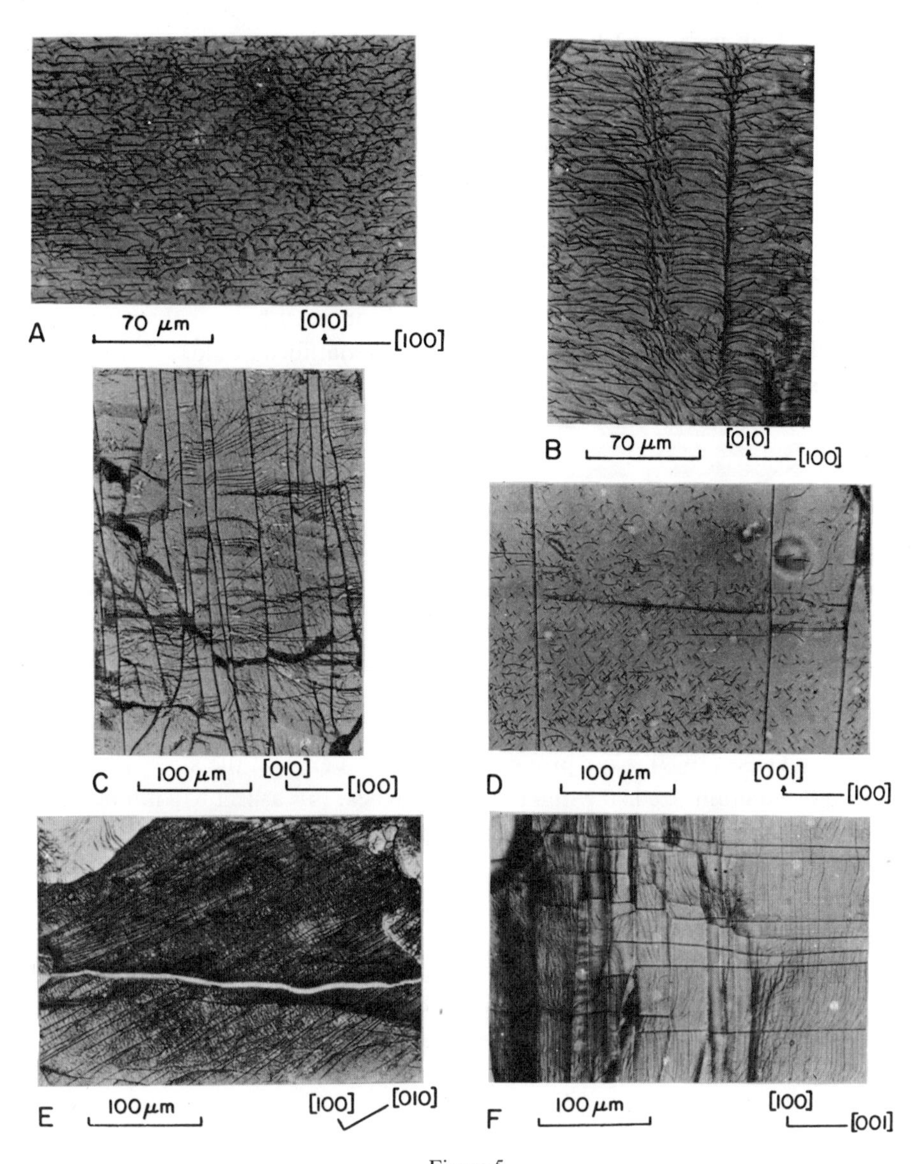

Figure 5

Dislocation networks in olivine as revealed in thin section by the decoration technique; the slip system is (010)[100]. A–D. Photomicrographs of different parts of a naturally deformed sample from San Quintin, Baja, California: (A) dislocation loops with dominant horizontal screw components; (B) formation of (100) walls of edge segments; (C) typical termination of subgrains in the [010] direction by coalescing (100) walls; (D) formation of (001) walls by cross-slip of screw dislocations. E. Experimentally formed subgrains viewed parallel to [001]. The narrow subgrains are bounded by northeast-trending bands normal to the slip direction and the dislocation density is high. F. Experimentally formed subgrains viewed normal to [010]. In this orientation the subgrains appear as subequant cells.

The results discussed in this section are of a preliminary nature and require careful evaluation which is merited because of their potential importance. There are two central problems in employing these techniques directly. The first is that dislocations and subgrains can be introduced during the sample preparation process if care is not taken and they can be introduced at any time during the emplacement process and hence may not reflect steady-state stresses at the depth of equilibration. The second problem is that, while the subgrain sizes decreased with increasing stress for AgCl (PONTIKIS and POIRIER 1975), they did not change upon lowering the stress. This observation, which may indicate that low angle boundaries are not as mobile as previously believed, is of obvious importance and must be investigated for rock-forming minerals. In any instance, the difficulties just outlined would lead to maximum stress estimates. Alternatively, low estimates would be obtained from free dislocation densities in xenoliths statically recovering while ascending in hot magmas.

Estimates from recrystallized grain size

The recrystallization process. It is observed also that the size of newly recrystallized grains is a function of stress only (e.g., JONAS *et al.* 1969), an observation apparently first made for rocks by POST (1973) who recognized this effect in experimentally deformed olivine. This is a most important result that bears close scrutiny because of the expected inherent stability of new grain sizes as compared to subgrain sizes and dislocation densities. LUTON and SELLARS (1969) and, more recently, MERCIER (1976a) and TWISS (1977), have proposed theories to explain the stress dependence of recrystallized grain size using different approaches; this aspect will not be emphasized here although we expect to test these theories later. Rather, we shall summarize briefly the main physical characteristics of the process as distinct from other recovery processes making use mainly of reviews by CAHN (1970) and JONAS *et al.* (1969) for metals and CARTER (1976) for rocks.

Primary annealing and syntectonic recrystallization involve the growth of new relatively strain-free grains with high-angle boundaries from old strained ones and may occur under hydrostatic stress (annealing) or during deformation (hot-working or syntectonic recrystallization). There are many similarities in the nature of the recrystallization process under static and dynamic conditions and because of these similarities, some workers have taken the view that recrystallization during hot-working might be interpreted as static recrystallization following the test or during breaks in intermittent tests, as discussed below. In both instances a critical strain, which is a function of stress, temperature, and purity, is required before the process can take place and the recrystallized grain size increases for decreases in the flow stress for cold-working (prior to anneal) and for hot-working. For both types, nucleation occurs in regions of highest strain energy, preferentially at grain boundaries, and the process of nucleation is thermally activated which accounts for the induction period frequently observed before annealing recrystallization becomes detectable.

Thus, the process is competitive with recovery kinetically in that recovery begins very rapidly and decreases with time whereas recrystallization starts slowly, builds to a maximum and then decreases with time. Increasing the impurity content increases the activation energy for both recovery and recrystallization, but affects recrystallization more, so that it may be retarded entirely. A dispersed second phase may accelerate nucleation if the particles are large (because of stress concentrations) but may retard it if they are small; in most instances dispersed particles impede the growth of new grains primarily by exerting a drag on the boundary.

There are three major models currently held for the nucleation process of recrystallization. The first is a form of the classical nucleation hypothesis in which new nuclei are generated preferentially at intersections of subgrain and grain boundaries (Fig. 6). Inasmuch as syntectonic recrystallization of both naturally and experimentally deformed rocks occurs mainly at grain boundaries, this model is preferred and is that used by MERCIER (1976a) to develop a theoretical expression, given below, relating grain size to stress. In the second model, nuclei form by cell growth in which

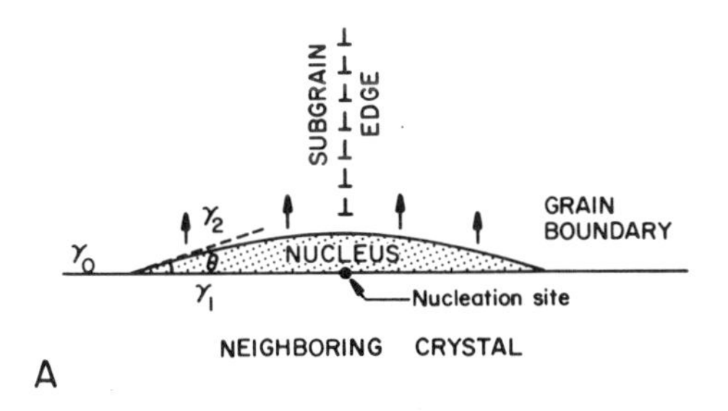

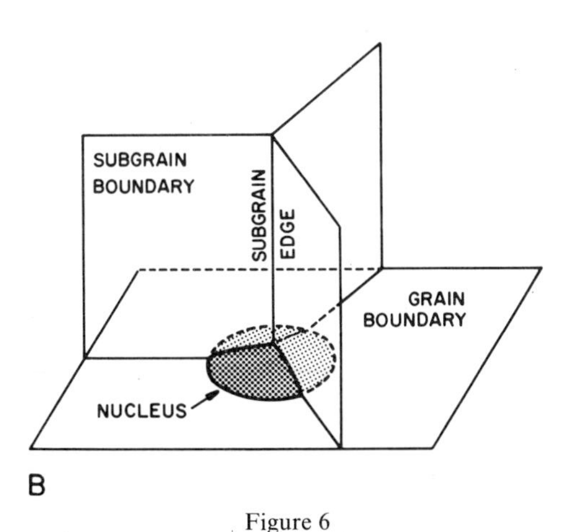

Figure 6

Formation of a nucleus at subgrain-grain intersections. Formation of a flattened nucleus increases the surface energy only slightly, by an amount $\gamma_1 + \gamma_2 - \gamma_0$(A), the main driving force for growth being normal to the grain boundary. Three-dimensional view of a growing nucleus is given in (B).

the center of a sharply bent region polygonizes and then grows into the strained crystal. Alternatively, such a cell could grow in a recovered structure by assimilating dislocations in adjacent subgrain boundaries so that, eventually, the wall becomes a mobile large-angle boundary (CAHN 1970). A variant of this hypothesis, proposed for olivine (POIRIER and NICOLAS 1975), is that subgrains are formed initially under the applied stress and with increasing strain the dislocations migrate into the fixed subgrain boundaries increasing the misorientation across them until a large-angle boundary is produced. The third model involves strain induced boundary migration (bulge nucleation; BECK and SPERRY 1950) in which grains with coarse substructure (low strain energy) sweep into those with fine cell structure. Boundary migrating rates depend, in addition to misorientation and impurity content, on their ability to emit, and in particular, absorb vacancies. In any instance, for annealing recrystallization, a growth selection mechanism seems to operate whereby, statistically, nuclei favorably oriented with respect to the host crystals, grow. For f.c.c. metals, a rotation of 38° about [111] (the Kronberg-Wilson rotation) is most common. A rotation of about 30° is also observed for silicates that have been annealed or stress-annealed provided that the new grain is totally contained within the host grain (CARTER 1976).

For annealing, when primary recrystallization is complete, most of the driving force for recrystallization is exhausted but the material still has surface energy which can be relieved only when converted to a single crystal. In an attempt to achieve this, initially large new grains of an orientation different from that of the primary fabric and with more than six sides (in two dimensions, 14 in three dimensions) grow at the expense of the others. This phase, termed secondary recrystallization, results in a fabric different from the primary one. Finally, the cube fabric in metals, (001)[100], can develop during tertiary annealing recrystallization from the ordinary (110)[001] fabric after prolonged treatment at high temperature (CAHN 1970). Fabrics analogous to cube fabrics have been produced experimentally in quartz (GREEN 1967) and in olivine (DOLLINGER; personal communication, 1975).

As noted above, the similarities in the recrystallization process under static and dynamic conditions have resulted in skepticism concerning the occurrence of dynamic recrystallization which has been relatively little studied in metals. However, JONAS *et al.* (1969) point out distorted annealing twins in recrystallized grains indicating subsequent deformation. Furthermore, the substructure in grains recrystallized during hot-working is identical to that produced in the initial grains during deformation. By contrast, grains recrystallized during annealing are virtually dislocation free. Finally, there are important differences in the stress-strain curves for two groups of metals, Al, α-Fe and their alloys and Cu, Ni, γ-Fe and their alloys (JONAS *et al.* 1969; LUTON and SELLARS 1969). The nature of curves characteristic of the former group is consistent with dynamic recovery whereas that of the latter group indicates dynamic recrystallization.

Differences in stress-strain characteristics have not been observed for dynamically recovered and recrystallized rocks but this lack may be due to the relatively low stress sensitivity of the equipment at the high pressures required. However, microstructural features are similar to those in dynamically recrystallized metals and it is important here to discuss the criteria available for distinguishing between syntectonic and annealing recrystallization in rock-forming materials. These criteria, which are as yet rather vaguely specified, include grain shape, dislocation density and configuration and preferred crystal orientations. Briefly, annealed grains are commonly equant and polyhedral (Fig. 7B) with 120° triple junctions whereas syntectonically recrystallized grains generally have irregular boundaries and are flattened normal to $\sigma_1(\varepsilon_1)$ and elongated parallel to $\sigma_3(\varepsilon_3)$. Unfortunately, this generalization can not be applied to olivine as, in some instances, tablet-shaped grains grow during annealing (Fig. 7A) and equant grains may be produced during natural syntectonic recrystallization (MERCIER and NICOLAS 1975). Annealed grains most commonly contain no subgrains and have low dislocation densities whereas GREEN and RADCLIFFE (1972) have shown that the densities and configurations are similar in host grains and syntectonically recrystallized ones. In some xenoliths from kimberlites, poorly recovered host crystals contain optically undeformed recrystallized grains, indicating annealing. Microstructures, such as kink bands, in host crystals are continuous in trend when being consumed by annealing whereas they are commonly displaced when being recrystallized syntectonically. Finally, the preferred orientations of annealed grains are related to the orientation of the host whereas those of syntectonically recrystallized grains at host grain boundaries and in totally recrystallized specimens are related to the principal stress and/or strain axes and the fabrics are generally stronger than are annealed fabrics.

Thus, while any single criterion may be insufficient for distinguishing between static and dynamic recrystallization, a combination of these criteria may suffice. However, the distinction is not always possible and additional and more definitive criteria are clearly needed.

Stress dependence of grain size. Figures 7C-F show a sequence of photomicrographs illustrating the increase of recrystallized grain size in olivine resulting from progressively decreasing applied differential stress in the steady state regime. The small (10 μm) grains, mainly at grain boundaries, in Fig. 7C have developed at 9 kb stress difference, enlarging to 25 μm at 5.4 kb (7D), to 80 μm at 2.3 kb (7E) and finally to 400 μm at 0.7 kb (7F). A similar sequence is observed in quartzite deformed in the steady state in the presence of H_2O (PARRISH *et al.* 1976) although through a smaller range in stress. Experiments at a given stress level but at different temperatures and strain rates on dunites of different composition and H_2O content yield the same new grain size indicating that the grain size is dependent mainly on stress. There appears to be no effect of total strain in the steady state regime although, as expected from the results on metals, minimum critical strains and temperatures are required for recrystallization to be initiated. When these conditions are met, steady state grain sizes are achieved in a matter of a few hours in steady-state laboratory experiments and hence kinetic effects are not expected to be of importance in natural deformations.

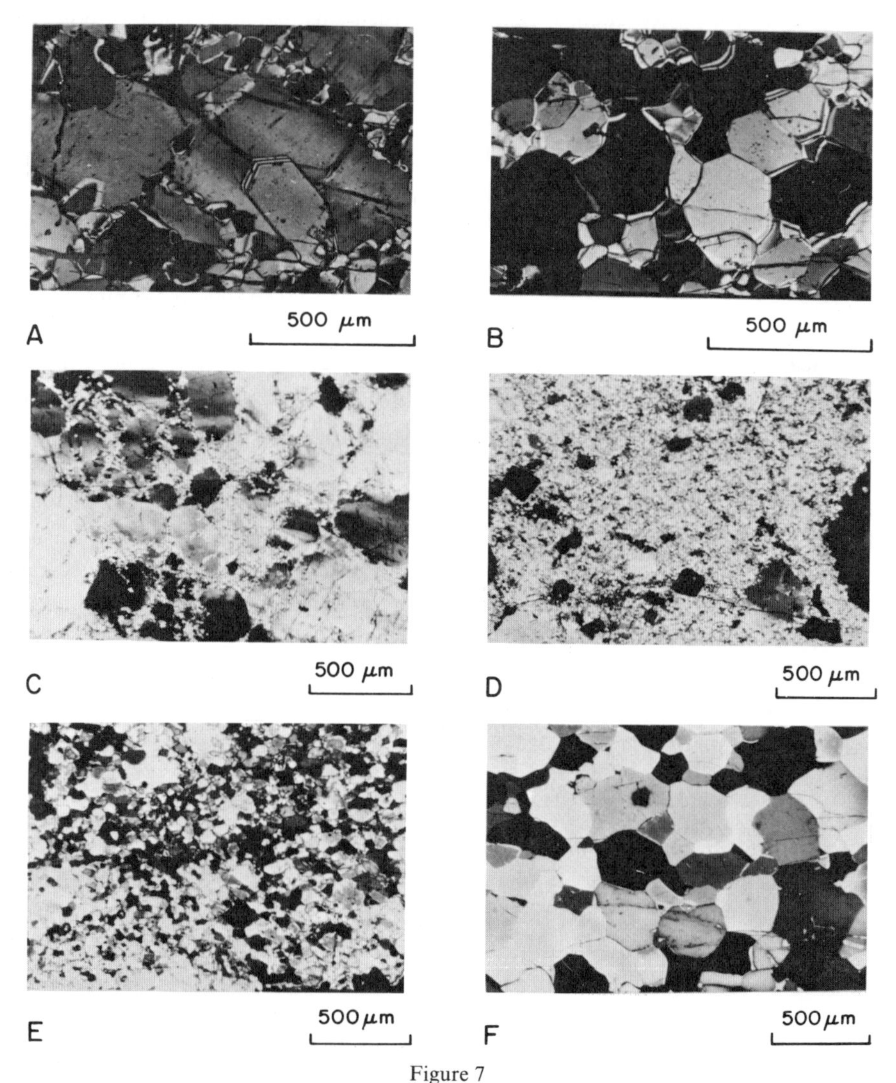

Figure 7
Grain growth and stress dependence of recrystallized grain size. Experimentally recrystallized samples under crossed polarizers. A. Primary annealing showing tablet shaped new grains growing into deformed host crystals. B. Secondary annealing recrystallization resulting from coarsening following primary annealing. Notice the slight curvature of grain boundaries, the center of curvature being on the side of the smaller grains which are being consumed. C–F. Recrystallized grain sizes increase with decreasing applied differential stress: (C) σ = 9.0 kb, D = 10 μm; (D) σ = 5.4 kb, D = 25 μm; (E) σ = 2.3 kb, D = 80 μm; (F) σ = 0.7 kb, D = 400 μm.

Figure 8 shows our results for wet quartzite and for both natural and synthetic dunites deformed in the presence and absence of H_2O released from the talc confining medium. Starred points indicate stresses calculated from flow laws for the materials (Table 1) for specimens whose temperature and strain rates were well established but for whose stress values were in question. For the wet quartzite

$$D = 6.5\,\sigma^{-1.4} \tag{6}$$

and for the dunite

$$D = 94\,\sigma^{-1.23} \tag{7}$$

POST (1973, 1976) obtained for Mt. Burnet dunite

$$D = 80\,\sigma^{-1.49} \tag{8}$$

which is somewhat different from our result, but which yields virtually the same stress level as a function of grain size. As noted above, MERCIER (1976a) has developed a theoretical expression relating steady-state grain size to stress on the basis of the model shown in Fig. 6 with the result

$$D = 1.13\,\sigma^{-1.4}\exp(13\,500/RT) \tag{9}$$

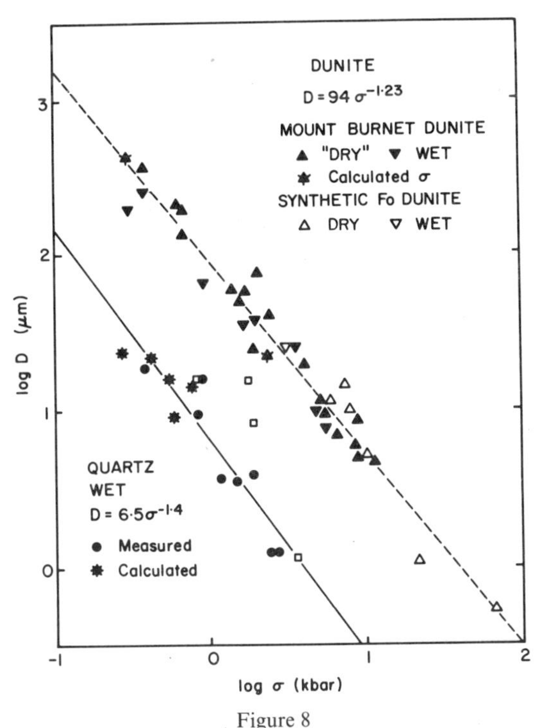

Figure 8

Log-plot of dynamically recrystallized grain size against differential stress for specimens of natural and artificial dunite and natural quartzite deformed in the steady state regime. Equations shown relating grain-size to stress are least squares fits to the data (see text). Open squares are wet Mount-Burnet dunite data not used in the fit: excess H_2O would enhance recovery more than recrystallization, as a steady state grain size was not yet achieved.

As is apparent from (9), there is a slight temperature dependence through the small positive energy term and this results from the reasonable assumptions that subgrain sizes are temperature-independent and that nucleation rates are more sensitive to temperature variations than are growth rates. Fits of the grain size data as functions of both stress and temperature also suggest a weak temperature dependence of the sort indicated by equation (9) and this problem is currently under investigation. In the meantime, we shall use equation (9) for the analyses that follow but we note that differences in stress estimates using (7), (8) or (9) are negligible for the physical conditions of interest here.

5. *Some applications of the results*

These preliminary observations, if substantiated by careful future work, may be used directly to determine stress magnitudes during crustal and upper mantle deformations, thus providing a powerful new and simple technique for estimating this important quantity. As an example, a map of stresses determined from (9) for grain sizes of the Lanzo massif (NICOLAS *et al.* 1971) is given in Fig. 9. The massif is divisible on the basis of petrogenetic and textural grounds, into three submassifs as is also indicated by the stress estimates. The central part of each submassif is characterized by large grain size and hence low stress. These low values may reflect relict stresses from steady-state flow in the lithosphere, as is also suggested by similarities in texture with xenoliths. The grain size decreases steadily towards the margins of each submassif indicating higher stresses, to 3 kb locally, these high stresses most likely having been generated during the emplacement process; both textures and asymmetrical fabrics (NICOLAS *et al.* 1971) support this interpretation. Similar results have been obtained for the Newfoundland ultramafics by MERCIER (1976a) and are expected for other highly deformed ophiolite sequences, such as the Vourinos Complex, Greece.

Of perhaps more interest and importance are stress estimates from peridotite from basalts and kimberlites, whose textures and microstructures have not been altered appreciably during their journey to the surface. That is, we are concerned here mainly with steady-state flow stresses estimated using equation (9) in specimens that have equilibrated at depth, the depth being given for suitable specimens by pyroxene geothermometry and geobarometry (e.g., BOYD 1973; MACGREGOR 1974; MERCIER and CARTER 1975). Specifically, the depths obtained for the stress profiles to be presented below are from a more general single-pyroxene technique (MERCIER 1976b). The profiles have generally been constructed on the basis of 15 to 20 data points randomly distributed in the depth interval shown (for details, see MERCIER 1976a) and the curvatures shown apparently are real.

Figure 10A shows the results obtained in this manner to date for the southwestern United States and Fig. 10B shows results for southern Africa. Depth in kilometers is

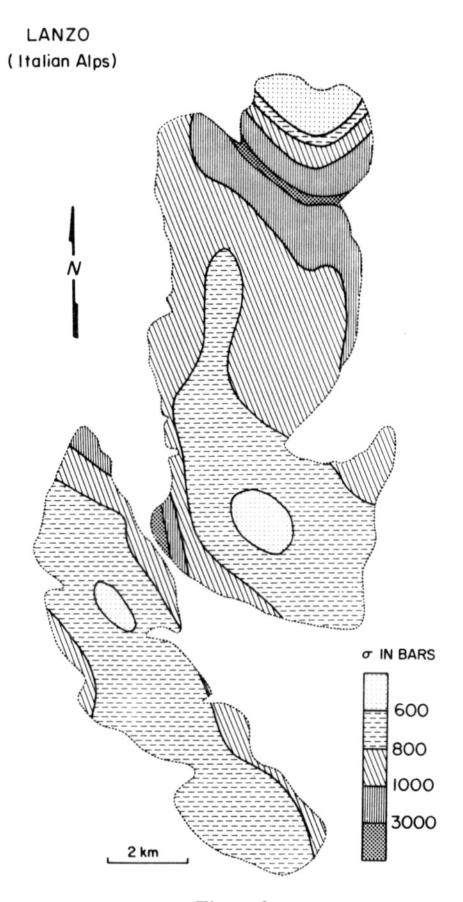

Figure 9
Stress estimates for various zones of the Lanzo massif based on average grain sizes (NICOLAS *et al.*, 1971) and equation (9). The massif is also divisible into three submassifs on petrogenetic grounds. The low stresses near the center of the submassifs are believed to reflect steady state flow in the lithosphere whereas the progressively higher stresses (smaller grain size) toward the margins are interpreted as having been generated during emplacement. The stresses shown on this map are only apparent figures (slightly over-estimated) since the average grain size used is a whole-rock cumulative average. However, the trends are real as the recrystallized grain size (related to stress) and the fraction of recrystallized material (related to stress and strain) decrease simultaneously.

indicated in each profile on the ordinate and stress difference, in hundreds of bars, is given on the abscissa. The hatched patterns indicate xenoliths in quaternary basalts or diatremes whereas the stippled pattern indicates xenoliths in older rocks. For the southwestern United States, the stresses indicated are lowest in Nevada, possibly indicative of an incipient spreading ridge in that area. Over most of the Colorado plateau region and to the west, the main features are stresses of 200 to 300 bars at depths of 40 to 50 km, decaying to about 100 bars at deeper levels. The only data available from west of the San Andreas fault are those from San Quintin for which

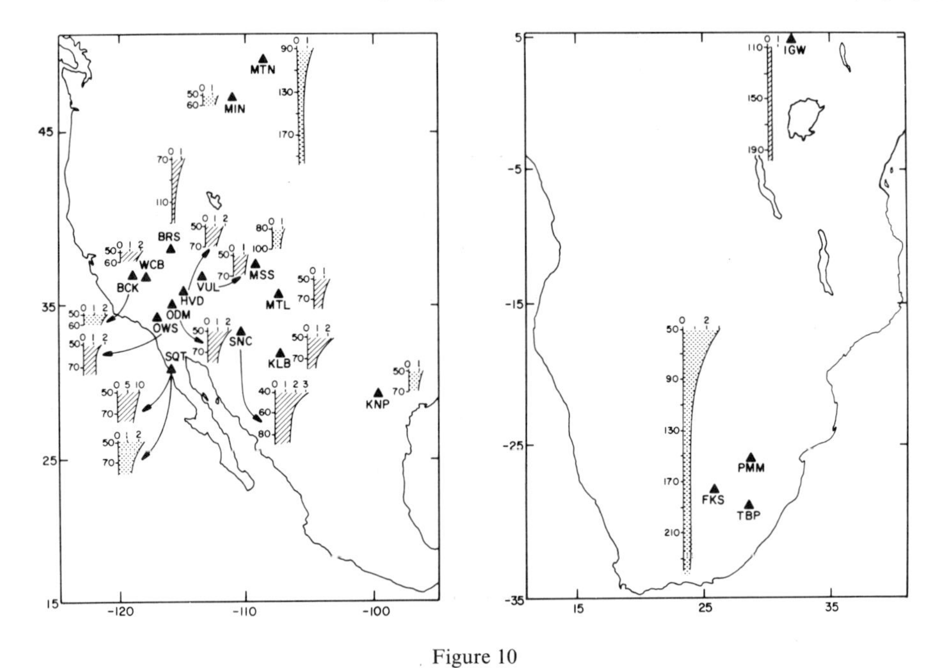

Figure 10
Profiles of stresses (absisscae, in hundreds of bars) as a function of depth (ordinate, in kilometers) for xenoliths from basalts and kimberlites whose localities are indicated by triangles in southwestern United States (A) and South Africa (B). Stresses are estimated from equation (9) and depths are estimated from pyroxene geothermometry and geobarometry MERCIER (1976b). Hatched profiles are for quaternary basalts and stippled patterns indicate older rocks (see text for discussion). Localities corresponding to initials are given in the caption for Figure 11.

two distinct levels of stress magnitudes are obtained. The granular xenoliths give stress profiles similar to those to the east but the sheared ones indicate stresses to about one kilobar. The latter, we believe, are associated with a shear zone (possibly a strike-slip fault); similar sheared xenoliths in basalts are associated with the North-Pyrenean fault zone (MERCIER, 1976a). MACGREGOR and BASU (1976) have interpreted the San Quintin associations as well as the Mt. Albert peridotite in reference to a diapir model but we believe the data are better fit by reactivated shear zone (San Quintin) and obduction (Mt. Albert) models, respectively (MERCIER 1976a).

Stresses estimated from specimens in kimberlites from Lesotho and from Frank Smith Mines, South Africa (Fig. 10B) decay from about 300 bars at 50 km to about 50 bars at 130 km, remaining constant thereafter to 240 km depth. Similarly, a low constant stress level of about 30 bars is recorded at depths from 110 km to 200 km in the Igwisi diatreme to the north. Additional data at hand are from the Massif Central region, Southern Europe and from Oahu and Kauai, Hawaii. For Kauai, the stresses drop from about 300 bars at 40 km depth to 80 bars at 120 km whereas the variation for Oahu is from 200 bars at 40 km to 80 bars at 100 km. Xenolith-bearing volcanics from Massif Central range in time from late Oligocene to present, and although the geotherm remained constant during this period, xenolith textures changed with time indicating variations in stress amplitudes throughout the Cenozoic era. Tertiary stress levels vary from less than 200 bars at 50 km to about 120 bars at 80 km with no observable lateral variation. Quaternary volcanics contain two additional contrasting series of xenoliths with fine or coarse-grained textures depending on the specific volcano. The finer-grained series yields stresses of about 300 bars at 50 km falling to nearly constant values of 100 to 150 bars at 70 to 90 km whereas the coarser-grained series gives about 170 bars at 50 km decaying to 30 bars at 120 km. These results for xenoliths from the Quaternary volcanics are interpreted as reflecting lateral heterogeneity of stress in the upper mantle during this period.

6. *Discussion and conclusions*

The data discussed above for xenoliths from basalts and kimberlites from four different regions indicate a rather similar stress variation with depth based on the steady-state recrystallized grain size technique employed. The stresses so obtained generally range between 200 and 300 bars at depths near 50 km, decaying to less than 100 bars at depths below about 100 km. There appear to be no kinetic problems associated with achieving steady state grain sizes at natural deformation rates and the occurrence of large (several cm) poikiloblastic olivine crystals in some of the specimens indicates that the presence of other phases apparently does not inhibit olivine growth.

The stress profiles obtained from the xenoliths are plotted for continental (infra-cratonic) and oceanic (high-temperature thermal regime) mantle in Fig. 11A and 11B, respectively, where they are compared with profiles calculated at a strain rate of 10^{-14}/sec from the flow parameters (Table 1) of CARTER and AVE'LALLEMANT (1970) and of POST (1976). For the continental data, the comparison is best with the extrapolation of CARTER and AVE'LALLEMANT's mechanical data and it is seen that the stresses level off for all curves at a depth of about 120 to 130 km; this depth may mark the lid of the asthenosphere in those regions. The oceanic data fall between the extrapolation of the two sets of mechanical equations at high levels and stresses but are clearly more in accord with POST's (1976) data at the lower stresses and depths. POST's Mt. Burnet specimens and assemblies were carefully dried at high temperature whereas CARTER and AVE'LALLEMANT report about 0.3 wt per cent H_2O, probably bound in serpentine, available in their 'dry' experiments. Thus, a plausible interpretation of the observation above is that the upper mantle beneath continents contains small but significant quantities of H_2O and more than that beneath oceans. This suggestion is supported by the presence of phlogopite, amphibole and other water-bearing minerals in some xenoliths from the upper mantle beneath continents.

For the oceanic data, there are two classes of stress profiles, one in which the stress decays more or less continuously with depth to very low values and the other

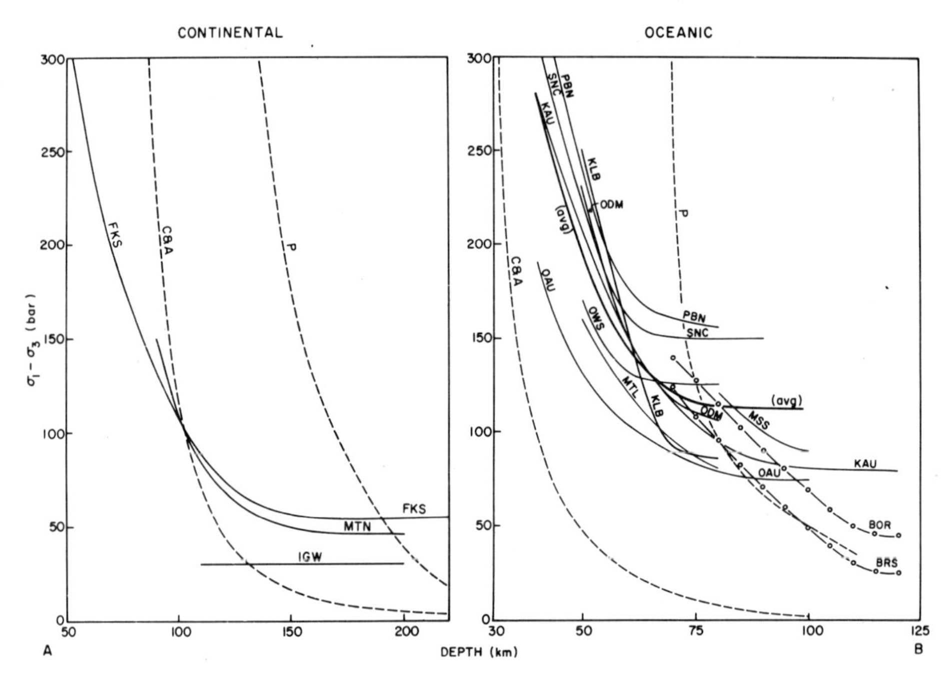

Figure 11
Stress profiles (solid curves) in the upper mantle beneath continents (A) and ocean basins (B) as estimated from recrystallized grain size equation (9) in xenoliths; depths are estimated by means of pyroxene geothermometry and geobarometry. Dashed curves are extrapolations, at $\dot{\varepsilon} = 10^{-14}$/sec, of POST's (1976) and CARTER and AVE'LALLEMANT's (1970) mechanical data using appropriate T/T_m values (Fig. 3B) and equation (2). Bold curve in 11B is an eyeball average of profiles for Kilbourne-type sequences. Locality designations in this figure and figure 10 are as follows: BCK, Big Creek, California; BOR, Borée, France; BRS, Black Rock Summit, Nevada; FKS, Frank Smith, South Africa; HVD, Hoover Dam, Arizona; IGW, Igwisi, Tanzania; KAU, Kauai, Hawaii; KLB, Kilbourne Hole, New Mexico; KNP, Knippa, Texas; MIN, Ming's Bar, Montana; MSS, Moses Rock, Utah; MTL, Mount Taylor, New Mexico; MTN, Montana kimberlites; OAU, Oahu, Hawaii; ODM, Old Dad Mountain, California; OWS, Old Woman Spring, California; PBN, Puy Beaunit, France; PMM, Premier Mine, South Africa; SNC, San Carlos, Arizona; SQT, San Quintin, Baja California; TBP, Thaba Putsoa, Lesotho; VUL, Vulcan's Throne, Arizona; WCB, Waucoba, California.

in which the stress levels off at fairly high values. The first type of behavior is typified by the Boree (France, BOR; BRS, Nevada) sequence and the second by the Kilbourne (New Mexico, KLB) sequence. There are differences in textures and phase chemistry between the two types of sequences although in both the textures range from a coarse-grained type at depths below 90 km to a granuloblastic type at shallower levels by progressive reduction of olivine grain size. In the Kilbourne-type sequence, the coarse-grained (*ca* 1 cm.) texture has equant olivine grains with curvilinear boundaries (protogranular texture of MERCIER and NICOLAS, 1975). For the Boree-type sequence, the deepest textures are poikiloblastic with tablet-shaped olivine crystals commonly larger than 10 cm and this texture grades into a coarse-grained texture at higher levels. Chemically, the Boree-type sequences are much more depleted than the

Kilbourne-type indicating that partial melt must have been removed as it formed. The Boree-type sequences are believed to represent conditions at incipient spreading ridges (Central Nevada; Southwestern Europe; MERCIER, 1976a) whereas the Kilbourne-type may reflect stresses in the upper mantle over major parts of oceanic plates, in accord with data for ophiolites (MERCIER, 1976a). It will be noted in Figure 11B that the stresses for the Kilbourne type level off to the range 75 to 150 bars at depths varying from 60 to 75 km, also possibly marking the lithosphere-asthenosphere boundary. These average depths suggested for the 'boundary' under continents and ocean basins are roughly in accord with those assigned on the basis of seismic information.

It is clear from Figs. 11A and 11B that the observed stress profiles do not coincide with those extrapolated from the mechanical data. A counter-clockwise rotation of Carter and Ave'Lallemant's extrapolation under continents, at the point of intersection with the observed profiles, would bring the curves into near coincidence. A small rotation of Post's curve for oceanic conditions, about an axis near the 100 bar stress level, would achieve the same result. This lack of coincidence (and accommodation for it) could be accounted for by varying the H_2O content as a function of depth and/or by adjusting the thermal gradient or melting temperature, and hence T/T_m values in equation (2). We suggest, however, that most of the discrepancy has arisen because the strain rate, which heretofore has been held constant at an average value of 10^{-14}/sec, may vary with depth.

In order to examine the possible nature of variation of strain rate with depth, we have assumed that the stress values estimated from recrystallized grain sizes in the xenoliths are correct. For the upper mantle beneath continents, the Frank Smith Mines (FKS) are regarded as representative. For the upper mantle beneath ocean basins only Post's data are employed along with the BRS (Black Rock Summit, Nevada) data for the Boree-type environment and the average (avg.) curve shown in Fig. 11B (bold) for Kilbourne-type conditions. Variations in strain rate and equivalent viscosity as a function of depth, evaluated from equation (2) using the stresses given in Fig. 11, are shown in Fig. 12A and 12B, respectively. For the continental upper mantle, the strain rate and viscosity drop rapidly at depths below about 100 km for both sets of data. Values obtained for Carter and Ave'Lallemant's data are perhaps more reasonable at depths shallower than 150 km whereas, at deeper levels, those obtained from Post's flow equation may be more nearly representative. The behavior is similar, although at shallower depths, for the Kilbourne-type (avg.) stress profiles under ocean basins. However, for the Boree-type (BRS), the strain rate drops to about 10^{-14}/sec at 80 km, and remains nearly constant thereafter to a depth of 125 km. This leveling off of the strain rate, which was anticipated because the BRS profile nearly coincides with that of Post's extrapolation, is due to the balancing effects of the decreasing stress level, which acts to lower the strain rate and the increasing T/T_m value which acts to increase it. In the same depth interval, the equivalent viscosity drops slowly but steadily from about 10^{22} poises to 10^{21} poises, this decay being the result of the steady drop in stress while the strain rate has remained constant ($\eta = \sigma/3\dot{\varepsilon}$). Thus, this analysis suggests that, in general, the steady state strain rate increases with depth with a concomitant decrease of viscosity but that the nature of the variation depends on the nature of the state of stress and on the thermal structure.

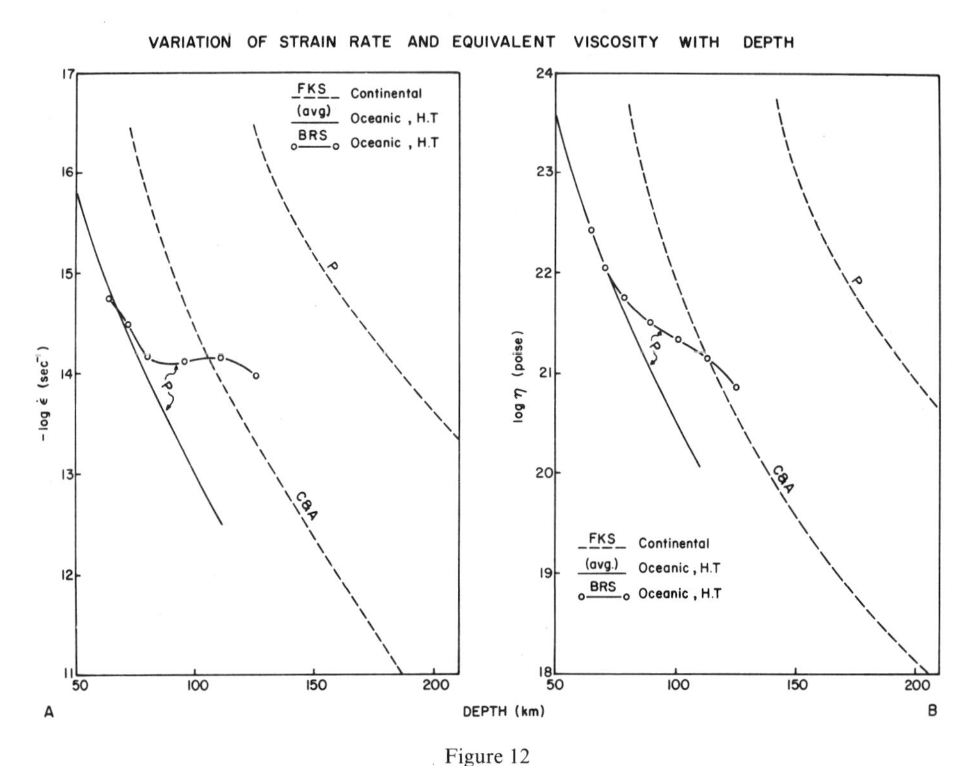

Figure 12
Variations in strain rate and equivalent viscosity with depth using POST's (1976) and CARTER and AVE'LALLEMANT's (1970) mechanical data, appropriate T/T_m values (Fig. 3B), equation (2) and assuming that the stress profiles obtained for the xenoliths (Fig. 11) are correct. The variations beneath continents are calculated from stresses for Frank-Smith xenoliths and under oceans for the average (avg.) of Kilbourne-type sequences and from the Black Rock Summit (BRS) data for Boree-type sequences.

In Fig. 13 we have plotted variations in plate velocities assuming homogeneous shearing to the depth indicated and that the strain rate profiles of Fig. 12A are representative. It is evident that the resulting surface velocities have meaning only if plates are dragged by flow in the material below (whether seismic lithosphere or asthenosphere) and are not resisted by a counterflow, and if small-scale convection normal to the axis of spreading is negligible. For the Boree-type (BRS) sequences, the surface velocity is nearly a linear function of the depth over which shearing takes place because the strain rate is nearly constant at 10^{-14}/sec. However, for the other oceanic sequences (avg.) and for the continental strain rate profiles (FKS), surface velocities increase rapidly with the depth of flow because of the very rapid increase in strain rate with depth. Thus, for these strain rate profiles, reasonable plate velocities are attained only if the depth over which flow takes place is restricted to about 50 km, for the model employed. Alternatively, such rapid increase in the strain rate, resulting in a nearly exponential increase in surface velocity as a function of depth of flow, might argue in favor of decoupling and the push-pull-counterflow hypothesis although there is as yet no direct evidence to support a mechanical discontinuity. In any instance, we do not have the answer to this important question and any interpretation of the present observations, especially as portrayed in Fig. 12 and 13, must be regarded as conjectural.

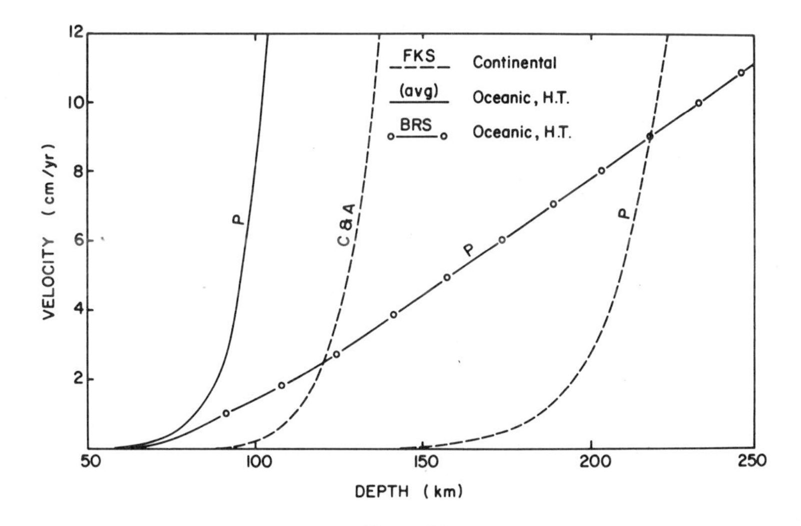

Figure 13
Plot of surface velocities assuming homogeneous strain over depths indicated and strain rate profiles of Figure 12A (see text).

In summary, we have obtained '*in situ*' stress estimates, on the basis of olivine steady state recrystallized grain sizes, from xenoliths in basalts and kimberlites from four different regions in the upper mantle beneath continents and ocean basins. The results are similar with flow stresses ranging from 200 to 300 bars at 40 to 50 km depth to a few tens of bars at depths below 100 km. With the exception of Boree-type sequences, these stresses attain nearly a constant value at depths of 60 to 75 km beneath ocean basins and at 120 to 130 km beneath continents, values interpreted as possibly marking the lid of the asthenosphere but there is no evidence for a mechanical discontinuity at the lithosphere-asthenosphere boundary. The flow stresses obtained are regarded as reasonable and are in fair accord with those estimated from steady-state mechanical data when account is taken of the likelihood of a general increase in strain rate with depth. While these results are of a preliminary nature, and the problem is currently under intensive investigation, we believe that they are qualitatively correct and should be incorporated into or used as constraints for attempts to model plate motions.

Acknowledgments

We wish to thank C. Archambeau for allowing us to make use of his projections in Fig. 10 and R. Twiss for critical comments on the manuscript. This work was supported by National Science Foundation Grant EAR 7604129.

References

BALL, C. J. (1957), Phil. Mag. *7*, 1011.

BECK, P. A. and SPERRY, P. R. (1950), *Strain-induced grain-boundary migration in high purity aluminum*, J. Appl. Phys. *21*, 150–152.

BOYD, F. R. (1973), *The pyroxene geotherm*, Geochem. Cosm. Acta. *37*, 2533–2546.

CAHN, R. W., *Recovery and recrystallization*, in *Physical Metallurgy*, (American Elsevier, New York, 1970), pp. 1129–1198.

CARTER, N. L. (1975), *High-temperature flow of rocks*, Rev. Geophys. Space Phys. *13*, 344–349.

CARTER, N. L. (1976), *Steady state flow of rocks*, Rev. Geophys. Space Phys. *14*, 301–360.

CARTER, N. L. and AVE'LALLEMANT, H. G. (1970), *High temperature flow of dunite and peridotite*, Geol. Soc. Amer. Bull. *81*, 2181–2202.

CATHLES, L. M., *The viscosity of the earth's mantle*, (Princeton Univ. Press, 1975), p. 386.

CLARK, S. P., JR. and RINGWOOD, A. E. (1964), *Density distribution and constitution of the mantle*, Rev. Geophys. *2*, 35–68.

CRITTENDEN, M. D., JR. (1967), *Viscosity and finite strength of the mantle as determined from water and ice loads*, Geophys. J. Roy. Astron. Soc. *14*, 261–279.

FROIDEVAUX, C. and SCHUBERT, G. (1975), *Plate motion and structure of the continental asthenosphere: A realistic model of the upper mantle*, J. Geophys. Res. *80*, 2553–2564.

GILLULY, J., *Tectonics involved in the evolution of mountain ranges*, in *The Nature of the Solid Earth*, (ed. E. C. Robertson) (McGraw-Hill, New York, 1972). pp. 406–439.

GLOVER, G. and SELLARS, C. M. (1973), *Recovery and recrystallization during high temperature deformation of α-iron*, Met. Trans. *4*, 765–775.

GEOTZE, C. (1975), *Sheared lherzolites: From the point of view of rock mechanics*, Geology, *3*, 172–173.

GOETZE, C. and KOHLSTEDT, D. L. (1973), *Laboratory study of dislocation climb and diffusion in olivine*, J. Geophys. Res. *78*, 5961–5971.

GREEN, H. W. (1967), *Extreme preferred orientation produced by annealing*, Science *157*, 1444–1447.

GREEN, H. W. and RADCLIFFE, S. V. (1972a), *Dislocation mechanisms in olivine and flow in the upper mantle*, Earth Planet. Sci. Lett. *15*, 239–247.

GREEN, H. W. and RADCLIFFE, S. V. (1972b), *Deformation Processes in the upper mantle*, Amer. Geophys. Un. Mono. *16*, 139–156.

GRIGGS, D. T., *The sinking lithosphere and the focal mechanism of deep earthquakes* in *Nature of the Solid Earth* (ed. E. C. Robertson) (McGraw-Hill, New York, 1972), pp. 361–384.

HASKELL, N. A. (1935), *Motion of viscous fluid under a surface load, 1*, Physics *6*, 265–269.

HEARD, H. C. (1972), *Steady-state flow in polycrystalline halite at pressure of 2 kilobars*, Amer. Geophys. Un. Mem. *16*, 191–210.

HEARD, H. C. (1976), *Comparisons of the flow properties of rocks at crustal conditions*, Phil. Trans. Roy. Soc. Lon. (in press).

HEARD, H. C. and CARTER, N. L. (1968), *Experimentally induced 'natural' intragranular flow in quartz and quartzite*, Am. J. Sci. *266*, 1–42.

HEARD, H. C. and RALEIGH, C. B. (1972), *Steady-state flow in marble at 500°–800°C*, Geol. Soc. Amer. Bull. *83*, 935–956.

HERRIN, E., *A comparative study of upper mantle models: Canadian shield and Basin and Range provinces* in *Nature of the Solid Earth* (ed. E. C. Robertson) (McGraw-Hill, New York, 1972), pp. 216–231.

HOUSTON, M. H. JR. and DE BREMAECKER, J-Cl (1975), *Numerical models of convection in the upper mantle*, J. Geophys. Res. *80*, 742–755.

JONAS, J. J., SELLARS, C. M. and MCG. TEGART, W. J. (1969), *Strength and structure under hot-working conditions*, Metallurgical reviews *130*, 1–24.

KIRBY, S. H. and RALEIGH, C. B. (1973), *Mechanisms of high-temperature, solid-state flow in minerals and ceramics and their bearing on the creep behavior of the mantle.* Tectonophysics *19*, 165–194.

KOHLSTEDT, D. L., GOETZE, C., DURHAM, W. B. and VANDER SANDE, J. (1976), *New technique for decorating dislocations in olivine*, Science *191*, 1045–1046.

KRAUT, E. A. and KENNEDY, G. C. (1966), New melting law at high pressures, *Phys. Rev. 151*, 668–675.

KUSHIRO, I., *Partial melting of garnet lherzolites in kimberlites at high pressures* in *Lesotho Kimberlites* (ed. P. H. Nixon) (L.N.D.C., Maseru, Lesotho, 1973), pp. 294–299.

LACHENBRUCH, A. H. (1968), *Preliminary geothermal model for the Sierra Nevada*, J. Geophys. Res. *73*, 6977–6989.

LANGSETH, M. G., JR, LE PICHON, X. and EWING, M. (1966), *Crustal structure of the mid-ocean ridges, 5, Heat flow through the Atlantic ocean floor and convection currents,* J. Geophys. Res. *73*, 5321.

LUTON, M. J. and SELLARS, C. M. (1969), *Dynamic recrystallization in nickel and nickel-iron alloys during high temperature deformation*, Acta Metallurgica *17*, 1033–1043.

MACGREGOR, I. D. (1974), *The system* MgO-Al_2O_3-SiO_2*: solubility of* Al_2O_3 *in enstatite for spinel and garnet peridotite assemblages*, Am. Min. *59*, 110–119.

MACGREGOR, I. D. and BASU, A. R. (1976), *Geological problems in estimating mantle geothermal gradients*, Am. Min. *61*, 715–724.

MCCONNELL, R. K., *Viscosity of the earth's mantle* in *The History of the Earth's Crust*, (ed. R. A. Phinney) (Princeton Univ. Press, Princeton, N.J., 1968), pp. 45–57.

MCKENZIE, D. P. (1969), *Speculations on the consequences and causes of plate motions*, Geophys. J. Roy. Astron. Soc. *18*, 1–32.

MCKENZIE, D. P., ROBERTS, J. M. and WEISS, N. O. (1974), *Convection in the earth's mantle: Toward a numerical simulation*, J. Fluid Mech. *62*, 465–538.

MERCIER, J-C. C. (1976a), *Natural peridotites: chemical and rheological heterogeneity of the upper mantle*, Ph.D. thesis S.U.N.Y. at Stony Brook.

MERCIER, J-C. C. (1976b), *Single pyroxene geothermometry and geobarometry*, Am. Min. *61*, 603–615.

MERCIER, J-C. C. and CARTER, N. L. (1975). *Pyroxene geotherms*, J. Geophys. Res. *80*, 3349–3362.

MERCIER, J-C. C. and NICOLAS, A. (1975), *Textures and fabrics of upper-mantle peridotites as illustrated by xenoliths from basalts*, J. Petrol. *16*, 454–487.

MINEAR, J. W. and TOKSÖZ, M. N. (1970), *Thermal regime of a downgoing slab and new global tectonics*, J. Geophys. Res. *75*, 1397–1419.

MOSES, P. L. (1961), *Geothermal gradients known in greater detail*, World Oil *152*, 79–82.

NICHOLS, E. A. (1947), *Geothermal gradients in mid-continent and Gulf Coast oil fields*, Trans. AIME Petrol. Dev. *170*, 44–50.

NICOLAS, A., BOUCHEZ, J-L, BOUDIER, F. and MERCIER, J-C. (1971), *Textures, structures and fabrics due to solid flow in some European lherzolites*, Tectonophysics *12*, 55–86.

OXBURGH, E. R. and TURCOTTE, D. L. (1968), *Mid-ocean ridges and geotherm distribution during mantle convection*, J. Geophys. Res. *73*, 2643.

PARRISH, D. K., KRIVZ, A. and CARTER, N. L. (1976), *Finite element folds of similar geometry*, Tectonophysics *32*, 183–207.

POIRIER, J-P. and NICOLAS, A. (1975), *Deformation-induced recrystallization due to progressive misorientation of subgrains with special reference to mantle peridotites*, Jour. Geol. *83*, 707–720.

PONTIKIS, V. and POIRIER, J-P. (1975), *Phenomenological and structural analysis of recovery-controlled creep, with special reference to the creep of single-crystal silver chloride*, Phil. Mag. *32*, 577–592.

POST, R. L., JR. (1973), *The flow laws of Mt. Burnet dunite*, Ph.D. thesis, Univ. Calif., Los Angeles.

POST, R. L., JR. (1976), *High temperature creep of Mt. Burnet dunite*, Tectonophysics (in press).

RALEIGH, C. B. and KIRBY, S. H. (1970), *Creep in the upper mantle*, Mineral. Sci. Amer. Spec. Pap. *3*, 113–121.

RALEIGH, C. B., KIRBY, S. H., CARTER, N. L. and AVE'LALLEMANT, H. G. (1971), *Slip and the clinoenstatite transformation as competing rate processes in enstatite*, J. Geophys. Res. *76*, 4011–4022.

ROSS, J. V. (1977), *Flow and recrystallization of orthopyroxene* (in press).

SCLATER, J. G. and FRANCHETEAU, J. (1970), *The implications of terrestrial heat flow observations on current tectonic and geochemical models of the crust and upper mantle of the earth*, Roy, Astron. Soc. Geophys. Jour. *20*, 509–537.

SHERBY, O. D. and BURKE, P. M. (1968), *Mechanical behaviour of crystalline solids at elevated temperature*, Prog. Met. Sci. *13*, 325–390.

STACEY, F. D., *Physics of the Earth*, (John Wiley and Sons, Inc., New York, 1969), p. 324.

TORRENCE, K. E. and TURCOTTE, D. L. (1971), *Structure of convection cells in the mantle*, J. Geophys. Res. *76*, 1154–1161.

TWISS, R. J. (1977), *Theory and applicability of recrystallized grain size paleopiezometer*, Pure appl. Geophys., this issue.

WALCOTT, R. I. (1970), *Flexural rigidity and viscosity of the lithosphere*, J. Geophys. Res. *75*, 3941–3954.

WEERTMAN, J. (1968), *Dislocation climb theory of steady-state creep*, Trans. ASME *61*, 681–694.

WEERTMAN, J. (1970), *The creep strength of the earth's mantle*, Rev. Geophys. Space Phys. *8*, 145–168.

WEERTMAN, J. and WEERTMAN, J. R., *Mechanical properties, strongly temperature-dependent* in *Physical Metallurgy* (ed. R. W. Cahn) (1970) pp. 983–1010.

WHITTEN, C. A. (1956), *Crustal movements in California and Nevada*, Am. Geophys. Un. Trans. *37*, 393.

(Received 3rd November 1976)

Pageoph, Vol. 115 (1977), Birkhäuser Verlag, Basel

Theory and Applicability of a Recrystallized Grain Size Paleopiezometer

By ROBERT J. TWISS[1])

Abstract – An approximate theoretical relation is derived which relates stress during steady state creep to both subgrain size and dynamically recrystallized grain size. The relation results from equating the dislocation strain energy in the grain boundary to that in the enclosed volume. Available data on metals and silicates are in excellent agreement with the theory. For paleopiezometry, the recrystallized grain size must be preserved by quenching, by cooling under stress, or by inhibition of grain growth by intimate mixture of two or more phases. In general, stress may be underestimated using rocks in which grain size has been reduced by dynamic recrystallization, especially if the grain size is very small. Stress may be overestimated using coarse grained rocks in which the grain size has increased toward the steady state value. Quantitative limits remain to be established. The theoretical relation can in principle be applied to any metal or mineral if only the effective isotropic elastic moduli and the Burgers vector are known. When used as a paleopiezometer, the technique indicates that high stresses on the order of 100 MPa are not infrequently associated with mantle diapirism and with large scale thrust faulting. Consideration of the Mt. Albert ultramafic body suggests that texturally inferred stresses from peridotite massifs and from ultramafic xenoliths in alkali olivine basalts might reflect either horizontal variations in stress across a rising diapir or else a vertical variation in stress as defined by the pyroxene geobarometer (MERCIER *et al.* 1977). In either case the stresses are probably characteristic of local diapirism. Stresses characteristic of global upper mantle flow might be inferred from xenoliths originating from above kimberlite-producing diapirs.

Key words: Paleopiezometer; Tectonic stress; Recrystallized grain size; Subgrain size.

1. *Introduction*

Rocks which have undergone natural ductile deformation at high temperatures are commonly referred to as tectonites. They are characterized by the presence of preferred orientations of crystallographic axes; by substructures indicative of the operation of dislocation mechanisms of deformation, such as polygonized grains, subgrains, deformation lamellae, dynamically recrystallized grains and high dislocation densities; and by textures indicative of dynamic recrystallization such as porphyroclastic[2])

[1]) Department of Geology, University of California, Davis, California 95616, USA.

[2]) This term has been objected to presumably because of its sedimentary connotations (BLACIC in MERCIER 1976). The term *blastogranular* used by MERCIER (1976), however, does not describe the characteristic bimodal grain size distribution. *Porphyroblastic* would be ideal except for its genetic connotation that the porphyroblast grew to a large size in the fine grained groundmass of the metamorphic rock. The only solution for describing a metamorphic rock having large *relict* metamorphic grains in a fine grained recrystallized groundmass would seem to be porphyroclastoblastic which becomes ridiculous. I retain the somewhat unsatisfactory term which has gained wide acceptance in the literature.

textures. Tectonites are characteristically found in the metamorphic cores of mountain belts and as xenoliths brought up from the mantle in alkali olivine basalts and kimberlites.

Although techniques have been developed by which, under good conditions, the geometry and the kinematics of deformation in high grade metamorphic terranes can be unraveled (TURNER and WEISS 1963, HANSEN 1971), the lack of adequate constitutive equations describing ductile behavior in rocks, and the difficulty in inferring the strain rates or stresses experienced by the rocks usually make dynamic analysis all but impossible. Progress is being made in the laboratory on defining constitutive equations, especially for olivine and quartz (for a review, see CARTER 1976). The present paper outlines a theory and discusses the application of a technique by which, under some circumstances, the paleostress may be inferred from the textures of tectonites.

In the metals literature there is ample evidence that dislocation density, subgrain size (BIRD *et al.* 1969), and dynamically recrystallized grain size (LUTON and SELLARS 1969, GLOVER and SELLARS 1973, BROMLEY and SELLARS 1973) all show a monotonic variation with steady state stress during hot creep, *independent of temperature and total strain.* GOETZE and KOHLSTEDT (1973), KOHLSTEDT and GOETZE (1974), GOETZE (1975) and KOHLSTEDT *et al.* (1976a) have proven that for olivine as well, the dislocation density varies monotonically with steady state stress, and they have used this relation as a paleopiezometer. POST (1973), CARTER and MERCIER (1976), and MERCIER *et al.* (1977) showed that the dynamically recrystallized grain size in olivine varies with stress, and the latter authors and TWISS (1976a) have used this relation to infer stresses from mantle xenoliths.

In principle, dislocation density, subgrain size and dynamically recrystallized grain size can all be used to infer stresses from deformed minerals if the appropriate relations are known. In practice, the determination of dislocation densities and subgrain sizes is difficult and expensive because it requires the use of the transmission electron microscope, although the decoration technique of KOHLSTEDT *et al.* (1976b) allows easy examination of dislocations in olivine. Optical methods of determining the dislocation density by etch pit counts, and of determining the subgrain size by optically visible misorientation have proven unreliable (e.g., BIRD *et al.* 1969). The determination of the dynamically recrystallized grain size, however, is easily done on a petrographic microscope and hence suggests itself as a simple and practical paleopiezometer.

At present, experimental data exist only for olivine (POST 1973, CARTER and MERCIER 1976, MERCIER *et al.* 1977) and quartz (MERCIER *et al.* 1977). A theory (TWISS 1976b) is outlined here by which the relation between stress and both recrystallized grain size and subgrain size may be predicted for other minerals. The theory is in good agreement with existing data from both metals and silicates. The conditions under which the technique may be used to determine paleostress are also discussed, and some applications are considered.

2. *Summary of theory*

When a material is subjected to hot creep, it is generally observed that dislocations tend to gather into planar walls which divide individual grains into subgrains. Under steady state conditions, the mean subgrain size is a function of the applied differential stress and is a steady state feature of the substructure. The subgrain walls coexist with a steady state density of dislocations in the volume, and that density is also a function of the steady state stress. It is not uncommon for crystalline materials to undergo recrystallization during creep. When first formed, the dynamically recrystallized grains are observed to be effectively dislocation free, and the mean size is a function of the steady state stress. It is the object of the theory to account for the variation with steady state stress of the mean subgrain size and the mean dynamically recrystallized grain size. It will appear that the dislocation density within the subgrain or the recrystallized grain is critically important in determining its size.

The fundamental assumption on which the theory is based (A. K. MUKHERJEE 1975, personal communication) is that the formation of subgrains and recrystallized grains must be energetically favorable processes. Thus the total strain energy of dislocations ordered into a closed surface, i.e. either a subgrain or a recrystallized grain boundary, must be less than or equal to the total energy of a steady state density of dislocations within the enclosed volume. Assuming the equality, leads directly to a unique grain dimension d for any given state: as the grain size changes, the total volume energy varies directly as the volume, and hence as d^3. The total grain boundary energy, however, varies as the surface area or only d^2. Thus there must exist a unique grain size at which the energies are equal.

The condition is expressed for an idealized cubic subgrain or recrystallized grain by

$$6\gamma d^2 = w\rho d^3 \tag{1}$$

where γ is the dislocation strain energy per unit area in the grain boundary, w is the dislocation strain energy per unit length in the grain volume, d is the grain dimension, and ρ is the steady state dislocation density in the grain volume. The stress is introduced into the relation through the equation relating differential stress σ to the dislocation density

$$\sigma = \alpha\Gamma b\rho^{1/2}, \qquad \Gamma \equiv \frac{\mu}{1 - \nu} \tag{2}$$

where b is the Burgers vector, μ the shear modulus, ν the Poisson ratio, and α an empirical parameter of order 1. This equation is theoretically justified by the assumption that at steady state creep, the back stress on a dislocation caused by its interaction with other nearby dislocations must equal the applied stress (e.g., BASINSKI 1959). The parameter α corrects for the complexity of the dislocation interactions.

By making the simplifying assumptions that all dislocations are edge dislocations, that all boundaries are simple tilt boundaries, and that crystals are elastically isotropic, relatively simple equations can be found for the volume and boundary energy densities (HIRTH and LOTHE 1968, equations 3–49, 19–84). All these assumptions are adequate for a first order approximation. The self energy of edge and screw dislocations differs only by a factor $(1 - \nu)$, and tilt boundaries are very common among subgrain boundaries. The low angle tilt boundary theory is good for misorientations up to 10° to 15°. This includes subgrain boundaries, which commonly have misorientations of 1° or less, and it is considered adequate for the initial stages of recrystallization before grain rotation has proceeded too far. After some manipulation (TWISS 1976b), equation (1) can be reduced to the forms

$$\frac{\sigma}{\Gamma} = K\left(\frac{d}{b}\right)^{-p} \quad \text{or} \quad \log\frac{\sigma}{\Gamma} = \log K - p\log\frac{d}{b} \tag{3}$$

where

$$p \equiv \frac{\phi}{2\phi - 1}, \qquad \log K = p\left[\log\frac{3e\alpha^2\beta}{\pi\phi} - \frac{1}{\phi}\log\frac{\alpha\beta}{2}\right] \tag{4}$$

and where ϕ is the ratio of total dislocation length in the boundary to that in the enclosed volume, β is a parameter which accounts for the energy of the dislocation core, and e is the Napierian base. Values of α and β are most accurately evaluated empirically, but known values vary only over a small range: $0.7 \lesssim \alpha \lesssim 2.3$ (BIRD *et al.* 1969, KOHLSTEDT *et al.* 1976); $1 \lesssim \beta \lesssim 8$ (HIRTH and LOTHE, 1968). The values of α here are appropriate for the use of Γ rather than just μ in equation (2).

The value of the parameter ϕ is the only difference between subgrains and recrystallized grains in this theory. Approximate theoretical values of ϕ_s for subgrains and ϕ_r for recrystallized grains can be obtained as follows: We assume that the subgrain volume contains a steady state density of dislocations. Equation (1) can be combined with equations (3–49) and (19–84) from HIRTH and LOTHE (1968) to yield a relation between dislocation energy per unit length in the boundary and in the volume (TWISS 1976b).

$$D\gamma = \frac{1}{\phi}w \tag{5}$$

where D is the spacing of dislocations in the boundary. From equations (3) and (4) at constant σ/Γ, an increase in ϕ can be shown to provide an increase in the predicted grain size. Assuming the subgrain size that forms is the smallest stable size, we must take $\phi_s = 1$. For lower values, dislocations would increase their energy by moving into the boundary (equation 5) and the boundary would not form.

To produce a dislocation-free recrystallized grain, imagine that the dislocations in the volume all move outward by expansion of loops to form the grain boundary.

Assuming again that the smallest stable grain size is the one that develops, ϕ is minimized by requiring the boundary to be formed only by expansion of those dislocations originally in the volume at the time of recrystallization. The increase in dislocation length can be represented by the expansion of a loop from some mean diameter to the grain diameter d. Assuming, for lack of a better approximation, that the mean diameter is between $0.5d$ and $0.7d$, the diameter of that loop that encloses one half the grain cross sectional area, then the ratio of the expanded length to the unexpanded length of the mean loop gives $1.4 \leq \phi_r \leq 2$.

Using the above values of ϕ in equation $(4)_1$ gives $p_s = 1$ and $0.78 \gtrsim p_r \gtrsim 0.67$.

The success of the theory is indicated in Fig. 1 in which data on the variation with stress of subgrain and recrystallized grain size is plotted on nondimensional coordinates for olivine, quartz, and a variety of metals. The dashed lines are the linear regressions on individual data sets. The solid lines represent best fits to the two groups of data assuming a slope which is a mean of the slopes for the relevant data sets. The slopes and intercepts are

$$p_s = 1 \pm 0.03 \qquad \log K_s = 0.91 \pm 0.01 + 3.85\Delta p_s \tag{6}$$

$$p_r = 0.68 \pm 0.02 \qquad \log K_r = 0.38 \pm 0.01 + 5.30\Delta p_r \tag{7}$$

where the error on the slopes is the 70% confidence limit (approximately 1 standard deviation) calculated by assuming the mean slopes are the best fits, and where the first error on each intercept is the 70% confidence limit for the mean value of $\log \sigma/\Gamma$, and the second error prescribes the dependence of the intercept on a change in slope (WINE 1964). The equations are, for subgrains:

$$\log \frac{\sigma}{\Gamma} = 0.91 - \log \frac{d}{b} \tag{8}$$

and for recrystallized grains:

$$\log \frac{\sigma}{\Gamma} = 0.38 - 0.68 \log \frac{d}{b} \tag{9}$$

The values of slopes and intercepts from (6) and (7) can be substituted into (4) to calculate values for ϕ, α, and β.

$$\phi_s = 1 \pm 0.03 \qquad \phi_r = 1.89 \pm 0.15 \qquad \alpha = 1.57 \pm 0.33 \qquad \beta = 0.89\,{}^{-0.55}_{+2.46} \tag{10}$$

where the errors indicate the extremes calculated using the extreme values from (6) and (7). If the values (10) are introduced into equations (3) and (4), the theoretical curves fit the empirical ones exactly. The values of the parameters ϕ, α, β are well within the rather narrow independently expected limits; therefore the theory, although an approximate one, is in excellent accord with the experimental data.

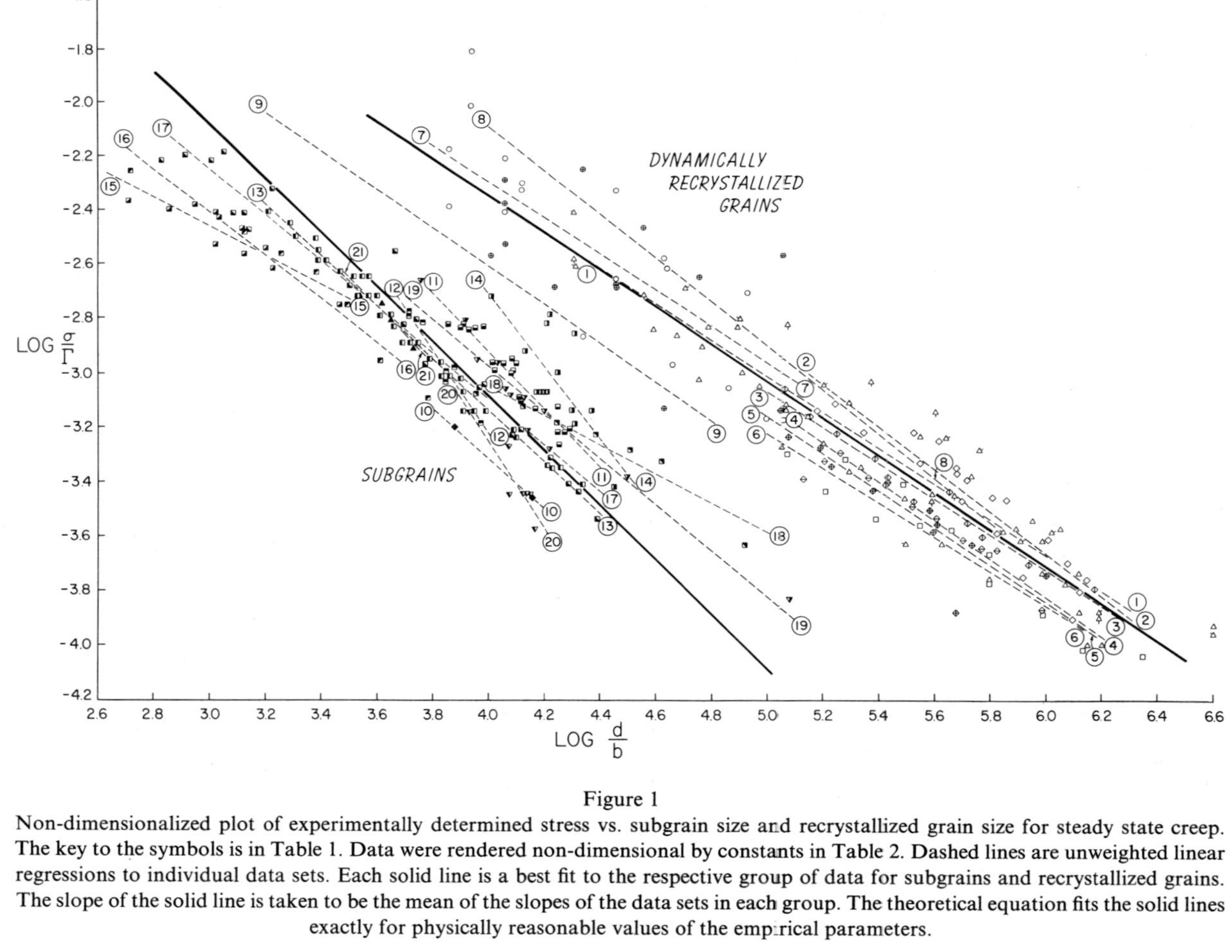

Figure 1

Non-dimensionalized plot of experimentally determined stress vs. subgrain size and recrystallized grain size for steady state creep. The key to the symbols is in Table 1. Data were rendered non-dimensional by constants in Table 2. Dashed lines are unweighted linear regressions to individual data sets. Each solid line is a best fit to the respective group of data for subgrains and recrystallized grains. The slope of the solid line is taken to be the mean of the slopes of the data sets in each group. The theoretical equation fits the solid lines exactly for physically reasonable values of the empirical parameters.

Table 1

Key to Figure 1

Curve No.	Material	Symbol	No. Data Points	Temperature Range °K	Slope*	Intercept*	Source
				RECRYSTALLIZED GRAIN SIZE DATA			
1	Cu	△	30	814–1220			
	Cu 1% Al	△	8	814–1220	−0.66	0.27	Bromley and Sellars (1973)
	Cu 4% Al	△	8	814–1220			
	Cu 8% Al	△	8	814–1220			
2	Ni	◇	19	1035–1552	−0.78	0.99	Luton and Sellars (1973)
3	Ni 5% Fe	◇	12	1207–1552	−0.66	0.21	Luton and Sellars (1973)
4	Ni 10% Fe	◇	11	1207–1552	−0.71	0.44	Luton and Sellars (1973)
5	Ni 20% Fe	◇	11	1207–1552	−0.68	0.25	Luton and Sellars (1973)
6	αFe	□	12	1023–1153	−0.63	−0.06	Luton and Sellars (1973)
7	Olivine $(Mg_{.92}Fe_{.08})_2SiO_4$						
	wet	○	18	1073–1423	−0.67	0.37	Post (1973)
	dry	⊕	12	1148–1699			
8	Olivine $(Mg_{.92}Fe_{.08})_2SiO_4$	NO DATA PLOTTED			−0.81	1.15	Mercier *et al.* (1977)
9	Quartz (SiO_2)	NO DATA PLOTTED			−0.71	0.23	Mercier *et al.* (1977)
				SUBGRAIN SIZE DATA			
10	Ni	◆	2	923	−0.93	0.40	Mitra and McLean (1967)
11	αFe, vacuum melted	⬓	21	773–1123	−1.06	1.33	Glover and Sellars (1973)
12	αFe, zone refined	⬒	7	773–1123	−1.53	2.89	Glover and Sellars (1973)
13	low C Steel	◧	35	719–870†)	−0.95	0.66	Goldberg (1966)
14	Austenitic Stainless Steel	◨	16	977–1103	−1.33	2.60	Garofalo *et al.* (1963)
15	316 Stainless Steel	◪	11	973–1173	−0.54	−0.84	Young and Sherby (1973)
16	Fe 25% Cr	◩	7	773–1073	−0.80	−0.01	Young and Sherby (1973)
17	Fe 3% Si	⬔	21	643–1373	−0.85	0.30	Young and Sherby (1973)
18	Fe 3% Si	⬕	7	916	−0.54	−0.87	Barrett *et al.* (1966)
19	Al	▽	12	528–889	−0.85	0.44	McQueen *et al.* (1967)
20	Al	▼	7	433–573	−1.62	3.22	Chen *et al.* (1975)
21	Cu	▲	5	273	−1.02	0.92	Pratt (1966)

*) Slopes and intercepts are from my unweighted linear regressions on the reported data.
†) Goldberg's data points are averages over the temperature range studied.

3. *Limitations of the technique*

Although the recrystallized grain size paleopiezometer is potentially very useful and is capable of providing heretofore unavailable data on the dynamics of tectonic processes, it is necessary to be aware of the limitations of the technique. The following discussion of these limitations is necessarily qualitative because the technique is new and incompletely studied. Such a discussion, however, can at least point out potential problems on which further work is needed.

The necessary requirement for the paleopiezometer to be useful is that the grain size measured represent the *steady state size* of *syntectonically recrystallized* mineral grains. There are several mechanisms that limit the extent to which this requirement can be met.

During steady state dynamic (syntectonic) recrystallization the steady state grain size can be viewed as a balance between two processes: (1) recrystallization, which diminishes the size of grains, and (2) grain growth which increases the size of grains. Evidence from the metals literature indicates that for a given set of conditions, a minimum critical strain is required in a grain before recrystallization can occur (e.g., LUTON and SELLARS 1969). Moreover, the energy considerations presented in the theory imply that the theoretical recrystallized grain size d is the smallest possible stable grain size. Therefore a grain theoretically must be at least $2d$ in diameter before it can recrystallize into two new grains.

In principle, then, the mean grain size $\overline{d}$ will always be somewhat larger than the newly recrystallized grain size d, and the relative rates of strain and of grain growth govern the final mean and the standard deviation of the grain size.

If the critical strain accumulates in newly recrystallized grains in a time less than that required for the grains to double their size, then recrystallization can occur as soon as grain growth permits two new grains each of diameter d to be produced. Thus the mean grain size $\overline{d}$ will be between d and $2d$, and the standard deviation will be relatively small. If, on the other hand, the critical strain accumulates in a time very long compared with the time required to double the size of a newly recrystallized grain, grain growth can continue beyond a grain size of $2d$, and the mean grain size $\overline{d}$ and the standard deviation will both be a larger multiple of d than in the previous case. In strict terms, the values presented in equations (10) were not obtained properly because equation (10) results from fitting a theory involving newly recrystallized grain size to data involving the mean grain size. The difference, however, is not very large, although if independently determined values for the material parameters were used instead of those in equation (10), the application of the resulting equations to mean grain size data could result in an underestimate of the stress.

After stress is removed from a tectonite, dynamic recrystallization ceases, and if the temperature remains high, annealing and grain growth can proceed. Thus any tectonite that has been annealed will yield stress estimates which are too low. The driving forces for recrystallization and grain growth during annealing are (1) the reduction of strain energy by the reduction of dislocation density as recrystallization and grain growth proceed, and (2) the decrease in grain boundary energy density through grain growth and the consequent reduction in grain boundary area per unit volume. The driving force, and therefore the rate of grain growth, is highest for a material having a high dislocation density and a small grain size. The textures associated with high stresses are therefore the most difficult to preserve, and stresses inferred from them will tend to be too low.

Post-deformational annealing must be a common phenomenon in nature, and it puts severe limitation on the rocks to which this paleopiezometer may be applied. There are several mechanisms, however, that can allow the preservation of dynamically recrystallized grain size, and there must be adequate evidence that one of these mechanisms has operated before inferred stresses can be considered meaningful. *First*, if a rock is rapidly cooled from a high temperature of deformation, or if the temperature of deformation has never been very high, the kinetics of grain growth can become slow enough to prevent a significant post-deformational increase in grain size. *Second*, if a constant tectonic stress is maintained on a tectonite during even a slow cooling history, the dynamically recrystallized grain size, being independent of temperature, will not change. *Third*, the grain size of a particular mineral may be preserved by the physical inhibition of its growth by adjacent grains of a different mineral.

Different problems must be considered when the approach to a steady state grain size involves an increase in the initial grain size, especially if the applied stress is very low and the theoretical steady state grain size is large. The problems are that: (1) a decrease in stress from an initial steady state requires a gradual reorganization of the internal structure to a new steady state condition; and (2) the driving forces at low stresses may be insufficient to cause grain growth to a steady state condition in the available time.

If an initial steady state stress applied to a rock is decreased, and if we ignore temporarily the problem of the kinetics of grain growth, it is necessary to know how rapidly the structure of the material will change so that new recrystallized grains will reflect the new low stress condition. Some data exist on the variation of dislocation density and subgrain size following a decrease in stress, and the process appears to be slow (Mitra and McLean 1967). In fact in aluminium, the reorganization of substructure to reflect a lower stress occurs gradually through a strain of approximately 5% to 10% (Sherby *et al.* 1956). If this is typical also of silicates, then for olivine at stresses near 5 MPa and at reasonable mantle temperatures, a time on the order of 10^6 years would be required to alter the internal structure to a new steady state. Such lengths of time are not negligible for certain geologic processes, and if recrystallized grain size behaves in a similar manner, the application of this paleopiezometer could yield excessive values of stress for the last tectonic event. Some experimental work on quartz and olivine indicates that the dynamically recrystallized grain size recovers quite rapidly after a decrease in stress (Mercier *et al.* 1977; Carter and MERCIER 1976, personal communication) so that this problem may not be troublesome. Quantitative data on the process are needed, and the matter is under investigation in Carter's laboratory.

In addition to this problem, however, there is the additional problem of the kinetics of grain growth. As mentioned before, grain growth is driven by a decrease in surface energy and a decrease in internal strain energy. As the grain size gets larger, the surface energy term becomes small. Moreover, at the low stresses for which the steady state grain size is large the internal strain energy is also small. For example for olivine at a stress of 5 MPa, corresponding roughly to a 5 mm recrystallized grain size (equation 11), the dislocation density is on the order of 10^5 cm^{-2}. This is 4 to 5 orders of magnitude smaller than dislocation densities typical of common laboratory deformation experiments, and Nicolas (1976) assumes densities of 10^5 to 10^6 cm^{-2} to be typical of an undeformed, statically annealed mantle. The driving forces for grain growth are therefore small, and if they are sufficiently small, the grain size may not reach steady state in the time available.

These problems are most important for coarse grained rocks in which the rates of grain growth and change in internal structure are small. If the grain size cannot reach steady state in the time available, the inferred stress will be an overestimate of the true stress. It may therefore be important to demonstrate whether the observed grain size in a coarse grained rock is a result of an increase or a decrease in the initial grain size, and in the former case, it is certainly important to be able to determine whether the increase was under conditions of static annealing or dynamic recrystallization.

4. *Applications of the technique*

The significance of equation (9) lies in the fact that simply by knowing the value of the Burgers vector and the effective isotropic elastic moduli for a mineral, a relation can be obtained that allows the steady state recrystallized grain size to be used as a paleopiezometer, within the error of the theoretical approximation (Fig. 1) and within the limitations discussed above.

Equation (9) may be written in dimensional form for the minerals anorthite, calcite, olivine and quartz using the data in Table 2.

$$\sigma = Bd^{-0.68} \qquad B \equiv K\Gamma b^{p} \tag{11}$$

where σ is in MPa, d is in mm, and B takes on the following values: 7.8 (anorthite), 7.5 (calcite), 14.6 (olivine), 5.5 (quartz). Thus for a given stress, the steady state recrystallized grain size for all of these minerals is predicted to be approximately the same, within a factor of two. Because the elastic moduli and the Burgers vectors for many other minerals do not differ greatly from those considered here, the prediction is considered quite general.

The ultramafic xenoliths found in alkali olivine basalts and kimberlites provide one

Table 2

Elastic parameters and Burgers vectors for polycrystalline materials*

Material	μ† 10^4 *MPa*	ν	Γ† 10^4 *MPa*	b 10^{-7} *mm*
Aluminium	2.55	0.34	3.86	2.86
Copper‡)	4.62	0.35	7.11	2.56
α Iron‡	7.65	0.28	10.6	2.48
γ Iron	7.65	0.28	10.6	2.52
Nickel‡	7.93	0.31	12.2	2.49
Anorthosite	3.8	0.31	5.51	7
Dunite (Twin Sisters)	7.8	0.24	10.3	6.98
Marble (Danby, Vt.)	3.3	0.31	4.8	8.08
Quartzite	4.2	0.15	4.9	5

*) Elastic parameters for metals are room temperature data from Smithells (1962). Elastic parameters for rocks were calculated from wave velocities at 400 MPa pressure using formulas for isotropic bodies (Birch, Table 7-16 in Clark, 1966).

†) 1 MPa = 10 bars.

‡) The same data were used for all alloys of each metal.

example of textures preserved by quenching. Although there is evidence that some annealing has occurred (Radcliffe and Green 1973, Goetze 1975), the presence of extremely fine grained rocks (~10 μm) especially among the samples from kimberlites argues that the annealing cannot have proceeded very far. An empirical equation similar to equation (11) was determined on experimentally deformed specimens by Post (1973). This equation has been used (Twiss 1976a) to interpret the texture of an ultramafic xenolith studied by Green and Boullier (referenced in Green 1976). The rock has a porphyroclastic texture, and the grain size of the groundmass is approximately 10 to 25 μm (Green 1976, personal communication). Using Post's equation, the grain size indicates stresses between 400 and 220 MPa. Equation (11) for olivine implies the stresses are approximately between 350 and 190 MPa. The agreement is good. Moreover, Goetze (1975) concluded that the dislocation density in the porphyroclasts of the rock indicates stresses between 400 and 200 MPa. Thus the two independent paleopiezometers are in excellent agreement.

The Mt. Albert ultramafic intrusive body has been studied in detail by MacGregor and Basu (1976a, b). They infer that it has intruded diapirically as a hot solid into cooler country rocks, and it possibly represents a situation in which the stress was maintained throughout at least the high temperature part of the cooling history. The result has been the preservation of a radial gradient in texture and grain size from a mosaic texture of 40 μm olivine grains around the margins, through a porphyroclastic texture, to a central area of coarse grained olivine having a mean diameter of about 3 to 4 mm (MacGregor and Basu 1976b). Taken at face value, interpretation of the grain size using equation (11) indicates a gradation in stress approximately from 130 MPa at the margin to 6 MPa in the center. This is qualitatively the type of stress distribution to be expected within a rising diapir. A suite of xenoliths from San Quintin, Baja California, showing similar characteristics to the Mt. Albert intrusive was studied by BASU (1975a, b). The grain size in these rocks varies approximately from 300 μm to 4 mm again suggesting a range of stresses approximately from 30 MPa to 6 MPa.

Mylonites are commonly associated with large thrust faults in metamorphic rocks (e.g., HSU 1955, CHRISTIE 1963, BELL and ETHERIDGE 1976). It is now generally recognized that most mylonites are the product of ductile deformation and recrystallization rather than cataclasis (CHRISTIE 1963, BELL and ETHERIDGE 1973). Indeed the examples of very fine grained dunite discussed above would classify as mylonites according to the revised definition of BELL and ETHERIDGE (1973).

The size of recrystallised grains in mylonites associated with large thrust faults is commonly in the range of 10 to 40 μm indicating that annealing has been prevented either by rapid cooling or by cooling under stress. In the San Gabriel Mountains, Southern California, HSU (1955) reported mylonites in a thrust zone with quartz and feldspar grain sizes of about 10 μm. Stresses inferred from equation (11) indicate 130 to 190 MPa. In the Moine thrust, CHRISTIE (1963) reported quartz grain sizes in the 10 to 40 μm range implying stresses between 50 and 130 MPa. Thrusting in the Arltunga Nappe Complex, Central Australia, produced quartz mylonites with a recrystallized grain size near 30 to 40 μm implying stresses of 50 to 60 MPa (HOBBS *et al.* 1976 referencing YAR KAHN 1972). BELL and ETHERIDGE (1976) also describe mylonites from the Woodroffe Thrust in Central Australia which contain dynamically recrystallized quartz grains in the range 10 to 100 μm indicating stresses between 30 and 130 MPa.

On the basis of evidence such as this, there can be little doubt that tectonic stresses on the order of 100 MPa are commonly associated with the emplacement of thrust sheets in crustal rocks. Accepting the inference of diapirism for Mt. Albert and the San Quintin xenoliths (MACGREGOR and BASU 1976) such stresses also appear to be associated with local mantle diapirism.

HOBBS *et al.* (1976, Fig. 2.19a) show an example in which the size of the quartz grains in a schist is a function of the amount of mica present. Where mica is abundant, the growth of quartz grains has been inhibited, and the grain size is markedly smaller. Unfortunately the texture in this example is ambiguous, as it is not possible to decide whether the fine grain size represents the preservation of a small dynamically recrystallized grain size or the preservation of a grain size that was prevented from increasing to the appropriate steady state size.

Recent investigations of ultramafic xenoliths have shown that textural variations are correlated with position on the chemically determined temperature-pressure curve (e.g., BOULLIER and NICOLAS 1973). MERCIER *et al.* (1977) have used this association to infer variation of stress with depth. This may prove to be a unique method of obtaining direct information about the dynamics of the upper mantle, but several problems suggest caution in the interpretation of the results.

It is common now to assume that the whole mantle is in a dynamic state. It is important, however, not to impose this preconception on available data for fear of dismissing significant information. Apart from any other considerations, an increase in grain size with depth can be explained qualitatively in strictly static terms. During static annealing, the grain size obtained depends on the kinetics of grain growth, the time of the anneal, and the homologous temperature T/T_m (T_m = melting temperature of a particular mineral). As homologous temperature increases, grain growth speeds up, so that for a given annealing time, larger grains will form at higher homologous temperatures. There is evidence (TOZER 1972, TWISS 1976a) that homologous temperature increases with depth in the mantle until it reaches a constant value below the asthenosphere. The grain size in a static portion of the mantle could therefore change in a similar way. Thus it becomes important to prove that textures of coarse grained mantle rocks are caused by dynamic recrystallization and not by static annealing. In some instances this may not be immediately apparent, as, for example, with the protogranular texture of MERCIER and NICOLAS (1975) and BASU (1975a, Plates 1–4). MERCIER and NICOLAS (1975) and NICOLAS (1976) have interpreted the protogranular texture as showing no evidence of deformation. The free dislocation density in olivine grains is approximately 10^5 to 10^6 cm^{-2} (GUEGUEN referenced in NICOLAS, 1976) and NICOLAS (1976) considers this typical of statically annealed mantle. MERCIER *et al.* (1977), however, have inferred that such textures indicate dynamic recrystallization. More work appears to be necessary on ways of distinguishing between these two possibilities.

To associate a texturally inferred stress with a chemically inferred pressure requires the assumption that the texture and the chemistry equilibrate at the same time. The kinetics of reaching a steady state recrystallized grain size, however, are not necessarily the same as the kinetics of equilibrating a pressure-sensitive chemical composition. The first process involves the operation of dislocation mechanisms of deformation and grain growth driven largely by internal strain gradients across grain boundaries. The chemical equilibration is driven by chemical potential gradients across grain diameters. Under constant or very slowly changing conditions such as horizontal shear by steady state creep in the mantle, the effect of the different kinetics is negligible. In a rapidly changing environment such as is found in a rising diapir, however, the different kinetics could become important.

It is useful to explore the possibility with reference to an exposed ultramafic body. MACGREGOR and BASU (1976a, b) have studied the Mt. Albert ultramafic body, and they concluded that it intruded as a hot solid into greenschist facies metamorphic rocks. The body displays a roughly radial variation in the chemically inferred temperature and pressure ranging approximately from 1120°K, 1000 MPa at the margins to 1470°K, 3000 MPa in the core. Variations in texture correlate well with the temperature-pressure data, varying from very fine grained ($\cong$40 μm) at the margin to coarse grained ($\cong$3 to 4 mm) in the core. These data, however, all come from the same level in the ultramafic body, and metamorphic assemblages in the contact aureole indicate maximum conditions of roughly 1000°K, 400 MPa.

The chemical compositions are demonstrably in disequilibrium with the pressure. From the same level in the body a variation of 2000 MPa in chemically recorded pressure is found. Moreover, none of the chemical data indicate equilibration to the ambient pressure at the depth of intrusion. The deformation, however, clearly must have continued up to the present structural level, and there is no independent evidence to indicate at what depth the textures ceased to record the applied stresses.

The temperature-pressure data have been interpreted as indicating diapiric upwelling in the mantle (BASU 1975b, MACGREGOR and BASU 1976a, b). The disequilibrium of the chemical compositions is easily explained in this model by assuming the rate of change of pressure due to the rise of the diapir was much more rapid than the rate of chemical equilibration. Taken at face value, the grain sizes indicate a variation of stress from 130 MPa at the margin to 6 MPa in the center. This is at least qualitatively in accord with the expected *horizontal* distribution of stresses in a rising diapir.

The correlation of texture with inferred temperature and pressure may be explained by an increase in the rate of equilibration in the highly stressed rocks. The high stresses produce high dislocation densities and small grain sizes which promote more rapid equilibration through dislocation pipe diffusion, through the increase in grain boundary diffusion provided by the increase in grain boundary area per unit volume, and through a decrease in the diffusive path length necessary for equilibration. The influx of water into the ultramafic body from the surrounding country rocks also undoubtedly assisted the re-equilibration process (MACGREGOR and BASU 1976a). Thus the more highly stressed marginal rocks were able to proceed further toward equilibration with the ambient conditions than were the rocks in the relatively unstressed core.

Thus there appear to be two equally viable hypotheses to account for the Mt. Albert data: first, that the texturally inferred stresses can be uniquely associated with the chemically inferred pressures to give a stress variation with depth (MERCIER *et al.* 1977), and second that the two inferences cannot be uniquely associated, and the stress data simply record the radial stress variation from the margin to the core across a rising diapir. Because the textural data from the Mt. Albert body are strikingly similar to data from other peridotite massifs (MACGREGOR and BASU 1976a) and to data gathered from suites of spinal peridotite xenoliths from alkali olivine basalts (BASU 1975b, MACGREGOR and BASU 1976a, b) the ambiguities associated with the interpretation from the Mt. Albert body also exist in the interpretation of those other ultramafic suites. In either case, it appears that the data are probably relevant to a local mantle process of diapiric upwelling and do not add directly to our knowledge of the dynamics of mantle-wide convective flow.

GREEN and GUEGUEN (1974) have argued that the texture-pressure-temperature systematics of xenoliths from kimberlites indicate that the xenoliths represent samples from within and above the top of a rising diapir. This is in contrast to the xenoliths from alkali olivine basalts which are interpreted to originate between the core of a rising diapir and its boundary (MacGregor and Basu 1976a). If this distinction is in fact correct, and if the mantle above a kimberlite-producing diapir is indeed undisturbed by immediately preceding diapirs, then the xenoliths which originate above a kimberlite-producing diapir may be considered to represent more 'normal' mantle. In this context, 'normal' mantle is characterized by an environment that does not change rapidly with time and that therefore allows processes having different kinetics to equilibrate. Thus if the grain size can be shown to be the result of dynamic recrystallization, the stress-depth curves inferred from such suites should be more reliable, within the limitations of the technique discussed above.

Carter and Mercier (1976, personal communication) and Moores (1976, personal communication) have suggested that the Mt. Albert massif can be interpreted as part of an ophiolite complex. In this case the temperature-pressure data would have to be interpreted as being in part a disequilibrium relict of an earlier history of diapiric (?) rise through the lithosphere. The mylonitic textures and the chemical alteration reflecting lower temperatures and pressures could then be interpreted as being associated with the process of ophiolite emplacement, and the roughly concentric geometry of textures, temperatures, and pressures as being due to folding. If this were the case, then the comparison to mantle diapirs is not as relevant. This does not, however, negate the arguments (1) that mantle diapirism is involved in the process that leads to partial melting and eruption of xenolith-bearing alkali-olivine basalts; (2) that textures do not necessarily reflect conditions at the depths recorded by the pyroxene geobarometer; and (3) that the observed textures record the dynamics of local processes such as diapirism rather than mantle-wide processes.

5. *Summary*

A theoretical relation has been derived which predicts the variation of subgrain size and dynamically recrystallized grain size with stress at steady state creep. The theory is in excellent agreement with available data on metals and silicates, and it provides the basis for a paleopiezometer using the dynamically recrystallized grain size of a wide variety of minerals.

The paleopiezometer can only be used on rocks which have undergone dynamic recrystallization and which subsequently have not been annealed. The recrystallized grain size can be preserved by quenching, by remaining under constant stress during cooling, or by physical inhibition of grain growth by different neighboring phases. In most cases, application of the paleopiezometer to fine grained recrystallized tectonites can be expected to yield a minimum value for the stress.

If the applied stress has decreased to a low value from a higher steady state value, or if the applied stress requires an increase in grain size to a large value, the approach to a steady state condition may be very slow. If steady state is not reached, the inferred stress will be an overestimate of the true stress. The accuracy of the paleopiezometer applied to natural rocks and the limits of its applicability, however, must still be evaluated by more rigorous analysis and more experimental data.

Under the same stress, the steady state recrystallized grain size for numerous minerals is predicted to be the same within a factor of 2. The application of the paleopiezometer to fine grained ultramafic rocks and to mylonites in large thrust faults indicates that maximum tectonic stresses on the order of 100 MPa are associated with rising mantle diapirs and with large scale thrust faults.

Suites of ultramafic xenoliths must be distinguished according to whether they represent samples from interiors of mantle diapirs, or from more normal mantle undisturbed by diapirs. Peridotite massifs and xenolith suites from alkali olivine basalts probably represent the interiors of mantle diapirs (MACGREGOR and BASU 1976). Estimates of stress from the recrystallized grain size in these rocks may correlate with chemically inferred pressure, but are equally well interpreted as representing a radial distribution of stress from margin to core in a rising diapir independent of the chemically inferred pressure. In either case the inferred stresses are probably associated only with local diapirism. Coarse grained xenoliths from kimberlites may represent normal mantle (GREEN and GUEGUEN 1974), in which case, if the textures are those of dynamic recrystallization, inferred stresses may be correlated with chemically inferred pressures (MERCIER *et al.* 1977) to give valuable information on the dynamics of large scale mantle deformation.

The recrystallized grain size paleopiezometer is a new and potentially powerful tool with which to gain insight into tectonic processes. Like any new technique, important problems remain to be solved, but if applied with circumspection, it can give critical information on tectonic stresses which heretofore has been unobtainable.

Acknowledgement

I am indebted to A. K. Mukherjee for stimulating discussions which led to the basic assumption on which the theory depends. W. Belfield and H. W. Green provided helpful comments on preliminary versions of the manuscript. The reviews by N. L. Carter and J.-C. C. Mercier were also beneficial, although these gentlemen do not subscribe to all the interpretations given in the paper. This research was supported by grant number EAR-75-21878 from the National Science Foundation.

REFERENCES

BARRETT, C. R., NIX, W. D. and SHERBY, O. D. (1966), *The influence of strain and grain size on the creep substructures of* Fe-3Si ASM Trans. Quart. *59*, 3–15.

BASINSKI, Z. S. (1959), *Thermally activated glide in face centered cubic metals and its application to the theory of strain hardening*, Phil. Mag., Ser. 8, *4*, 393–432.

BASU, A. R. (1975a), *Petrogenesis of xenoliths from the San Quintin volcanic field, Baja California*, Ph.D. thesis, University of California, Davis, 229 pp.

BASU, A. R. (1975b), *Hot spots, mantle plumes, and a model for the origin of ultramafic xenoliths in alkali basalts*, Earth Planet. Sci. Letts. *28*, 261–274.

BELL, T. H. and ETHERIDGE, M. A. (1973), *Microstructure of mylonites and their descriptive terminology*, Lithos *6*, 337–348.

BELL, T. H. and ETHERIDGE, M. A. (1976), *The deformation and recrystallization of quartz in a mylonite zone, Central Australia*, Tectonphys. *32*, 235–267.

BIRD, J. E., MUKHERJEE, A. K. and DORN, J. F., *Correlations between high temperature creep behavior and structure*, in *Quantitative Relation Between Properties and Microstructure* (eds. D. G. Brandon and R. Rosen), (Proceedings of an International Conference, Haifa, Israel, 1969), pp. 255–342.

BOULLIER, A. M. and NICOLAS, A., *Texture and fabric of peridotite nodules from kimberlite at Mothae, Thaba, Putsoa, and Kimberley*, in *Lesotho Kimberlites* (ed. P. H. Nixon), (Lesotho National Development Corporation, Maseru, 1973), pp. 57–66.

BROMLEY, R. and SELLARS, C. M., *High temperature deformation of copper and copper-aluminum alloys*, in *The Microstructure and Design of Alloys*, (Proceedings of the Third International Conference on the Strength of Metals and Alloys, v. *1*, 1973), pp. 380–385.

CARTER, N. L. (1976), *Steady state flow of rocks*, Rev. Geophys. Space Phys. *14*, 301–360.

CLARK, S. P., *Handbook of Physical Constants* (rev. ed.), (The Geological Society of America, Inc., New York, 1966), 587 pp.

CARTER, N. L. and MERCIER, J.-C. C. (1976), *Stress dependence of olivine neoblast grain sizes*, EOS Trans. AGU *57*, 322.

CHEN, P. W., YOUNG, C. T. and LYTTON, J. L., *Effect of dislocation substructure on the primary creep behavior of aluminum at elevated temperatures*, in *Rate Processes in Plastic Deformation of Materials* (eds. J. C. M. Li and A. K. Mukherjee), (American Society of Metals, 1975), pp. 605–628.

CHRISTIE, J. M. (1963), *Moine Thrust Zone in the Assynt Region, N. W. Scotland*, Univ. Calif. Publ. Geol. Sci. *40*, 345–439.

GAROFALO, F., *Fundamentals of Creep and Creep Rupture in Metals* (MacMillan, New York, 1965), 258 pp.

GAROFALO, F., RICHMOND, O., DOMIS, W. F. and VON GEMMINGEN, F. (1963), *Joint International Conference on Creep*, London, The Inst. of Mech. Eng., pp. 1–31, referenced in Garofalo (1965).

GLOVER, G. and SELLARS, C. M. (1973), *Recovery and recrystallization during high temperature deformation of α-iron*, Metallurgical Trans. *4*, 765–775.

GOETZE, C. (1975), *Sheared lherzolites: from the point of view of rock mechanics*, Geology *3*, 172–173.

GOETZE, C. and KOHLSTEDT, D. L. (1973), *Laboratory study of dislocation climb and diffusion in olivine*, J. Geophs. Res. *78*, 5961–5971.

GOLDBERG, A. (1966), *Influence of cold work on creep resistance and microstructure*, J. Iron and Steel Inst. *204*, 268–277.

GREEN, H. W., II, *Plasticity of olivine in peridotites*, in *Electron Microscopy in Mineralogy* (ed. H.-R. Wenk), (Springer-Verlag, New York, 1976), pp. 443–464.

GREEN, H. W., II and GUEGUEN, Y. (1974), *Origin of kimberlite pipes by diapiric updwelling in the upper mantle*, Nature *249*, 617–620.

HANSEN, E., *Strain Facies* (Springer-Verlag, New York, 1971), 207 pp.

HIRTH, J. P. and LOTHE, J., *Theory of Dislocations* (McGraw-Hill, New York, 1968), 780 pp.

HOBBS, B. E., MEANS, W. D. and WILLIAMS, P. F., *An Outline of Structural Geology*, (Wiley & Sons, 1976), 571 pp.

HSU, K. J. (1955), *Granulites and mylonites of the region about Cucamonga and San Antonio Canyons, San Gabriel Mountains, California*, Univ. California Publ. Geological Sciences *30*, 223–352.

KOHLSTEDT, D. L. and GOETZE, C. (1974), *Low stress high temperature creep in olivine single crystals*, J. Geophys. Res. *79*, 2045–2051.

KOHLSTEDT, D. L., GOETZE, C, and DURHAM, W. B., *Experimental deformation of single crystal olivine with application to flow in the mantle*, in *The Physics and Chemistry of Minerals and Rocks* (ed. R. G. J. Strens), (John Wiley and Sons Ltd., London, 1976a), pp. 35–49.

KOHLSTEDT, D. L., GOETZE, C., DURHAM, W. B. and VANDERSANDE, J. (1976b), *New technique for decorating dislocations in olivine*, Science *191*, 1045–1046.

LUTON, M. J. and SELLARS, C. M. (1969), *Dynamic recrystallization in nickel and nickel-iron alloys during high temperature deformation*, Acta Metallurgica *17*, 1033–1043.

MACGREGOR, I. D. and BASU, A. R. (1976a), *Geological problems in estimating mantle geothermal gradients*, Amer. Min. *61*, 715–724.

MACGREGOR, I. D. and BASU, A. R. (1976b), *Petrogenesis of the Mt. Albert ultramafic intrusion*, Bull. Geol. Soc. Amer. (in press).

MCQUEEN, H. J., WONG, W. A. and JONAS, J. J. (1967), *Deformation of aluminum at high temperatures and strain rates*, Canad. Jour. Phys. *45*, pp. 1225–1234.

MERCIER, J.-C. C. (1976), *Single pyroxene geothermometry and geobarometry*, Amer. Min. *61*, 603–615.

MERCIER, J.-C., ANDERSON, D. A. and CARTER, N. L. (1977), *Stress in the lithosphere: inferences from steady state flow of rocks*, Pure appl. Geophys. (this issue).

MERCIER, J.-C. C. and NICOLAS, A. (1975), *Textures and fabrics of upper-mantle peridotites as illustrated by xenoliths from basalts*, J. Petrol. *16*, 454–478.

MITRA, S. K. and MCLEAN, D. (1967), *Cold work and recovery in creep at ostensibly constant structure*, Metals Sci. Jour. *1*, 192–198.

NICOLAS, A. (1976), *Flow in upper mantle rocks: some geophysical and geodynamic consequences*, Tectonophysics *32*, 93–106.

PRATT, J. E. (1966), *Dislocation substructure in strain cycled copper*, J. Materials *1*, 77–88.

POST, R. L., Jr. (1973), *The flow laws of Mt. Burnett dunite*, Ph.D. thesis, University of California, Los Angeles, 272 pp.

RADCLIFFE, V. M. and GREEN, H. W., II (1973), *Substructural changes during annealing of deformed olivine: implications for xenolith tenure in basaltic magma*, Trans. Am. Geophys. Union *54*, 453.

SHERBY, O. D., TROZERA, T. A. and DORN, J. G. (1956), *Effects of creep stress history at high temperatures on the creep of aluminum alloys*, Proc. Am. Soc. Test. Mat. *56*, 789–804.

SMITHELLS, C. J., *Metals Reference Book* (3rd ed.), vol. 2 (Butterworths, Washington, 1962), 1087 pp.

TOZER, D. C. (1972), *The present thermal state of the terrestrial planets*, Phys. Earth Planet. Interiors *6*, 182.

TURNER, F. J. and WEISS, L. E., *Structural Analysis of Metamorphic Tectonites*, (McGraw-Hill, New York, 1963), 545 pp.

TWISS, R. J. (1976a), *Structural superplastic creep and linear viscosity in the earth's mantle*, Earth Planet. Sci. Letts. *33*, 86–100.

TWISS, R. J. (1976b), *Static theory of the variation of subgrain and recrystallized grain size with stress*, in preparation.

WINE, R. L., *Statistics for Scientists and Engineers* (Prentice-Hall, Englewood Cliffs, N.J., 1964), 671 pp.

YAR KAHN, M. (1972), *The Structure and Microfabric of a Part of the Arltunga Nappe Complex, Central Australia*, Ph.D. thesis, Australian National University, Canberra, 113 pp.

YOUNG, C. M. and SHERBY, O. D. (1973), *Subgrain formation and subgrain-boundary strengthening in iron-based materials*. J. Iron Steel Inst. *211*, 640–647.

(Received 1st November 1976)

Pageoph, Vol. 115 (1977), Birkhäuser Verlag, Basel

State of Stress in the Lithosphere: Inferences from the Flow Laws of Olivine

By STEPHEN H. KIRBY[1])

Abstract – The experimental flow data for rocks and minerals are reviewed and found to fit a law of the form

$$\dot{\varepsilon} = A' [\sinh (\alpha\sigma)]^n \exp[-(E^* + PV^*)/RT]$$

where $\dot{\varepsilon}$ = strain rate,
σ = differential stress,
P = hydrostatic pressure,
T = temperature, and
A', α, n, E^*, and V^* are material constants.

This law reduces to the familiar power-law stress dependency at low stress and to an exponential stress dependency at high stress. Using the material flow law parameters for olivine, stress profiles with depth and strain rate are computed for a representative range of temperature distributions in the lithosphere. The results show that the upper 15 to 25 km of the oceanic lithosphere must behave elastically or fail by fracture and that the remainder deforms by exponential law flow at intermediate depths and by power-law flow in the rest. A model computation of the gravitational sliding of a lithospheric plate using olivine rheology exhibits a very sharp decoupling zone which is a consequence of the combined effects of increasing stress and temperature on the flow law, which is a very sensitive function of both.

Key words: Stress in lithosphere; Flow of rocks.

1. *Introduction*

One of the methods of determining the stresses in the lithosphere is by physical modelling using the major geophysical observables of heat flow, gravity, seismic wave data, and plate velocities as constraints. A flow law is a kinetic equation of state which relates rates of strain to the applied stress, temperature, pressure, and any other parameters which influence strain rates. Fundamentally and dimensionally, the link between the geophysical observables and the non-hydrostatic stress is through the flow laws appropriate to rocks in the lithosphere. In this paper I review the general empirical laws of flow for rock-forming minerals and then attempt to compute stress in the lithosphere as a function of strain rate and depth. Finally, as an illustration of effects of the form of the flow law, I compute the depth of decoupling for gravitationally sliding lithospheric slabs of uniform thickness and slope.

[1]) U.S. Geological Survey, 345 Middlefield Road, Menlo Park, California 94025, USA.

2. *Flow laws for the lithosphere*

My purpose here is not to provide a comprehensive review of the flow law data for rocks and minerals, since several excellent modern reviews are already published (WEERTMAN 1970, WEERTMAN and WEERTMAN 1975, KIRBY and RALEIGH 1973, STOCKER and ASHBY 1973, CARTER 1976, CHRISTIE and ARDELL 1976, HEARD 1976). Rather, I shall focus on the data which has bearing on the rheology of rocks and minerals representative of the crust and mantle parts of the lithosphere.

Under conditions which promote ductility in rocks, all experiments show an initial transient regime in which the relationship between stress and strain rate is time dependent. It is generally assumed that, given the long times available for flow in the earth, transient effects are unlikely to be important in large scale flow processes in the lithosphere and asthenosphere. This simplification is assumed hereafter.

Most experiments are conducted on right cylinders of rock which are subject to a stress state of uniaxial symmetry: $\sigma_{11} \neq \sigma_{22} = \sigma_{33} = P$ where the x_1 direction is parallel to the cylinder axis and P is the confining pressure. The differential stress $\sigma = |\sigma_{11} - P|$ and is related to the strain rate ($\dot{\varepsilon}$), temperature (T) and pressure (P) by a flow law of the general form

$$\dot{\varepsilon} = A' [\sinh (\alpha\sigma)]^n \exp [-(E^* + PV^*)/RT] \tag{1}$$

where A', α, n, E^*, and V^* are material properties and R is the gas constant. This flow law is followed by most metals (GAROFALO 1965), ice (BARNES *et al.* 1971), alkali halides (HEARD 1972, BURKE 1968), and olivine (POST 1973). At low stresses, the law of equation (1) reduces to the familiar power law stress dependency:

$$\dot{\varepsilon} = A\sigma^n \exp [-(E^* + PV^*)/RT] \tag{2}$$

where $A = A' \alpha^n$. At high stresses, equation (1) reduces to the exponential stress dependency:

$$\dot{\varepsilon} = A'' \exp (\beta\sigma) \exp [-(E^* + PV^*)/RT] \tag{3}$$

where $A'' = A'/2^n$ and $\beta = n\alpha$. The stress at the transition $\sigma_t = 1/\alpha$.

The ratio σ_t/μ (where μ is the average shear modulus) varies from 1 to 4×10^{-3} in non-metallic crystalline materials: 1.04×10^{-3} in halite (HEARD 1972), 1.08×10^{-3} in ice I (BARNES *et al.* 1971), 3.2×10^{-3} in olivine (POST 1973), and 3.7×10^{-3} in calcite (HEARD and RALEIGH 1972).

In addition to the variation of the flow law constants of equation (1) between various minerals and rocks, which represent variations of bulk chemistry and crystal structure, variations of the trace concentrations of dissolved impurity water have dramatic weakening effects on some silicates, such as quartz (GRIGGS and BLACIC 1964, 1965, BLACIC 1971, 1975, KIRBY 1975). Recent experiments on synthetic quartz (KIRBY 1978) indicate that increasing water concentration decreases the activation energy for creep E^* (Fig. 1) and that there is no significant effect on n. This implies

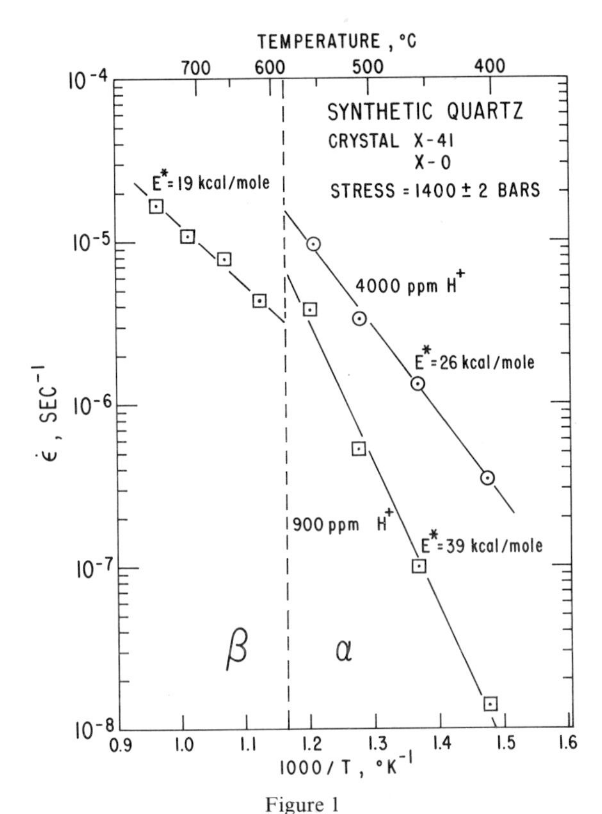

Figure 1

The effect of temperature and the α–β phase change on creep rate in synthetic quartz crystals X-41 (with 900 atomic ppm H^+) and X-0 (4000 ppm H^+). Strain rates are those at 4% strain. Note the difference in activation energy between X-41 and X-0 in the α quartz stability field and the decrease in activation energy from 39 to 19 kcal/mole on transforming to β-quartz for crystal X-41.

that the greatest weakening effect occurs at the lowest temperatures. Since water is an abundant volatile in the crust, water weakening may be dominant in determining rock strength there.

Although the flow laws for some crustal rocks [such as calcite marble (HEARD and RALEIGH 1972) and halite (HEARD 1972) are very well known, the steady flow laws for *representative* crustal rocks (such as granite, metamorphic rocks, diorite, and gabbro) are not. What are needed are not only the 'dry' steady state flow laws but also the flow laws under controlled water pressure.

Fairly reliable flow data are now available for olivine, the volumetrically dominant mineral in the upper mantle. POST (1973) and CARTER and AVÉ LALLEMENT (1970) have determined the flow law for polycrystalline olivine in the kilobar range of stresses both dry and in presence of water. KOHLSTEDT and GOETZE (1974) have determined the flow law for dry single crystal olivine at high temperatures and stresses between 200 and 1000 bars. KOHLSTEDT *et al.* (1976) showed that the high stress polycrystalline dry data are consistent with the low stress single crystal data when the effects of temperature are taken into account. Post's 'wet' olivine data give:

$$A = 8.16 \times 10^{-21} \text{ cgs}$$
$$n = 3.2$$
$$E^* = 91.0 \text{ kcal/mole}$$
$$\alpha = 4 \times 10^{-10} \text{ cm}^2/\text{dyne}$$
$$A' = 9.66 \times 10^{9} \text{ sec}^{-1}$$

with an associated transition stress σ_t of 2.5 kb.

My interpretation of the dry single crystal data of DURHAM and GOETZE (1978) results in the following parameters for the 'average' orientation:

$$A = 1.22 \times 10^{-24} \text{ cgs}$$
$$n = 3.6$$
$$E^* = 95 \text{ kcal/mole}$$

Comparison between these single crystal data and the higher stress data of PHAKEY *et al.* (1972) indicates a transition stress between 1 and 4 kb. For convenience I arbitrarily set σ_t at 2.5 kb, the same as that for Post's wet data results. Thus for dry olivine:

$$\alpha = 4 \times 10^{-10} \text{ cm}^2/\text{dyne}$$
$$A' = 8.30 \times 10^{9} \text{ sec}^{-1}$$

The activation volume for creep, V^*, is the only parameter which has not been determined directly. KIRBY and RALEIGH (1973) estimated V^* from the empirical law of SHERBY and SIMNAD (1961) which replaces the Boltzmann factor $\exp[-(E^* + PV^*)/RT]$ by $\exp[-gT_m/T]$, where g is a constant and T_m is the melting temperature of olivine at pressure. $V^* = 11.0$ cm^3/mole from the experimental pressure effect on T_m. This is the only estimate of V^* which currently has any experimental validity and agrees with estimates of V^* based on defect models for 0^{2-} diffusion, which controls creep rates (KIRBY and RALEIGH 1973).

The flow laws of equations (1), (2), and (3) relate one component of the stress tensor to one component of the strain rate tensor. NYE (1953) has shown how equation (2) can be generalized to general states of stress and strain rate. Shear stress (τ) and shear strain rate ($\dot{\gamma}$) are conceptually the easiest to relate to plate movements. NYE (1953) shows that the basic form of equation (2) can be transformed to

$$\dot{\gamma} = B\tau^n \exp[-(E^* + PV^*)/RT] \tag{2a}$$

where $B = 3^{(n+1)/2} A$. Similarly equation (3) can be transformed to

$$\dot{\gamma} = B'' \exp(\beta''\tau) \exp[-(E^* + PV^*)/RT] \tag{3a}$$

where $B'' = \sqrt{3}\, A''$ and $\beta'' = \sqrt{3}\, \beta$. Finally,

$$\dot{\gamma} = B' [\sinh (\alpha'\tau)]^n \exp [-(E^* + PV^*)/RT] \quad (1a)$$

where $\alpha' = \beta''/n = \sqrt{3}\, \beta/n$ and $B' = B/\alpha'^n$. Using the above relationships, the flow law constants for dry olivine are:

$$B = 1.53 \times 10^{-23} \text{ cgs}$$
$$\beta'' = 2.49 \times 10^{-9} \text{ cm}^2/\text{dyne}$$
$$\alpha' = 6.93 \times 10^{-10} \text{ cm}^2/\text{dyne}$$
$$B' = 1.44 \times 10^{10} \text{ sec}^{-1}$$

where n, E^*, and V^* are as before.

The transformed flow law constants from the wet olivine data of POST (1973) are:

$$B = 8.20 \times 10^{-20} \text{ cgs}$$
$$\beta'' = 2.22 \times 10^{-9} \text{ cm}^2/\text{dyne}$$
$$\alpha' = 6.93 \times 10^{-10} \text{ cm}^2/\text{dyne}$$
$$B' = 1.67 \times 10^{10} \text{ sec}^{-1}$$

where n, E^*, and V^* are as in equation (1) for 'wet' olivine.

3. *Stresses in the lithosphere*

Olivine is the only rock-forming mineral which is representative of the lithosphere and for which reliable flow law data are available. I assume that the creep behavior of olivine dominates the flow of the mantle part of the lithosphere, which is composed predominantly of olivine. The lack of steady state creep data for representative crustal rocks, especially under hydrous conditions, precludes estimating creep stresses in the crust. Physical modelling of plate interactions, utilizing the appropriate flow laws, is the most powerful approach in calculating the stresses in the lithosphere. One can, however, estimate the range of stresses in the lithosphere by estimating the temperature, pressure, and strain rate and then calculating the stress from the flow law. This is attempted in the section that follows.

3.1 Estimates of the controlling parameters

The pressure distribution is firmly established from inversions of seismic wave data and other geophysical data to density profiles. An average lithosphere density of 3.45 gm/cm^3 is a good fit to the density profiles of PRESS (1972) and was used to calculate pressure as a function of depth. Since pressure has a small effect on calculated stresses, this simplification produces insignificant error.

Thermal transport in the lithosphere takes place by conduction and, to a lesser extent, radiation, since convective transfer is, by definition, confined to the asthenosphere. The temperature distribution can be calculated, given the surface heat flow, surface temperatures, the material thermal constants, the distribution of heat sources, and information on the boundary and initial conditions. I have selected a range of published solutions of the temperature distribution which bracket most estimates of temperature as a function of depth for 'normal' lithosphere (excluding ridge areas and subducting lithosphere). The continental shield geotherm of GREEN and RINGWOOD (1967) was used as the lower bound on the temperature distribution (Fig. 2).

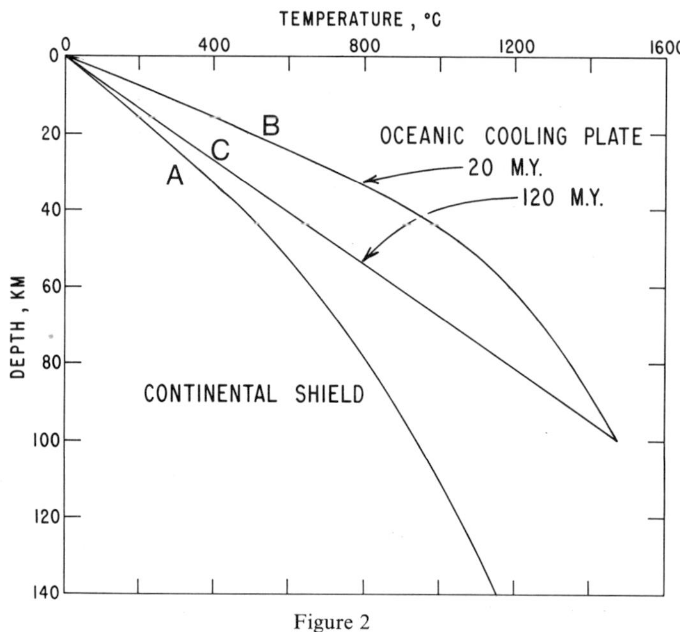

Figure 2

Representative temperature distributions with depth in the lithosphere. Continental shield geotherm from GREEN and RINGWOOD (1967) and cooling oceanic plate geotherms from SCLATER and FRANCHETEAU (1970) for plate ages of 20 and 120 M yr.

SCLATER and FRANCHETEAU (1970) and SCLATER *et al.* (1972) showed that the falloff of heat flow and ocean bottom bathymetry away from the ridges could be accounted for by a single oceanic cooling plate model. The temperature distributions associated with their preferred model for the Northern Pacific at plate ages of 20 M yr and 120 M yr (distances from ridge of 1000 and 6000 km, respectively) are shown in Fig. 2. The geotherms shown in Fig. 2 bracket most of the estimates of the geotherm based on more sophisticated models.

Despite the conceptual view of the lithosphere as being rigid, there are regions where clearly it must be deforming. At boundaries between plates, such as at transform faults, a shear zone must exist with a finite width. Geodetic measurements indicate that the magnitude of the relative velocity between the North America plate and the Pacific plate along the San Andreas fault is 3.2 ± 0.5 cm/yr (SAVAGE and BURFORD 1973). A remarkable fact is that with the exception of the region of the Transverse Ranges, the zone associated with San Andreas tectonic features is only 110 km wide at most and recent geodetic measurements indicate that locally it may be only 100 meters wide (R. O. BURFORD, personal communication). This range of shear zone widths corresponds to shear strain rates between 10^{-11} and 10^{-14} sec^{-1}. Simple calculation reveals that the maximum shear strain rate associated with bending of the lithosphere in subduction zones is about 10^{-14} sec^{-1} and an average is 10^{-15} to 10^{-16} sec^{-1} (SEGAWA and TOMODA 1976, T. C. HANKS, personal communication). Given the range of plate velocities, the time scales of plate motion, and a reasonable range of plate dimensions, a minimum shear strain rate of 10^{-16} sec^{-1} is probably appropriate. Thus a range of 10^{-11} to 10^{-16} sec^{-1} spans the probable range of strain rates in the lithosphere where significant flow is taking place.

3.2 Stress profiles in the lithosphere

The flow stresses were calculated as functions of depth from the hydrous and anhydrous olivine flow laws using the above estimates of the controlling creep parameters. These stress profiles are presented in Figs. 3, 4, and 5 for the three representative geotherms of Fig. 2.

Several preliminary comments should be made to aid discussion of the results. The first is that the stresses are appropriate to the mantle part of the lithosphere where significant solid state flow takes place. Clearly lower stresses can be supported elastically. Thus the flow stresses illustrated in Figs. 3, 4, and 5 represent upper limits on tectonic stresses. It is also expected that solid state ductile flow should be supplanted by brittle fracture at shallow depths where the temperatures are low. BYERLEE (1968) has shown that the brittle-ductile transition at room temperature can be accounted for by assuming that failure occurs when the shear stress τ_c on a potential fracture plane is just high enough to form a fault and overcome static friction. BYERLEE (1968) also showed that the static frictional shear stress τ_f at room temperature obeys the following law for a wide variety of rocks:

$$\tau_f \text{ (kb)} = 0.8 + 0.6\,\sigma_n \text{ (kb)} \tag{4}$$

where σ_n is the normal stress on the fault plane. The brittle-ductile transition has not been systematically explored at higher temperatures. However, the stresses achieved under ductile conditions in the published experimental results at intermediate and high temperatures for a variety of rock types (see review by HEARD 1976) are all in the ductile field of BYERLEE (1968), as expected. If we assume that σ_n is dominated by the hydrostatic component P of the stress field due to the lithostatic load, then:

$$\tau_c \text{ (kb)} = 0.8 + 0.6\,P \text{ (kb)} \tag{4a}$$

and the stress τ_c above which brittle behavior is to be expected is given by equation (4a). τ_c is plotted in Figs. 3, 4, and 5 as a guide to the transition from ductile to brittle behavior at low pressure and high calculated flow stresses. Since the tectonic component of the stress field can contribute to variations in σ_n and thus τ_c, the brittle-ductile transition can be affected by the particular state of stress. For example, in the bending of the lithosphere in subduction zones, the normal stresses can be significantly lower than lithostatic in the upper half of the plate where extension has to take place. Hence equation (4a) overestimates τ_c, the stress above which brittle behavior can be expected. Also, equation (4a) overestimates τ_c when fluids lower the effective normal stress on potential fracture planes (HUBBERT and RUBEY 1959). A final preliminary comment is that viscous flow can significantly affect the temperature distribution by viscous heating and thermal feedback (SCHUBERT *et al.* 1976, MELOSH 1976). Viscous heating can be expected to lower the calculated flow stresses and hence the stress profiles which do not incorporate viscous heating sources may overestimate

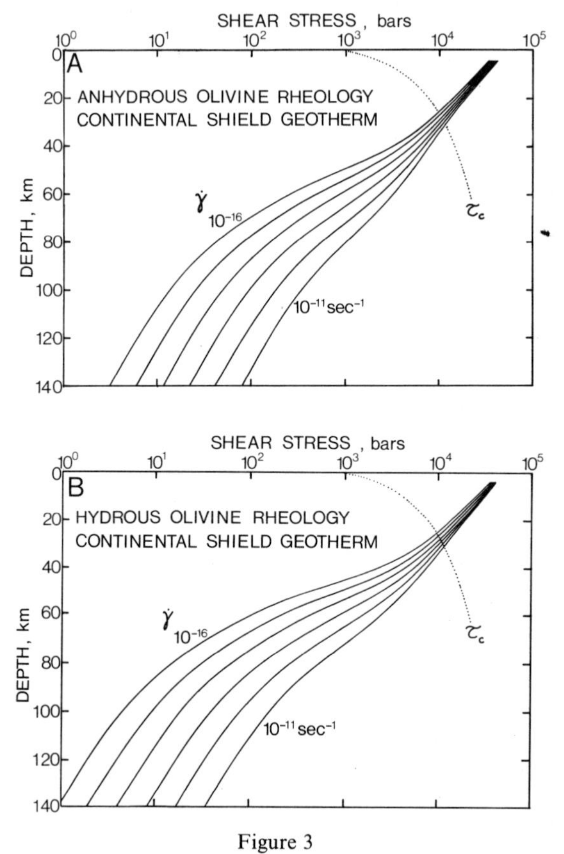

Figure 3

Stress profiles with depth calculated from olivine flow laws using the continental shield geotherm of GREEN and RINGWOOD (1967): (A) Anhydrous olivine rheology, (B) Hydrous olivine rheology.

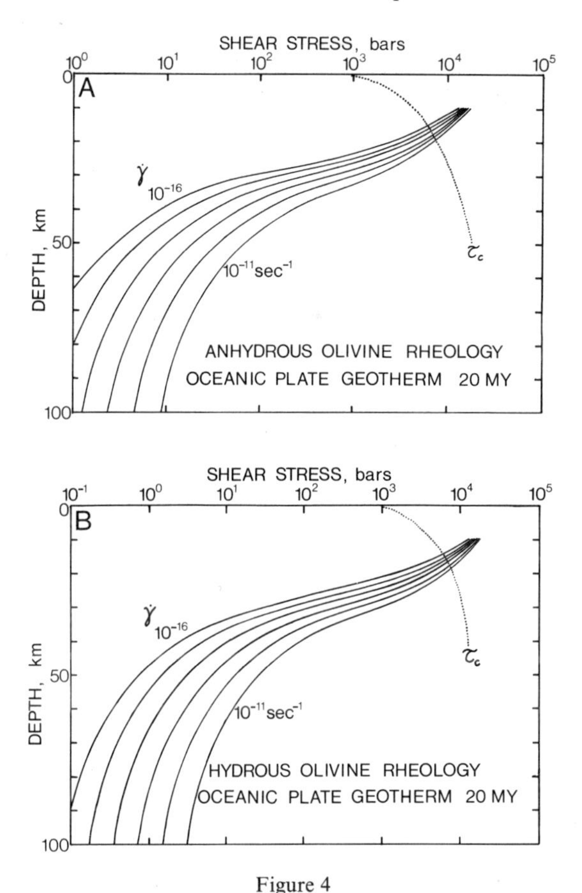

Figure 4
Stress profiles with depth calculated from olivine flow laws using the cooling oceanic plate geotherm of SCLATER and FRANCHETEAU (1970) at a plate age of 20 M yr. (A) Anhydrous olivine rheology, (B) Hydrous olivine rheology.

the flow stress. In summary of the points discussed above, the stress profiles illustrated in Figs. 3, 4, and 5 are upper limits on the stresses which can be sustained by the mantle part of the lithosphere.

Mindful of the above points, several conclusions can be drawn from the illustrated stress profiles:

(1) *A significant portion of the upper lithosphere should be following the exponential stress dependency* of equation (3a). This is evident from the converging of the stress profiles at stresses above 1.4 kb, which is the transition stress τ_t between power law and exponential law flow. As exponential law flow becomes more important at the higher stresses, the uncertainties in calculated flow stresses due to uncertainties in the strain rate are diminished, since the calculated flow stress becomes less and less sensitive to variations in strain rate. Thus model

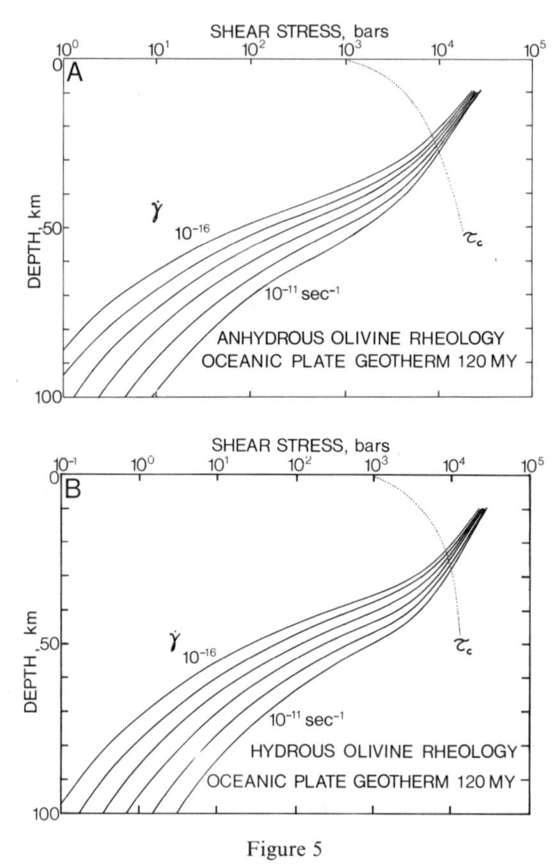

Figure 5
Stress profiles with depth calculated from olivine flow laws using the cooling oceanic plate geotherm at a plate age of 120 M yr. (A) Anhydrous olivine rheology, (B) Hydrous olivine rheology.

computations which use the power law of equation (2) may be in considerable error in the upper part of the lithosphere where $\tau > 1.4$ kb.

(2) *The effect of water on the calculated flow stresses is relatively small.* This is primarily due to the high stress sensitivity of the flow law, and to the higher sensitivity of flow stress in the anhydrous olivine rheology to changes in temperature. This causes the calculated flow stresses for the hydrous and anhydrous olivine rheologies to converge at higher temperatures and greater depths.

(3) *Depending on the geotherm, the upper 15 to 30 km of the lithosphere should deform in the brittle regime.* This conclusion is relatively insensitive to the tectonic strain rate since ductile flow at the brittle-ductile transition is well into the exponential flow law regime. In particular, the bending of the lithosphere in the subduction zones must involve brittle deformation, probably involving faulting, in the upper 25 km of the plate (see Fig. 5). Contributions of the

tectonic stresses to lowering σ_n and thus τ_c in equation (4a) as discussed above would increase this depth interval of brittle behavior.

(4) *The calculated flow stresses at the base of the lithosphere vary between tenths and tens of bars*. This is evident from the profiles and the current estimates of the thicknesses of the lithospheric plates.

I believe that stress profiles based on the rheology of rocks appropriate to the lithosphere are our best constraints on the magnitudes of the boundary tractions on the lithospheric plates. Studies which consider the mechanical equilibrium of the plates in search of the driving mechanisms should more effectively incorporate these constraints.

4. *Gravitational sliding of the lithosphere*

It has been suggested that a significant driving force for plate motion is the movement of the lithosphere away from the ridges by gravitational sliding (FRANK 1972). A simple calculation is made (similar to that of a sliding glacier (PATERSON 1969)), which illustrates the effects of the form of the olivine flow law and also explores aspects of the driving force. Frank showed that the free sliding on a flat earth of a flat plate with uniform thickness H and average density $\bar{\rho}$ produces an average basal shear stress

$$\tau_{x'z'} = -g(\bar{\rho} - \rho_w) H (h/L) \tag{5}$$

where ρ_w is the average density of seawater, L is the distance between ridge and an arbitrary point near the trench and h is the elevation difference between those two points (see Fig. 6). The quantity (h/L) is approximately equal to the slope. It is a reasonable question to ask what combinations of slopes and plate thicknesses are necessary to produce a strain rate from the flow law which is appropriate to the observed plate velocity. If we regard the plate thickness as a variable with depth, then

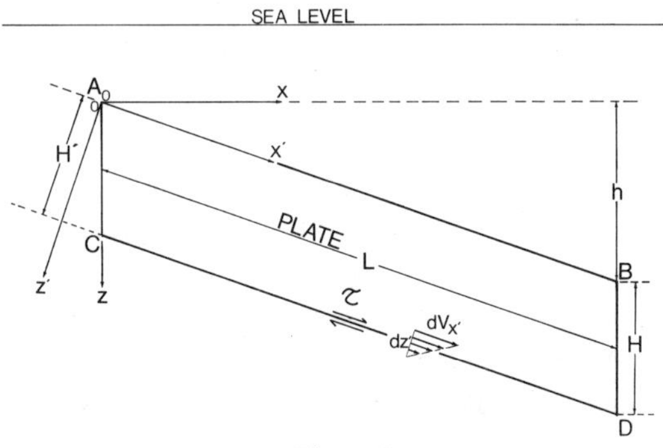

Figure 6
Sketch showing the geometry used in the gravitational sliding computation. After FRANK (1972, Fig. 1a).

the shear strain rate, referred to the x'–z' coordinates, can be calculated with depth from the flow law of equation (1a), the stress given by equation (5), and the temperature distribution. I assume the steady-state temperature distribution of SCLATER and FRANCHETEAU (1970) (C in Fig. 2), which is a conservative estimate of the geotherm for the Pacific plate. Since $\dot{\gamma}_{x'z'} \equiv \partial V_{x'}/\partial z'$ where $V_{x'}$ is the sliding velocity, the cumulative plate velocity $\Delta V_{x'}$ can be calculated as a function of H by integrating $\dot{\gamma}_{x'z'}$ from zero to $z' = H' \simeq H$. The cumulative plate velocity versus plate thickness or depth is shown in Fig. 7 for a range of slopes of 10^{-1} to 10^{-5}. The actual distribution of plate velocity is $V_{x'} = V_{x'}^{\text{tot}} - \Delta V_{x'}$ where V_{x}^{tot} is the observed velocity, in this

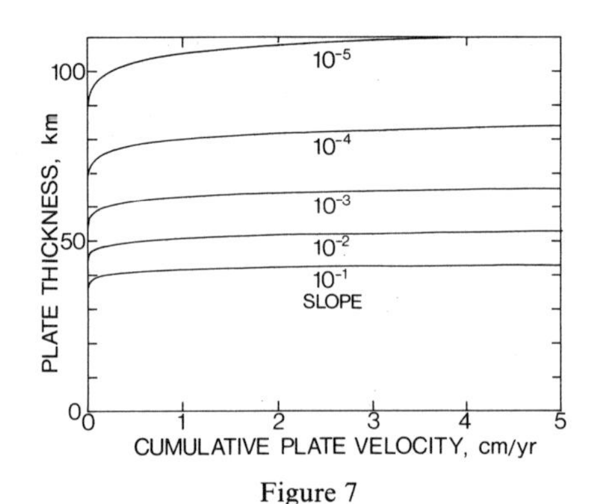

Figure 7
Cumulative plate velocity due to gravitational sliding as a function of plate thickness and plate slope, assuming the geotherm C of Fig. 2 and a mean plate density of 3.45 gm/cm^3. Anhydrous olivine rheology.

case 5 cm/yr for the North Pacific. Presumably, the plate thickness for a given slope is given by the value H where $\Delta V_{x'} = V_{x}^{\text{tot}}$. Several comments should be made on the results:

(1) *The zone of decoupling is sharply defined* with zone thicknesses varying between 2 and 20 km (depending on the slope) necessary to achieve the observed plate velocity of 5 cm/yr. This sharpness of the decoupling reflects the high sensitivity of strain rate to changes in stress and temperature, both of which increase approximately linearly with depth. The plate is essentially rigid above the decoupling zone.

(2) *The depth to the decoupling zone varies from about 80 to 105 km for slopes between 10^{-4} and 10^{-5}*, which are admissible, judging from the current refinement of gravity models for the deep ocean basins. The above range of plate thicknesses are consistent with the estimates of average plate thickness of 75 to 100 km for the Northern Pacific which best fit the observed bathymetry and heat flow distribution in the cooling plate calculations.

(3) *No thermal feedback is incorporated in the calculation but such feedback, if it exists, is likely to increase the sharpness of the decoupling.*

There are many obvious problems with this highly simplified model, many of which are discussed by FRANK (1972), but it clearly shows that the Northern Pacific plate is capable of moving by gravity sliding at its observed rate with a rather modest slope and a conservative temperature distribution and that the depths to the zone of decoupling are reasonable in light of current estimates of plate thickness.

REFERENCES

BARNES, P., TABOR, D. and WALKER, J. C. F. (1971), *The friction and creep of polycrystalline ice*, Proc. Roy. Soc. Lond. *A234*, 127–255.

BLACIC, J. D. (1971), *Hydrolytic weakening of quartz and olivine*, Thesis, University of California, Los Angeles, California, 205 p.

BLACIC, J. D. (1975), *Plastic deformation mechanism in quartz: the effect of water*, Tectonophysics *27*, 271–294.

BURKE, P. M. (1968), *High temperature creep of polycrystalline sodium chloride*, Thesis, Stanford University, Stanford, California, 122 p.

BYERLEE, J. D. (1968), *Brittle-ductile transition in rocks*, J. Geophys. Res. *73*, 4741–4750.

CARTER, N. L. (1976), *Steady state flow of rocks*, Rev. Geophys. Space Phys. *14*, 301–360.

CARTER, N. L. and AVÉ LALLEMANT, H. G. (1970), *High temperature flow of dunite and peridotite*, Geol. Soc. Amer. Bull. *81*, 2181–2202.

CHRISTIE, J. M. and ARDELL, A. J., *Deformation structures in minerals* in *Electron Microscopy in Mineralogy* (ed. H. R. Wenk) pp. 374–403 (Springer, Berlin-Heidelberg 1976), 564 p.

DURHAM, W. B. and GOETZE, C. (1978), *Plastic flow of oriented single crystals of olivine*, J. Geophys. Res., in press.

FRANK, F. C., *Plate tectonics, the analogy with glacier flow, and isostasy* in *Flow and Fracture of Rocks* (ed. H. C. Heard, I. Y. Borg, N. L. Carter and C. B. Raleigh) pp. 285–292 (American Geophysical Union Monograph, Washington 1972), 352 p.

GAROFALO, F., *Fundamentals of Creep and Creep Rupture in Metals* (McMillan, New York, 1965), 258 p.

GREEN, D. H. and RINGWOOD, A. E. (1967), *The stability fields of aluminous pyroxene peridotite and garnet peridotite and their relevance in upper mantle structure*, Earth Planetary Sci. Letters *3*, 151–160.

GRIGGS, D. T. and BLACIC, J. D. (1964), *The strength of quartz in the ductile regime*, Trans. Amer. Geophys. Union *45*, 102–103.

GRIGGS, D. T. and BLACIC, J. D. (1965), *Quartz: anomalous weakness of synthetic crystals*, Science *147*, 292–295.

HEARD, H. C., *Steady flow in polycrystalline halite at pressure of 2 kilobars* in *Flow and Fracture of Rocks* (ed. H. C. Heard, I. Y. Borg, N. L. Carter and C. B. Raleigh) pp. 191–209 (American Geophysical Union Monograph 19, Washington 1972), 352 p.

HEARD, H. C. (1976), *Comparison of the flow properties of rocks at crustal conditions*, Phil. Trans. R. Soc. Lond. *A283*, 173–186.

HEARD, H. C. and RALEIGH, C. B. (1972), *Steady-state flow in marble at 500°C to 800°C*, Geol. Soc. Amer. Bull. *83*, 935–956.

HUBBERT, M. K. and RUBEY, W. W. (1959), *Role of fluid pressure in mechanics of overthrust faulting*, Geol. Soc. Am. Bull. *70*, 115–166.

KIRBY, S. H. (1975), *Creep of synthetic alpha quartz*, Thesis, University of California, Los Angeles, California, 193 p.

KIRBY, S. H. (1978), *The alpha-beta inversion in quartz: the effect of temperature on creep rates*, Geophys. Res. Letters, in press.

KIRBY, S. H. and RALEIGH, C. B. (1973), *Mechanisms of high temperature solid-state flow in minerals and ceramics and their bearing on the creep behavior of the mantle*, Tectonophysics *19*, 165–194.

KOHLSTEDT, D. L. and GOETZE, C. (1974), *Low stress high-temperature creep in olivine single crystals*, J. Geophys. Res. *79*, 2045–2051.

KOHLSTEDT, D. L., GOETZE, C. and DURHAM, W. B., *Experimental deformation of single crystal olivine with application to flow in the mantle* in *The Physics and Chemistry of Minerals and Rocks* (ed. R. G. J. Strens) pp. 35–49 (Wiley, London 1976), 697 p.

MELOSH, H. J. (1976), *Plate motion and thermal instability in the asthenosphere*, Tectonophysics *35*, 363–390.

NYE, J. F. (1953), *The flow law of ice from measurements in tunnels, laboratory experiments and the Jungfraufirn borehole experiments*, Proc. Roy. Soc. Lond. *A219*, 477–489.

PATERSON, W. S. B., *The Physics of Glaciers* (Pergamon, Oxford-London, 1969), 250 p.

PHAKEY, P., DOLLINGER, G. and CHRISTIE, J. (1972), *Transmission electron microscopy of experimentally deformed olivine crystals* in *Flow and Fracture of Rocks* (ed. H. C. Heard, I. Y. Borg, N. L. Carter and C. B. Raleigh) pp. 117–138 (American Geophysical Union Monograph, Washington 1972), 352 p.

POST, R. (1973), *The flow laws of Mt. Burnett dunite*, Thesis, University of California, Los Angeles, California, 271 p.

PRESS, F., *The earth's interior inferred from a family of models* in *The Nature of the Solid Earth* (ed. E. C. Robertson) pp. 147–171 (McGraw-Hill, New York 1972), 677 p.

SAVAGE, J. C. and BURFORD, R. O. (1973), *Geodetic determination of relative plate motion in central California*, J. Geophys. Res. *78*, 833–845.

SCHUBERT, G., FROIDEVAUX, C. and YUEN, D. A. (1976), *Oceanic lithosphere and asthenosphere: thermal and mechanical structure*, J. Geophys. Res. *81*, 3525–3540.

SCLATER, J. G., ANDERSON, R. N. and BELL, M. L. (1972), *Elevation of ridges and evolution of the central eastern Pacific*, J. Geophys. Res. *76*, 7888–7915.

SCLATER, J. G. and FRANCHETEAU, J. (1970), *The implications of terrestrial heat flow observations on current tectonic and geochemical models of the crust and upper mantle of the earth*, Geophys J. Roy. Astron. Soc. *20*, 509–542.

SEGAWA, J. and TOMODA, Y., *Gravity measurements near Japan and study of the upper mantle near the oceanic trench-marginal sea transition zones* in *The Geophysics of the Pacific Ocean Basin and Its Margin* (ed. G. Sutton, M. Manghnani and R. Moberly) pp. 35–52 (American Geophysical Union Memoir *19*, Washington 1976), 480 p.

SHERBY, O. D. and SIMNAD, M. T. (1961), *Prediction of atomic mobility in metallic systems*, Amer. Soc. Metals Trans. *54*, 227–240.

STOCKER, R. L. and ASHBY, M. F. (1973), *On the rheology of the upper mantle*, Rev. Geophys. Space Phys. *11*, 391–426.

WEERTMAN, J. (1970), *The creep strength of the earth's mantle*, Rev. Geophys. Space Phys. *8*, 145–168.

WEERTMAN, J. and WEERTMAN, J. R. (1975), *High temperature creep rock and mantle viscosity*, Ann. Rev. Earth Plan. Sci. *3*, 293–315.

(Received 5th January 1977)

Pageoph, Vol. 115 (1977), Birkhäuser Verlag, Basel

On Correlation Between Seismic Velocity Anisotropy and Stresses *In Situ*

By I. A. TURCHANINOV, V. I. PANIN, G. A. MARKOV, V. I. PAVLOVSKII, N. V. SHAROV and G. A. IVANOV[1])

Abstract – Seismic velocity anisotropies measured in underground mines are compared with *in situ* stress measurements in these mines. These underground data are also compared with seismic velocity anisotropies observed by large scale seismic sounding conducted from the earth's surface. The velocity anisotropies are about 10 % and the data obtained by different methods on different scales and frequencies agree with each other. The directions of largest and smallest velocities coincide with the largest and smallest horizontal stresses, respectively. These results suggest that the direction and magnitude of stresses in potential mining areas could be estimated from velocity anisotropies observed in seismic prospecting of the area.

Key words: Stress *in situ*; Seismic velocity anisotropy; Stress in mines.

1. Introduction

There are two types of correlation between stresses and physical, particularly mechanical, properties of rocks *in situ*. On one hand, higher stresses are observed in harder rock zones, which form the so-called 'carrying construction' of the earth's crust; on the other hand, stresses have a significant effect on rock properties, thus causing certain anomalies in corresponding geophysical fields.

Strain relief and ultrasonic velocity measurements show that the more competent rock formations contain higher stresses in the apatite-nepheline mines in the Khibiny Mountains. Figure 1 shows the dependence of P-velocity and Young's modulus on the largest principal stress measured *in situ*. The stress was determined by the strain relief method *in situ*. The velocity and Young's modulus in the direction of greatest principal stress were found in the laboratory at atmospheric pressure on samples taken from points of stress determinations. Statistical analysis of these results shows a rather distinct correlation between stress magnitudes *in situ* and longitudinal wave velocities and the Young's modulus, respectively.

The most obvious effect of the stress state on physical rock properties manifests itself in velocity anisotropy. It is well known that stress anisotropy causes velocity anisotropy in rocks. This difference can be significant and is supported by numerous

[1]) Kola Branch of the USSR Academy of Sciences, Apatity, Murmanskaja Oblast Phersman Street 24, USSR.

laboratory experiments (TOCHER 1957, TURCHANINOV and PANIN 1976), and by respective field measurements (TURCHANINOV and PANIN 1976, ALEINIKOV *et al.* 1971). This interrelation can be established distinctly and unambiguously by results of measurements on samples of dimensions of 0.1 m or on small scales from 1 m to 3 m in mines.

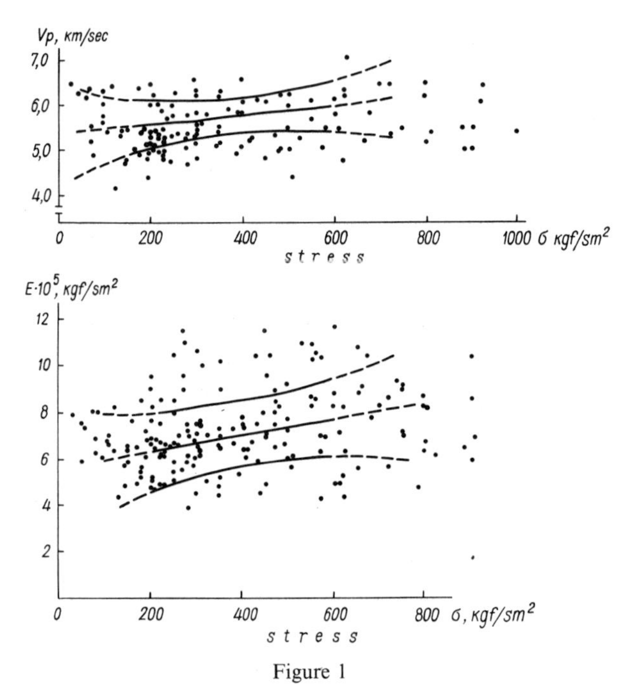

Figure 1

Correlation between stress magnitudes *in situ* and longitudinal wave velocities (a) and the Young's modulus (b), respectively.

However, the estimate of the state of stress of a massif derived from velocity measurements remains rather problematic. To find out whether the state of stress can be estimated from large-scale seismic velocity measurements, it is reasonable to conduct seismo-acoustic and seismic observations on different scales and in different directions using elastic waves in a wide band of frequencies. If enough different frequencies and scales are used, one may estimate the degree of anisotropy in the medium on the basis of volumetric indicatrixes of the velocities V_p and V_{sv}. Below we will compare *in situ* stress relief measurements in the apatite-nepheline mines of the Khibiny, with velocity measurements obtained on scales of 1 m, 25 m, and 10–110 km.

2. *Velocity anisotropy and stress measurements*

Data on anomalously high stresses obtained by studying the stress state at Khibiny apatite mines have been published fully enough (TURCHANINOV and MARKOV 1966,

TURCHANINOV *et al.* 1973). Stress relief measurements were conducted in the southern part of the Khibiny (Fig. 2). Measurements were obtained in the area of about 10 km^2, opened by mining to the depth of 0.5 km. The main components of complete stress tensor and characteristics of stress field anisotropy were determined by statistical analysis of more than 5000 measurements.

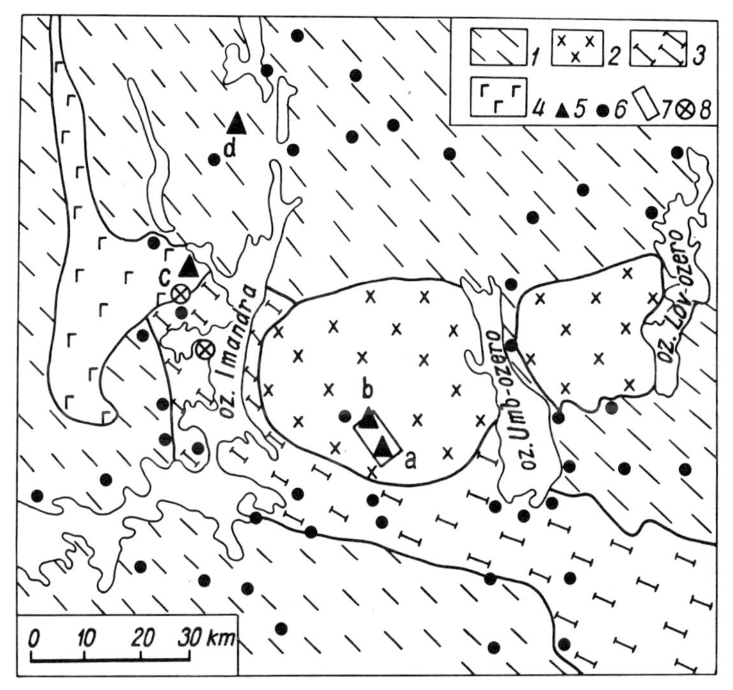

Figure 2

Schematic map of the investigated region. 1 – gneiss, diorite gneiss, granite gneiss; 2 – nepheline sienite complex (khibinite, ristschorrite, lavochorrite, foyaite, ijolite-urtite); 3 – Imandra-Varzuga effusive-sedimentary complex; 4 – gabbro, gabbro-norite, gabbro-anorthosite; 5 – industrial explosion points, mines respectively: (a) Central and Rasvumchorr, (b) Kirovsk and Udspor, (c) Monchegorsk, (d) Olenegorsk; 6 – registration points of seismostation 'Poisk-KMPV'; 7 – region of experimental determination of stress *in situ*; 8 – sections of observation by reflection method.

The largest compression is statistically horizontal, equal to 200 to 600 kg^f/cm^2 and oriented (also statistically) in approximately an EW direction. Stress compression in the NS direction is characterized by average magnitude from 100 to 300 kg^f/cm^2. The compression anisotropy coefficient along these directions varies inside the limits of 0.1 to 0.6.

Results of ultrasonic measurements at apatite-nepheline mines have also been published (TURCHANINOV and PANIN 1976, TURCHANINOV and PANIN 1970). Mine measurements were conducted mainly on scales from 0.5 to 1.0 m by ultrasonic transducers with frequency of about 70 kc/sec. On the whole, the ultrasonic results agree statistically in value with stress relief data, i.e., one can observe statistical correlation between magnitudes of stress *in situ* and longitudinal wave velocities.

Seismic investigations by the transmitted wave method have been conducted at a section of the Kirovskii mine (measurement scale from 20 to 30 m, frequency about 500 c/sec) (PANIN and RITTER 1974). The results of these investigations corroborated the presence of a distinct velocity anisotropy *in situ*, corresponding to the stress state of rocks in this section.

Seismic sounding investigations (measurement scale from 10 to 110 km, for prevailing number of points – from 40 to 60 km, frequency from 5 to 25 c/sec) have been made in the area outlining the Khibiny and Lovozero massifs (Fig. 2). Industrial explosions in quarries and mines in the area near Kirovsk (in the southern part of the Khibiny massif), and Olenegorsk and Monchegorsk (in the north-western part of the investigated area) were used as oscillation sources. On the whole, 144 industrial explosions were fixed at 43 points. An aerial variation of the seismic sounding method (HALEVIN 1975) with two serial seismostations 'Poisk-KMPV' and seismographs SV2-10 with own frequency 10 c/sec has been applied.

The report contains results of the first stage of data handlings: hodographs of the first arrivals and arrival velocities of the first P and S waves from different directions around the same sources. The waves in the first arrivals at distances from 10 to 30 km from the source are characterized by a progressive increase in values of apparent velocity (V_k) from 5.7–5.8 to 6.0–6.4 km/sec. With a further increasing of the distances up to 80–90 km, the value of V_k remains practically unchanged. At distances from 30 to 100 km, first arrivals were registered which were weakly refracted on a discontinuity at depths from 1 to 2 km (LITVINENKO and NEKRASOVA 1963, ANKUDINOV 1974). Apparent velocities of these waves are close to the velocities on the depths of elasticity parameters stabilization. For the region under investigation at distances from 30 to 100 km from the point of explosion, the obtained magnitudes of apparent velocities of the first arrivals are practically equal to the velocities estimated by division of the distance from the center of the explosion to the point of observation by the time of the first arrivals.

Figure 3 shows the estimated velocities V_p and V_{sv} as a function of azimuth, determined for the points around the Khibiny massif (points of explosions near Kirovsk). It also shows the orientation of the horizontal stress ellipse, obtained by the stress relief method. From Fig. 3 it is clear that a distinct velocity anisotropy exists in the Khibiny massif (anisotropy coefficient reaches 10%, error of velocity determination does not exceed 1%). The major axes of both the velocity anisotropy and ellipse of elasticity are oriented mainly in the west-north-western and the east-south-eastern directions. From this correlation we conclude that the stress field causes the observed velocity anisotropies.

Seismic observations by reflection method were conducted on gabbroids of the Monchegorsk region over two disconnected sections (Fig. 2). The length of hodograph varies from 1175 to 2350 m, oscillation frequency ranges from 30 to 50 c/sec, observational spacing equals 25 m, and the depth of the first refraction horizon amounts to first tens of meters. Observations on each section have been conducted on three mutually intersected profiles. Azimuth profile locations of both sections are the same. Gabbroids of the investigated region have no significant distinctions in physical properties.

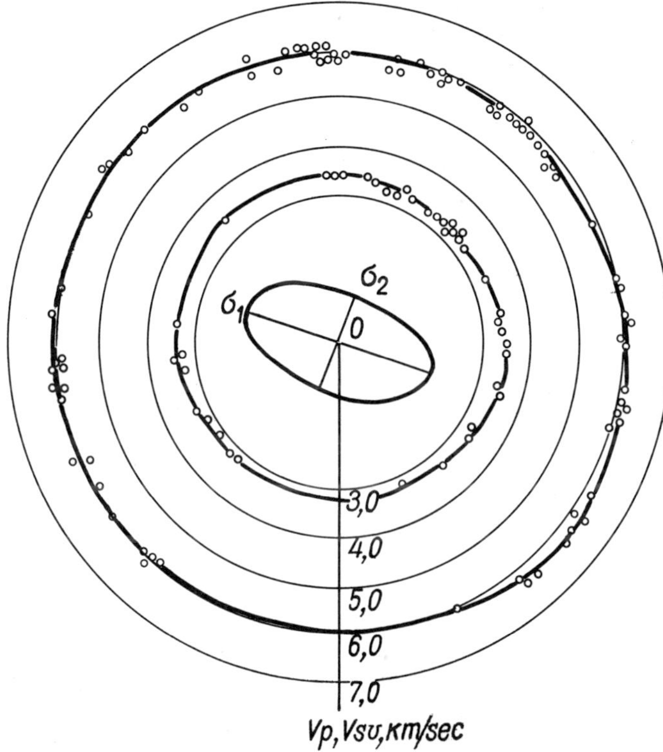

Figure 3
Velocities V_p and V_{sv} are shown as a function of azimuth and compared with the compressive stress for rocks of the Khibiny massif.

The values of first arrival velocities on reversed profiles of both directions are practically equal. This enables us to plot a common clock diagram of longitudinal and transverse wave velocities for the gabbroids from the sections under consideration (Fig. 4). This figure shows in another scale the indicatrixes of the estimated velocities V_p and V_{sv}, determined by a seismic sounding method for the points in the southern and eastern directions from industrial explosions in Olenegorsk and Monchegorsk regions.

The character of the wave velocities as a function of azimuth, obtained by measurements on different scales and frequencies, points to the existence of a velocity anisotropy (anisotropy coefficient approaches 8 % for V_p, and 20 % for V_{sv}) in the Monchegorsk region. The major anisotropy axes obtained by the different methods coincide and are oriented mainly in south-south-western and north-north-eastern directions (Fig. 5).

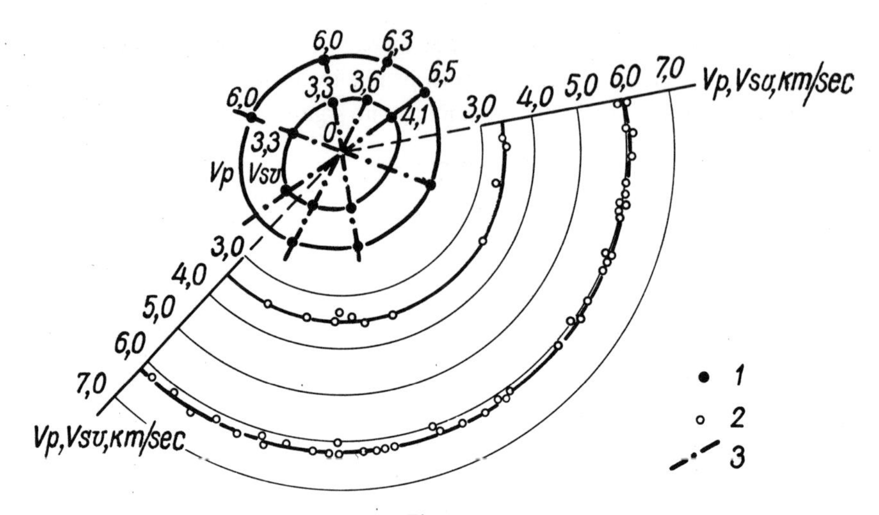

Figure 4
Velocities V_p and V_{sv} as a function of azimuth. Data obtained by reflection and seismic sounding for rocks of Monchegorsk and Olenegorsk regions. 1 – reflection data; 2 – seismic sounding data; 3 – azimuthal profile location.

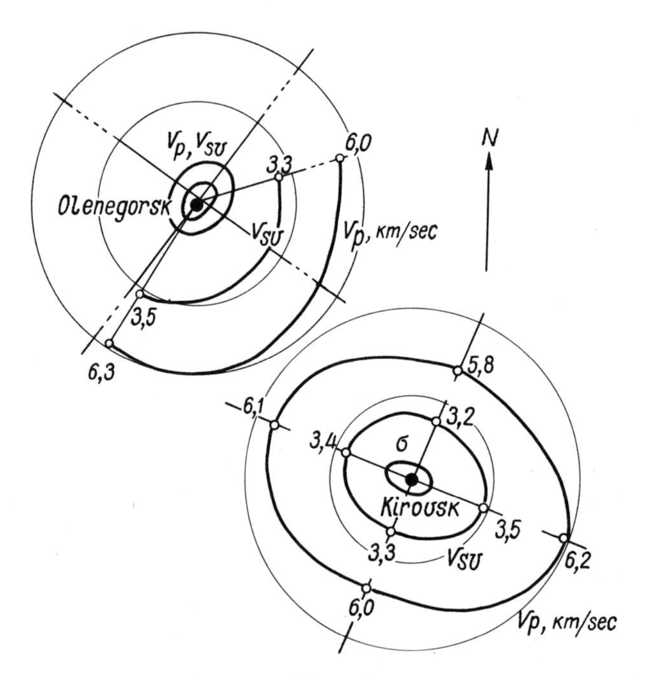

Figure 5
Scheme of the first arrival velocity distribution V_p and V_{sv} for the whole region under study. The upper and lower diagram refers to the northwestern and southeastern part of Fig. 2, respectively.

Unfortunately, no direct stress measurements have yet been conducted in the Monchegorsk region. But extrapolating the results from the Khibiny massif to the gabbroid massif in Monchegorsk region, one can suggest *a priori* that the velocity anisotropy in this region is also due to stress states.

No doubt seismic anisotropy of rocks can to a large extent be caused by respective structure and jointing. But the structure and jointing of rocks are formed by relative stress fields, thus carrying certain information on stress state functioning for some time (in geological scale) until the moment of measurements, which has been experimentally supported at one of the polymetallic mines (PANIN *et al.* 1970).

3. Conclusions

The stress state of rocks *in situ* is one of the significant factors determining the relative distribution of seismic wave velocities, and one should take into account its effect on these fields. On the other hand, seismic factor distribution analysis can serve as an important source of information on the present stress state of rocks *in situ*. The use of seismic methods for evaluation of the stress state of rocks provides information on stress field parameters in a deposit at the preliminary stage of prospecting. This will enable engineers to plan in advance the details of mining operations because they will know in advance what stresses to expect.

REFERENCES

ALEINIKOV, A. L., ZUBKOV, A. V. and HALEVIN, N. I. (1971), *On possible correlation between elastic wave velocity anisotropy and earth crust stress state*, Dokl. Akad. Nauk SSSR, *1971*, 78.

ANKUDINOV, S. A. (1974), *On nature and particularities of waves registered at the first arrivals by seismic sounding in the Baltic shield*, Zap. LGI, *66* (2), 31.

HALEVIN, N. I., *Explosion Seismology in the Urals M.* (Izd. Nauka, 1975), 133 pp.

LITVINENKO, I. V. and NEKRASOVA, K. A. (1963), *Crystalline rocks differentiation on elastic properties in Northern Kareliya*, Zap. LGE, *46* (2), 22.

PANIN, V. I. and RITTER, F. A., *Estimation of seismic methods efficiency for determination of stress state in situ, in Izmerenie napryazhenii v massive gornyh porod* (ed. Zbornik) (Novosibirsk, Izd. Nauka, 1974), p. 138.

PANIN, V. I., TOMA, K. KNOLL, P. and ADLER, A. (1970), *Stress state determination in a polymetallic mine*, Phyzico-technicheskie razrabotki polezynh iskopaemyh *2*, 1970, 128.

SAVICH, A. I., KOPTEV, V. I., NIKITIN, V. N. and YASCHENKO, Z. G., *Seismo-acoustic Methods of Rock Mass Studying* (M. Izd. Nedra, 1969), 240 pp.

SILAEVA, O. I. and BAYUK, E. I. (1967), *On anisotropy nature of elastic rock properties*, Izv. Akad. Nauk SSSR, Physica Zemli, *12*, 22.

TOCHER, D. (1957), *Anisotropy in rocks under simple compression*, Trans. Amer. Geophys. Union 38.

TURCHANINOV, I. A. and MARKOV, G. A. (1966), *Latest tectonics influence on stress rock state in Khibiny apatite mines*, Izv. Akad. Nauk SSSR, Physica Zemli *8*, 83.

TURCHANINOV, I. A., MARKOV, G. A., IVANOV, V. I. and KOZYREV, A. A. (1973), *Tectonic stress field from measurements in the Khibiny massif*, Kn.: Napryazhennoe Sostoyanie Zemnoi Kory, Izd. Nauka.

(Received 9th November 1976)

Pageoph, Vol. 115 (1977), Birkhäuser Verlag, Basel

Simultaneous Monitoring of Stress and Strain in Massive Rock

By MICHAEL T. GLADWIN[1])

Abstract – Continuous measurements of ultrasonic pulse transit times over 10 m baselines between fixed points in each of three mutually perpendicular directions are made routinely in solid rock. The most significant current experiment is in a large underground support pillar (30 m × 30 m × 200 m high) 700 m below ground level in an active mine. Velocity measurements to a precision of 2 parts in 10^6 allow stress changes of order 1 kPa to be monitored, and compared with simultaneous strain measurements (for which a capacitance strain sensor is used) to examine the mechanisms of large scale stress relief processes. Characteristic stress relief cycles (of magnitude 200–3000 kPa) are found to migrate through the pillar as impressed loads are accommodated by the rock mass.

Key words: Stress in situ; Seismic velocity anisotropy; Creep.

1. Introduction

Numerous strain monitoring techniques have been developed for geophysical use. In most instances strain measurements are used to infer stress conditions, and an in situ elastic modulus must be observed or assumed. However, in many cases observed strain is not directly related to the accumulation of elastic energy, since stress is almost invariably accompanied by anelastic creep and redistribution of block boundaries. These relief processes are particularly relevant on the long time scales of the monitoring programmes for shallow earthquake studies and are important also in mining situations.

In the present experiment, the stress dependence of acoustic velocity, which accompanies the phenomenon of pore closure at low ambient stress levels (<2 k bar), is used to monitor stress in a large underground mining support pillar as part of a cooperative programme (BRIDGES *et al.*, 1976). The stress dependence of acoustic velocity has not yet been widely used for monitoring, probably because early workers (KANAMORI, 1970; FEDOTOV *et al.*, 1970; EISLER, 1969) concentrated on long base line measurements which necessitated the use of explosive sources, and which have a low total path sensitivity to stress since for most of the ray path the conditions of

[1]) Department of Physics, University of Queensland, Brisbane, Australia 4067.

hydrostatic pressure ensure that pore closure is essentially complete. Short baseline measurements are not subject to these limitations, and routine continuous monitoring at much higher sensitivity, using boreholes up to 100 m apart is quite feasible.

REASENBERG and AKI (1974) presumably recognized this in their use of timed pulse trains from an air gun in water. However, in this experiment and in the previous work of DE FAZIO, AKI and ALBA (1973), implausibly high apparent sensitivity of velocity to stress (20 % per bar) was obtained, quite out of line with laboratory measurements and our own better controlled observations. Their very high sensitivities are due partly to the fact that observations are made at the surface at low ambient stress levels, partly to the ambiguity in separating the chosen Rayleigh wave arrival maximum from the *p*- and *s*- wave codas but more particularly because their observation (which is essentially a phase velocity measurement) is subject to amplitude variations on all of the multiple paths which reach the detection point (GLADWIN and STACEY, 1974). Their observations appear to be limited to ± 0.15 ms in 170 ms (≈ 0.1 %) and, as inferred in their summary, resolution at this level is of little significance for geophysical observations at reasonable depths. Observation of *p*-wave first arrivals is not subject to such limitations.

Ultimate stress sensitivity in such a system is determined by the accuracy of timing of individual pulse transits. This is limited by the degradation of pulse rise time with distance from the transmitter, but it can be shown [GLADWIN and STACEY, 1974] that this rise time is actually proportional to travel time, so that the relative precision of a velocity measurement is independent of travel time provided no instrumental limitation is imposed by the timing electronics. Demonstrations that multiple sampling of acoustic pulse arrivals reproducibly improves the timing precision in an almost perfect statistical ($\sqrt{N}$) manner have been reported (GLADWIN, 1973; GLADWIN and STACEY, 1974) and are here confirmed. By averaging travel times of 10^4 pulses at 20 ms intervals, the averaged readings (obtained at 200 sec. intervals) are found to be reproducible to several nanoseconds although individual pulses over 5 m paths (about 900 μs travel time) can be timed only to about 0.2 μsec.

Experiments of this kind in tunnels of the Snowy Mountains Hydroelectric Authority and in both lead and copper ore bodies at the Mount Isa mine in North Queensland, in which tunnelling and mining operations have produced stresses which were believed to be approximately known, have all yielded velocity changes corresponding satisfactorily with the velocity stress sensitivities established by laboratory measurements.

The most recent experiment uses an orthogonal array of velocity monitors in a pillar in the copper ore body at Mount Isa. Simultaneous measurements of strain are being made with a borehole capacitance strain meter developed for the project from the highly linear displacement transducer previously reported (STACEY *et al.*, 1969; GLADWIN and WOLFE, 1975). Apart from providing routine analysis of the state of stress of the pillar (of which a reasonably large volume (10^4 m^3) is sampled), comparison with the independent strain measurements allows both the elastic and inelastic components of the loading sequence to be isolated. The pillar, which is to be taken through an extended stage of vertical uniaxial compression as part of a mining sequence, provides a very large scale quasi-laboratory experiment which should allow conclusions to be drawn on the mechanisms of stress relief and redistribution in massive rock.

2. *Ultrasonic stress monitoring*

Changes in pulse travel time due to changes in compressional wave velocity in the direction of propagation are proportional over a limited range to changes of stress in the rock. For most sites in massive rock the stress sensitivity of pulse velocity is of order 1 part in 10^6 per kilopascal (1 part in 10^4 per bar) allowing stress change of 5 kPa (1/20 bar) to be resolved easily. The system is sensitive to changes of either hydrostatic or deviatoric stress. Each installation consists of a magnetostrictive transmitting transducer (resonant at 37.5 kHz) and two piezoelectric detectors set in three parallel 90 mm boreholes. For spot monitoring (required by mining personnel) a portable measurement package is used to produce a mean travel time averaged over any number of pulses chosen by the operator (typically 10^3 or 10^4). For the current continuous monitoring experiment in the Mt. Isa copper ore body, a number of installations are multiplexed into a single timing module and the travel times are digitally serialized via a telephone line to a 7 track incremental magnetic tape recorder (in an air conditioned control room). Four paths are being monitored, two in the vertical direction and two at right angles to each other in the horizontal plane. A significant improvement and simplification of the technique, as originally described (GLADWIN 1973), has been obtained by timing the first zero crossing of the p-wave arrival instead of using an arbitrary discriminator level on the leading pulse. A histogram of standard deviations of hourly readings over a six day period for a particular station (travel time 905 μs) shown as Fig. 1, indicates the reproducibility of observations. More than 50 % of all the data show hourly standard deviations of 2 nanoseconds or less, i.e., a timing precision of 2 parts in 10^6. It should be noted that these data have been obtained in a site with very high ambient noise, in an active mine, within 100 m of routine drilling operations. This is accomplished because the rapid spatial attenuation of mine-generated high frequency noise allows a good signal to noise ratio at the chosen frequency. This precision could not in principle be achieved using techniques such as those of REASENBERG and AKI (1974), who do not use first arrivals, but is necessary for stress monitoring in sites which have realistic velocity-stress sensitivities.

Calibration of the ultrasonic stress monitor for each particular site is achieved by direct comparison of observed velocity changes with the long term stress changes observed by over-coring of soft borehole inclusion strain meters every few months and by direct comparison with strain observations. Laboratory observations on the elastic properties of the rock which is a strong recrystallized breccia of compressive strength 170 MPa have also been made.

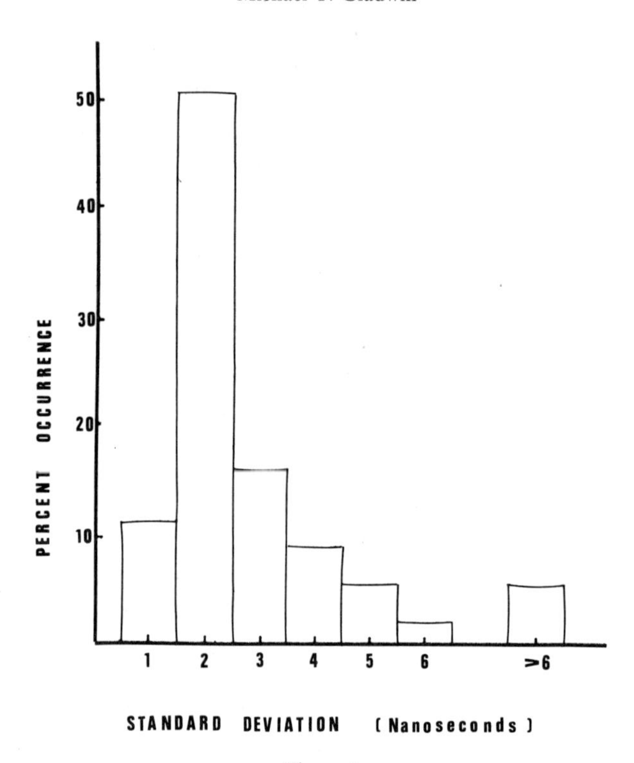

Figure 1
Histogram of hourly standard deviations of time of flight readings for a 5 m path ($t = 905$ μs) in the Mt. Isa copper ore body over a six day period in February 1976.

Recordings during 1974 in the Mt. Isa lead ore body yielded the following results:

(1) Changes of velocity of predictable magnitudes, totalling up to 7% at one site, occurred as loading was increased through approximately 40 MPa (400 bars).

(2) Velocity anisotropies of up to 3% were observed in extreme cases.

(3) Stress relief processes occur by episodal relief events. These are characterized by gradual increase of velocity in the direction of increased stress toward a reasonably constant level representing the maximum which the rock can withstand, followed by a more dramatic decrease in velocity accompanying local failure and stress relief. An example is shown in Fig. 2.

(4) Relief events, when they occur, typically give stress drops in the range 500–3000 kPa (5–30 bars). These values are estimated from laboratory measurements of stress sensitivity and are consistent with known long term total loading figures and overcore measurements.

(5) The stress relief events are not instantaneous but typically occur over periods of 4 to 6 hours.

(6) Stress relief events in the direction of maximum principal stress are not simultaneously transmitted to the minor stress axes in nearby areas.

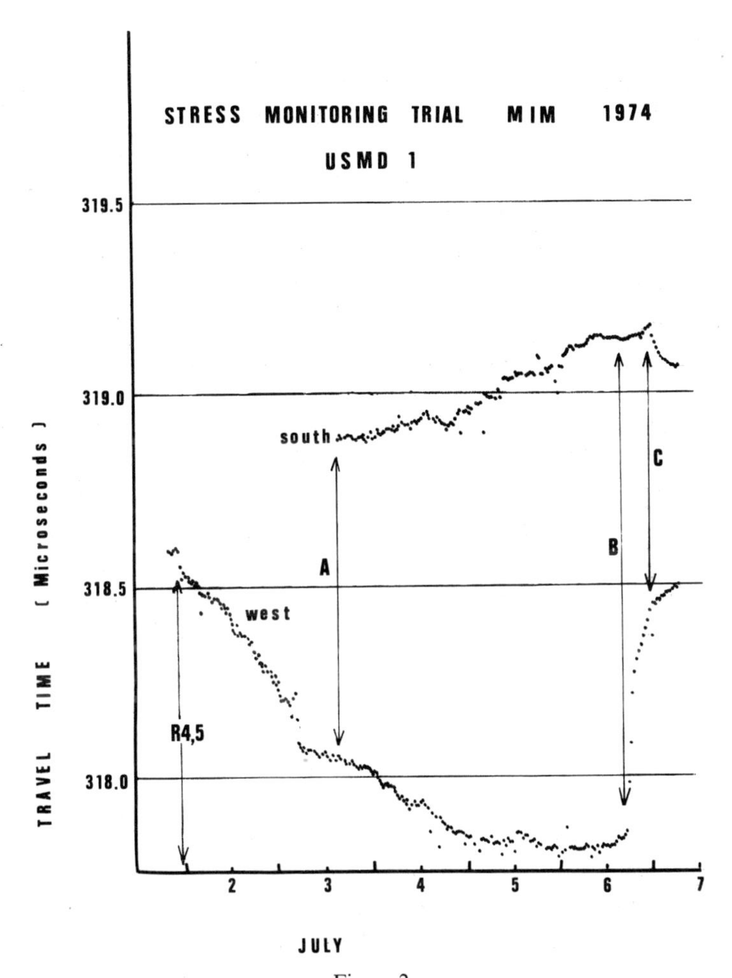

Figure 2
A stress relief event in the Mt. Isa lead ore body. Raw travel times for two short paths at right angles to each other are shown over a six day period. The stress release initiated at B on the lower plot has a magnitude of approximately 30 bars.

(7) The number of stress-relief events per unit time tends to increase prior to massive failure of the whole rock body.

Measurements of stress (as distinct from strain) are notoriously difficult to make in a mining environment. Thus the stress relief events observed in this experiment are of particular interest. The one reproduced in Fig. 2 is representative. Two nearly perpendicular paths are shown. The one marked WEST coincided with the near-vertical maximum principal stress and showed an increase in stress initiated by firing at the time marked (R4,5) of two rings of explosive which removed adjacent supporting rock. The second path (SOUTH) was displaced by 5 m from the West path, but in the absence of inelastic effects would be expected to show a simultaneous reduction in stress. At the time of firing no marked changes in velocity occurred. However, the expected velocity changes occurred over the next few days. At the point marked B, after the west path had reached what seemed to be a steady state, a stress relief event (increase in travel time) occurred at a time unrelated to mining activity. At this time the south path showed no response, but about 7 hours later (C in Fig. 2) an increase in stress is indicated, evidently when the relieved stress on the west path was transferred to the region sampled by the south path. The inferred stress relief of the west path was 3000 kPa (30 bars).

3. *Simultaneous strain monitoring*

The 1974 velocity observations in the Mt. Isa lead ore body were not accompanied by well controlled strain measurements. In the current programme in the copper ore body, use is made of a new borehole strain meter based on the capacitance displacement technique developed in this department (STACEY *et al.*, 1969); MCKAVANAGH and STACEY, 1973; GLADWIN and WOLFE, 1975; DAVIS and STACEY, 1976). The strain meter measures plane strain perpendicular to a 90 mm borehole approximately 8 m long. Each instrument has three independent sensors with strain sensing axes at 120° to each other. On each axis the change in diameter of the borehole is monitored using a three plate capacitor with centre plate mounted from one end of the diameter and outer plates mounted from the opposite end. Strain is routinely recorded to 1 part in 10^8. The strain meter probe is located close to the ultrasonic stress monitor.

During 10 months of continuous operation a systematic strain rate of about 3×10^{-8} per day has been observed. During periods of stress accumulation (as measured by increase in ultrasonic pulse velocity) this strain rate tends to decrease. An example of simultaneous stress and strain data is shown in Fig. 3. The mining extraction sequence which will ultimately put the pillar into uniaxial compression is in its early stages, so that only small loading events are available for study so far. The strain step (magnitude 9×10^{-7}) at midnight of 30 July, was triggered by a major blast at that time. A decrease in travel time over an ultrasonic path with the same orientation as the strain sensor occurred over the next five days, as the load was redistributed in the pillar. At the end of this time, (August 5), a stress relief event occurred without a strain increment and the strain trend returned to the normal long term drift rate. The July 31 strain step (on the upper plot) would, if elastic, correspond to a stress change of order 50 to 70 kPa (0.5 to 0.7 bars). The total associated velocity change over the next 5 days was 0.44 microseconds. With the approximate velocity-stress sensitivity of 2×10^{-6} per kPa (2×10^{-4} per bar), the corresponding stress change would be 240 kPa (2.4 bar). It may be possible that the stress sensitivity is increased to approximately 7 parts in 10^4 per bar by microcracks, but in any case the observations are obviously not explicable in terms of elastic deformation. The stress relief event on August 5 is a striking example of effects observed many times in both the lead and copper ore bodies at Mt. Isa. None of the observed events appear as straightforward elastic deformation.

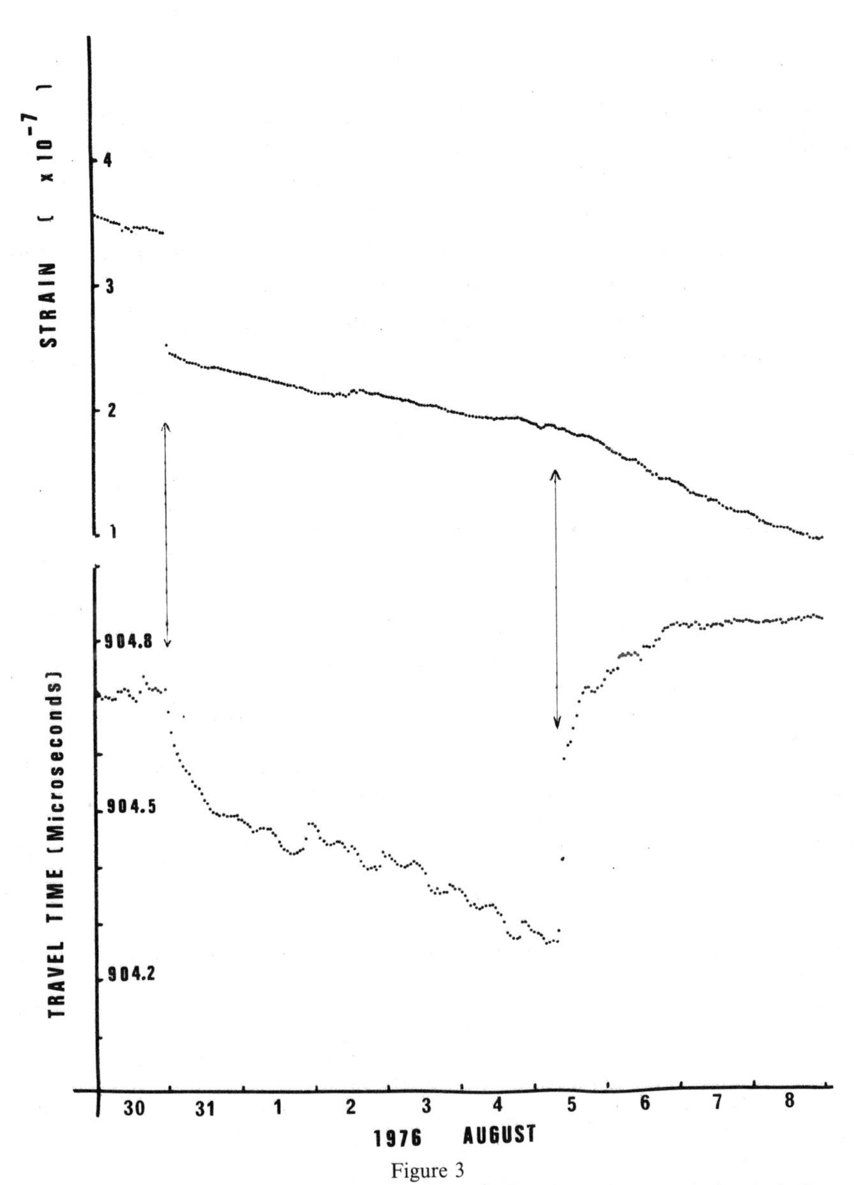

Figure 3

Hourly average values of strain and of ultrasonic travel time (inferred stress) at a particular site in the copper ore body at Mt. Isa. The increase in velocity associated with the strain step on July 31 occurred over a period of five days during which an anomalous strain rate was also observed. A typical stress relief event, accompanied by a change in the strain rate, occurred on August 5.

4. *Summary*

Simultaneous monitoring of stress and strain allows the separation of elastic and inelastic processes of rock deformation. In the experiments reported here, in situ deformation of rock is found to be so dominated by inelastic processes that we are doubtful of the validity of stresses inferred from strain observations. Use of ultrasonic stress monitoring would be a valuable extension to many current strain monitoring programmes.

Acknowledgements

Our collaboration with Mt. Isa Mines Limited was originally arranged by Dr. K. McCracken of the CSIRO, Division of Mineral Physics. The experiment has been made effective by the help of B. H. B. Brady, M. C. Bridges, J. Dwyer, V. Stampton, J. Spreadborough and T. van Neilson of M.I.M.

The equipment was fabricated by Messrs. T. van Doorn, A. Roth, F. Kingshott and C. Croskell at the University of Queensland, and financed by the Australian Research Grants Committee through a grant to Professor F. D. Stacey, from whose earth strain research programme this work developed.

REFERENCES

BRIDGES, M. C., GLADWIN, M. T., GREENWOOD-SMITH, R., and MUIRHEAD, K. (1976). *Monitoring of stress strain and displacement in and around a vertical pillar at Mount Isa Mine*, ISRM Symposium 1976, The Institution of Engineers, Australia, *76*, 44–48.

DAVIS, P. M. and STACEY, F. D. (1976). *Direct measurement of magnetostriction in rocks using capacitance micrometry*. Geophys. J. R. Astron. Soc. *44*, 1–6.

DE FAZIO, T. L., AKI, K. and ALBA, J. (1973). *Solid earth tide and change in the in situ seismic velocity*. J. Geophys. Res. *78*, 1319–1322.

EISLER, J. D. (1969). *Investigation of a method for determining stress accumulation at depth II*. Bull Seismol. Soc. Am. *59*, 1, 43–58.

FEDOTOV, S. A., DOLBILKINA, W. A., MOROZOV, V. N., MYACHKIN, V. I., PREOBRAZENSKY, V. B., and SOBOLEV, G. A. (1970). *Investigation on earthquake prediction in Kamchatka*, Tectonophysics, *9*, 249–258.

GLADWIN, M. T. (1973). *Precise in situ measurement of acoustic velocity in rock*, Rivista Italiana di Geofisica, *22*, 5/6, 283–286.

GLADWIN, M. T. and STACEY, F. D. (1974). *Ultrasonic velocity as a rock stress sensor*, Tectonophysics, *21*, 39–45.

GLADWIN, M. T. and WOLFE, J. (1975). *Linearity of capacitance displacement transducers*, J. Sci. Instrum. *46*, 8, 1099–1100.

KANAMORI, H. (1970). *Recent developments in earthquake prediction research in Japan*, Tectonophysics, *9*, 291–300.

MCKAVANAGH, B. and STACEY, F. D., (1974). *Mechanical hysteresis in rocks at low strain amplitudes and seismic frequencies*, Phys. Earth and Plan. Int. *8*, 246–250.

REASENBERG, P. and AKI, K., (1974). *A precise, continuous measurement of seismic velocity for monitoring in situ stress*. J. Geophys. Res. *79*, 399–406.

(Received 28th September 1976)

Pageoph, Vol. 115 (1977), Birkhäuser Verlag, Basel

Stress Distribution near Lake Jocassee, South Carolina

By PRADEEP TALWANI[1])

Abstract – Seven months of seismic monitoring near Jocassee reservoir (impounded 1974) resulted in the detection of four clusters of earthquake activity. Composite fault plane solutions (CFPS) for each cluster indicate strike slip faulting for shallow earthquakes (< 1.0 km) on the shores of the reservoir and normal faulting for deeper events (1–3 km) in the middle of the reservoir. The directions of the axes of maximum and minimum compression inferred from the CFPS were found to be NW and NE respectively and contrast with the NE and NW directions obtained by hydrofracture in a shallow well (230 m) at Bad Creek, about 10 km from the epicentral region.

Key words: Fault plane solutions; Hydrofracture; Seismicity at reservoir.

1. Introduction

The Jocassee hydro station in Oconee County, South Carolina (inset Fig. 1) was built in the early 1970's. The lake, with a capacity of 1430×10^6 m^3 was filled in January 1974 and achieved a maximum head of 102 m on March 15, 1975. Lake Jocassee lies in the Inner Piedmont geologic belt with the predominant rock unit being Henderson augen gneiss. This region had been regarded as aseismic until a MM intensity III–IV event occurred near the dam on October 18, 1975. This was followed by felt events on November 6 and November 25, 1975. This latter event ($M_L = 3.2$) was felt over 3000 sq km and was assigned a MM intensity IV–V.

We have monitored the region continuously from November 1975 to June 1976 using four to seven portable seismographs. The preliminary results of those investigations have been discussed elsewhere (see TALWANI *et al.* 1976). In this paper I will attempt to infer stress directions from composite fault plane solutions of events in the vicinity of Lake Jocassee.

2. Location of microearthquake activity

Over one thousand events ($-1.2 < M_L < 2.5$) were recorded between November 8, 1975 and May 31, 1976. Of these, over 300 were located (with an accuracy of

[1]) Geology Department, University of South Carolina, Columbia, South Carolina 29208, USA.

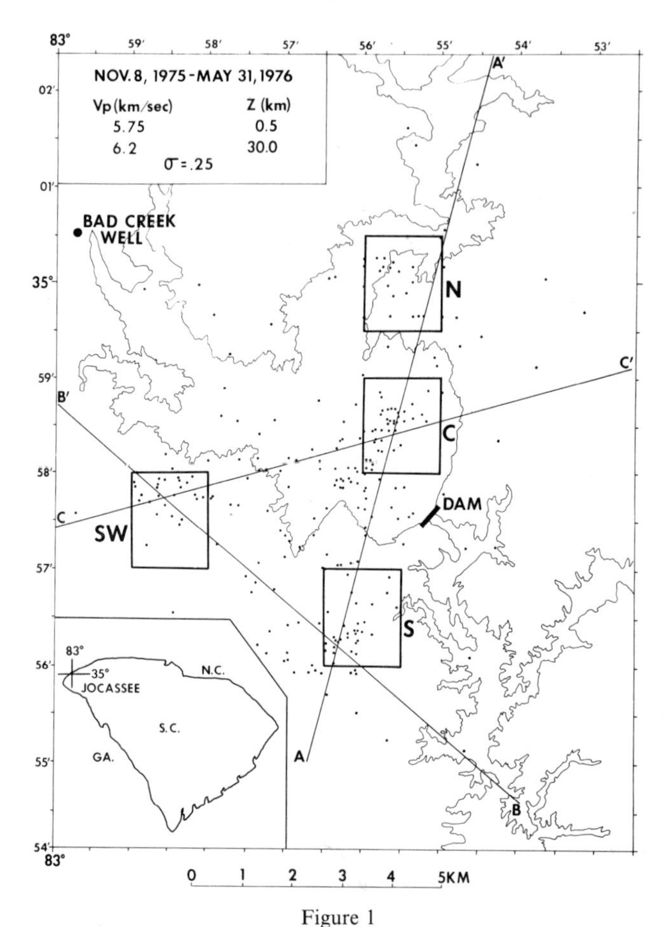

Figure 1

Shows the location of epicenters of events between November 8, 1975 and May 31, 1976 near the Jocassee dam in northwest South Carolina (inset). The activity is seen to lie in four clusters – north (N), central (C), south (S) and southwest (SW). Events within these clusters were used to obtain CFPS. The velocity model used to locate the epicenters is shown in another inset. AA′, BB′ and CC′ show location of profiles along which hypocenters were plotted in Fig. 2. *In situ* stress measurements were made at the Bad Creek well.

± 200 m) and are shown in Fig. 1. The activity occurred in four clusters: in the middle of the lake and to its north, south and southwest. These locations are referred to as central, north, south, and southwest clusters. The earliest activity (November, 1975) was concentrated in the central cluster; subsequently it spread to the outer clusters.

Three hypocentral cross sections are shown in Fig. 2. In AA′, which passes through three clusters the activity is deepest under the lake. (Computed depths have an accuracy of ± 0.4 km). In CC′ also, the zone of activity deepens into the central cluster. In BB′ passing through the southern and southwestern clusters there appears to be a lack of activity between the clusters. The activity is perhaps associated with steeply dipping joints that are present in the area and trend NE and NW. The joints appear to be concentrated in the Henderson gneiss unit.

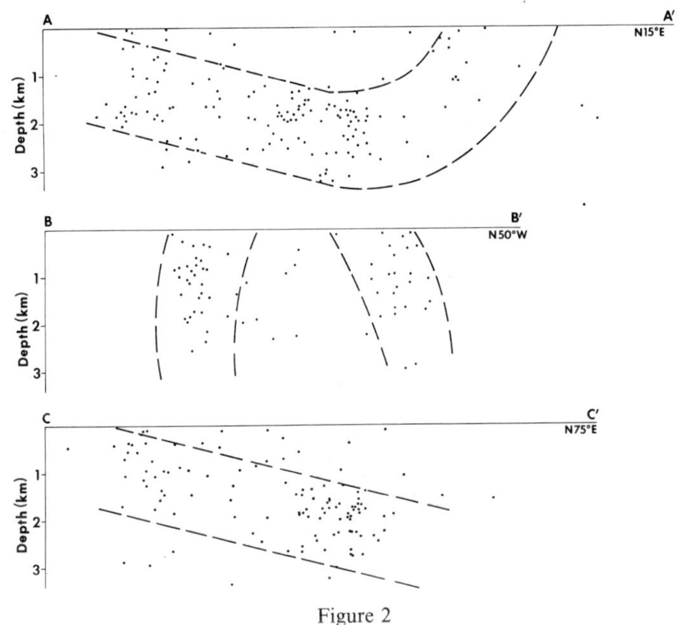

Figure 2
Hypocentral cross sections along AA′, BB′ and CC′ in Fig. 1.

3. *Composite fault plane solutions and stress distribution*

Composite fault plane solutions (CFPS) were made for 85 events that occurred during the period December 1975–March 1976. Only P-wave first motions were used. CFPS were obtained using 10, 20, 20, and 35 events respectively for the N, S, SW and C clusters. For the first three, only events with depths <1 km were used. The results (Fig. 3) indicate strike slip faulting is associated with the shallower N, S, and SW clusters, and normal faulting in the deeper C cluster.

The P (compressional) axis and T (tensional) axis, located in the center of the dilatational and compressional quadrants, respectively, and oriented 45° to the nodal planes are also shown in Fig. 3, and are listed in Table 1. The slip vector corresponding to any fault plane is obtained at the intersection of that plane with a plane through P and T axes.

Following RALEIGH *et al.* (1972) and SBAR and SYKES (1973), σ_1, the maximum compressive principal stress was chosen 30° from the slip vector in the direction of the P axis and in the plane of the slip vector and the P and T axes. Similarly, σ_3, the least compressive principal stress was chosen 60° from the slip vector in the direction of the T axis. The null axis (that is, the intersection of the two nodal planes) corresponds to σ_2.

Two nodal planes were obtained from each cluster from the CFPS. Since it was not clear which of the two nodal planes is the fault plane, slip vectors corresponding to both nodal planes were obtained. The directions of σ_1 and σ_3 corresponding to both nodal planes were calculated. The differences in the direction of σ_1 (or σ_3) obtained from the two nodal planes was found to be less than 36° (except in one case), (see Table 2). This difference was small enough for us to assume that the mean values were $\pm 30°$ of the 'true' stress directions. This result agrees with the observation of RALEIGH *et al.* (1972), who found that the 'true' directions of σ_1 and σ_3 lie within $\pm 40°$ of the direction of P and T axes respectively. (The mean directions of σ_1 and σ_3 calculated above correspond to the directions of P and T axes).

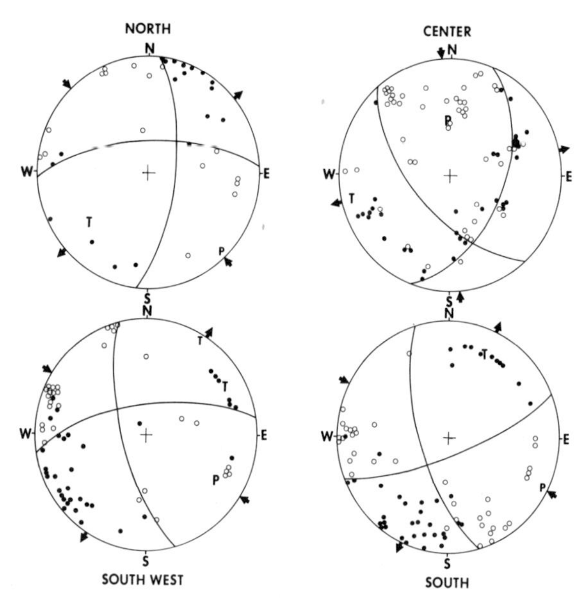

Figure 3
Equal area lower hemisphere projections for composite fault plane solutions for four clusters. Solid dots are for compressions and open dots for dilatations. P (compressional) and T (tensional) axes are shown for each cluster.

Table 1
Geometric data for the CFPS of events in different clusters

Cluster	Nodal plane			P axis		T Axis	
Depth (km)	No.	Strike	Dip	Azim	Plunge	Azim	Plunge
N	1.	N6°E	70E°	S44°E	4°	N48°E	32°
≤1	2.	N87°E	67°N				
S	1.	N17°W	70°W	S63°E	4°	N26°E	24°
≤1	2.	N68°E	76°S				
SW	1.	N15°W	72°W	S58°E	26°	N32°E	0°
≤1	2.	N81°E	70°N				
C	1.	N20°E	50°E	N5°W	52°	S77°W	8°
1–3	2.	N42°W	62°W				

To compare the stress directions obtained in the different clusters, σ_2 and the mean directions (and plunges) of σ_1 and σ_3 were plotted in Fig. 4. We note that for the three shallow clusters—N, S and SW—σ_1 and σ_3 are horizontal, trending NW and NE respectively. For deeper earthquakes in the central cluster, σ_1 is vertical and σ_3 is horizontal, trending NE.

Table 2

Computed directions of principal stresses for different clusters

Cluster	Nodal	Slip vector		σ_1		σ_2		σ_3	
	Plane	Azim	Plunge	Azim	Plunge	Azim	Plunge	Azim	Plunge
N	1.	S4°E	26°	S31°E	12°	N41°E	59°	S66°W	32°
	2.	N84°W	20°	N56°W	6°			S30°W	34°
S	1.	N21°W	12°	N49°W	2°	S34°W	64°	N42°E	24°
	2.	N74°E	20°	S77°E	10°			N9°E	22°
SW	1.	N10°E	20°	S42°E	28°	N52°W	61°	N45°E	8°
	2.	N74°E	20°	S75°E	27°			S19°W	7°
C	1.	N47°E	29°	N17°E	47°	S19°E	38°	S85°W	20°
	2.	N70°W	40°	N28°W	52°			N68°E	3°

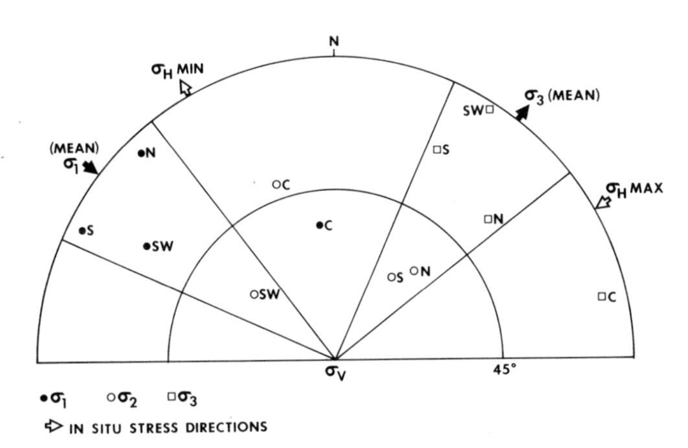

Figure 4

Mean values of σ_1, σ_2 and σ_3 for the four clusters plotted on the northern hemisphere of an equal area net, and compared with *in situ* stress directions at Bad Creek well. Stresses with a plunge of less than 45°, are considered to be horizontal, while those with a plunge greater than 45°, inner circle, are considered to be vertical. The letters N, S, etc., pertain to the different clusters.

4. *Discussion*

In situ stress measurements were made by HAIMSON (1975) at the Bad Creek well, located on a hillside about 100 m higher than, and about 10 km NW of the dam crest. By hydrofracturing at different depths to 230 m, he obtained 62 bars for the vertical stress, 159 bars for the least horizontal stress at N30°W, and a maximum horizontal stress of 228 bars at N60°E. In order to compare these with directions obtained from CFPS we recall that for strike slip motion on a fault both σ_1 and σ_3 are horizontal, while for a normal fault σ_1 is vertical and σ_3 is horizontal. We find that the direction of σ_3 obtained from the CFPS of each cluster is NE (see Table 3 and Fig. 4). The direction of σ_1 for earthquakes associated with strike slip faulting (N, S and SW clusters) is NW. Thus, we infer that the directions of maximum and minimum horizontal stresses are NW and NE respectively, which are *opposite* of the directions obtained by hydrofracture at the Bad Creek well. Since the maximum and minimum horizontal stress directions obtained from the CFPS for all the clusters are consistent with each other, we suggest that they represent the tectonic directions in the epicentral region rather than those obtained by hydrofracture.

Table 3

Variation of stress as a function of depth near Lake Jocassee

Location	Depth (km)	σ_{HMax} (σ_1)	σ_{Hmin} (σ_2)	σ_{vert} (σ_3)
Bad Creek Well*	0.23	N60°E	N30°W	Vert
N, S, SW Clusters	1.0	N52°W ± 20°	Vert	N38°E ± 20°
C Cluster	1–3	Vert	N19°W ± 20°	N77°E ± 20°

*) *In situ* stress measurements by HAIMSON (1975).

At a depth of 230 m in the Bad Creek well, the vertical stress is the least principal stress, at a depth of about 1 km, near the dam, it is the intermediate stress, while at greater depths it is the maximum principal stress. This suggests that the increase of vertical stress (with depth) is greater than that of horizontal stresses.

We conclude that in the vicinity of the Jocassee dam, the directions of maximum and minimum horizontal stresses are NW and NE respectively. The vertical stress is less than the horizontal stresses at shallow depths (<1 km). At a depth of about 1 km the vertical stress is the intermediate principal stress, and at greater depths, it is the greatest principal stress.

Acknowledgements

This study was partially funded by U.S. Geological Survey Contract No. 14-08-0001-14553.

REFERENCES

HAIMSON, B. C. (1975), *Design of underground powerhouses and the importance of preexcavation stress measurements*, preprint, Sixteenth Symposium on Rock Mechanics, Minneapolis, Minnesota.

RALEIGH, C. B., HEALY, J. H. and BREDEHOEFT, J. D. (1972), *Faulting and crustal stress at Rangely, Colorado*, Geophys. Monogr. Am. Geophys. Union *16*, 275–284.

SBAR, M. L. and SYKES, L. R. (1973), *Contemporary compressive stress and seismicity in Eastern North America: An example of Intra-Plate Tectonics*, Geol. Soc. Am. Bull. *84*, 1861–1882.

TALWANI, P., STEVENSON, D., CHIANG, J. and AMICK, D. (1976), *The Jocassee Earthquakes – A Preliminary Report*, Third Technical Report, Contract No. 14-08-0001-14553, U.S. Geological Survey, 92 pp.

(Received 13th October 1976)

Pageoph, Vol. 115 (1977), Birkhäuser Verlag, Basel

Focal Mechanism of Earthquakes in the Kamchatka–Commander Region and Heterogeneities of the Active Seismic Zone

By VYACHESLAV M. ZOBIN and IRINA G. SIMBIREVA[1])

Summary – Focal mechanisms of 74 shallow and 16 intermediate earthquakes (1964–1970) in the Kamchatka–Commander region are discussed. Regional stress systems and the nature of faulting are analyzed. Complex stress fields in the Kamchatka–Commander region and variable aftershock processes of strong Kamchatka earthquakes show the existence of heterogeneities of the active seismic zone in the Region.

Key words: Fault plane solutions; Tectonics of Kamchatka.

1. Introduction

Mechanism studies of earthquakes have become more interesting because of their relevance to the main hypotheses of the new global tectonics (ISACKS *et al.* 1968). Numerous focal mechanism solutions for a region give us the possibility of the reconstruction of the regional stress system and the nature of faulting.

This paper discusses the heterogeneities of the seismic focal zone in the Kamchatka–Commander region. The heterogeneities of the regional stress systems reflected by foci of moderate earthquakes and particularities of aftershock processes of strong earthquakes are considered.

Mechanisms of 74 shallow and 16 intermediate moderate earthquakes with magnitudes $m_b = 4.5$–6.5 for the period 1964–1970, and four strong ($M_s > 7.0$) earthquakes and their aftershocks for the period 1969–1973, in the Kamchatka–Commander region were investigated using the data of the first motion of P- and S-waves. Focal mechanism solutions were determined according to the A. V. VVEDENSKAYA (1969) method. The initial materials used were the signs of P- and S-wave first motion recorded at seismic stations of Kamchatka as well as the data of other stations published in the Operative Seismological Bulletin of the Institute of the Physics of the Earth, USSR; Earthquake Data Reports of CGS, USA and BISG issues, Edinburgh. The catalogue of 74 shallow earthquakes and 16 intermediate earthquakes for which focal mechanisms were determined is listed in Table 1. Hypocenters of these earthquakes were taken from annuals 'Earthquakes in the USSR'.

[1]) Institute of Volcanology, Pobeda Ave., 19, Petropavlovsk-Kamchatsky, USSR.

Table 1

Focal mechanism of earthquakes (1964–1970)

								Compression		Tension		Intermediate stress		Fault plane I				Fault plane II			
	Date	Origin time GCT	Latitude °N	Longitude °E	depth km	$K_{S1,2}^{F68}$	m_b	$Az°$	$e°$	$Az°$	$e°$	$Az°$	$e°$	dip direction	dip angle	dip component	strike component	dip direction	dip angle	dip component	strike component
(1)	(2)	(3)	(4)	(5)	(6)	(7)	(8)	(9)	(10)	(11)	(12)	(13)	(14)	(15)	(16)	(17)	(18)	(19)	(20)	(21)	(22)
1.	02.01.64	05–21	54 31	161 30	30–50	12.1	4.9	350	18	110	54	250	30	30	38	+0.56	+0.83	148	72	+0.85	−0.53
2.	14.01.64	01–11	52 56	159 53	40	11.5	4.9	242	51	228	18	118	36	15	75	−0.64	+0.77	266	41	−0.56	−0.83
3.	15.03.64	09–49	53 05	157 56	180	12.0	4.8	140	20	48	13	292	70	95	70	−0.70	−0.99	186	84	−0.41	+0.91
4.	22.03.64	00–52	52 41	160 39	25	12.4	5.0	58	20	316	26	175	56	275	84	+0.53	+0.84	360	56	+0.035	−0.99
5.	23.04.64	21–08	52 39	161 02	5–10	11.6	4.8	240	42	142	10	42	46	288	70	−0.64	+0.77	180	54	−0.41	−0.91
6.	24.04.64	00–42	52 42	161 04	10	12.2	4.9	88	8	358	52	195	40	308	64	+0.70	+0.72	58	52	+0.53	−0.85
7.	31.05.64	10–24	53 26	159 09	130	11.3	4.8	50	10	315	22	160	65	270	82	+0.37	+0.93	360	66	+0.14	−0.99
8.	21.06.64	01–33	50 39	157 30	60–70	13.3	5.7	225	1	135	54	136	35	193	54	+0.64	−0.77	75	56	+0.70	+0.70
9.	14.07.64	13–58	53 13	159 59	45	12.6	5.5	46	7	305	55	318	35	10	50	+0.62	−0.79	252	60	+0.74	+0.67
10.	04.11.64	03–37	54 44	163 04	5	11.8	5.0	220	20	314	20	264	64	267	64	0.0	+1.0	178	90	+0.42	−0.91
11.	05.12.64	23–51	53 43	161 42	0	12.5	5.3	278	4	188	4	246	86	324	88	+0.09	+0.99	235	86	+0.03	−1.0
12.	05.12.64	23–55	53 42	161 48	0	12.3	5.0	224	6	316	20	300	20	270	74	+0.17	+0.98	23	80	+0.28	−0.96
13.	07.12.64	18–30	51 31	158 16	50	11.8	5.1	153	7	244	20	230	70	292	80	+0.34	−0.94	200	72	+0.17	+0.98
14.	26.12.64	14–30	51 36	157 20	140	14.1	5.1	106	28	228	46	176	35	260	80	+0.82	−0.57	155	38	+0.34	+0.54
15.	29.01.65	09–35	54 37	162 06	30	12.3	5.8	22	8	194	7	217	80	148	88	+0.14	+0.99	240	82	+0.03	−1.0
16.	07.02.65	19–29	54 57	162 23	20–40	12.6	5.2	348	40	130	42	60	20	150	90	+0.94	−0.34	60	20	0.0	+1.0
17.	08.02.65	15–46	54 59	165 32	20–30	13.1	5.6	355	45	162	4	342	48	295	58	−0.60	−0.80	44	68	−0.59	+0.81
18.	08.02.65	17–37	55 00	165 29	20–40	13.1	5.8	350	30	90	14	5	54	45	60	−0.26	+0.97	302	72	−0.53	−0.85
19.	04.03.65	02–23	52 12	160 49	10–20	11.1	4.6	245	22	350	34	310	50	300	50	+0.10	+0.99	30	84	+0.64	−0.77
20.	01.04.65	17–52	54 39	161 54	50–60	11.4	5.0	288	4	23	40	194	50	324	62	+0.48	+0.87	72	70	+0.54	−0.84
21.	13.04.65	17–45	51 26	159 48	0–30	11.9	4.9	220	6	122	62	312	28	194	46	+0.77	−0.64	65	56	+0.81	+0.59
22.	24.07.65	11–45	54 35	163 10	10–20	10.5	4.8	222	55	317	5	50	36	284	50	−0.67	+0.74	168	60	−0.7	−0.7
23.	16.10.65	20–01	55 55	164 36	0–40	13.1	5.4	304	30	54	30	178	46	86	90	+0.70	−0.72	358	46	0.0	+1.0
24.	21.12.65	00–32	52 25	159 04	10–40	12.0	5.1	336	26	74	12	254	55	296	84	−0.44	−0.90	27	64	−0.21	+0.98
25.	22.12.65	00–28	52 20	160 43	0–10	13.0	5.5	270	40	140	38	25	32	114	90	−0.50	+0.87	205	32	−0.67	−0.74
26.	22.12.65	03–22	52 20	160 44	0–10	11.2	5.1	45	30	312	4	210	60	90	74	−0.39	+0.92	355	66	−0.30	−0.95
27.	22.12.65	07–27	52 26	160 32	0–20	12.0	5.4	40	14	132	15	275	70	85	70	0.0	+1.0	355	90	+0.34	−0.94
28.	07.01.66	07–45	52 40	160 13	30	11.9	5.2	355	5	252	62	88	38	330	46	+0.81	−0.59	200	55	+0.80	+0.60
29.	16.01.66	19–44	54 51	165 33	10–30	12.4	5.4	323	22	57	10	175	64	280	84	−0.40	−0.91	12	64	−0.17	+0.98

30.	28.01.66	22–38	51 21	157 13	130–140	14.1	5.7	152	4	62	5	287	85	107	85	0.0	−1.0	17	90	+0.12	+0.99
31.	05.02.66	14–24	52 29	159 12	10–40	12.5	5.3	296	2	205	2	60	86	250	86	+0.03	−1.0	340	88	+0.05	+1.0
32.	20.02.66	05–58	52 57	159 57	30	11.8	5.1	54	26	314	17	198	58	94	82	−0.53	+0.85	360	60	−0.14	−0.99
33.	08.04.66	01–46	51 58	158 10	75	14.3	6.0	240	22	10	56	142	24	280	30	+0.60	+0.80	40	70	+0.89	−0.45
34.	19.04.66	20–26	52 55	159 33	60	12.6	5.0	336	30	146	4	164	60	290	68	−0.34	−0.94	28	70	−0.42	+0.91
35.	25.04.66	23–03	52 07	160 50	5–10	11.2		145	14	143	26	138	60	290	64	+0.21	+0.98	28	80	+0.50	−0.87
36.	10.07.66	01–40	55 58	164 43	20–30	12.4	5.3	174	12	85	4	350	78	128	80	+0.10	−0.99	220	84	+0.17	+0.98
37.	22.09.66	00–04	52 23	159 40	25	11.5	4.9	350	15	255	15	122	70	302	70	0.0	−1.0	212	90	+0.34	+0.94
38.	28.09.66	15–15	53 03	160 05	40	11.2		315	16	204	53	58	32	280	40	+0.52	−0.86	160	70	+0.81	+0.59
39.	06.10.66	13–48	51 19	159 43	15–25	11.1	4.9	145	50	45	7	310	40	280	54	−0.59	−0.80	195	63	−0.70	+0.70
40.	22.10.66	12–47	54 58	162 26	30	13.3	5.5	195	5	105	5	320	84	240	90	−0.03	+1.0	152	85	0.0	−1.0
41.	14.12.66	01–59	56 41	161 28	10–15	11.4	4.6	4	2	93	13	297	76	140	80	+0.19	−0.98	48	80	+0.17	+0.98
42.	08.01.67	05–02	56 07	163 06	10–20	12.1	5.1	20	18	286	18	143	64	332	86	+0.40	+0.91	244	65	+0.03	−1.0
43.	08.01.67	06–43	55 53	163 14	0	12.0	5.1	316	24	220	12	102	64	265	66	−0.14	−0.99	360	84	−0.41	+0.91
44.	08.01.67	08–32	56 08	163 17	0–5	11.8	4.9	18	15	282	15	145	74	240	88	+0.34	+0.94	328	74	+0.03	0.0
45.	28.01.67	22–27	54 44	160 34	110	11.6	5.0	10	4	274	50	103	40	224	60	+0.64	+0.77	338	54	+0.57	−0.82
46.	22.03.67	20–47	52 23	160 51	20	11.7	4.8	30	37	296	6	198	54	80	70	−0.5	+0.87	337	60	−0.44	−0.90
47.	02.03.67	23–03	58 30	160 35	25	12.6	5.2	32	8	290	50	130	40	240	66	+0.70	+0.70	355	52	+0.59	−0.81
48.	23.05.67	05–53	52 01	160 33	10	11.1	4.3	87	45	254	45	350	8	80	90	+0.99	+0.10	170	8	+0.09	−1.0
49.	23.05.67	20–56	52 01	160 33	10	11.4	4.5	235	34	140	10	42	52	280	74	−0.59	+0.81	182	56	−0.34	−0.94
50.	04.06.67	05–26	51 16	159 40	0–10	12.3	4.8	46	16	140	5	245	71	64	62	+0.62	−0.79	304	47	+0.67	+0.74
51.	04.06.67	06–23	51 16	159 36	5	11.9		320	40	218	14	110	40	2	72	−0.64	+0.77	257	43	−0.42	−0.91
52.	05.06.67	16–38	51 16	159 40	0	12.0	4.5	350	17	90	35	230	60	130	84	+0.50	+0.87	40	60	+0.09	−1.0
53.	07.10.67	14–37	52 10	160 26	10	11.2	4.7	26	25	120	2	207	60	78	70	−0.34	+0.94	338	70	−0.37	−0.93
54.	14.12.67	18–25	54 20	160 49	80	12.8	5.5	152	40	265	25	15	40	112	80	−0.77	−0.64	218	40	−0.26	+0.97
55.	06.02.68	09–47	54 48	162 16	20	11.7	4.8	120	14	214	15	345	70	78	90	+0.34	−0.94	167	80	0.0	+1.0
56.	15.04.68	17–25	58 12	159 48	50	11.3	4.7	55	14	145	26	295	58	100	60	+0.17	+0.98	193	84	+0.47	−0.88
57.	30.04.68	01–43	54 12	159 18	110	11.5	5.1	335	8	243	10	116	70	296	70	0.0	−1.0	25	90	+0.34	+0.84
58.	20.05.68	11–53	51 48	158 36	60	13.0	5.3	338	25	70	8	164	64	290	74	−0.34	−0.94	22	70	−0.28	+0.96
59.	14.06.68	13–23	51 28	159 53	20–30	12.5	5.0	270	24	3	6	101	66	318	70	−0.24	+0.97	228	76	−0.34	−0.94
60.	14.06.68	15–19	51 30	159 40	5–10	11.4	4.4	290	20	28	20	160	60	340	60	0.0	+1.0	68	90	+0.50	−0.87
61.	14.06.68	23–03	51 28	159 48	0	11.3	4.9	75	2	344	18	167	70	300	78	+0.24	+0.97	28	74	+0.19	−0.98
62.	15.06.68	00–43	51 27	159 55	30	11.1	4.4	290	22	198	10	90	68	335	80	−0.34	+0.94	245	70	−0.17	−0.98
63.	23.07.68	21–12	55 36	166 24	30–50	13.2		175	10	85	8	310	78	130	78	0.0	−1.0	40	90	−0.21	+0.98
64.	04.09.68	10–34	52 58	159 50	40	12.8	4.7	318	22	50	8	156	65	273	80	−0.34	−0.94	6	68	−0.21	+0.98
65.	19.12.68	15–15	53 16	160 22	40–50	13.4	5.4	343	13	208	70	78	12	332	32	+0.94	−0.34	174	60	+0.96	+0.27
66.	02.01.69	14–07	53 43	160 43	30	12.5	14.9	84	6	76	26	345	65	126	72	+0.24	+0.97	228	76	+0.34	−0.94
67.	21.01.69	23–12	55 48	163 22	10	11.6	4.8	96	8	185	3	182	80	50	86	−0.10	−0.99	142	84	−0.99	+1.0

								Compression		Tension		Intermediate stress		Fault plane I				Fault plane II			
	Date	Origin time GCT	Latitude °N	Longitude °E	depth km	$K_{S1,2}^{F68}$	m_b	$Az°$	$e°$	$Az°$	$e°$	$Az°$	$e°$	dip direction	dip angle	dip component	strike component	dip direction	dip angle	dip component	strike component
(1)	(2)	(3)	(4)	(5)	(6)	(7)	(8)	(9)	(10)	(11)	(12)	(13)	(14)	(15)	(16)	(17)	(18)	(19)	(20)	(21)	(22)
68.	22.01.69	00–42	55 49	163 02	10	11.8	5.5	270	10	178	10	58	80	315	90	+0.21	+0.98	225	80	0.0	−1.0
69.	22.01.69	03–17	55 47	163 21	10	11.8	5.0	12	3	280	20	107	68	322	70	+0.26	−0.97	235	76	+0.28	+0.96
70.	22.01.69	03–54	55 47	163 10	10	11.8	4.8	275	16	178	25	20	62	140	85	+0.45	+0.89	225	64	+0.09	−1.0
71.	25.01.69	12–10	55 50	163 16	10	11.8	4.9	56	6	322	34	130	40	260	50	+0.59	+0.81	14	64	+0.37	−0.93
72.	26.01.69	15–05	55 50	163 03	20	12.5	5.5	315	20	52	14	170	64	275	85	−0.42	−0.91	5	66	−0.12	+0.99
73.	29.01.69	05–18	55 44	163 03	10	11.4		310	40	202	18	94	44	352	76	−0.69	+0.72	244	50	−0.34	−0.94
74.	31.01.69	04–10	53 18	159 03	140	12.4	5.2	284	35	18	6	112	56	335	62	−0.37	+0.93	237	70	−0.47	−0.88
75.	12.02.69	15–39	55 46	163 10	0–10	11.0	5.1	222	16	116	40	330	48	75	78	+0.67	+0.74	176	50	+0.31	−0.95
76.	28.02.69	13–47	51 30	158 08	30	10.3	5.0	88	30	324	44	195	30	292	80	+0.85	+0.58	29	32	+0.17	−0.98
77.	17.04.69	12–48	54 58	166 57	20	12.2		18	40	115	6	210	50	328	70	−0.59	−0.81	74	60	−0.47	+0.88
78.	08.06.69	14–49	53 10	159 54	40	12.5	5.0	130	4	225	65	42	24	290	54	+0.88	−0.47	157	48	+0.81	+0.59
79.	19.06.69	18–56	53 02	160 04	40	12.8	5.2	60	12	160	50	317	40	95	50	+0.50	+0.87	210	70	+0.72	−0.69
80.	16.07.69	08–16	52 09	159 07	60	12.5	5.8	277	15	12	25	160	56	322	60	+0.17	+0.98	56	80	+0.47	−0.88
81.	03.10.69	01–51	51 45	157 52	120	12.7	5.2	62	2	330	40	152	50	278	65	+0.54	+0.84	26	62	+0.53	−0.85
82.	22.11.69	23–09	57 45	163 45	30	14.4	6.3	356	16	98	42	252	45	142	74	+0.64	−0.77	38	50	+0.37	+0.93
83.	02.12.69	04–12	57 18	163 48	0–20	11.9	5.1	265	40	160	15	55	45	305	75	−0.64	+0.77	210	50	−0.26	−0.97
84.	03.12.69	12–34	54 30	161 32	30–40	11.8	4.9	106	12	356	42	203	44	68	55	+0.50	−0.87	318	68	+0.64	+0.77
85.	08.12.69	05–18	57 07	163 05	0	12.2	4.6	288	26	170	43	40	40	139	80	+0.17	+0.98	238	40	+0.26	−0.97
86.	23.12.69	13–22	57 12	162 53	0	12.1	5.4	277	3	186	18	13	68	140	76	+0.26	+0.97	233	73	+0.24	−0.97
87.	06.02.70	00–12	54 31	163 39	40	13.4	5.6	210	8	120	8	325	80	74	88	+0.17	+0.98	168	80	+0.03	−1.0
88.	19.04.70	04–42	51 22	157 54	90	11.3	5.6	46	35	310	8	205	50	88	70	−0.56	+0.83	348	56	−0.37	−0.98
89.	19.06.70	18–53	57 28	163 30	10	12.1	5.2	276	10	185	10	52	76	143	90	+0.24	+0.97	232	76	0.0	−1.0
90.	30.08.70	00–38	52 04	159 41	20–30	13.0	5.3	348	27	80	5	182	62	300	76	−0.42	−0.91	36	68	−0.26	+0.97
91.	28.09.70	17–22	53 08	159 08	120	12.6	5.4	258	45	358	12	100	44	320	50	−0.50	+0.87	214	68	−0.67	−0.74

Note: $Az°$ = azimuth; $e°$ = the angle with horizontal surface;
dip-component: (+) reverse type of faulting;
(−) normal type of faulting
strike-component: (+) right-lateral strike-slip;
(−) left-lateral strike-slip.

The error of hypocenter determinations is not more than ± 5–10 km. Energetic class of earthquakes $K_{S_{1,2}}^{F_{68}}$ ($K = \lg E_{\text{joule}}$) was calculated from Fedotov's nomogram $K_{S_{1,2}}^{F_{68}}$ (FEDOTOV 1972) connected to magnitude m_b by the empirical formula

$$K_{S_{1,2}}^{F_{68}} = 2.1 + 2.0\, m_b$$

2. *Seismicity*

Kamchatka and the Commander Islands are located at the junction of the Kurile–Kamchatka and Aleutian island arcs. The region is characterized by high intensity of seismic activity. Most earthquakes of the region are concentrated in the focal zone dipping beneath Kamchatka and representing a part of the Kurile–Kamchatka seismofocal zone. Detailed seismological investigations by Prof. S. A. Fedotov and Dr. P. I. Tokarev with colleagues made it possible to study the characteristic features of the seismofocal zone (FEDOTOV *et al.* 1969, 1974).

The regional catalogues of earthquakes for the period 1964–1970 resultant of these detailed investigations were published in annuals 'Earthquakes in the USSR'. Using these catalogues maps of density of earthquake epicenters were constructed with $K_{S_{1,2}}^{F_{68}} \geq 8.5$ for shallow (0–50 km) and intermediate (more than 50 km) earthquakes, respectively (Fig. 1).

While constructing the maps, the region was divided into a set of squares 25 $\times$ 25 km^2. The number of earthquakes within a single square was attributed to its center. It is interesting to notice some similarity between patterns of isobath lines and isolines of earthquake density.

A band of epicenters of shallow earthquakes stretching along the Kamchatka coast and including the ends of east peninsulas of Kamchatka demonstrates a place of intersection of the Kurile–Kamchatka focal zone with the Earth's surface. The Kurile–Kamchatka focal zone joins the Aleutian seismofocal zone (Commander Islands) in the region of Kamchatsky Cape Peninsula. The eastern boundary of the Kamchatka seismoactive zone is the Kurile–Kamchatka deep-seated trench, earthquakes beyond which occur very seldom (FEDOTOV *et al.* 1974).

FEDOTOV *et al.* (1974) consider most of the earthquake foci of the region to be concentrated at depth of 0–50 km.

The foci of intermediate earthquakes are concentrated in the narrow seismofocal zone dipping beneath Kamchatka. Seismic activity of the focal zone at depths of more than 50 km (see Fig. 1) is essentially heterogeneous.

During a short interval of time (1964–1970) seismic activity in Kamchatka was distributed unevenly. Zones of intense seismic activity can be recognized to the south-east of Kamchatsky Cape Peninsula, near Kronotsky Peninsula, to the north-east of Shipunsky Cape and on its submarine extension.

The discussed peculiarities of the distribution of eathquake foci in space are caused by a complicated stress system.

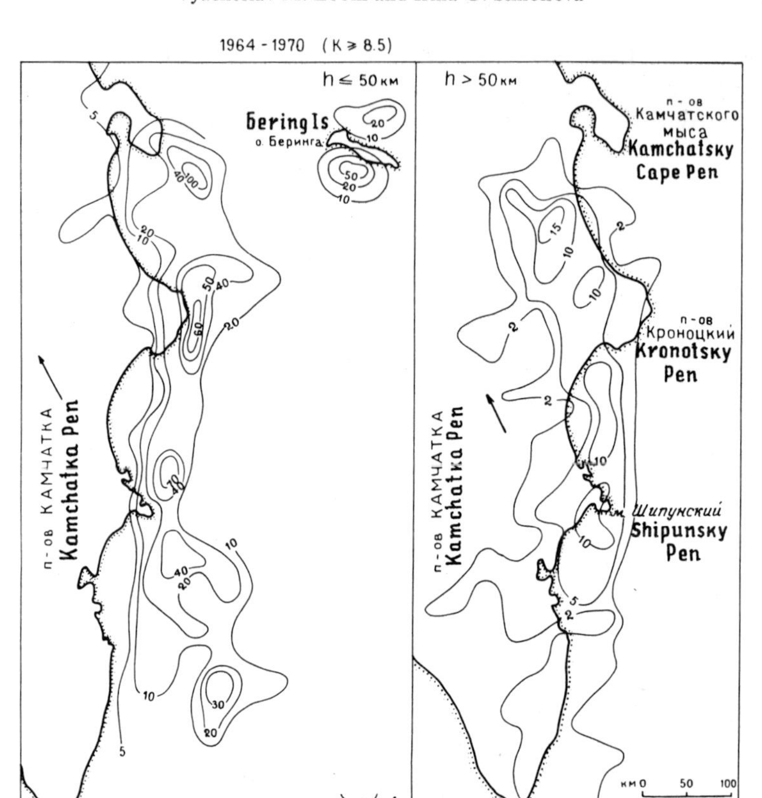

Figure 1
Map of density of earthquake foci in the Kamchatka–Commander region during 1967–1970. 1 – isolines of density of earthquake foci.

3. Focal mechanism of moderate earthquakes in the Kamchatka–Commander region during 1964–1970

Focal mechanism solutions for 91 moderate earthquakes are listed in Table 1. Their epicenters are shown in Fig. 2.

The following parameters have been determined for each earthquake: the axes of compression tension and intermediate stresses, and two possible fault planes. The errors in the determination of the position of axes and nodal planes are estimated to be about 5–15°.

The main features of the local and regional stress systems and the nature of faulting in the foci earthquakes were analyzed.

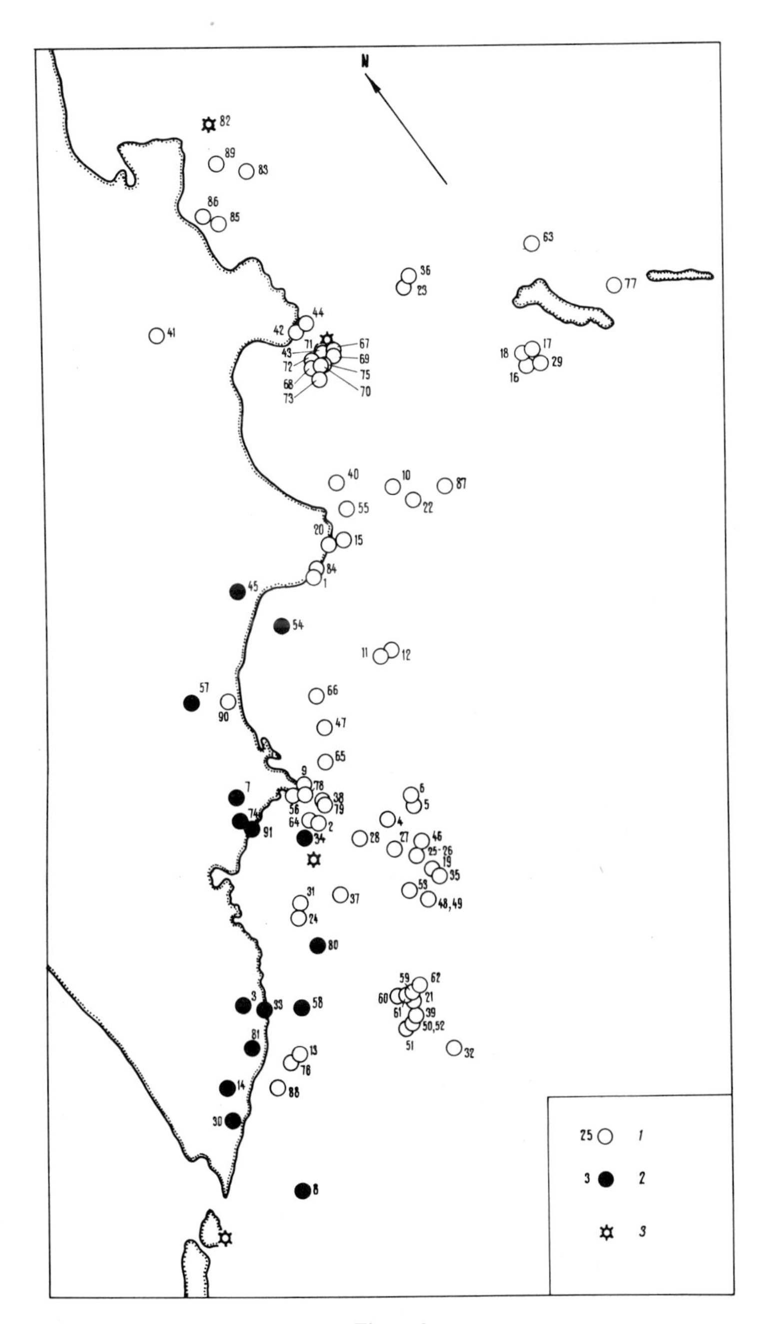

Figure 2
Map of epicenters of earthquakes discussed. 1 – epicenters of shallow earthquakes; 2 – epicenters of intermediate earthquakes; 3– epicenters of strong earthquakes ($M > 7.0$).

4. *Shallow earthquakes*

The nature of faulting was studied from the point of comparison of predomination on the one hand of vertical or horizontal component in foci, and on the other hand of the normal or reverse type of faulting for vertical component.

The horizontal component predominates for both possible nodal planes in 77 events of 91. The vertical component predominates only for 4 events.

The analysis of the sign of relative motion of the hanging wall side allowed us to differentiate the local particularities of the nature of faulting. The foci with the component of normal faulting and those with the component of reverse faulting were distinguished. The component of reverse faulting is present when the compression stress axis is more horizontal than the tension stress axis, and the component of normal faulting is present when the tension stress axis is more horizontal.

Figure 3 shows the spatial distribution of the zones of earthquake foci with different natures of faulting.

The main Kurile–Kamchatka focal zone near Kamchatka is characterized by predominant strike-slip type of faulting with a reverse faulting component. The Aleutian, Kronotsky and Shipunsky transverse structures cross the main focal zone and are characterized by predominant strike-slip type of faulting with normal faulting component.

Regional and local stress systems were analyzed by the reconstruction of principal normal elastic stresses according to the summary method of GUSHCHENKO (1975). Summary diagrams of stereographic projection of the upper hemisphere of pressure and tension axes of individual focal mechanism solutions for each quasihomogeneous region are drawn in Fig. 3.

According to the data of moderate earthquakes the regional stress field is heterogeneous. The local stress fields are stable only at short distances (100–200 km). It is interesting that the regions of homogeneous local stress systems coincide in space with those of homogeneous nature of faulting.

The main Kurile–Kamchatka focal zone is characterized by gradual evolution of regional stress system from south to north. The compression stresses for the entire zone are horizontal and nearly normal to the trend of the Kurile–Kamchatka arc; this supports the previous data by BALAKINA (1962). The tension stresses are also nearly horizontal near the southern and northern ends of Kamchatka, but nearly vertical in the center of the zone (see Fig. 3, zone II).

The *Aleutian* transverse zone (IV) includes the Commander Islands, Kamchatsky trough and Kamchatsky Cape Peninsula. The principal compression stress for this zone is nearly horizontal and quasiparallel.

The *Kronotsky* transverse zone (V) is characterized by nearly horizontal principal compression and tension stresses.

The *Shipunsky* transverse zone (IV^{a-b}) is divided into two subzones with little difference in the stress systems. Zone VI^{a} crosses the main Kurile–Kamchatka focal

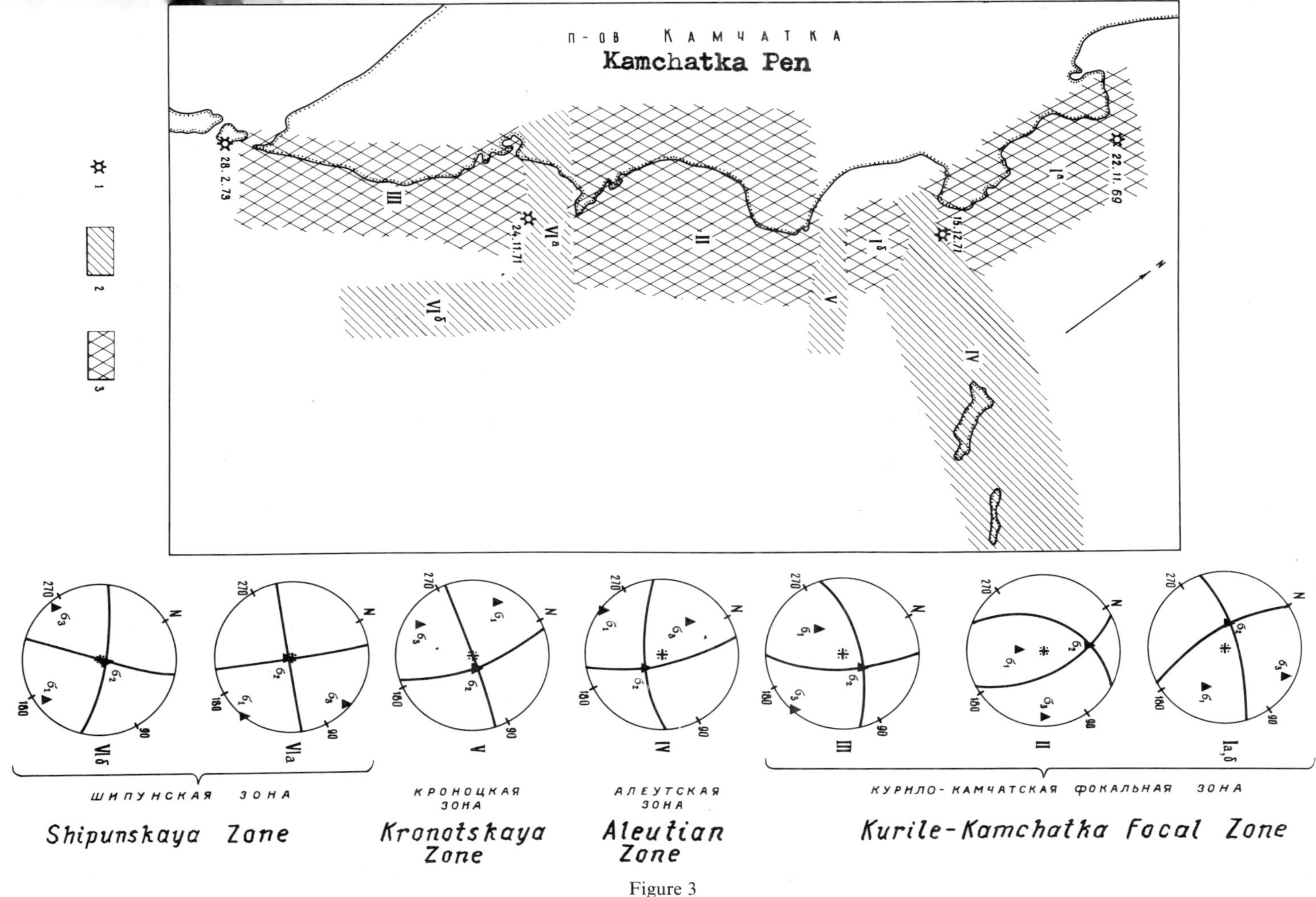

Figure 3

Nature of faulting and principal normal elastic stresses acting in the foci of moderate earthquakes. 1 – epicenters of strong earthquakes ($M > 7.0$); 2–3 – regions with predominance of strike-slip type of faulting with normal (2) and reverse (3) component of faulting. Right: stereographic projections of upper hemisphere of principal normal elastic stress $\sigma_1, \sigma_2, \sigma_3$ – the principal elastic tension, intermediate stresses and compression, respectively. I–IV – numbers of homogeneous regions.

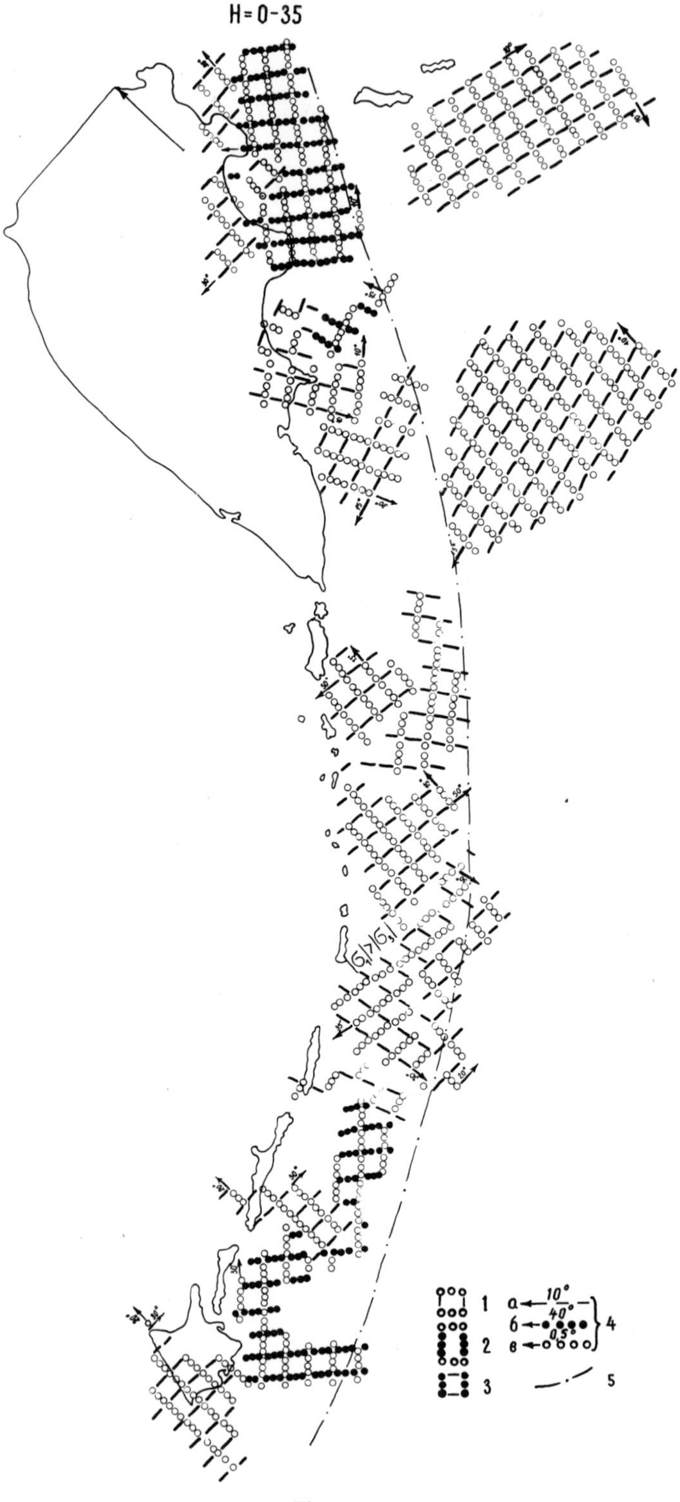

H=0-35

Figure 4

Principal stress axes orientation in the Kurile–Kamchatka region (after SIMBIREVA *et al.* 1976). 1 – pressure regions; 2 – tension regions; 3 – maximum gradient regions; 4 – principal stress axes orientations. Numerals indicate angle between stress axis and Earth surface: (a) – pressure, (b) – tension, (c) – intermediate stress. 5 – axis of oceanic trench.

zone near Shipunsky Peninsula, turns to the south and then runs along the main focal zone (VI^b). The two subzones are characterized by nearly horizontal principal compression and tension stresses.

The same investigation has been done by SIMBIREVA *et al.* (1976) for the Kurile–Kamchatka region according to more strong earthquake data (mostly $M \geq 6$). Simbireva's scheme is illustrated in Fig. 4. (Focal mechanism solutions which are in Table 1 do not participate in this scheme construction.)

Comparing Figs. 3 and 4 we can see that essential heterogeneity of the regional stress field is presented both on the high energetic level ($M \geq 6$) and on the moderate one ($4 \leq M < 6$). It should be noticed that the same parts distinguished in Figs. 3 and 4 are differed on their parameters, i.e., stress field reconstructed according to data of moderate earthquakes is not similar to stress field reconstructed under stronger earthquakes.

5. *Intermediate earthquakes*

The number of intermediate earthquakes is not enough for detailed constructions, but we can conclude that a single stress system and a single nature of faulting for intermediate earthquakes is absent under Kamchatka.

6. *Focal mechanisms and the development of aftershock activity of strong Kamchatka earthquakes* ($M > 7.0$)

This part of the paper is devoted to the comparative investigation of the focal processes of four strong Kamchatka earthquakes from 1969 to 1973. Focal mechanisms and space–time distribution of aftershocks are discussed.

Four strong earthquakes occurred after a long period of seismic silence following the Great Kamchatka earthquakes of 1952 and 1959. The seismic phone of moderate earthquakes for the period 1964–1970 was discussed in a previous part of the paper.

Epicenters of strong earthquakes are shown in Fig. 4.

The first in this series of strong shallow earthquakes was Ozernovskoe (22 Nov. 69), which occurred at the northern end of the Kurile–Kamchatka focal zone (Fig. 5). The epicenter of the main shock is situated on the northern end of the fault area. The aftershock activity spreads south of the main shock.

The next shallow strong earthquake (Ust'–Kamchatskoe 15 Dec. 71) took place in the region of Kamchatsky Cape Peninsula at the junction of the Aleutian and Kurile–Kamchatka structures (Fig. 6). This earthquake is characterized by a more intensive aftershock sequence than the Ozernovskoe earthquake and by a more complicated process in the fault area.

The last shallow strong earthquake (Severo–Kurilskoe, 28 Feb 73) occurred at the

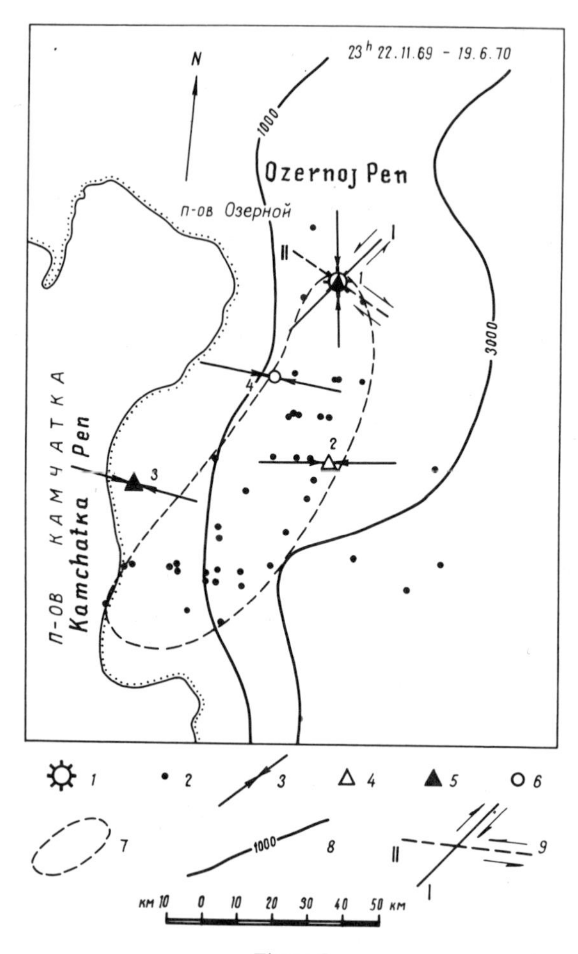

Figure 5
Focal process of Ozernovskoe earthquake. 1 – epicenter of the main shock; 2 – epicenters of aftershocks during the first day of activity; 3 – compression stresses acting in the foci of earthquakes; 4–6 – nature of faulting (4 – strike-slip type with normal faulting; 5 – strike-slip type with reverse faulting, 6 – strike-slip type); 7 – fault area; 8 – isobaths; 9 – possible fault planes.

southern end of Kamchatka near Paramushir Island (Fig. 7). In this case the development of the fault area spread north of the epicenter of the main shock.

The data on these earthquakes are presented in Table 2. Detailed investigation of their focal mechanism, macroseismic data and aftershock sequence was presented in a set of papers (FEDOTOV *et al.* 1973, GUSEV *et al.* 1975b, OSKORBIN *et al.* 1976).

All strong earthquakes had a seismic intensity up to 8 (MSK-64 scale).

The focal process of these earthquakes was investigated on the basis of data of the focal mechanism of main shocks and aftershocks and the temporal and spatial distribution of aftershocks.

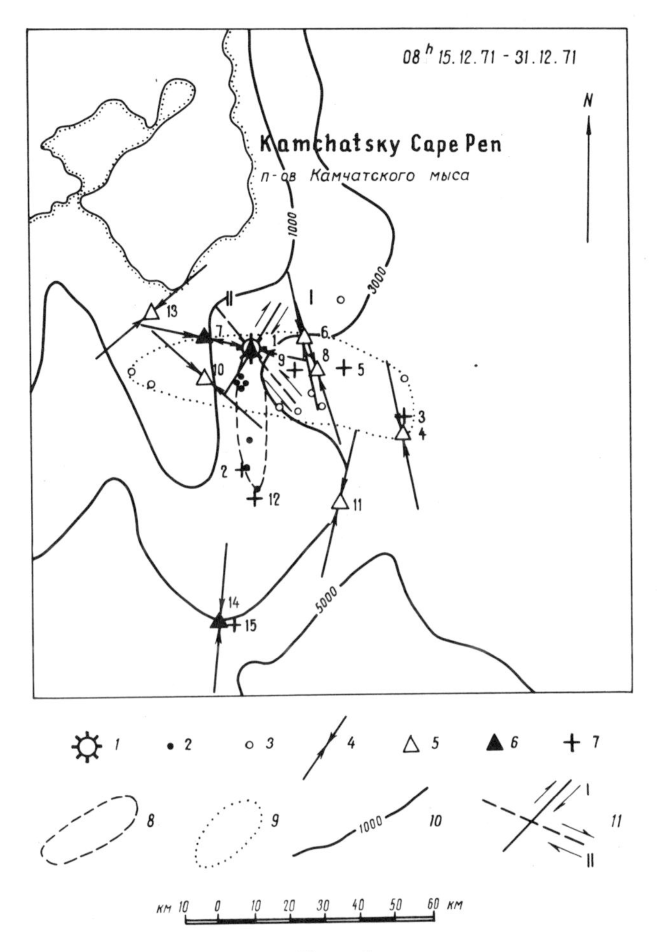

Figure 6
Focal process of Ust'–Kamchatskoe earthquake. 1 – epicenter of the main shock; 2, 3 – epicenters of aftershocks during the first day of activity (2 – the first 12 hours, 3 – the last 12 hours); 4 – compression stresses acting in the foci of earthquakes; 5–6 – nature of faulting (5 – strike-slip with normal faulting, 6 – strike-slip type with reverse faulting); 7 – epicenters of aftershocks for which focal mechanism was not determined; 8 – fault area; 9 – area of aftershock activity during the second half of the first day of activity; 10 – isobaths; 11 – possible fault planes.

The source dimensions were analyzed from the spatial distribution of the aftershock swarm during the first day of aftershock activity.

The aftershocks are shown in Figs. 5–7 by black circles delineating the fault area. The next period of the focal process was investigated using only the data of strong aftershocks ($m_b \geq 5.0$). Each strong aftershock is designated by an ordinal number in Figs. 5–7.

A comparative analysis of the main focal and source parameters reveals the difference in the nature of strong earthquakes.

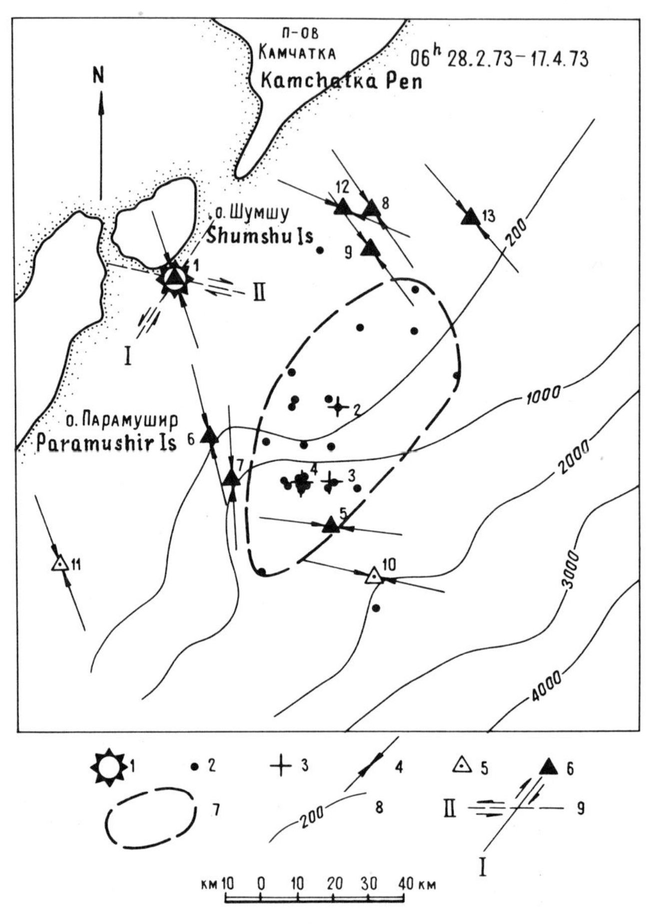

Figure 7

Focal process of Severo–Kurilskoe earthquake. 1 – epicenter of the main shock; 2 – epicenters of aftershocks during the first day of activity; 3 – epicenters of aftershocks for which focal mechanism was not determined; 4 – compression stresses acting in the foci of earthquakes; 5–6 – nature of faulting (5 – strike-slip type with normal component, 6 – strike-slip type with the reverse component); 7 – fault area; 8 – isobaths; 9 – possible fault planes.

Focal mechanisms of these earthquakes are different. The focal mechanism of the Ozernovskoe earthquake is characterized by a predominance of strike-slip type of faulting along the Kurile–Kamchatka focal zone with a little component of underthrusting beneath Kamchatka. The focal mechanisms of the Ust'–Kamchatskoe and the Severo–Kurilskoe earthquakes are characterized by a predominance of dip-slip type of faulting. The fault planes extend along the Kurile–Kamchatka focal zone and underthrust beneath Kamchatka.

The aftershocks of strong earthquakes can be divided into two types. Ust'–Kamchatskoe and Severo–Kurilskoe earthquakes (the first type) are characterized

Table 2

Data on strong Kamchatka earthquakes

Name	Date	Origin time	Lat. °N	Long. °E	depth, km	M_s	m_b	k	L length, km	dip-component of faulting	strike-component of faulting
Ozernovskoe	22.11.69	23.09	57.70	163.56	30	7.7	6.3	14.4	120	+0.64	−0.77
Petropavlovskoe	24.11.71	19.35	52.70	159.60	100	7.2	6.3	15.0	65	−0.47	+0.88
Ust'–Kamchatskoe	15.12.71	08.29	55.85	163.35	20–30	7.7	6.1	14.7	50	+0.92	+0.37
Severo-Kurilskoe	28.02.73	06.37	50.60	156.40	40–70	7.4	6.3	15.5	95	+0.85	−0.53

by the development of the focal process at the ends of the faults (see Fig. 6,7). Ozernovskoe earthquake (the second type) is characterized by the development of the focal process along the fault (see Fig. 5). It is interesting that the number of aftershocks with magnitude $m_b \geq 5.0$ is different for each strong earthquake (more than 10 for Ust'–Kamchatskoe and only 4 for Ozernovskoe).

Only one intermediate depth strong earthquake (Petropavlovskoe, 24 Nov. 71) occurred near Shipunsky Peninsula (Fig. 8) (GUSEV *et al.* 1975a). This shock took place at a depth of 100–120 km and is characterized by the absence of strong aftershocks. Only the nearest seismic stations registered five weak aftershocks. Four of them presented a chain along the coast of Avachinsky Bay. The focal mechanism of this

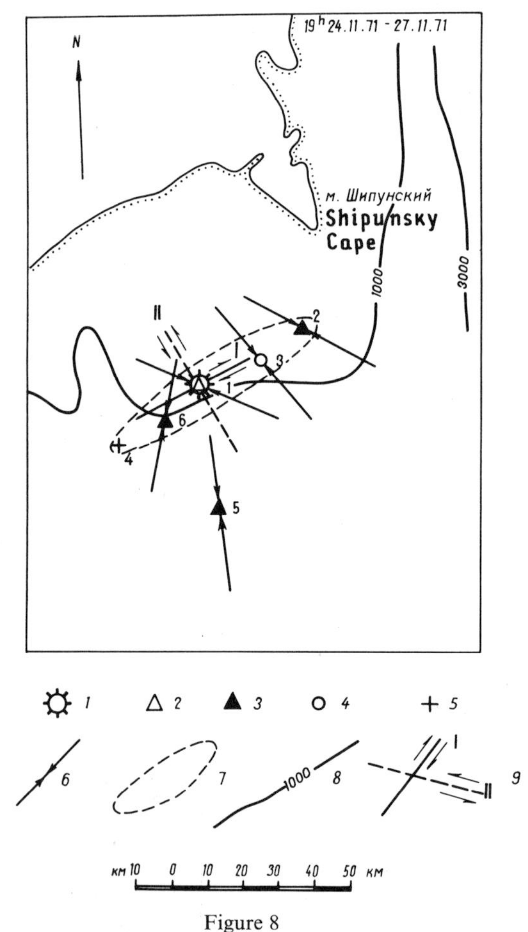

Figure 8

Focal process of Petropavlovskoe earthquake. 1 – epicenter of the main shock; 2–4 – nature of faulting (2 – strike-slip type with normal component, 3 – strike-slip type with reverse component, 4 – strike-slip type); 5 – epicenter for which focal mechanism was not determined; 6 – compression stresses acting in the foci of earthquakes; 7 – fault area; 8 – isobaths; 9 – possible fault planes.

strong earthquake is a right-lateral strike-slip along the Kurile–Kamchatka focal zone.

The difference in the focal processes of strong earthquakes with similar magnitude testifies to the existence of heterogeneities of the active seismic zone near Kamchatka.

7. *Conclusions*

1. Seismic activity of shallow and intermediate earthquakes in the Kamchatka–Commander region is unevenly distributed along the focal zone based on data of 1964–1970.

2. According to the moderate earthquake data the regional stress field in the Kamchatka–Commander region is complex and consists of the main Kurile–Kamchatka regional stress-field and several local transverse stress systems.

The nature of faulting in the region is characterized by a predominance of strike-slip type of faulting with normal or reverse faulting components.

3. The nature of aftershock activity and focal mechanism of the strong Kamchatka earthquakes are different and specified for each separate earthquake.

References

BALAKINA, L. M. (1962), *General regularities in the directions of the principal stresses effective in the earthquake foci of the seismic belt of the Pacific Ocean*, Izv. Acad. Sci. USSR, Geophys. Ser., No II (in Russian).

FEDOTOV, S. A. (1972), *Energetical classification of the Kurile–Kamchatka earthquakes and the magnitude problem*, Nauka (in Russian).

FEDOTOV, S. A., GUSEV, A. A. and ZOBIN, V. M. *et al.* (1973), *Ozernovskoe earthquake and tsunami 22.XI.1969* in *Earthquakes in USSR in 1969*, Nauka (in Russian).

FEDOTOV, S. A., TOKAREV, P. I. and BAGDASAROVA, A. M. (1969), *Seismicity of Kamchatka and Commander Islands according to detail seismic investigations* in *Structure and development of the earth crust in Soviet Far East*, Nauka (in Russian).

FEDOTOV, S. A., TOKAREV, P. I., GODZIKOVSKAY, A. A. and ZOBIN, V. M. (1974), *Detail data about seismicity of Kamchatka and Commander Islands (1965–1968)* in *Seismicity, seismic prediction, upper mantle properties and their relations to volcanism in Kamchatka*, Nauka (in Russian).

GUSEV, A. A., ZOBIN, V. M., KONDRATENKO, A. M. and SHUMILINA, L. S. (1975a), *Petropavlovskoe earthquake, 24.XI.71* in *Earthquakes in USSR in 1971*, Nauka (in Russian).

GUSEV, A. A., ZOBIN, V. M., KONDRATENKO, A. M. and SHUMILINA, L. S. (1975b), *Ust'-Kamchatskoe earthquake, 15.XII.71* in *Earthquakes in USSR in 1971*, Nauka (in Russian).

GUSHCHENKO, O. I. (1975), *Kinematic principle of a principal stress direction (based on geological and seismic data)*, Doklady Acad. Sci. USSR *225*:3 (in Russian).

ISACKS, B., OLIVER, J. and SYKES, L. R. (1968), *Seismology and the new global tectonics*, J. Geophys. Res. *73*.

OSKORBIN, L. S., LEONOV, N. N. and ZOBIN, V. M. (1976), *Earthquake and tsunami, 28.II.73* in *Earthquakes in USSR in 1973*, Nauka.

SIMBIREVA, I. G., FEDOTOV, S. A. and FEOFILAKTOV, V. D. (1976), *Heterogeneities of the stress field in the Kurile–Kamchatka arc in the light of seismological data*, Geol. Geophys. *1* (in Russian).

VVENDENSKAYA, A. V. (1969), *Investigations of stress and fractures in earthquake foci by use of dislocation theory*, Nauka (in Russian).

(Received 30th December 1976)

Pageoph, Vol. 115 (1977), Birkhäuser Verlag, Basel

Implications of Stress-Drop Models of Earthquakes for the Inversion of Stress Drop from Seismic Observations

By RAUL MADARIAGA[1])

Abstract – We discuss the inversion of stress drops from seismic observations on the basis of crack or stress-drop models of earthquake mechanism. Since a formal inverse problem cannot be posed at present we discuss implications of solutions to direct problems. We first discuss the static approximations used to obtain stress drop from seismic moment and source dimensions. We show that the geometrical effects are quite significant if only one source dimension has been retrieved from seismic observations. The effect of variable stress drop is discussed and we show that the inverted stress drop is not a simple average of the actual stress drops on the fault. We discuss the energy release during faulting and show that the apparent stress has a complicated relation to the stress drop on the fault. We also show that the static stress drops obtained by seismologists are a lower bound to the actual dynamic stress drops on the fault. This may in part explain disagreements with laboratory results. Finally, we discuss the inversion of source dimensions from the far-field radiation. We analyse two extreme, simple dynamical source models, a circular fault and a rectangular fault and show that geometry has a much more pronounced effect on radiation than is usually acknowledged.

Key words: Stress drop; Earthquake source theory; Earthquake source dimension.

1. Introduction

The number of stress drop determinations from seismic observations of earthquakes has increased substantially in recent years thanks both to developments in the study of the velocity and density structure of the earth and to the general acceptance of earthquake models based on the slip (dislocation) on a fault plane. Reviews of some of the available data may be found in KANAMORI and ANDERSON (1975) for large earthquakes $M_s > 7$ and in THATCHER and HANKS (1973) for determinations of stress drops for California earthquakes. These studies determine average stress drop, average effective stress or average apparent stress. A question naturally arises as to the exact meaning of these averages and how well can they be determined in the presence of complicating factors such as variable stress, variable rupture velocity, source geometry, etc. These questions properly pertain to the inverse problem of the earthquake source. Suppose we have been able to completely correct for the effects of propagation and attenuation (a very difficult task itself) and we want to find the geometry, stress drop distribution and rupture history of the earthquake. Such a formal inverse problem may not be considered at the present time because we do

[1]) Department of Earth and Planetary Sciences, Massachusetts Institute of Technology, Cambridge, MA 02139, USA.

not know enough about the direct problem. In fact, only a few extremely simplified three dimensional fault models have been numerically solved for the direct problem (MADARIAGA, 1976, 1977, CHERRY *et al.*, 1976, ARCHULETA and FRAZIER, 1976). We will therefore content ourselves with a discussion of inferences made from the experience obtained solving these very simplified source models.

In the first section we discuss the usual method to find stress drop from seismic moment and source area observations. We shall discuss the influence of source geometry and what kind of average is usually found when the stress drop is variable on the fault. Next we discuss seismic energy radiation which is a convenient way to introduce boundary conditions at the fault and discuss the relation between the frictional law on the fault and the seismic energy. In the last section we consider the effect of geometry on the seismic radiation and spectra in the far-field. The discussion of two simple models; a circular and a rectangular fault, serves to emphasize the significant effect of geometry on the seismic radiation and the inversion of stresses.

2. *Static parameters*

The most important static parameter that may be obtained from seismic observations is the seismic moment. In the far-field, and at periods much longer than the source size, a fault appears as a point double couple source. The scalar value of the moment of one of these couples is the seismic moment (AKI, 1966). The seismic moment is related to the final slip D at the fault by the following relation (BURRIDGE and KNOPOFF, 1964).

$$M_0 = \mu \int_S dS\, D(x, y) \tag{1}$$

where μ is the rigidity of the elastic medium surrounding the fault and S is the final area of the fault. It is frequently rewritten in the form

$$M_0 = \mu \bar{D} S \tag{2}$$

where $\bar{D}$ is the average final slip at the fault.

In order to obtain the stress drop from the seismic moment we need a relation between $\bar{D}$ and the stress drop $\Delta\sigma$ on the fault. Seismologists usually assume that the stress drop on the fault is constant and then compute the average slip for different geometries (circular or two-dimensional plane and antiplane). Posed in this form the problem is that of a shear crack under uniform stress drop, i.e., uniform difference between tectonic and frictional stresses. The solution to this problem may be written in the general form

$$\bar{D} = C \frac{\Delta\sigma}{\mu} W \tag{3}$$

where W is a characteristic length associated with the *narrowest* dimension of the fault. For circular faults it is the radius, for elliptical faults the semi-minor axis, for long faults its half width. The coefficient C is a form factor that depends on the geometry of the fault and the direction of slip.

For an elliptical fault of semi-major axis L and semi-minor axis W we may find C using some results, of ESHELBY (1957). For transversal slip, i.e., for slip in the direction W we find

$$C = 4/\left[3E(k) + \frac{w^2}{L^2}(K(k) - E(k))/k^2\right] \tag{4}$$

where $k = (1 - W^2/L^2)^{1/2}$ and $K(k)$ and $E(k)$ are complete elliptical integrals. Similarly, for longitudinal slip (along the semi-major axis) we find

$$C = 4/\left[3E(k) + (E(k) - \frac{w^2}{L_2}K(k))/k^2\right] \tag{5}$$

Both in (5) and (4) we have assumed that the Poisson ratio $v = 0.25$. When $L = W = R$ (circular fault) we find $C = 16/7\pi = 0.728$ a result obtained by KEILIS BOROK (1959). This is the minimum value of C from (4) or (5). On the other hand for a very long, rectangular, longitudinal slip fault $C = \pi/2 = 1.57$ as shown by KNOPOFF (1958). This value of C is larger than C for any elliptical fault and it is quite probably the maximum possible value for any fault geometry. For a transversal slip long fault we find the intermediate value $C = 3\pi/8 = 1.18$. Thus it appears that C varies at most by a factor of two for different source geometries.

Combining (3) and (1) we find the stress drop

$$\Delta\sigma = \frac{M_0}{CSW} \tag{6}$$

Thus, if we know M_0, S and W, the stress drop $\Delta\sigma$ may be found with a precision better than a factor of at most two. However, in most cases, only one size parameter is obtained from seismic observations via the corner frequency (see section 4). In that case (6) is frequently replaced by the rough approximation

$$\Delta\sigma = \frac{M_0}{CS^{3/2}} \tag{7}$$

which is obviously a poor approximation for long thin faults and may produce errors much larger than a factor of two.

Thus far we have assumed a constant stress drop on the fault, which is quite unlikely in earthquakes since we expect substantial variation of both the initial tectonic stress field and the final frictional stress. In such cases we have to replace $\Delta\sigma$ in (3), (6) or (7) by a certain weighted average $\Delta\sigma^*$ of the stress drop on the fault. In order to find an expression for $\Delta\sigma^*$ for faults of general shape we would have to solve a difficult crack problem with variable stress. However, we can get an estimate of the effect of variable stress drop studying a simpler plane fault model. A solution for the slip on a plane fault due to a general stress drop was found by SNEDDON and LOWENGRUB (1969, pp. 25–35). Averaging their results yields the following expression for the average slip on the fault

$$\bar{D} = C \frac{2}{\mu\pi W} \int_{-W}^{W} \Delta\sigma(x) \sqrt{W^2 - x^2}\, dx$$

where x is the coordinate along the fault. Thus, the weighted average stress drop is

$$\Delta\sigma^* = \frac{2}{\pi W^2} \int_{-W}^{W} \Delta\sigma(x) \sqrt{W^2 - x^2}\, dx \tag{8}$$

We have designated this average with a star to indicate that it is not a straight average but an average with a weighting function that emphasizes the stress drops near the center of the fault. Thus, an identical concentrated stress applied near the center produces larger slip than it would produce if it were applied near the borders. We may say that the fault is stiffer near the edges and softer near the center.

Using (6) or (7) to obtain the stress drop will yield a weighted average stress drop $\Delta\sigma^*$ which in general will differ from the unweighted average stress drop $\overline{\Delta\sigma}$. However, the weighting function $4/\pi(W^2 - x^2)^{1/2}$ has an elliptical shape which does not deviate much from a constant average. Only in the case of strong stress drop concentration near the edges will $\Delta\sigma^*$ differ significantly from $\overline{\Delta\sigma}$.

3. *Seismic energy*

Traditionally, seismologists have defined the seismic energy E_s as a fraction of the potential energy released by faulting W,

$$E_s = \eta W \tag{9}$$

where η is the seismic efficiency. This relation synthesizes all the dynamic processes of energy release in the efficiency η, which cannot be obtained by purely seismological observations. An expression for E_s in terms of the stresses and slip at the fault was obtained by KOSTROV (1974) and extended recently by DAHLEN (1976) to include the effect of non-elastic prestress fields and rotation of the earth. Here we will derive these relations by a slightly simpler method trying to emphasize the dependence of E_s on the rupture process at every point on the fault.

Let us consider a fault surface $\Sigma(t)$ that grows with time until it reaches its final area Σ_f. The seismic energy is the total flow of elastic energy through a surface S that surrounds the source (Fig. 1). S is sufficiently far from the source that the static field due to the source is negligible. Then the seismic energy is (KOSTROV, 1974).

$$E_s = \int_0^{t^*} dt \int_S dS\, \dot{u}_i\, \Delta\sigma_{ij} n_j \tag{10}$$

where n_j is the normal to S, $\dot{u}_i$ is the particle velocity and $\Delta\sigma_{ij} = \sigma_{ij}(\mathbf{r}, t) - \sigma_{ij}^0(\mathbf{r})$ is the *change* in stress due to the elastic waves. The integral over time is taken until no significant motion arrives from the source.

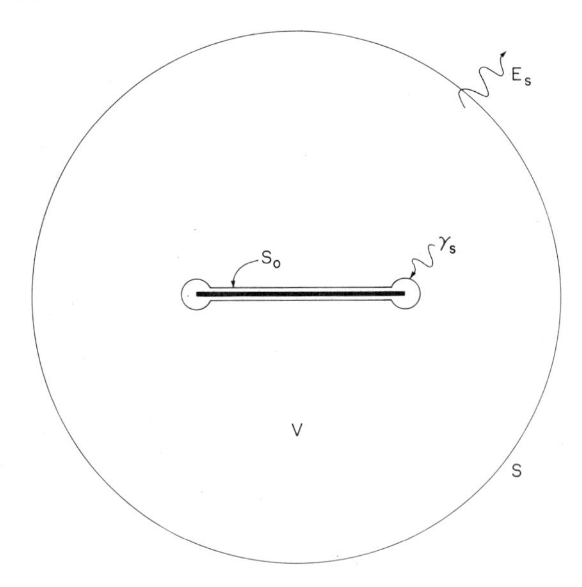

Figure 1
Geometry used to study seismic energy radiation E_s. S is a sphere in the far-field. S_0 is a surface that surrounds all the inelastic zone in the vicinity of the fault. γ_s is the fracture energy absorbed per unit extension of the rupture front.

The definition of E_s in (10) is in terms of stresses and velocities at S and we want to express it in terms of the field at the source. Let us surround the source by a boundary $S_0(t)$ than includes all the non-elastic zone near the fault as shown in Fig. 1. Most of the significant non-linear (plastic) effects occur in a small zone near the fault's edge where large stress concentrations occur. These zones are indicated by the circular indentations near the edges of the fault in Fig. 1. As the fault grows there is irreversible non-elastic energy dissipation in these zones. This energy is used in plastic deformation and in breaking new fault surface. KOSTROV *et al.* (1970) and FREUND (1972) have calculated the energy absorbed and expressed it in terms of the stress concentrations at the crack tip. Here we shall simply designate with γ the energy dissipated per unit advance of the rupture front.

Simple energy balance indicates that the energy flow E_s through S is equal to the energy flow out of the source minus the potential energy change W_s in the elastic volume V and minus the energy absorbed by the fault edge during rupture. Since the problem is linear we proceed by ignoring the pre-stress field and write an energy balance for the dynamic field only. The effect of pre-stress may be included later through interaction energy terms as shown by ESHELBY (1957). We find

$$E_s = \int_0^{t^*} dt \int_{S_0} \dot{D}_i \, \Delta\sigma_{ij} n_j \, dS - W_S - \int_{S_0} \gamma \, dS \tag{11}$$

Where $\dot{D}$ is the slip velocity at the fault, $\Delta\sigma_{ij}$ is the dynamic stress drop and W_s is the potential energy change due to the total stress drop in the absence of pre-stress

$$W_s = \frac{1}{2} \int_{S_0} D_i^f \, \Delta\sigma_{ij}^f n_j \, dS \tag{12}$$

where D_i^f is the final slip at the fault and $\Delta\sigma_{ij}^f$ the final stress drop. In (12) we used the fact that S is sufficiently far from the source so that the static stress field is negligible.

Inserting (12) into (11) we find

$$E_s = \int_0^{t^*} dt \int_{S_0} dS \, \dot{D}_i [\overline{\sigma_{ij}} - \sigma_{ij}(t)] n_j - \int_{S_0} dS \gamma \tag{13}$$

where $\overline{\sigma_{ij}} = 1/2(\sigma_{ij}^0 + \sigma_{ij}^1)$ is the average of the initial stress σ_{ij}^0 and the final stress σ_{ij}^1 at the fault, and $\sigma_{ij}(t)$ is the frictional stress at the fault while slip proceeds ($\dot{D} \neq 0$).

If we assume that the inelastic zones at the edge of the rupture front are small compared to the source dimension then we can rewrite E_s in the form

$$E_s = \int_{S_0} dS \int_0^{D_i^f} dD_i \, [\overline{\sigma_{ij}} - \sigma_{ij}(D)] n_j - E_{\text{rupt}} \tag{14}$$

where E_{rupt} is the total energy dissipated inelastically at the rupture front. $\sigma(D)$ is the frictional law for slip at each point on the fault. For simplicity we have excluded viscous friction in which case $\sigma(D, \dot{D})$ is also a function of slip velocity. In Fig. 2 we sketch a possible friction law $\sigma(D)$ at a point on the fault and we also indicate the initial (σ^0), final (σ^1) and average ($\bar{\sigma}$) stresses. Also indicated in this figure is the static friction or upper yield point σ^y of the constitutive relation. As suggested by the figure the static friction σ^y may be higher than the initial stress. The increase in stress necessary to produce rupture is provided by dynamic stress concentration ahead of the crack. The history of stress at a point is as follows: first, with the arrival of P and S waves radiated from the nucleation point the stress departs from the initial σ^0 level and as the rupture front gets closer the stress increases. When the stress reaches the level σ^y rupture starts and the stress $\sigma(D)$ decreases as D increases. During the slip, part of the potential energy in the elastic body is released and used to overcome friction, to produce additional fracture and to radiate seismic waves. As indicated by (14) the contribution to E_s from one point on the fault is

$$e_s = \int dD_i [\overline{\sigma}_{ij} - \sigma_{ij}(D)] n_j - \gamma \tag{15}$$

which we may call specific available seismic energy. As seen from Fig. 2 the contribution to e_s is negative for $\sigma(D) > \bar{\sigma}$ and positive for $\sigma(D) < \bar{\sigma}$. The negative contribution is a dissipation and should perhaps, be incorporated into the fracture energy.

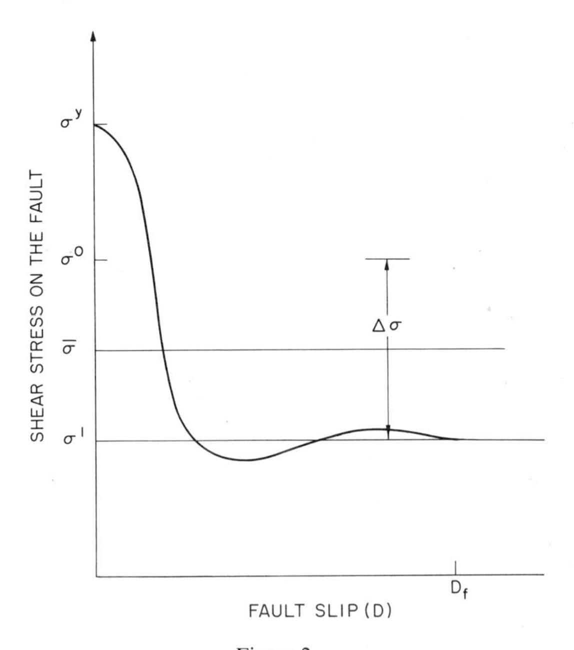

Figure 2

Sketch of a possible frictional low at the fault. σ^y is the yield stress or static friction, σ^0 the initial stress before the nucleation of rupture, σ^1 is the final stress after all motion has ceased, $\bar{\sigma} = \frac{1}{2}(\sigma^0 + \sigma^1)$ is the average stress on the fault. D_f is the total slip.

In fact, IDA (1972) and ANDREWS (1975) have proposed models where there is no inelastic dissipation outside of the fault itself. In this case energy is dissipated at the rupture front only when $\sigma(D) > \bar{\sigma}$. In order to study dynamic faulting problems we have to specify the friction law $\sigma(D)$. This can only be obtained from experiments or, perhaps, eventually inverted from seismic observations. Since we do not know much about $\sigma(D)$ in many dynamical models of source mechanisms a simplified frictional law is used. The assumption is that $\sigma(D)$ drops abruptly to a constant kinetic friction σ^k immediately after slip starts. Some basic considerations of elasticity demonstrate that in this case the stress ahead of the crack has a typical square root singularity and σ^y is necessarily infinite. However the square of the stress intensity factor may be shown to be proportional to the total fracture energy dissipation (KOSTROV *et al.*, 1970, FREUND, 1972). This approximation is well known in fracture mechanics and has been proved to be adequate in most instances.

The frictional law in Fig. 2 is similar to the frictional law proposed by BYERLEE (1970) in his study of the stick-slip instability since, as proposed by BRACE and BYERLEE (1969), the same, or similar, frictional instability may be active in earthquakes as in stick-slip. We can think of an earthquake as a stick-slip instability that propagates on the fault. But there are significant differences between both phenomena due to the dynamic stress concentration that control the spreading of the instability in earthquakes. An important consequence is that the stress drop measured in stick-slip events is the difference $\delta\sigma = \sigma^y - \sigma^1$ in Fig. 2. In earthquakes, on the other hand, the stress drop $\Delta\sigma$ has to be equal to $\delta\sigma$ only near the rupture nucleation point or at isolated asperities. At other points the stress is driven to the static friction σ^y by the dynamic stress concentrations ahead of the rupture front so that the dynamic stress drop $\delta\sigma$ may be larger than the static stress drop $\Delta\sigma$ determined from (6). This may explain, at least in part, that stress drops measured in the laboratory are significantly larger than the average stress drops obtained for earthquakes.

The seismic energy has been used to obtain the apparent stress (AKI, 1966)

$$\sigma_{\text{app}} = \eta\bar{\sigma} = \mu E_s/M_0 \tag{16}$$

where $\bar{\sigma}$ is the average stress and η the efficiency. This relation has the advantage it does not require an estimate of the source dimensions. E_s is estimated from the energy-magnitude relations so that the estimate (16) may be very rough. Assuming that E_s is well determined we may find an expression for σ_{app} using (14) and (2)

$$\sigma_{\text{app}} = \eta\bar{\sigma} = \frac{1}{\bar{D}S}\int_{S_0} dS \int_0^{D_s} dD_i[\bar{\sigma}_{ij} - \sigma_{ij}(D)]n_j - \frac{\gamma}{\bar{D}} \tag{17}$$

So that the apparent stress is an average of $\bar{\sigma} - \sigma(D)$ during the fault process. In the case of constant friction $\sigma(D) = \sigma^k$ and (17) reduces to

$$\eta\bar{\sigma} = \frac{1}{S}\int dS[\bar{\sigma} - \sigma^k] - \frac{\gamma}{\bar{D}} \tag{18}$$

a result found also by HUSSEINI (1976). Thus, $\eta\bar{\sigma}$ is not really a measure of absolute stress but, as remarked by SAVAGE and WOOD (1971) and HANKS and THATCHER (1972), $\eta\bar{\sigma}$ is really a measure of stress drop on the fault.

4. *Seismic radiation*

In order to recover stress drops (6 or 7) from seismic observations we need the seismic moment and estimates of the source area and its geometry. For larger earthquakes the area may be estimated from aftershock zones, but for many earthquakes the fault dimensions may only be recovered from the seismic radiation in the time or frequency domain.

Let us consider the radiation of body waves in the far-field and assume that we have removed the effect of the path so that we can treat the source as if it were embedded in an infinite, homogeneous medium. The far-field radiation may be obtained from the representation theorem (BURRIDGE and KNOPOFF, 1964) assuming a planar fault and slip in only one direction

$$u(\mathbf{R}, t) = \frac{\mu}{4\pi\rho c^3} R_{\theta\phi} \frac{1}{R} U_{\text{ff}}(\hat{R}, t - R/c) \tag{19}$$

where, ρ is the density, μ the rigidity and c is either v_p, the P-wave velocity or v_s, the S-wave velocity, $R_{\theta\phi}$ is the double couple radiation pattern, $\hat{R}$ the direction of the observer with respect to a reference point on the source and R is the distance to the observer.

The shape of the far-field pulse is given by

$$U_{\mathrm{ff}}(\hat{R}, \tau) = \iint_S dS\, \dot{D}\left(\mathbf{r}, \tau + \frac{\hat{R}\cdot\mathbf{r}}{c}\right) \tag{20}$$

where $\mathbf{r}$ is a position vector on the fault plane and $\dot{D}(\mathbf{r}, t)$ is the slip velocity distribution on the fault plane. As (20) indicates the far-field pulse is a delayed average of the slip velocity at the source. Since the fault area S_0 is finite and the duration of slip on the source (rise time) is also finite the pulse U_{ff} will have a finite duration. This property is shared by all source models and is used either in the time domain (HASKELL'S (1964) model) or in the spectral domain (BRUNE'S (1970) model) to retrieve the source dimensions. In order to discuss the spectral properties of the far-field pulse it is convenient to introduce the space and time Fourier transform of the slip velocity function

$$\dot{D}(\mathbf{k}, \omega) = \int_0^{\infty} dt\, e^{-i\omega t} \int_{S_0} dS\, e^{-i\mathbf{k}\cdot\mathbf{r}}\, \dot{D}(\mathbf{r}, t) \tag{21}$$

Then, transforming (20) we find the following relation between the far-field spectrum and the source slip-velocity spectrum

$$U_{\mathrm{ff}}(\hat{R}, \omega) = \dot{D}(\mathbf{k}, \omega)|_{\mathbf{k}=(R/c)\omega} \tag{22}$$

Expressions similar to this one were obtained by AKI (1967) for linear faults and MADARIAGA (1976) for circular faults.

Many properties of the far-field spectrum may be obtained from (22). For instance, at low frequencies the spectrum is flat and proportional to the seismic moment M_0 (AKI, 1967). Also, as noted by MOLNAR *et al.* (1973b), if $\dot{D}(\mathbf{r}, t)$ is everywhere on the fault of the same sign (unipolar) then the spectral amplitude $|U_{\mathrm{ff}}(\hat{R}, \omega)|$ is maximum at zero frequency, a property disputed by ARCHAMBEAU (1972, 1975). But the property of Fourier transform that interests us the most is that any pulse of finite duration (say ΔT) has a spectrum of finite width (Δf in Hz) such that $\Delta f\,\Delta t \approx 1$. In seismology it is more common to use the corner frequency in a logarithmic plot of the spectrum (see Fig. 4 and 6) rather than the width of the spectrum. Roughly, the corner frequency, f_0, is one half the spectral width so that $f_0\,\Delta t \approx 0.5$. Thus, the corner frequencies are an inverse measure of the pulse width and, therefore, inversely proportional to some combination of the source size and the rise time. The exact relationship between corner frequency and the source dimensions has to be determined for specific fault models. It will depend on the geometry of the source, the distribution of stress drop on the fault, the presence of multiple sources, vicinity to the free surface, the structure in the vicinity of the source, etc. Little is known about the effects of these complicating factors. In the following we will concentrate on the discussion of two relatively simple, but quite different geometries, a circular and a rectangular fault.

5. *The circular fault model*

We consider a circular fault that grows symmetrically from a point at a constant rupture velocity v_R and finally stops suddenly at a certain radius a. A kinematic analysis, with arbitrary specification of $\dot{D}(r, t)$ was made by SAVAGE (1966). Later, BRUNE (1970) approximated the dynamic solution assuming that rupture occurred instantaneously on the fault. Recently, we were able to solve this problem numerically (MADARIAGA, 1976). The high degree of symmetry makes this the easiest fault model to study when we specify a constant stress drop on the fault. In order to study the radiation we have to find first the slip function at the source which is the main result of the numerical solution. In Fig. 3 we show the slip history as a function of radius of the fault. This slip function is very approximately cylindrically symmetric, i.e., independent of azimuth. The most important feature of the slip function is that the slip is concentrated near the center of the fault, a result expected from static solutions of circular faults (ESHELBY, 1957). The rise time is also longer near the center of the fault. The slip function shown in Fig. 3 was calculated assuming a constant rupture velocity $v_R = 0.87\, v_S$, v_S = shear velocity. The effect of the rupture velocity is not too severe in the far field radiation of this model because of the great symmetry of

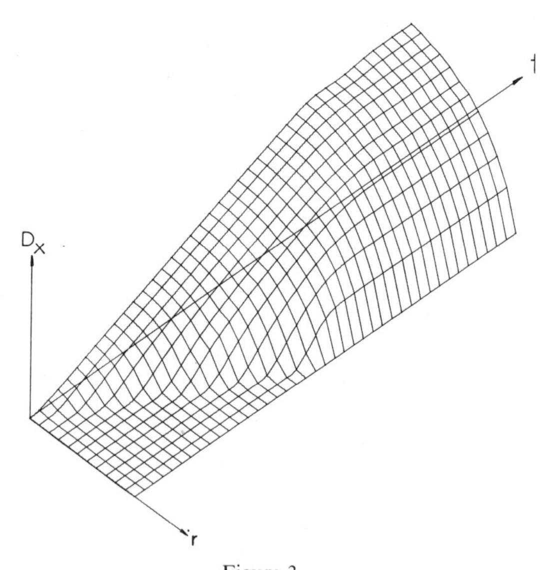

Figure 3
Source slip function for a circular fault with rupture velocity $v_R = 0.87\, v_s = 0.5\, v_p$. This plot shows the slip history as function of radius on the fault.

the fault and the fact that most of the radiation comes from stopping phases generated during the abrupt stopping of rupture at the final radius (a). Thus varying the rupture velocity within reasonable limits – say $v_R = 0.6$–$0.9\ v_S$ – does not affect the radiation significantly.

Far-field radiation from the circular model was obtained by a series of Fourier transforms and a relation of the type of (22). In Fig. 4 we show an example of the far-field displacement pulse radiated in a direction $\theta = 60°$ from the normal to the fault. Radiation is symmetrical about this normal. The shape of the far-field pulse and spectra change very slowly as a function of θ; $\theta = 60°$ may be considered as an 'average' radiation. From Fig. 4 we see that the spectral amplitude has a characteristic corner frequency f_0 defined by the intersection of the low and high frequency asymptotes. We find that the corner frequency, averaged for different rupture velocities and azimuth, has the following relation to the source radius

$$f_0^P = 0.32\ v_s/a$$
$$f_0^S = 0.21\ v_s/a \qquad (23)$$

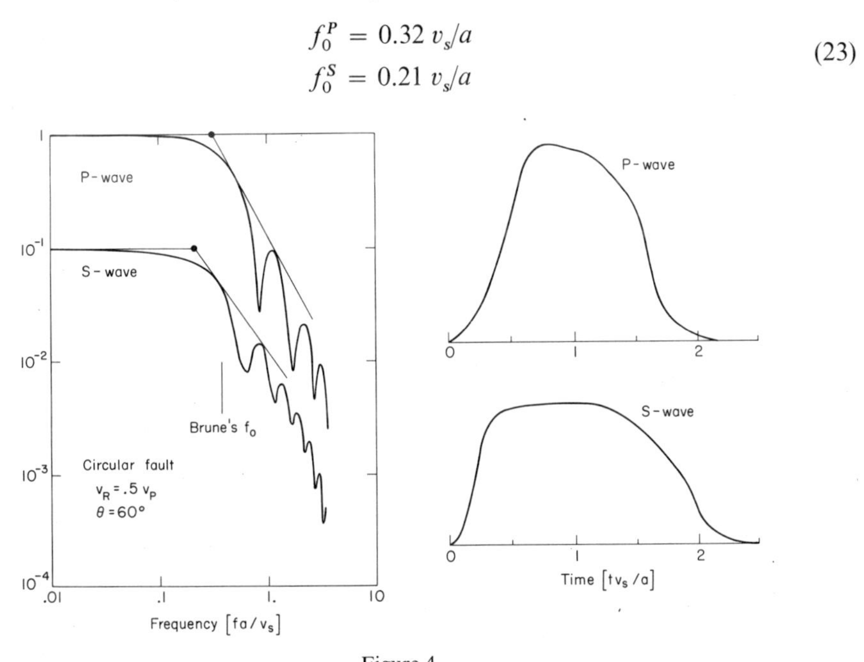

Figure 4
Far-field P and S pulses and their spectra radiated by the circular fault whose source slip function is shown in Fig. 3. The spectrum of P waves is normalized to 1 at low frequencies. The spectrum of S waves is normalized to 0.1 at low frequencies. Both pulses are normalized to the same seismic moment (area).

for P and S waves respectively. BRUNE (1970) proposed a coefficient 0.375 for S waves. We notice that the P-wave corner is at higher frequency than the S-corner frequency, a relation that has been observed in many earthquake spectra as shown by MOLNAR *et al.* (1973a). Higher P-corner frequencies are a property of faults of low length-to-width ratio. It is interesting to note that the corner of the displacement amplitude spectrum is very approximately the peak frequency of the energy density spectrum. That is, the corner defines the frequency at which the source is more efficient in transferring power to the far-field radiation. The wavelengths associated with the corner frequencies are $\lambda^P = 5.41a$ for P-waves and $\lambda^S = 4.76a$ for S-waves. Thus, the circular source has a maximum efficiency to transfer energy at wavelengths of about 2.5 times the source diameter.

6. *Rectangular fault model*

The rectangular fault model was the first source model studied by seismologists (BEN-MENAHEM, 1961, HASKELL, 1964). Rupture nucleates instantaneously across the width of the fault and then extends longitudinally at a constant rupture velocity v_R either unilaterally or bilaterally until it stops abruptly. In this model the nucleation and stopping of rupture are not considered in detail so that the model is valid only down to wavelengths of the order of the width of the fault. The original studies of this model were purely kinematical, i.e., stress plays no role, assuming a propagating ramp dislocation. We have studied this problem numerically (MADARIAGA, 1977) assuming a constant stress drop on the fault in order to find a dynamically acceptable source time function and slip distributions across the fault. In Fig. 5 we show the source time function along the longitudinal axis of a rectangular fault model of length-to-width ratio $L/W = 3$. Rupture velocity is $v_R = 0.9\ v_S$ and slip is longitudinal (strike slip). The source time function is quite different from that of the

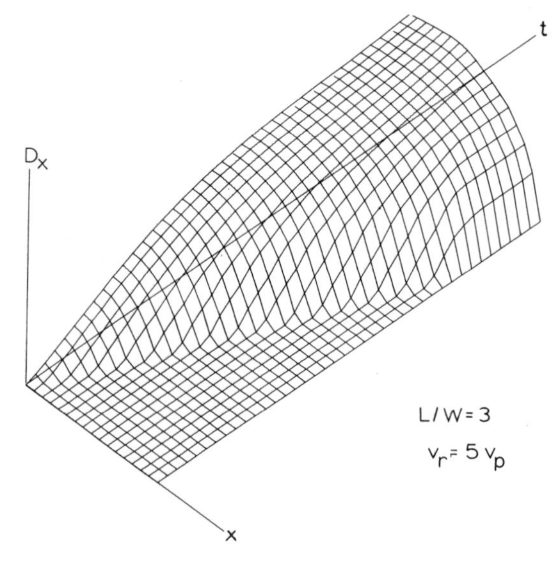

Figure 5
Source slip function for a rectangular fault with bilateral rupture velocity $v_R = 0.87\ v_s = 0.5\ v_p$. This plot shows the slip history for several points along the longitudinal axis of the fault. The length to width ratio of this fault is $L/W = 3$.

circular fault, the slip history is approximately the same along the fault substantiating the kinematic models in which the source was modelled as a propagating dislocation. The slip function at a point, however, differs markedly from a ramp, being sharply rounded near the rupture front. The rise time is approximately a constant along the fault, $T_R = 1.3\ w/v_S$ in contrast to the circular fault where the rise time was longer at the center. In a long fault two patches of rupture move away from the nucleation point leaving a healed fault in their wake. This produces strong focussing of the radiation in directions along the axis of the fault. In contrast a circular fault focuses energy in the direction normal to its plane.

In Fig. 6, we show far-field pulses and spectra radiated by a bilateral crack ($L/W = 3$, $v_R = 0.5\ v_S$) for a direction that makes an angle $\psi = 58°$ with the longitudinal axis of the fault. This is a representative or 'average' direction of radiation. The most noticeable difference with radiation from a circular fault is that the far-field pulse is more complex. There seem to be two distinct pulses, the longer one, of smaller amplitude is due to radiation from the rupture front that moves away from the observer, while the shorter, stronger initial pulse is associated with the rupture moving towards the observer. The spectra are correspondingly more complicated and in Fig. 6 there seem to be two obvious corner frequencies. The lower corner is associated with longitudinal propagation, while the higher corner is associated with the width and rise time of the fault. It is difficult to establish a simple relationship between corner frequency and source length because the rupture velocity affects the corner frequencies markedly and the spectra are very sensitive to the angle of

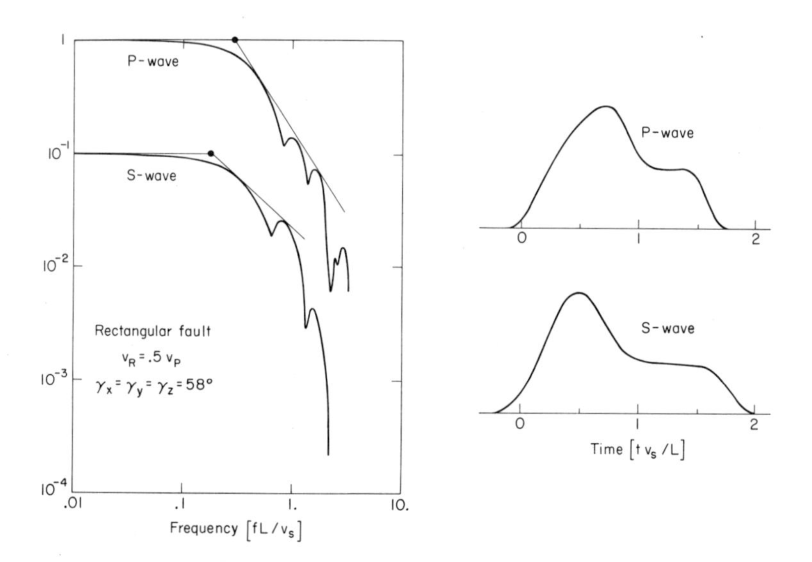

Figure 6
Far-field P and S pulses and their spectra radiated by the rectangular fault of Fig. 5. The spectra and pulses are normalized in the same manner as those in Fig. 4.

radiation. Tentatively, for $v_R = 0.9\, v_S$ we propose, on the average,

$$f_0^P = 0.29\, v_s/L$$
$$f_0^S = 0.20\, v_s/L \tag{24}$$

This problem is further discussed in MADARIAGA (1977).

The rectangular fault model is much more complex than the circular fault model in its dynamics. The two scales of the model, length and width affect different regions of the spectrum and, as a consequence, we expect at least two corner frequencies. In order to obtain an estimate of the stress drop from (6) it is necessary to find both the length and the width from the far-field radiation. This is a very difficult observational problem which may be feasible for only a few large earthquakes. For smaller earthquakes at most one corner frequency may be retrieved from the spectrum or, equivalently, not more than the pulse width may be obtained from the far-field radiation. It appears that information not contained in the spectra, like aftershock distribution, is necessary to establish the width of the fault and then estimate the stress drop. Many authors assume that the fault is circular and proceed to find the stress drop from (7) rather than (6). If the fault happened to be elongated, this procedure would produce large errors because the stress drop in (6) is inversely proportional to the square of the fault width.

7. *Discussion*

We have reviewed the results of attempts to model earthquake sources as mechanical instabilities caused by a drop in the ability to sustain stress on a plane fault. The models studied are highly idealized because the mathematical problems that arise even with the simplest fault models make it necessary to use expensive numerical methods. Moreover, fault radiation requires the use of three dimensional models because of the significant geometrical effects that appear in elongated faults. In a discussion of the estimate of stress drop (equation 6) by usual methods we have emphasized that the fault geometry is an important variable that is not usually incorporated in the inversion of stress drop from seismic observations. These geometrical effects were shown to dominate the slip history and radiation from the two simple source models that we discussed – a circular fault and a rectangular fault. Using the wrong fault model for a given earthquake may generate large errors in the inversion of stress drop.

We have also discussed the role of the frictional law in the modelling of earthquake sources. This frictional law is largely unknown but we have argued that it should be similar to the frictional law in laboratory stick-slip experiments. One significant property of dynamical source models is the presence of stress concentrations ahead of the rupture front. These stress concentrations are purely dynamical effects which are absent from frictional experiments in the laboratory. As a consequence the dynamic stress drop (which should be equal to stress drop in stick-slip) may be larger than the static stress drops obtained by seismologists. The seismic stress drops are thus lower bounds to the actual dynamic stress drops.

Acknowledgements

This work was supported by the Division of Earth Sciences of the National Science Foundation under Grant EAR75–14808 A01.

References

AKI, K. (1966), *Generation and propagation of G waves from the Niigata earthquake of June 16, 1964, 2, Estimation of earthquake moment, released energy, and stress-strain drop from G-wave spectrum,* Bull. Earthquake Res. Inst. Tokyo Univ. *44*, 73–88.

AKI, K. (1967), *Scaling law of seismic spectrum,* J. Geophys. Res. *72*, 1217–1231.

ANDREWS, D. J. (1976), *Rupture propagatian with finite strain in antiplane strain,* J. Geophys. Res. *81*, 3575–3582.

ARCHAMBEAU, C. B. (1972), *The theory of stress wave radiation from explosions in prestressed media,* Geophys. J. *29*, 329–366.

ARCHAMBEAU, C. (1975), *Developments in seismic source theory,* Rev. Geophys. Space Phys. *13*, 304–306.

ARCHULETA, R. J. and FRAZIER, G. A. (1976), *A three dimensional, finite element solution in a half space for the near field motion of a propagating stress drop* (abstract), Earthquake Notes *47*, 37–38.

BEN-MENAHEM, A. (1962), *Radiation of seismic body waves from a finite moving source in the earth,* J. Geophys. Res. *67*, 345–350.

BRACE, W. F. and BYERLEE, J. D. (1966), *Stick-slip as a mechanism for earthquakes,* Science, *153*, 990–992.

BRUNE, J. N. (1970), *Tectonic stress and the spectra of seismic shear waves from earthquakes,* J. Geophys. Res. *75*, 4997–5009.

BURRIDGE, R. and KNOPOFF, L. (1964), *Body force equivalents for seismic dislocations,* Bull. Seism. Soc. Am. *54*, 1875–1888.

BYERLEE, J. D. (1970), *The mechanics of stick-slip,* Tectonophysics, *9*, 475–486.

CHERRY, J. T., BACHE, T. C. and MASSO, J. F. (1976), *A three dimensional finite-difference simulation of earthquake faulting* (abstract), EOS, Trans. Am. Geophys. Un. *57*, 281.

DAHLEN, F. A. (1976), *Seismic faulting in the presence of a large compressive stress,* Geophys, Res. Lett. *3*, 245–248 and 506.

ESHELBY, J. D. (1957), *The determination of the elastic field of an ellipsoidal inclusion, and related problems,* Proc. Roy. Soc. *A241*, 376–396.

FREUND, L. B. (1972), *Energy flux into the tip of an extending crack in an elastic solid,* J. Elasticity, *2*, 341–348

HANKS, T. C. and THATCHER, W. (1972), *A graphical representation of seismic sources,* J. Geophys. Res. *77*, 4393–4405.

HASKELL, N. A. (1964), *Total energy and energy spectral density of elastic wave radiation from propagating faults,* Bull. Seism. Soc. Am. *54*, 1811–1841.

HUSSEINI, M. I. (1976), *Energy balance for motion along a fault,* submitted to Geophys. J. Roy. astr. Soc.

IDA, Y. (1972), *Cohesive force across the tip of a longitudinal shear crack and Griffith's specific surface energy,* J. Geophys. Res. *77*, 3796–3805.

KANAMORI, H. and ANDERSON, D. L. (1975), *Theoretical basis of some empirical relations in seismology,* Bull. Seism. Soc. Am. *65*, 1073–1095.

KEILIS-BOROK, V. I. (1959), *On the estimation of the displacement in an earthquake source and of source dimensions*, Ann. Geofis. *12*, 205–214.

KNOPOFF, L. (1958), *Energy release in earthquakes,* Geophys. J. *1*, 44–52.

KOSTROV, B. V. (1974), *Seismic moment and energy of earthquakes, and seismic flow of rock,* Izv. Fizika Zemli, *1*, 23–40.

KOSTROV, B. V. and NIKITIN, L. V. (1970), *Some general problems of mechanics of brittle fracture*, Arch. Mech. Stosow., *22*, 749–776.

MADARIAGA, R. (1976), *Dynamics of an expanding circular fault,* Bull. Seism. Soc. Am. **65**, 163–182.

MADARIAGA, R. (1977), *A dynamic stress drop model of a rectangular fault,* in preparation.

MOLNAR, P., TUCKER, B. and BRUNE, J. N. (1973a), *Corner frequencies of P and S waves and models of earthquake sources,* Bull. Seism. Soc. Am. *65*, 2091–2104.

MOLNAR, P., JACOB, K. H. and MCCAMY, K. (1973b), *Implications of Archambeau's earthquake source theory for slip on fault,* Bull. Seism. Soc. Am. *63*, 101–104.

SAVAGE, J. C. (1966), *Radiation from a realistic model of faulting,* Bull. Seism. Soc. Am. *56*, 577–592.

SAVAGE, J. C. and WOOD, M. D. (1971), *The relation between apparent stress and stress drop,* Bull. Seism. Soc. Am. *61*, 1381–1388.

SNEDDON, I. N. and LOWENGRUB, M. (1962), *Crack problems in the classical theory of elasticity,* (John Wiley and Sons, New York.

THATCHER, W. and HANKS, T. (1973), *Source parameters of southern California earthquakes,* J. Geophys. Res. *78*, 8547–8576.

(Received 4th November 1976)

Pageoph, Vol. 115 (1977), Birkhäuser Verlag, Basel

Apparent Stress and Stress Drop for Intraplate Earthquakes and Tectonic Stress in the Plates

By Randall M. Richardson[1]) and Sean C. Solomon[1])

Abstract – The magnitude of shear stress in the lithosphere is bounded from below by the apparent stress and stress drop during intraplate earthquakes. Apparent stresses and stress drops for a number of mid-plate earthquakes are calculated from the earthquake magnitude, SH wave amplitude spectra, and estimates of the length of the fault zone. Apparent stresses vary between 0.1 and 2 bars, if m_b is used as a measure of seismic energy, and stress drops lie between 2 and 70 bars. There is no systematic difference in either apparent stress or stress drop between these intraplate events and typical plate boundary earthquakes. These bounds on intraplate shear stresses are consistent with the inference from current models of plate tectonic driving forces that regional stress differences in the plates are typically on the order of 100 bars. The highest stress drops measured for midplate earthquakes under this model represent nearly total release of local tectonic stress.

Key words: Stress drop; Intraplate stress; Apparent stress.

1. Introduction

The state of stress in the lithosphere is an important though poorly known quantity. The magnitudes and directions of principal stresses are closely related to the forces that produce them, including the tectonic forces associated with the driving mechanisms for plate tectonics. Knowledge of the level of stress in the earth is clearly vital for proper evaluation of potential seismic hazards, especially in areas of only occasional seismic activity. In this paper bounds on the magnitude of deviatoric stress in the lithosphere are determined from the apparent stresses and stress drops associated with midplate earthquakes. A consistency argument is then used to develop a model for the general magnitude of intraplate stress from these bounds and from independent evidence on the nature of tectonic forces acting on the plates.

A variety of methods, all with some limitations, exist for inferring the state of stress in the lithosphere. Fault plane solutions of midplate earthquakes indicate the faulting mechanism and, by inference, the orientation but not the magnitude of principal stresses in the lithosphere. *In situ* measurements can determine both the orientation and magnitude of stress in the crust. Indirect methods such as strain relief require additional information about the material properties of the medium to infer stresses. Hydrofracturing measures stresses directly and has the potential of

[1]) Department of Earth and Planetary Sciences, Massachusetts Institute of Technology, Cambridge, Massachusetts 02139, USA.

economically providing both the orientation and magnitude of stresses in continental regions. Hydrofracturing data from several continents show average values for maximum stress differences on the order of a few hundred bars with the largest values less than one kilobar (HAIMSON 1975, 1976, RANALLI 1975). Both direct and indirect *in situ* methods, however, are restricted to the top few kilometers of the earth's crust.

Stress in the lithosphere may be caused by a number of mechanisms, and regional stresses may locally be masked by large stress due to heterogeneities in structure or history. Nonetheless, there are large regions of the earth characterized by consistent orientations of principal stresses as determined both by earthquake mechanisms and by a variety of *in situ* measurements (SYKES and SBAR 1973, 1974, SBAR and SYKES 1973, AHORNER 1975). Such consistency argues strongly for a tectonic origin for such stresses.

Several additional arguments may be used to bound the magnitudes of shear stress in the lithosphere. Heat flow measurements across the San Andreas fault place an upper limit of about a hundred bars on shear stress across the fault, but do not constrain well the orientation of stresses (BRUNE *et al.* 1969). Laboratory measurements of shear stresses required to fracture rocks or to initiate rupture on a pre-existing break are on the order of several kilobars (STESKY *et al.* 1974). If such measurements can be extrapolated to fault dimensions, stresses of up to several kilobars may exist along active faults and presumably within the lithosphere.

Lower bounds on shear stress magnitudes in the lithosphere are provided by the apparent stress and stress drop acting during an intraplate earthquake. In this paper, we compute the apparent stress and stress drop for a number of earthquakes far from plate boundaries and we relate the results to our earlier plate tectonic models of intraplate stress (RICHARDSON *et al.* 1976, SOLOMON *et al.* 1975, 1977a) based only on the directions of principal stresses as inferred from midplate earthquake mechanisms. A quantitative model of intraplate stress is completely consistent with plate tectonic constraints on possible driving forces if the general level of shear stress within the plates is about 100 bars.

2. *Apparent stress and stress drop*

Both the apparent stress and stress drop represent measures of some fraction of the shear stress acting during an earthquake and are derivable from seismic observations. The apparent stress is the product of the average shear stress $\bar{\sigma}$ on the fault before and after faulting and an unknown seismic efficiency factor η, and is given by (AKI 1966):

$$\eta\bar{\sigma} = \mu \frac{E_S}{M_0} \tag{1}$$

where μ is the shear modulus, taken as 3×10^{11} dyne/cm^2, E_S is the radiated seismic energy in dyne-cm, and M_0 is the seismic moment in dyne-cm.

The stress drop $\Delta\sigma$ is the difference between the initial and final shear stress on the fault. For a vertical strike-slip earthquake the stress drop is given by (KNOPOFF 1958):

$$\Delta\sigma = \frac{2M_0}{\pi L w^2} \tag{2}$$

and for a dip-slip earthquake it is (AKI 1966):

$$\Delta\sigma = \frac{4}{\pi} \frac{(\lambda + \mu)}{(\lambda + 2\mu)} \frac{M_0}{L w^2} \tag{3}$$

where L and w are fault length and width, respectively, for a rectangular fault, and λ is the Lamé constant. The quantities E_S and M_0 can be determined or estimated from seismic wave amplitude measurements, and the parameters L and w can be determined either from field observations (surface rupture, aftershock locations) or from seismic measurements (corner frequency, directivity) and a model of the earthquake source.

3. *Calculation of apparent stress*

The midplate earthquakes for which determinations of apparent stress and stress drop have been made are listed in Table 1 and their epicenters are shown in Fig. 1.

Table 1
Intraplate earthquakes used in this study

Event	Date	Time	Location	Reference for fault plane solution
1	14 May 1964	13:52:14	65.3N, 86.5W	HASHIZUME (1974)
2	21 Jan 1972	14:43:43	71.4N, 74.7W	HASHIZUME (1973)
3	21 Oct 1965	02:04:36	37.6N, 90.9W	MITCHELL (1973)
4	2 Dec 1970	11:03:10	68.4N, 67.4W	HASHIZUME (1973)
5	2 Oct 1971	03:19:29	64.4N, 86.5W	HASHIZUME (1974)
6	9 Nov 1968	17:01:38	38.0N, 88.5W	STAUDER and NUTTLI (1970)
7	25 Nov 1965	10:50:50	17.1S, 100.2W	MENDIGUREN (1971)
8	29 Sept 1969	20:03:32	33.2S, 19.3E	FAIRHEAD and GIRDLER (1971)
9	20 Oct 1972	04:33:50	20.6N, 29.7W	This paper (Figure 2)
10	14 Oct 1968	02:58:52	31.7S, 117.0E	FITCH *et al.* (1973)
11	26 Apr 1973	20:26:27	20.0N, 155.2W	UNGER and WARD (1974)
12	10 Dec 1967	22:51:23	17.6N, 73.8E	SINGH *et al.* (1975)
13	24 Mar 1970	10:35:17	22.1S, 126.6E	FITCH *et al.* (1973)
14	9 May 1971	08:25:01	39.8S, 104.9W	FORSYTH (1973a)
15	10 Oct 1970	08:53:05	3.6S, 86.2E	FITCH *et al.* (1973)
16	23 Oct 1964	01:56:05	19.8N, 56.1W	MOLNAR and SYKES (1969)

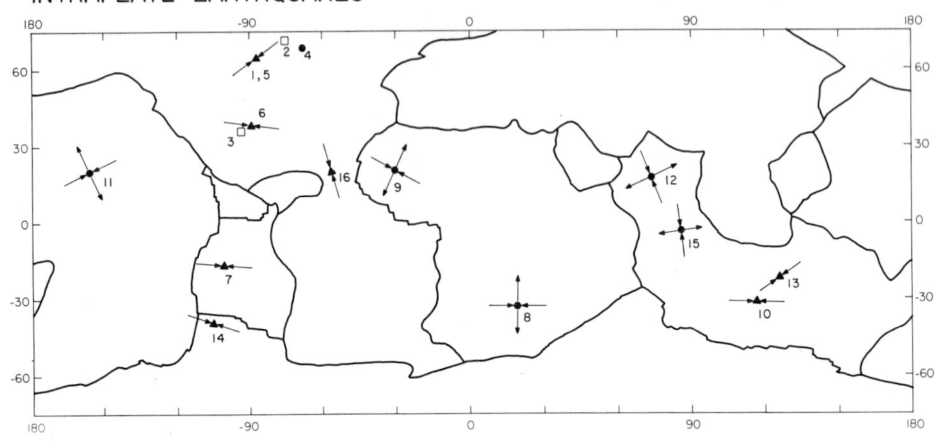

Figure 1
Epicenters of intraplate earthquakes used in this study. Event numbers are from Table 1. Solid lines are plate boundaries. Triangles denote thrust faulting, squares normal faulting, and circles strike-slip mechanisms. The horizontal projections of P and T axes are shown, where well determined from the fault plane. Cylindrical equidistant projection.

Earthquakes from all the major plates except the Eurasian plate are included. Fault plane solutions for all but one event appear in the literature; the solution for event 9 is shown in Fig. 2.

To determine the apparent stress from equation (1), both the seismic moment

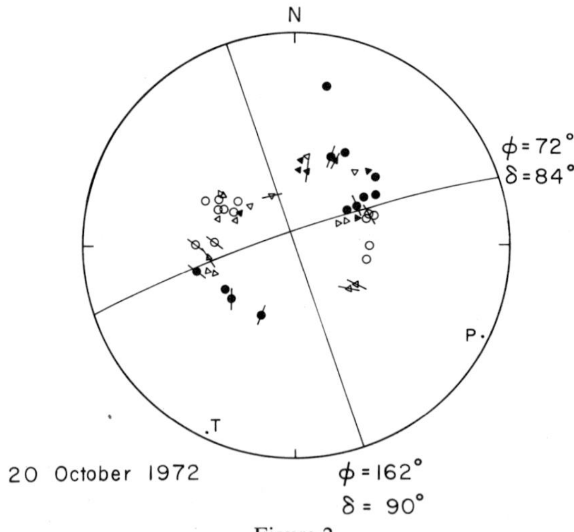

Figure 2
Fault plane solution for event 9. Equal-area projection of the lower focal hemisphere; closed and open circles represent clear compressional and dilatational P wave first motions, respectively. Triangles represent less reliable determinations. S wave polarizations are represented by solid lines through symbols. Strike and dip of nodal planes are marked. P and T denote the inferred axes of maximum and minimum compressive stress, respectively.

and seismic energy must be known. The seismic moment M_0 may be calculated either from long-period surfaces waves or from long-period body waves.

In this study, long period amplitude spectra of SH waves are used to calculate the moment when no published values are available. Long period horizontal components of shear waves recorded at WWSSN stations in the epicentral distance range of 32 to 67° are digitalized and rotated to isolate the SH component. The digitized signal is Fourier analyzed to obtain an amplitude spectrum; window lengths are generally 60 sec. The resulting spectrum is corrected for the instrument response of WWSSN stations (HAGIWARA 1958), and the spectral amplitude Ω_0 of the long period end of the spectrum is estimated (HANKS and WYSS 1972). The seismic moment may be derived from Ω_0 by the relation (KEILIS-BOROK 1960):

$$M_0 = \frac{4\pi\rho V^3}{RG}\Omega_0 \tag{4}$$

where ρ and V are the density and shear wave velocity, respectively, at the source, R is a correction for radiation pattern, and G is a correction for geometrical spreading, attenuation and the free surface. The radiation pattern correction, including superposition of the sS wave of appropriate source amplitude and free surface reflection coefficient, is taken from HIRASAWA (1966) given the fault plane parameters of each earthquake. The propagation correction term is given by (JULIAN and ANDERSON, 1968):

$$G = 2\left[\frac{\rho V^3\, dt/d\Delta\, |d^2t/d\Delta^2|}{\rho_0 V_0\, r_0^2 r^2 \sin\Delta \cos i \cos i_0}\right]^{1/2} \exp\left[-\pi f \int \frac{ds}{QV}\right] \tag{5}$$

where r is the radius from the earth's center, Δ is the epicentral distance, t is the travel time, i is the angle of incidence, the subscript 0 denotes the earth's surface, f is the frequency, $1/Q$ is the attenuation, and the integral is taken over the entire ray path. The Q model of TSAI and AKI (1969) and the shear wave velocity model SLUTD1 of HALES and ROBERTS (1970) are used in computing the attenuation correction and the travel time derivatives. The factor 2 in equation (5) accounts for the effect of the free surface at the station.

Values for moments calculated at each of the several stations for seven intraplate earthquakes are listed in Table 2. The scatter in the data is fairly large, indicative of uncertainties in the procedure. The radiation pattern correction probably introduces the greatest uncertainty, so stations are selected away from SH nodal planes since for such directions the radiation pattern corrections are largest and most sensitive to errors in fault plane geometry. Inaccuracies in the attenuation model have the least effect at low frequencies where Ω_0 is measured. Considering these uncertainties, the seismic moment calculated from equation (4) is probably accurate to about a factor of 2.

The mean moments, weighted by the quality of the determination, for these events are listed in Table 3. As a check on our procedure for calculating seismic

Table 2
Seismic moments calculated from long-period SH *wave spectra*

Event	Station	Δ, degrees	Ω_0, 10^{-2} dyne-sec	M_0, 10^{25} dyne-cm	Quality*)
8	NAI	35.7	3.0	3.3	A
	AAE	45.9	4.0	6.0	A
	EIL	64.1	1.3	4.2	A
	JER	66.3	2.0	6.8	B
9	BHP	49.4	1.5	2.1	A
	TAB	66.8	1.5	4.7	A
	TRN	32.0	1.1	1.8	C
	BOG	45.9	1.7	2.2	C
	HLW	55.4	1.3	3.5	B
11	COL	45.2	2.5	5.1	A
	GUA	57.5	3.0	8.2	C
	GIE	66.7	2.9	8.2	A
	TUC	41.4	1.0	6.1	B
	GOL	46.8	1.9	5.3	C
	MAT	59.9	3.1	6.7	B
12	MUN	63.7	2.1	23.0	A
	ATU	48.2	2.7	6.0	A
	KRK	58.7	1.2	4.5	B
13	GUA	39.6	1.7	8.3	B
	AFI	58.2	1.2	3.5	A
	MAT	59.2	1.0	3.3	A
15	LAH	36.7	2.8	8.3	B
	NIL	38.7	2.6	8.6	B
	SHI	46.2	3.8	6.3	A
	JER	59.9	2.4	5.2	B
	MAT	62.8	2.3	5.4	B
16	LPB	38.0	2.5	4.8	B
	TUC	50.3	1.4	3.6	B
	ARE	39.1	1.5	2.9	B
	TOL	48.6	3.4	14.0	A
	STU	59.1	3.5	13.0	B
	GIE	39.3	1.6	10.0	B

*) Subjective quality of determination of long period spectral level Ω_0.

moment from SH wave spectra, we may compare the computed M_0 for the 10 December 1967 Koyna Dam earthquake in India (event 12) with values obtained by SINGH *et al.* (1975) from surface waves and by LANGSTON (1976) from body waves. Our method gives $M_0 = 8.2 \times 10^{25}$ dyne-cm while Singh *et al.* and Langston obtained 8.2×10^{25} and 3.2×10^{25} respectively. MAASHA and MOLNAR (1972) estimated the moment for the 29 September 1969 Ceres, South Africa, earthquake (event 8) to be 2.0×10^{25} dyne-cm from long-period P waves. The moment determined here from SH waves is 4.8×10^{25} dyne-cm. The two determinations agree to within the errors in both calculations.

The radiated seismic energy E_S is not in general a simple function of a single measurable parameter. Two estimates of E_S in common use are from the energy-magnitude relations of GUTENBERG and RICHTER (1956):

$$\log_{10} E_S = 5.8 + 2.4\, m_b \tag{6}$$

and, for large events,

$$\log_{10} E_S = 11.8 + 1.5\, M_S \tag{7}$$

where M_S and m_b are the surface and body wave magnitudes, respectively. Because equations (6) and (7) are based upon an implied similarity of seismic spectra and because the Gutenberg-Richter relations were derived from magnitudes determined by techniques or on instruments somewhat different from those used currently, their quantitative utility to predict energy is doubtful. However, to the extent that $\eta\bar{\sigma}$ in

Table 3

Apparent stresses for intraplate earthquakes

Event	m_b[a])	M_S[b])	M_0[c]) 10^{25} dyne-cm	$(\eta\bar{\sigma})_b$, bars	$(\eta\bar{\sigma})_S$, bars
1	4.4	3.9	0.0018[e])	0.38	
2	4.2		0.0093[f])	0.02	
3	4.9	4.1[g])	0.017[g])	0.64	
4	4.9	4.3	0.051[f])	0.21	
5	5.0	4.5	0.013[e])	1.5	
6	5.5[h])	5.2[h])	0.20[h])	1.5	
7	5.3	4.9	0.18[i])	0.55	
8	5.6	6.3[d])	4.8	0.11	11
9	5.7	5.8[d])	2.8	0.32	
10	5.9	6.8[d])	6.1[j])	0.45	49
11	5.9	6.1[d])	6.4	0.43	4.2
12	5.9	6.3[k])	8.2[k])	0.33	6.5
13	6.1	5.9[d])	4.3	1.9	
14	6.0	6.0[d])	9.0[l])	0.53	2.1
15	5.8	6.3[d])	6.7	0.23	8.0
16	6.2	6.3	8.4	1.7	6.4

[a]) ISC values for m_b unless otherwise referenced.

[b]) Unless otherwise referenced, M_S calculated for this study using $M_S = \log(A/T) + 1.66 \log \Delta + 3.3$, where A is the maximum zero to peak amplitude of the long period vertical component of the Rayleigh wave in microns in the period T range of 18 to 22 sec and Δ is the distance in degrees.

[c]) Moment calculated for this study unless otherwise referenced.

[d]) USCGS.

[e]) HASHIZUME (1974).

[f]) HASHIZUME (1973).

[g]) MITCHELL (personal communication 1976).

[h]) BHATTACHARYA (1975).

[i]) MENDIGUREN (1971).

[j]) FITCH *et al.* (1973).

[k]) SINGH *et al.* (1975) and Table 2.

[l]) FORSYTH (1973b).

equation (1) is regarded only as some measure of short-period spectral level divided by a measure of long-period spectral level, then equations (6) and (7) may still be useful indicators of short-period level for comparison of relative differences in $\eta\bar{\sigma}$ within a population of seismic events. We use equations (6) and (7) below to measure an E_S and we caution that this quantity may not be closely related to actual seismic energy. The apparent stresses thus calculated are comparable only to those determined in other studies using the same procedure to estimate E_S.

4. *Apparent stress for intraplate earthquakes*

The apparent stresses calculated using E_S estimated from body-wave magnitudes are listed in Table 3. For events with $M_S \geq 6$, the apparent stress has also been estimated from the surface wave magnitude.

Considering the uncertainties associated with the calculation of magnitude and seismic moment, apparent stresses measured by a single procedure are accurate to about half an order of magnitude. Apparent stresses $(\eta\bar{\sigma})_b$ obtained from body wave magnitudes vary from 0.02 to 1.9 bars with most values near 0.5 to 1 bar. Apparent stresses $(\eta\bar{\sigma})_S$ obtained from surface wave magnitudes ($M_S \geq 6$) vary from 2 to 50 bars, with 5 to 10 bars a representative value.

KANAMORI and ANDERSON (1975) have suggested that, for events with M_S 6 to 8, intraplate earthquakes have higher apparent stresses (~ 50 bars) than interplate earthquakes (~ 10 to 20 bars). Their population of 'intraplate' earthquakes, however, consists only of events very near plate boundaries and only in restricted areas of high activity (Japan, California-Nevada, Azores-Gibraltar). The values of apparent stress for midplate earthquakes obtained from M_S in Table 3 are comparable to values obtained on plate boundaries (KANAMORI and ANDERSON 1975), and do not support a general distinction in stress level between intraplate and interplate events. The large M_S events in Table 3 range only from M_S 6.0 to 6.8, however.

As a more comprehensive comparison of apparent stress for intraplate versus interplate earthquakes, we plot in Fig. 3 M_0 versus m_b for the events in Table 3 and for a wide range of plate margin events. The plate boundary earthquakes include post-1963 events from the lists of intermediate-depth and deep focus South American and Tonga earthquakes from WYSS (1970a) and WYSS and MOLNAR (1972), shallow Aleutian and Tonga island arc earthquakes from WYSS (1970b) and MOLNAR and WYSS (1972), and ridge crest and transform earthquakes from WYSS (1970b). All apparent stresses for interplate events were calculated from equations (1) and (6), using M_0 from the original authors and m_b from the ISC.

The main conclusion to be drawn from Fig. 3 is that for all tectonic environments, including midplate regions, the apparent stresses are very similar; see also Table 4. Moment-magnitude relations *do not* support the contention that intraplate earthquakes are characterized in general by higher stress than plate boundary earthquakes.

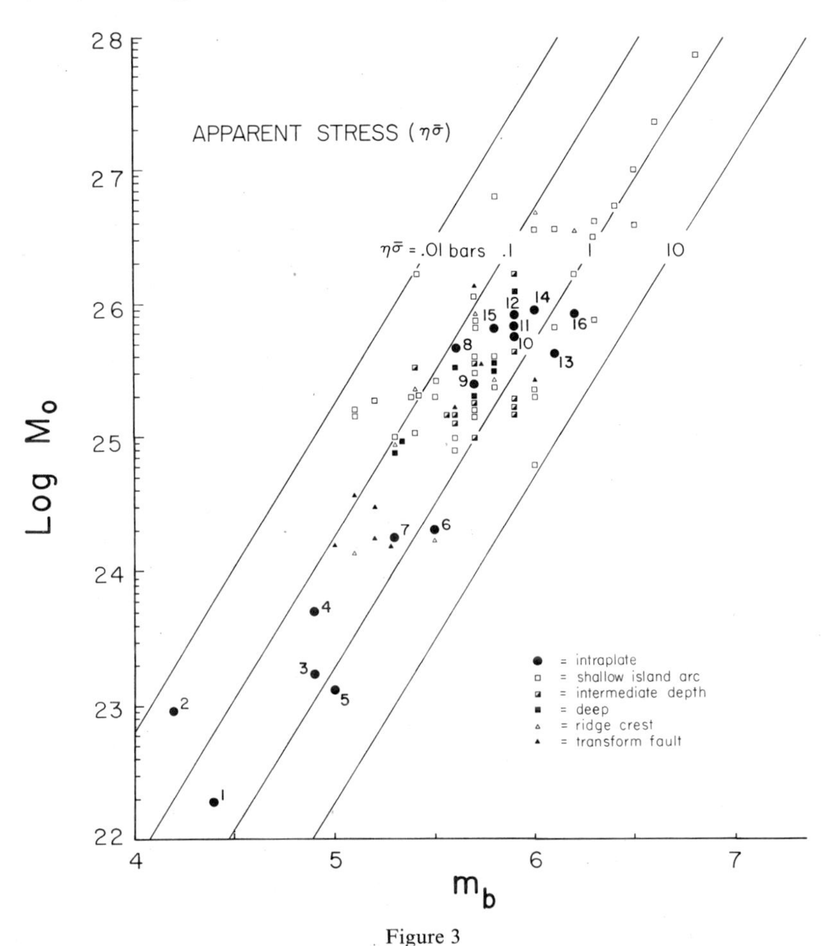

Figure 3
Relationship between seismic moment M_0 and body wave magnitude m_b for intraplate and interplate earthquakes. Straight lines represent constant apparent stress $\eta\bar{\sigma}$. Intraplate earthquake data are from Table 3. Interplate earthquake moments from WYSS (1970a, b), WYSS and MOLNAR (1972), and MOLNAR and WYSS (1972). All m_b are from the ISC.

There is no significant difference in Table 3 between the apparent stresses for oceanic earthquakes and those for intra-continental events.

5. *Stress drops for intraplate earthquakes*

The calculation for stress drop (equations 2–3) for midplate earthquakes is difficult because of the scarcity of information on fault dimensions and average fault displacement. Intraplate events are commonly small, usually occur far from seismological stations (which tend to be concentrated in more active areas) are not always accompanied by surface faulting, and rarely have aftershocks locatable teleseis-

Table 4
Summary of apparent stresses for various tectonic environments

Environment	No. of Events	Apparent Stress, bars Range	Geometric Mean	Arithmetic Mean	Median
Intraplate	16	0.02–1.9	0.42	0.68	0.44
Shallow island arc	34	0.01–7.6	0.27	0.70	0.30
Intermediate depth	14	0.05–1.6	0.34	0.54	0.37
Deep	8	0.10–0.52	0.25	0.29	0.27
Ridge crest	8	0.08–1.7	0.37	0.55	0.43
Transform fault	10	0.07–1.7	0.23	0.37	0.21

mically. For these reasons, observation of surface faulting and accurate determination of aftershock locations, the most reliable measures of fault dimensions, are not always available.

The values of stress drop for 5 intraplate earthquakes for which estimates of fault length are available are given in Table 5. Fault widths were estimated from an assumed geometric similarity $L/w \sim 2.5$, weakly supported by limited seismic data on fault dimensions (ELLSWORTH 1975, GELLER 1976). The range in $\Delta\sigma$ is 2 to 70 bars. The largest stress drop occurs for the single earthquake in oceanic lithosphere (event 14). The fault plane dimensions were estimated from aftershock epicenters relocated using a master event technique. Because of the distance of this earthquake sequence

Table 5
Stress drops for intraplate earthquakes

Event	M_0, 10^{25} dyne-cm	Length, km	$\Delta\sigma$, bars [a]
5	0.013	7 (Rayleigh wave amplitude spectra)[b]	2
8	3.8	27 (aftershocks)[c]	8
			25[d]
10	6.1	37 (surface cracks)[e]	7
			45–120[e]
12	8.2	40 (finite rupture velocity)[f]	5
			6–20[f]
14	9.0	8 (aftershocks)[g]	70

[a]) Stress drop calculated from equations (2) and (3), assuming $L/w = 2.5$.
[b]) HASHIZUME (1974).
[c]) FAIRHEAD and GIRDLER (1971).
[d]) MAASHA and MOLNAR (1972).
[e]) FITCH *et al.* (1973).
[f]) SINGH *et al.* (1975).
[g]) Main shock used as master event to relocate aftershocks and estimate $L = 8$ km, $w = 12$ km.

from seismic stations, the stress drop for event 14 probably has the greatest uncertainty of the listed values.

Independent estimates of $\Delta\sigma$ are available for events 8, 10, and 12. MAASHA and MOLNAR (1972) determined a stress drop of 25 bars for event 8 from the P-wave corner frequency. FITCH *et al.* (1973) estimated $\Delta\sigma$ to be 'about 100 bars' for event 10 using a somewhat lesser fault length and a width derived from the observed displacement across surface cracks in soil and weathered rock layers. SINGH *et al.* (1975) estimated a stress drop between 6 and 20 bars for event 12 using a length determined from rupture velocity and a width determined from an empirical relationship between fault area and earthquake magnitude. Given the sensitivity of $\Delta\sigma$ to fault dimensions, especially the uncertain fault width, the fair agreement of previous determinations with those reported here is perhaps acceptable.

Several workers have suggested that intraplate earthquakes may be characterized by higher apparent stresses than interplate events. MOLNAR and WYSS (1972) noted than an earthquake in the Indian plate had a stress drop (50 bars) higher than those of shallow Tonga arc earthquakes (1 to 25 bars) and that $\Delta\sigma$ was generally higher in environments where new faulting might be expected than where rupturing on older fault surfaces is likely. MAASHA and MOLNAR (1972) suggested that stress drops for African earthquakes are higher (8 to 25 bars) for those not associated with the East African rifting than for those events (3 to 18 bars) that are. KANAMORI and ANDERSON (1975), in their study of source parameters of earthquakes with M_S 6 to 8, contended that large intraplate earthquakes have generally higher stress drops (*c.* 100 bars) than do large interplate earthquakes (*c.* 30 bars). The stress drops in Table 5, from a few bars to a few tens of bars, are generally in the range reported for interplate events. Thus while some intralithosphere regions may be characterized by relatively high stress drops and high inferred stresses, particularly regions near plate boundaries such as those studied in the works mentioned, we find no compelling evidence from the midplate earthquakes considered in this study that stress drops or stresses are generally greater in intraplate environments than along plate boundaries.

6. *Relation to plate driving forces*

The reported measurements of apparent stress and stress drop for midplate earthquakes may be related to a considerable body of work on the application of equilibrium mechanics to the question of the relative magnitudes of possible driving and resistive forces acting on the plates. FORSYTH and UYEDA (1975) and TULLIS and CHAPPLE (1973) have studied the inverse problem of determining simply parameterized edge and surface forces from the relative angular velocities of the plates. SOLOMON and SLEEP (1974) and SOLOMON *et al.* (1975, 1977a) and RICHARDSON *et al.* (1976) have used the directions of principal stresses in the lithosphere inferred from midplate earthquake mechanisms (SYKES and SBAR 1973, 1974), *in situ* measurements and stress-sensitive geologic features to constrain the relative magnitude of the various plate driving and resistive forces.

A broadly consistent model of plate forces has emerged from all of these studies. The effective driving forces appear to be concentrated at plate boundaries, with viscous shear at the base of the plates acting primarily as a resistive force (RICHARDSON *et al.* 1976). At subduction zones there is both a large driving force component due to the negative buoyancy of the downgoing slab and a nearly compensating resistive force opposing motion of the slab into the mantle (SMITH and TOKSÖZ 1972, FORSYTH and UYEDA 1975). The net driving force exerted at subduction zones is comparable to the net driving force exerted at spreading centers (SOLOMON *et al.* 1975, RICHARDSON *et al.* 1976). Resistance to these forces is exerted at transform faults and by drag at the base of the plates. Drag may show significant regional variations but is probably not consistently much higher beneath continents than beneath oceans (SOLOMON *et al.* 1977b).

Driving forces at the ridge are among the easiest forces in the system to model quantitatively. Heat flow, gravity, and topography across the ridge suggest the presence of warm, low density material of minimal strength at shallow mantle depths. Thus forces adjacent to the ridge are due to density and topographic contrasts, and not to forces transmitted from great distances. Buoyancy forces at the ridge have been estimated to be equivalent to 100 to 200 bars compressive deviatoric stress across a 100 km thick lithosphere (HALES 1969, FRANK 1972, ARTYUSHKOV 1973).

Though the negative buoyancy of the subducted slab is potentially capable of transmitting tensional deviatoric stresses of several kilobars to the subducting plates (MCKENZIE 1969, JACOBY 1970, TURCOTTE and SCHUBERT 1971, SOLOMON and PAW U 1975), most of the buoyancy force is offset by local resisting forces (SMITH and TOKSÖZ 1972, FORSYTH and UYEDA 1975). That some net pulling force generally acts at subduction zones is implied by the focal mechanisms of intermediate depth earthquakes indicating down-dip extension within the slab (ISACKS and MOLNAR 1971).

The stress drops and apparent stresses of midplate earthquakes are entirely consistent with the conclusion that stress differences in the lithosphere are generally comparable to the compressive stress adjacent to spreading centers, or on the order of 100 bars. This conclusion is not meant to imply that stresses cannot locally be much greater or significantly less than such a value. At continent–continent collision zones, for instance, there is strong evidence for horizontal compressive stress of up to 1 kilobar (BIRD 1976). Along a major plate boundary fault, the stress may vary considerably depending on the recent rupture history. Because of large local driving and resistive forces (FORSYTH and UYEDA 1975), shear stresses within downgoing lithosphere may be considerably larger than within the plates; stress differences of 500 to 1,000 bars appear possible (SMITH and TOKSÖZ 1972, TOKSÖZ *et al.* 1973, ENGDAHL and SLEEP 1976). The deviatoric stress levels of a few hundred bars is suggested here for the typical regional stress field in most plate interiors.

Stress differences of about 100 bars in the plate are considerably less than the several kilobars required in the laboratory to fracture or fault dry crystalline rock (e.g., STESKY *et al.* 1974). Several possibilities exist that may account for the reduced stress needed to initiate rupture on faults in the earth. Fluid pore pressure for shallow midplate earthquakes may exceed hydrostatic pressure, thus reducing the friction which must be overcome for faulting to occur. Also, the presence of low strength material in the fault may lower the stress required for fracture.

7. *Conclusions*

The apparent stresses of midplate earthquakes are between 0.1 and 50 bars, depending on the definition of seismic energy, and the stress drops are in the range of 2 to 100 bars. Both sets of measurements are consistent with the inference from the fitting of global tectonic stress models to principal stress directions as indicated from midplate earthquake mechanisms that intraplate stress differences are on the order of the compressive stress adjacent to mid-ocean ridges, about 100 bars. One consequence of such a level for regional stresses is that the highest stress drops observed for midplate earthquakes represent nearly complete release of local stress for such events.

Intraplate earthquakes do not as a general rule have either apparent stresses or stress drops systematically larger than those for interplate earthquakes.

Acknowledgements

We have benefited from discussion with A. T. Smith, R. Madariaga, and N. H. Sleep. We thank N. C. Burr for programming assistance, R. North for help with master event relocations and P. Molnar and M. Wyss for critical reviews. This research was supported by the Earth Sciences section of the National Science Foundation, NSF Grant DES74–21894.

REFERENCES

AHORNER, L. (1975), *Present-day stress field and seismotectonic block movements along major fault zones in central Europe*, Tectonophys. *29*, 233–249.

AKI, K. (1966), *Generation and propagation of G waves from the Niigata earthquake of June 16, 1964. Part 2. Estimation of earthquake moment, released energy, and stress-strain drop from the G wave spectrum*, Bull. Earthquake Res. Inst., Tokyo Univ. *44*, 73–88.

ARTYUSHKOV, E. V. (1973), *Stresses in the lithosphere caused by crustal thickness inhomogeneities*, J. Geophys. Res. *78*, 7675–7708.

BHATTACHARYA, B. (1975), *Excitation and attenuation of Love waves in North America from the November 9, 1968 south central Illinois earthquake*, J. Phys. Earth *23*, 173–187.

BIRD, G. P. (1976). *Thermal and mechanical evolution of continental convergence zones: Zagros and Himalayas*, Ph.D. Thesis, Mass. Inst. Technol., Cambridge, Mass. 423 pp.

BRUNE, J. N., HENYEY, T. L. and ROY, R. F. (1969), *Heat flow, stress, and rate of slip along the San Andreas fault, California*, J. Geophys. Res. *74*, 3821–3827.

ELLSWORTH, W. L. (1975), *Bear Valley, California, earthquake sequence of February-March, 1972*, Bull. Seismol. Soc. Am. *65*, 483–506.

ENGDAHL, E. R. and SLEEP, N. H. (1976), *Seismicity and stress beneath the central Aleutian arc* (abstract), EOS, Trans. Amer. Geophys. Un. *57*, 329.

FAIRHEAD, J. D. and GIRDLER, R. W. (1971), *The seismicity of Africa*, Geophys. J. Roy. astro. Soc. *24*, 271–301.

FITCH, T. J., WORTHINGTON, M. H. and EVERINGHAM, I. B. (1973), *Mechanisms of Australian earthquakes and contemporary stresses in the Indian Ocean plate*, Earth Planet. Sci. Lett. *18*, 345–356.

FORSYTH, D. W. (1973a), *Compressive stress between two mid-ocean ridges*, Nature *243*, 78–79.

FORSYTH, D. W. (1973b), *Anisotropy and the structural evolution of the oceanic upper mantle*, Ph.D. Thesis, Mass. Inst. Technol., Cambridge, 253 p.

FORSYTH, D. W. and UYEDA, S. (1975), *On the relative importance of driving forces of plate motion*, Geophys. J. Roy. astro. Soc. *43*, 163–200.

FRANK, F. C. (1972), *Plate tectonics, the analogy with glacier flow and isostasy*, in *Flow and Fracture of Rocks*, Amer. Geophys. Un. Mon. *16*, 285–292.

GELLER, R. J. (1976), *Scaling relations for earthquake source parameters and magnitude*, Bull. Seismol. Soc. Am. *66*, 1501–1523.

GUTENBERG, B. and RICHTER, C. F. (1956), *Earthquake magnitude, intensity, energy and acceleration*, Bull. Seismol. Soc. Am. *46*, 105–145.

HAGIWARA, T. (1958), *A note on the theory of the electromagnetic seismograph*, Bull. Earthquake Res. Inst., Tokyo Univ. *36*, 139–164.

HAIMSON, B. C. (1975), *The state of stress in the earth's crust*, Rev. Geophys. Space Phys. *13*, 350–352.

HAIMSON, B. C. (1976), *Crustal stress measurements through an ultra deep well in the Michigan basin* (abstract), EOS, Trans. Amer. Geophys. Un. *57*, 326.

HALES, A. L. (1969), *Gravitational sliding and continental drift*, Earth Planet. Sci. Lett. *6*, 31–34.

HALES A. L. and ROBERTS, J. L. (1970), *Shear velocities in the lower mantle and the radius of the core*, Bull. Seis. Soc. Am. *60*, 1427–1436.

HANKS, T. C. and WYSS, M. (1972), *The use of body wave spectra in the determination of seismic source parameters*, Bull. Seismol. Soc. Am. *62*, 561–589.

HASHIZUME, M. (1973), *Two earthquakes on Baffin Island and their tectonic implications*, J. Geophys. Res. *78*, 6069–6081.

HASHIZUME, M. (1974), *Surface wave study of earthquakes near northwestern Hudson Bay, Canada*, J. Geophys. Res. *79*, 5458–5468.

HIRASAWA, T. (1966), *A least square method for the mechanism determination from S wave data;* Part 1., Bull. Earthquake Res. Inst., Tokyo Univ. *44*, 901–918.

ISACKS, B. and MOLNAR, P. (1971), *Distribution of stresses in the descending lithosphere from a global survey of focal-mechanism solutions of mantle earthquakes*, Rev. Geophys. Space Phys. *9*, 103–174.

JACOBY, W. B. (1970), *Instability in the upper mantle and global plate movements*, J. Geophys. Res. *75*, 5671–5680.

JULIAN, B. R. and ANDERSON, D. L. (1968), *Travel times, apparent velocities, and amplitudes of body waves*, Bull. Seismol. Soc. Am. *58*, 339–366.

KANAMORI, H. and ANDERSON, D. L. (1975), *Theoretical basis of some empirical relations in seismology*, Bull. Seismol. Soc. Am. *65*, 1073–1095.

KEILIS-BOROK, V. I. (1960), *Investigation of the mechanism of earthquakes*, Sov. Res. Geophys. (English transl.) *4*, 29.

KNOPOFF, L. (1958), *Energy release in earthquakes*, Geophys. J. Roy. astro. Soc. *1*, 44–52.

LANGSTON, C. A. (1976), *A body wave inversion of the Koyna, India earthquake of December 10, 1967, and some implications for body wave focal mechanisms*, J. Geophys. Res. *81*, 2517–2529.

MAASHA, N. and MOLNAR, P. (1972), *Earthquake fault parameters and tectonics in Africa*, J. Geophys. Res. *77*, 5731–5743.

MCKENZIE, D. P. (1969), *Speculations on the consequences and causes of plate motions*, Geophys. J. Roy. astro. Soc. *18*, 1–32.

MENDIGUREN, J. A. (1971), *Focal mechanism of a shock in the middle of the Nazca plate*, J. Geophys. Res. *76*, 3861–3879.

MITCHELL, B. J. (1973), *Radiation and attenuation of Rayleigh waves from the southeastern Missouri earthquake of October 21, 1965*, J. Geophys. Res. *78*, 886–899.

MOLNAR, P. and SYKES, L. R. (1969), *Tectonics of the Caribbean and middle America regions from focal mechanisms and seismicity*, Bull. Geol. Soc. Am. *80*, 1639–1684.

MOLNAR, P. and WYSS, M. (1972), *Moments, source dimensions and stress drops of shallow-focus earthquakes in the Tonga-Kermadec arc*, Phys. Earth Planet. Inter. *6*, 263–278.

RANALLI, G. (1975), *Geotectonic relevance of rock stress determinations*, Tectonophys. *29*, 49–58.

RICHARDSON, R. M., SOLOMON, S. C. and SLEEP, N. H. (1976), *Intraplate stress as an indicator of plate tectonic driving forces*, J. Geophys. Res. *81*, 1847–1856.

SBAR, M. L. and SYKES, L. R. (1973), *Contemporary compressive stress and seismicity in eastern North America: an example of intra-plate tectonics*, Geol. Soc. Am. Bull. *84*, 1861–1882.

SINGH, D. D., RASTOGI, B. K. and GUPTA, H. K. (1975), *Surface wave radiation pattern and source parameters of Koyna earthquake of December 10, 1967*, Bull. Seismol. Soc. Am. *65*, 711–731.

SMITH, A. T. and TOKSÖZ, M. N. (1972), *Stress distribution beneath island arcs*, Geophys. J. Roy. astro. Soc. *29*, 289–318.

SOLOMON, S. C. and PAW U, K. T. (1975), *Elevation of the olivine-spinel transition in subducted lithosphere: seismic evidence*, Phys. Earth Planet. Interiors *11*, 97–108.

SOLOMON, S. C. and SLEEP, N. H. (1974), *Some simple physical models for absolute plate motions*, J. Geophys. Res. *79*, 2557–2567.

SOLOMON, S. C., SLEEP, N. H. and RICHARDSON, R. M. (1975), *On the forces driving plate tectonics: inferences from absolute plate velocities and intraplate stress*, Geophys. J. Roy. astro. Soc. *42*, 769–801.

SOLOMON, S. C., SLEEP, N. H. and RICHARDSON, R. M. (1977a), *Implications of absolute plate motions and intraplate stress for mantle rheology*, Tectonophys. *37*, 219–231.

SOLOMON, S. C., SLEEP, N. H. and JURDY, D. M. (1977b), *Mechanical models for absolute plate motions in the early Tertiary*, J. Geophys. Res. *82*, 203–212.

STAUDER, W. and NUTTLI, O. W. (1970), *Seismic studies; south central Illinois earthquake of November 9, 1968*, Bull. Seismol. Soc. Am. *60*, 973–981.

STESKY, R. M., BRACE, W. F., RILEY, D. K. and ROBIN, P.-Y. F. (1974), *Friction in faulted rock at high temperature and pressure*, Tectonophys. *23*, 177–203.

SYKES, L. R. and SBAR, M. L. (1973), *Intraplate earthquakes, lithospheric stresses and the driving mechanism of plate tectonics*, Nature *245*, 298–302.

SYKES, L. R. and SBAR, M. L. (1974), *Focal mechanism solutions of intraplate earthquakes and stresses in the lithosphere* in *Geodynamics of Iceland and the North Atlantic Area* (ed. L. Kristjansson), 207–224.

TOKSÖZ, M. N., SLEEP, N. H. and SMITH, A. T. (1973), *Evolution of the downgoing lithosphere and the mechanisms of deep focus earthquakes*, Geophys. J. Roy. astro. Soc. *35*, 285–310.

TULLIS, T. E. and CHAPPLE, W. M. (1973), *What makes the plates go* (abstract), EOS, Trans. Amer. Geophys. Un. *54*, 468.

TURCOTTE, D. L. and SCHUBERT, G. C. (1971), *Structure of the olivine-spinel phase boundary in the descending lithosphere*, J. Geophys. Res. *76*, 7980–7987.

TSAI, Y. B. and AKI, K. (1969), *Simultaneous determination of the seismic moment and attenuation of seismic surface waves*, Bull. Seismol. Soc. Am. *59*, 275–287.

UNGER, J. D. and WARD, P. L. (1974), *Travel time delays and tectonic stress from a subcrustal Hawaiian earthquake* (abstract), EOS, Trans. Amer. Geophys. Un. *56*, 1150.

WYSS, M. (1970a), *Stress estimates for South American shallow and deep earthquakes*, J. Geophys. Res. *75*, 1529–1544.

WYSS, M. (1970b), *Apparent stresses of earthquakes on ridges compared to apparent stresses of earthquakes in trenches*, Geophys. J. Roy. astro. Soc. *19*, 479–484.

WYSS, M. and MOLNAR, P. (1972), *Source parameters of intermediate and deep focus earthquakes in the Tonga arc*, Phys. Earth Planet. Inter. *6*, 279–292.

(Received 4th October 1976)

Pageoph, Vol. 115 (1977), Birkhäuser Verlag, Basel

Source Parameters for the January 1975 Brawley – Imperial Valley Earthquake Swarm

By Stephen H. Hartzell and James N. Brune[1]

Abstract – The source parameters, moment, stress drop and source dimension are estimated for 61 events from the January 1975 Brawley earthquake swarm. Earthquakes studied range in local magnitude from 1.0 to 4.7. Stress drops range from 1 to 636 bars and increase with source depth. It is estimated that the sedimentary structure of the Imperial Valley amplifies shear waves by a factor of 2 to 3 in addition to the free surface amplification of 2. Estimates of moment from 10 sec surface waves are 4 to 6 times larger than the moment estimated from the relatively flat part of the local body wave spectrum at 1 sec. This may be due to after-slip on the fault, a long thin fault, or partial stress drop. It is shown that the experimentally determined ratio of stress drop to apparent stress should be approximately 4.0 when spectrum integration is used to obtain *S*-wave energy and the *P*-wave energy is 1/3 the *S*-wave energy.

Key words: Source parameters; Brawley swarm; Imperial Valley.

Introduction

The 4 March 1966 earthquake located on the Imperial fault between Brawley and El Centro with a local magnitude of 3.6 was evidently characterized by a low stress drop (~1 bar) with a large fault length (~10 km) (Brune and Allen 1967). This raised the following question concerning Imperial Valley earthquakes: Are all Imperial Valley earthquakes low stress drop events (because of particular structural and tectonic characteristics) or is the 4 March 1966 event relatively unique? The Brawley earthquake swarm of January 14 through January 31, 1975 provides an opportunity to study in some detail the source parameters for Imperial Valley earthquakes over a range in magnitudes and depths.

The Imperial Valley and upper Gulf is a seismically active region with seismic energy release often taking the form of a swarm of small earthquakes. A notable exception was the 1940 El Centro earthquake which was a more typical mainshock-aftershock sequence. The time distribution of cumulative moment for the 1976 Brawley activity is very similar to that obtained by Thatcher and Brune (1971) for an earthquake swarm in the northern Gulf of California. Johnson and Hadley (1976) have located 288 events associated with this swarm, ranging in local magnitude from 1.0 to 4.7. These locations are plotted in Fig. 1. The majority of the epicenters lie along the southern extension of the Brawley fault and two short faults at high

[1]) Scripps Institution of Oceanography, Institute of Geophysics and Planetary Physics, University of California, San Diego, La Jolla, California 92093, USA.

angles to the Brawley fault. These faults are about 6 km east of the trace of the 1940 break. The swarm is characterized by a period of low level 'foreshock' activity from January 14 to January 23, a short period of intense activity (including the main shock of $M_L = 4.7$) starting at 0348Z January 23 and continuing through January 25 and a period of lower level 'aftershock' activity lasting through January 31.

The Imperial Valley is a unique geological province. The area of the Brawley swarm is a region of crustal spreading covered by an accumulation of alluvial sediments up to 6 km deep. Shear wave velocities vary over a wide range, from 0.98 km/sec in the upper most layer to 2.71 km/sec just above basement and 3.70 km/sec in the basement (BIEHLER *et al.* 1964). The overlying low velocity layers cause amplification of the seismic waves. Thus, for a variety of reasons it is of great interest to study the spectra and source mechanism of the Brawley earthquake swarm.

Equipment and data

Figures 1 and 2 show the local and more distant stations, respectively, used in this study. Six Canadian short-period stations operating one second Wood–Anderson instruments were used to estimate m_b. Long-period WWSSN Press–Ewing records from south-western US stations were used to obtain estimates of moment at surface wave periods of 10 s covering a range in azimuths of 150 degrees. Near source data was analyzed from two stations operating in the Imperial Valley: (1) a portable digital recorder placed at two sites near the swarm (Fig. 1), (2) a strong motion accelerometer at the Brawley airport. Location of the digital recorder was chosen on the basis of preliminary Pasadena locations, and as a result not as close to the epicenters as we would have liked. The Canadian and US WWSSN stations all use standard instruments whose characteristics are well known. Below we describe the instrumentation used at the near source sites.

Digital recorder

The digital recording system has a sample rate of 128 samples/sec, data storage on 1/4-inch magnetic tape, and is similar to the system described by PROTHERO (1976). Two stage antialiasing filters are used with system response down by 6 db at 50 Hz. Relative time is provided by a temperature compensated crystal oscillator. Absolute timing to better than $\frac{1}{10}$ sec is obtained by reading WWV time pulses onto the beginning and end of each field tape. A horizontal seismometer having a 5-second natural period was used with the digital recorder. Response curves for the combined recorder-seismometer system are shown in Fig. 3 for various gain settings.

The digital recorder was operated at two sites during the course of the swarm: (1) 32° 55.17′N, 115° 31.18′W from 0656Z to 1629Z on January 24 and (2) 32° 55.17′N, 115° 32.04′W from 1749Z January 24 to 1830Z January 27 (Figure 1). Hundreds of

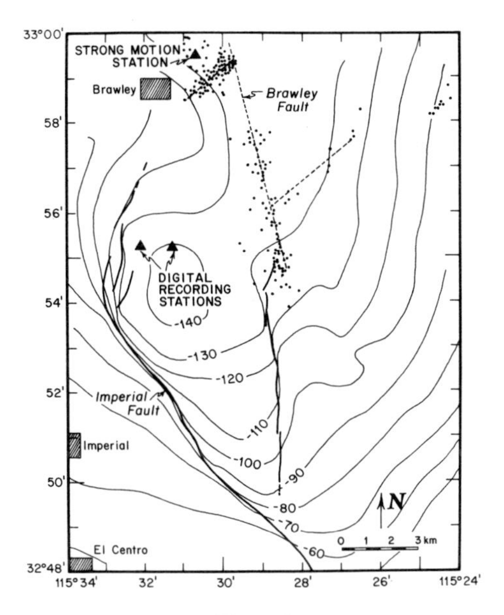

Figure 1

Brawley area map after JOHNSON and HADLEY (1976) showing location of instruments (triangles), swarm events (dots), observed surface faulting (heavy lines), faults inferred from seismicity (dashed lines), and topographic contours in feet (light lines).

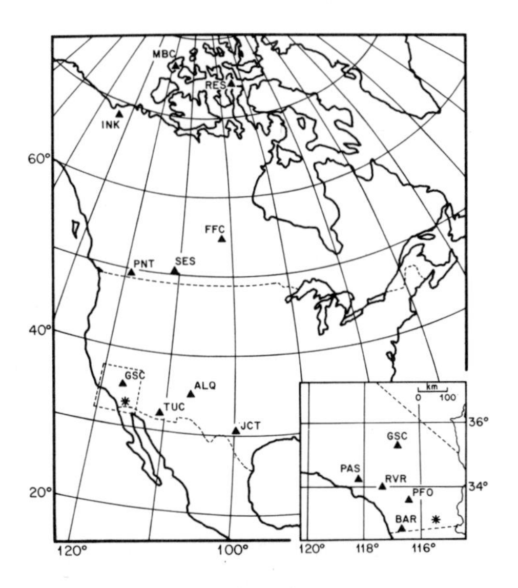

Figure 2

United States and Canadian stations used to estimate M_s and m_b.

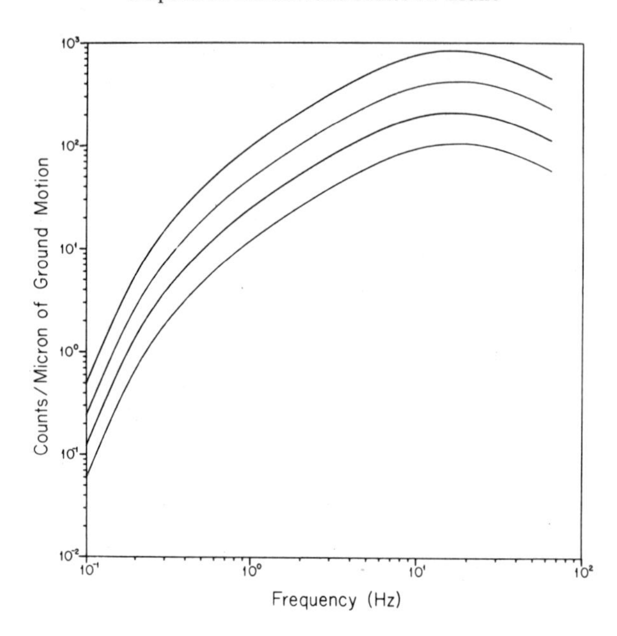

Figure 3
Combined seismometer ($T = 5$ sec) and digital recorder response for four amplifier settings.

events were recorded, with seismometer orientations of both NS and EW used to avoid possible radiation pattern nodes. Because of high gain setting, the amplifiers saturated for events with local magnitude greater than about 2.5. Thirty-seven of the larger events that did not saturate are included in this study.

The digital system operates in a triggered mode, recording for 16 seconds whenever the input level reaches 7 times ground noise. Two seconds of data prior to the trigger time are also recorded by means of a shift register delay line, thereby capturing the beginning of the triggering event. Figure 4 shows a sample of the digital data. The low P wave amplitudes are due to the near vertical incidence of rays from the hypocenter and the fact that a horizontal seismometer was used. In comparison, S phases are much larger in amplitude and of lower frequency content. A prominent shear wave free surface reflection is also visible on many of the records (Fig. 4). The impulsive nature of the S-wave indicates that most of the energy travels the direct path from hypocenter to station and that scattering has not greatly altered the signal character.

Strong motion accelerometer

Near source data for events larger than local magnitude 2.5 comes from a standard 3-component model SMA-1 strong motion instrument which records on 70 mm film. The accelerograph was installed by Dennis Johnson and Thomas Hanks (JOHNSON

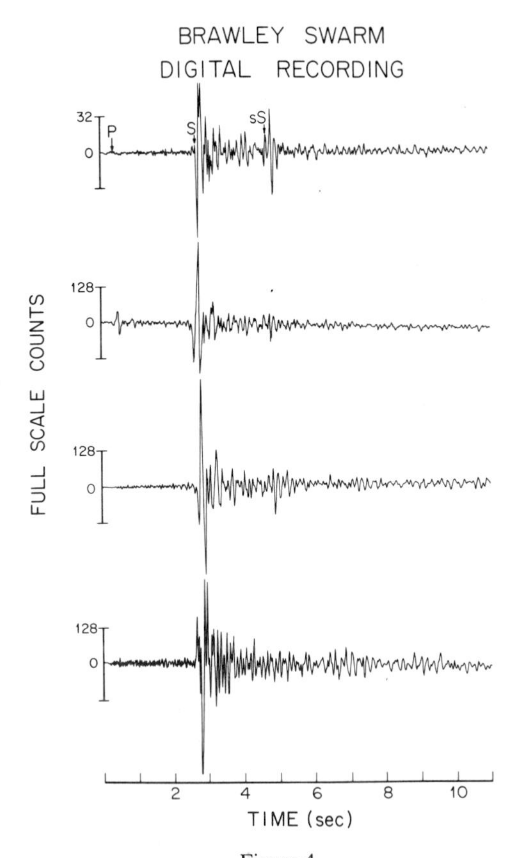

Figure 4
Examples of digital recording for four Brawley swarm events. Small P-wave amplitudes and generally simple wave forms are common.

and HANKS, 1976), and oriented such that upward trace deflection on the record corresponds to ground motion in the direction N45W (longitudinal component), up (vertical component), and S45W (transverse component). This instrument triggered by the vertical component, hence one to two tenths of a second of the initial *P*-wave motion is lost, but the entire *S*-wave is recorded. Calibration runs were performed on all three components to obtain the natural frequency ω_0, fraction of critical damping λ_0, and sensitivity (cm/G). Averages for the three-components are 25.0 Hz, 0.61, and 1.80 cm/G, respectively, with very little variation between components. For frequencies lower than the fundamental frequency, ω_0, the simple theory of a single-degree-of-freedom oscillator gives a good approximation to the instrument response (TRIFUNAC *et al.* 1973). Frequencies of interest to this study lie well below the natural frequency of 25 Hz. Therefore, the relative motion, x, of the transducer mass can be described by the viscosity damped harmonic oscillator equation:

$$\ddot{x} + 2\omega_0\lambda_0\dot{x} + \omega_0^2 x = -A \qquad (1)$$

where A is the absolute ground acceleration. This equation has been used to correct for instrument response.

The accelerometer was located at 32° 59.49′N, 115° 30.80′W and operated from 0508Z to 1918Z on January 25. During this period, 24 earthquakes were recorded with amplitudes large enough for detailed analysis (ranging in local magnitude from 2.5 to 4.3).

Analysis of the data

Body-wave spectra

Amplitude spectra were calculated for the 61 records described above and corrected for attenuation. The spectra were then interpreted in terms of a flat low frequency asymptote and a high frequency fall-off of ω^{-2}. The far field spectral parameters were then related to the source parameters seismic moment (M_0), source dimension (r), and stress drop ($\Delta\sigma$) using the relationships given by BRUNE (1970, 1971)

$$M_0 = 4\pi\mu\beta\Omega_0 R/0.85 \tag{2}$$

$$r = 2.34\beta/2\pi f_c \tag{3}$$

$$\Delta\sigma = (7/16)(M_0/r^3) \tag{4}$$

Here f_c is the spectral corner frequency, μ the rigidity, β the shear wave velocity, and Ω_0 the zero frequency spectral amplitude. The factor of 0.85 in the expression for moment is the produce of several correction terms. It allows for equal amplitude on the orthogonal horizontal component, $\sqrt{2}$, corrects for free-surface SH amplification, 2.0, and for the rms average of the shear displacement radiation pattern, 0.6 (THATCHER and HANKS 1973). Values of moment were also estimated for seven of the larger events using 10 sec surface waves recorded at US WWSSN stations and the Pinyon Flat Observatory 120 km NW of the swarm. Hypocentral locations were taken from JOHNSON and HADLEY (1976) (estimated accurately to ± 1 km). Rigidity, μ, was taken as 2.0×10^{11} dyne/cm, and shear wave velocity, β, as 2.6 km/sec.

Attenuation corrections

The attenuation factor Q^{-1} is an important factor in the correction of seismic spectra, and therefore deserves some discussion. Aside from isolated volcanic intrusions, the Imperial Valley is a 6 km thick accumulation of Cenozoic sediments (DIBBLEE, 1954). Since most of the events for which amplitude spectra were computed lie no deeper than 6 km, the ray paths to the Imperial Valley stations are predominantly within the sedimentary column. Unfortunately, measured values of Q for travelling shear waves in sedimentary rocks are very limited. Furthermore, Q of the unconsolidated surface layers is certainly much less than that of the highly compacted and metamorphosed layers just above the sediment-basement interface.

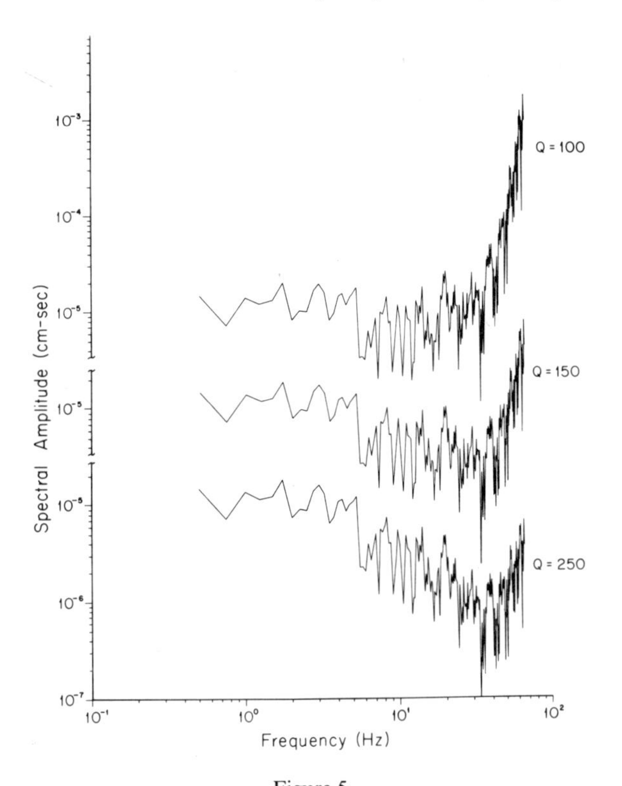

Figure 5
Typical shear wave spectrum corrected for three different values of $Q = 100, 150, 250$, A Q correction of 150 has been applied in this study.

In an effort to compensate for attenuation, an estimated average Q over the ray path was chosen as follows. Initial choices for Q of 100, 150, and 250 were used to correct the spectra of several typical events. Figure 5 shows the results for one earthquake recorded on the digital system (hypocentral distance about 9 km). The spectrum corrected for a Q of 100 is nearly flat out to a frequency of about 30 Hz (where the signal is dominated by noise), whereas values of 150 and 250 for Q produce the expected high frequency fall-off of about ω^{-2}. We have somewhat arbitrarily chosen a Q of 150. Possible errors due to the choice of Q are discussed later.

Digital recorder

The digital system has a sample rate of 128 samples/sec giving a Nyquist frequency of 64 Hz, well above the 8 Hz average corner frequency for the events studied. Spectral analysis was also done on several noise samples (taken from digitizer autoruns) and compared with the earthquake spectra. Domination of the spectrum by noise begins at 20 to 25 Hz, also above the observed corner frequencies. Typically the first two seconds of the S phase were Fourier transformed. This interval is long enough to encompass the major S wave motion without including significant scattered energy or any other distinct phases. Varying sample lengths from 1 to 4 seconds were tried without changing the general shape of the spectrum. The spectra are not considered reliable for periods greater than 1 sec.

Strong motion instrument

Shear wave phases from the strong motion records were digitized and corrected for instrument response using Equation (1), following the procedure in Routine Computer Processing of Strong-Motion Accelerograms by M. D. TRIFUNAC and V. LEE (1973). Enlargements of the 70 mm negatives were first made, and then digitized at evenly spaced intervals of 0.005 inches along the trace of the accelerogram. The trend and mean were removed by a least squares regression, and the curve resampled at evenly spaced intervals of 0.01 sec. To check for possible aliasing effects, selective events were resampled at 0.005 seconds, resulting in virtually identical spectra. The length of record transformed ranged from 2 to 3 sec, depending on the size of the event.

Displacement amplitude spectra were computed by two different methods and compared. The first method follows the analysis in Routine Computer Processing of

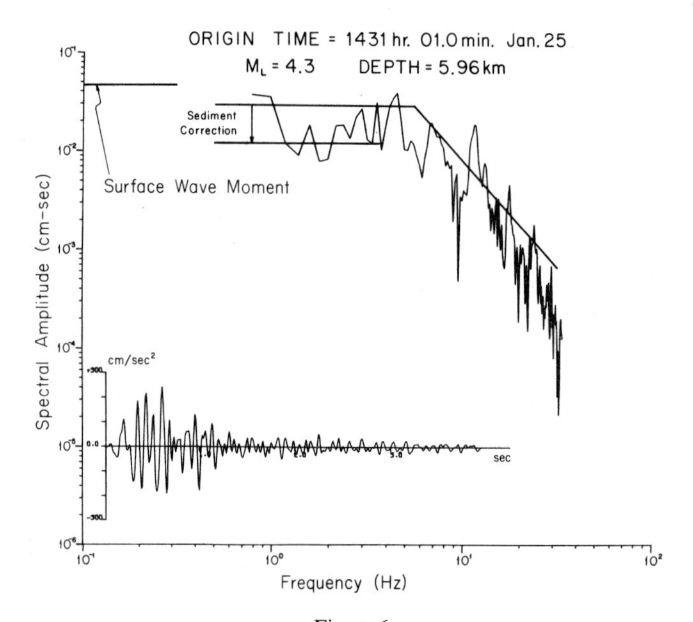

Figure 6
Shear wave displacement spectrum for a $M_1 = 4.3$ event digitized from the Brawley airport strong motion record. S-wave accelerogram (S45W component) is also shown with time scale in seconds. Low and high frequency asymptotes (heavy lines) yield a stress drop of 636 bars. Sediment amplification correction of 2.5X and spectral level determined from synthetic surface waves are also shown.

Strong-Motion Accelerograms. The accelerogram is twice integrated to obtain the displacement record which is then transformed giving the displacement amplitude spectrum. The second method transforms the accelerogram directly, avoiding the integration process. Division by ω^2 then gives the displacement amplitude spectrum. Final spectra obtained by both methods are nearly identical and differ only slightly at the low frequency end. Figure 6 shows the digitized accelerogram (*S* phase only) for one of the larger Brawley events and its displacement amplitude spectrum obtained by the second method.

Correction for local structure

The assumption of a flat-lying layered structure is a reasonable one for the alluvial environment of the Imperial Valley provided the recorders are near the epicenters. The Westmorland profile (BIEHLER *et al.* 1964) offers the best data to date on the seismic velocities in the area of the Brawley swarm. Results from this profile model the area as five layers over a half-space, with compressional and shear wave velocities increasing by nearly a factor of 3 from the surface layer to just above basement. Body waves incident on the near surface velocity layers encounter a significant impedance mismatch and a resulting amplification. To estimate this amplification (necessary in the determination of moment), an averaged Biehler–Westmoreland structure of two layers over a half space was used. This problem can be solved analytically for plane waves. Figure 7 is a plot of the amplification ratio for plane SH waves (amplitude

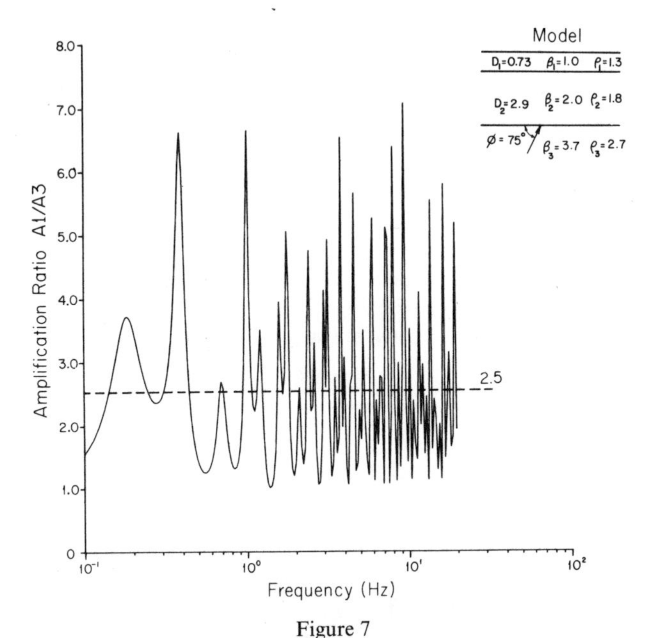

Figure 7
Amplitude of plane SH wave in top layer divided by amplitude in underlying halfspace (A_1/A_3) plotted as a function of frequency for the model shown.

in the topmost layer divided by the amplitude in the half space) over the frequency range of interest to this study. The specific structure used is shown in the upper right hand corner of Fig. 7. The angle of incidence used, $\phi = 75$ degrees, is an average of the angles of incidence for the digital and strong motion stations. A mean SH amplification factor of 2.5 is then obtained by taking a simple average over all frequency values. (If energy flux is considered, by taking the square of the amplitudes, the average amplification is 2.8.) All estimates of moment based on the apparently flat, low frequency part of the spectrum of Imperial Valley records have been adjusted down by a factor of 2.5.

Source parameters from local stations

Source parameters estimated from the digital station and strong motion instrument are summarized in Fig. 8. The cutoff in digitally recorded data at a moment of 6×10^{19} dyne-cm is artificial, being due to saturation of the instrument. Corner frequencies range from 2.3 Hz to 14 Hz with a wide range in stress drops, from 1 bar to 636 bars. The apparent gap in corner frequencies between 3 Hz and 5.7 Hz for the strong motion data is not believed significant and attributed to limited sampling.

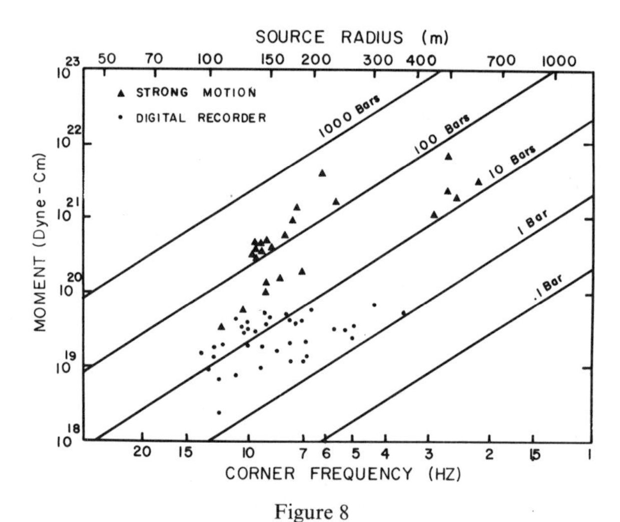

Figure 8
Summary of source parameters based on strong motion records and digital recordings.

By correcting the body wave spectra for a range in Q values of 100 to 250, it is estimated that the chosen corner frequencies may be off by a factor of 1.5, due to our poor knowledge of Q. This leads to an error of 3 to 4 in stress drop. Thus, the high stress drop value of 636 bars may be as low as about 200 bars or as high as several kilobars. However, since the same criterion was applied in choosing each individual spectral corner, the relative values of stress drop are considered to be much more accurate.

Variation of stress drop with depth

WYSS and BRUNE (1971) found an increase in apparent stress with depth for the Borrego Mountain region, and attributed it to greater hydrostatic pressure causing greater frictional resistance at depth. Figure 9 is a plot of stress drop versus depth for all Brawley swarm events of local magnitude between 3 and 4 included in this study as well as the $M_L = 4.3$ high stress drop event. A limited magnitude range was

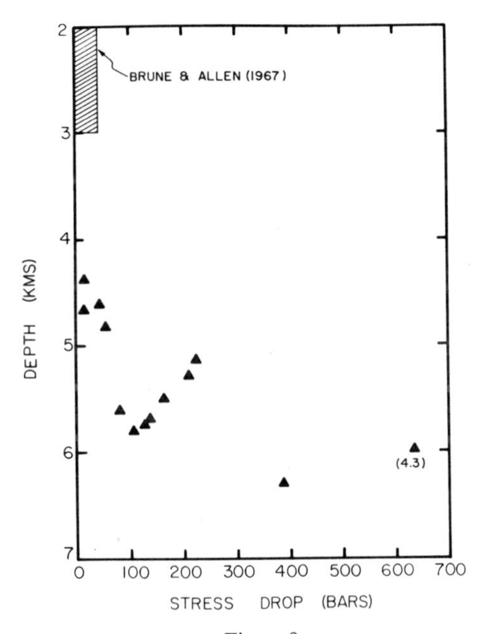

Figure 9

Stress drop as a function of depth for all magnitude 3 events included in this study plus the magnitude 4.3 event whose spectra is plotted in Fig. 6.

considered for this plot in order to minimize any variations in stress drop with magnitude. There is a clear increase in stress drop with depth. This result is consistent with the low stress drop of 1.1 bars reported by BRUNE and ALLEN (1967) for a small ($M_L = 3.6$) earthquake which occurred on the Imperial Fault a few kilometers west of the Brawley swarm. Seismic evidence indicated that that earthquake occurred very near the surface, less than 3 km deep. It has been plotted in Fig. 9 as a cross-hatched region, since no accurate hypocenter is available. It should be noted that plotting the Brawley events as point sources may be misleading. They could have ruptured upward from the hypocenter.

Surface wave moments

Surface wave excitation was studied in detail for two events with origin times 0701 hr 49.4 min and 1431 hr 01.0 min, both on January 25. These events (hereafter referred to as event A and event B, respectively) were chosen for analysis since both were recorded by the strong motion accelerograph at the Brawley airport and also produced readable traces on the Press–Ewing 15 second seismometers at Pasadena and Goldstone. Synthetic surface waves were generated using a synthetic seismogram program provided by David Harkrider (HARKRIDER, 1964, 1970).

Model parameters

Choice of the proper earth model and dissipation characteristics are important considerations for Brawley swarm events, since the sources lie in one structure, namely the Imperial Valley, and the stations are at sites outside the valley on a more rigid structure. Synthetic surface waves were generated for two different earth models and three different dissipation models. Elastic parameters for the two earth models (labelled KHC2 and KHC2-IV) are given in Table 1. Model KHC2 is an average

Table 1

Model KHC2-IV				Model KHC2			
H(km)	α(km/sec)	β(km/sec)	ρ(gm/cc)	H(km)	α(km/sec)	β(km/sec)	ρ(gm/cc)
0.18	1.70	0.98	1.28	1.0	2.50	1.44	2.50
0.55	1.96	1.13	1.36	3.0	5.50	3.14	2.60
0.98	2.71	1.57	1.59	23.4	6.30	3.63	2.70
1.19	3.76	2.17	1.91	5.0	6.80	3.92	2.90
2.68	4.69	2.71	2.19	8.6	8.37	4.73	3.50
21.82	6.30	3.63	2.70	20.0	8.37	4.71	3.52
5.0	6.80	3.92	2.90	20.0	8.08	4.62	3.47
8.6	8.37	4.73	3.50	20.0	7.93	4.35	3.43
20.0	8.37	4.71	3.52	20.0	7.78	4.18	3.40
20.0	8.08	4.62	3.47	25.0	7.75	4.22	3.38
20.0	7.93	4.35	3.43	25.0	7.78	4.30	3.36
20.0	7.78	4.18	3.40	25.0	7.97	4.44	3.35
25.0	7.75	4.22	3.38	25.0	8.19	4.56	3.34
25.0	7.78	4.30	3.36	25.0	8.39	4.61	3.34
25.0	7.97	4.44	3.35	25.0	8.52	4.58	3.37
25.0	8.19	4.56	3.34	25.0	8.55	4.57	3.41
25.0	8.39	4.61	3.34				
25.0	8.52	4.58	3.37				
25.0	8.55	4.57	3.41				

continental structure of Kanamori (personal communication), and is assumed to represent the structure outside the Imperial Valley. Model KHC2-IV replaces the top 5.58 km of KHC2 with the Imperial Valley structure of BIEHLER *et al.* (1964). Phase and group velocity curves for these two models are shown in Fig. 10. The common deep structure of KHC2 and KHC2-IV results in the convergence of the phase and group velocity curves for periods greater than 20 sec. Three different models

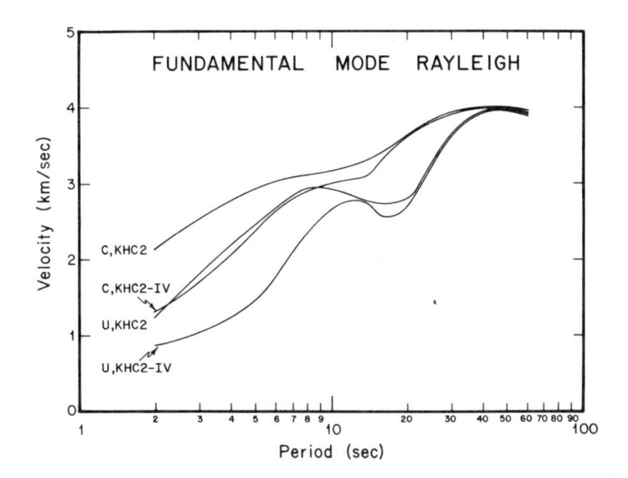

Figure 10
Fundamental mode Rayleigh group (U) and phase velocity (C) curves for the two earth models, KHC2 and KHC2-IV, of Table 1.

of attenuation as a function of period were considered. They are: (1) the continental attenuation model of ALEWINE (1974), which assumes a period dependent Q of several hundred and is based on accumulated observational data; (2) a constant Q of 100, with an amplitude attenuation coefficient ($\gamma = \pi f/QU$) of 2.5×10^{-4} km^{-1} at 10 sec; and (3) a constant Q of 30. Both attenuation models (2) and (3) set Q to 130 for periods greater than 50 sec, but this has no effect on the interpretation made here.

Fault parameters of strike, dip, and slip direction as well as source depths for events A and B were taken from JOHNSON and HADLEY (1976). Event A has a local magnitude of 3.7, a source depth of 5.48 km, and is primarily a thrust event with a dip of 63°NW. Event B has a local magnitude of 4.3, a source depth of 5.96 km, and represents primarily normal faulting at a dip of 72°SE. The source function used is a finite ramp with 1 sec rise time, but is not critical in the calculation of excitation of surface waves – a step function would have been equivalent.

Discussion of surface wave moments

Figure 11 summarizes the moment estimates for events A and B for the different structures and Q models considered. The estimates are based on the predominant pulse on the Press–Ewing records ($T \sim 10$ sec). Note that the results for the Alewine dissipation model are plotted at a Q of 300 for convenience. Actually, a period dependent Q is used in this model, but the average Q is near 300. Models KHC2 and KHC2-IV affect the estimated moments for events A and B quite differently. In going from model KHC2 to model KHC2-IV, the calculated moment for event B increases by three fold, whereas the moment estimate for event A increases by only 50%. This is due to the fact that surface wave excitation is a function of both the

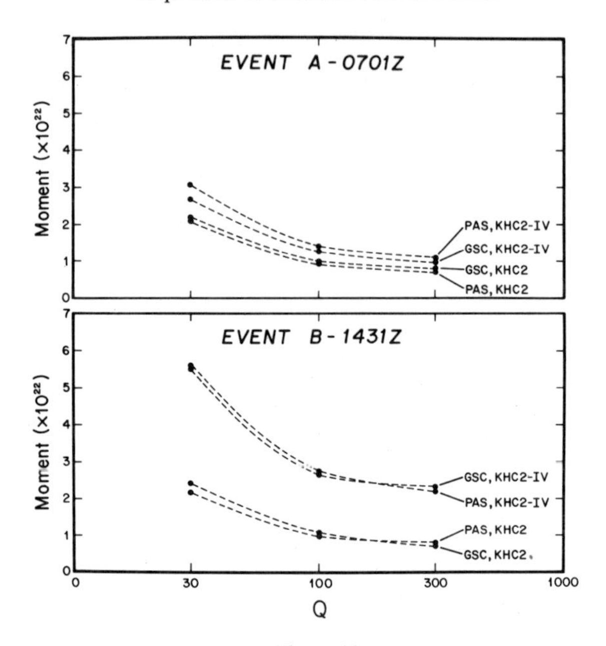

Figure 11
Estimates of moment versus Q for events A and B. Calculations are based on surface wave synthetics for stations PAS and GSC for the two earth models KHC2 and KHC2-IV.

source depth within the layered structure and the fault mechanism. If model KHC2 is choosen as the 'correct' earth structure, nearly the same moment is obtained for events A and B regardless of Q. This result is in conflict with independent observational data. The M_L (PAS) for event B is 4.3, in contrast with M_L 3.7 for event A, indicating a significantly higher moment for event B. On the other hand, if model KHC2-IV is used, the moment obtained for event B is approximately a factor of two greater than that obtained for event A in better agreement with the local magnitudes. This suggests that model KHC2-IV gives a more accurate prediction of surface wave excitation for Imperial Valley earthquakes, i.e., it suggests that the source structure yields a more accurate prediction of surface wave excitation than the receiver structure, even though the majority of the path lies in the receiver structure. This probably, however, does not apply to the shorter period energy ($T < \sim 5$ sec) which may not be efficiently transmitted across the boundary between the two structures.

For the attenuation model, the Alewine dissipation model is preferred. Synthetics for the lower Q values of 100 and 30 were generated in an effort to fit the low amplitude higher frequency content of the PAS and GSC records using model KHC2-IV. Visually, the best fit is obtained with $Q = 30$. However, this approach does not take into consideration the scattering of periods less than 5 sec at the boundary of the Imperial Valley. We believe the most reasonable procedure is to use the longer period 10 second energy and the Imperial Valley structure in conjunction with a higher, more reasonable Q.

We can estimate the effect of the boundary of the Imperial Valley on longer wavelength surface waves by modeling the edge of the Imperial Valley as a vertical boundary between two quarter spaces. Transmission and reflection coefficients can then easily be estimated for this model. The results will be a long-wavelength-limit since structure within the quarter spaces, which would affect wavelengths on the order of the thickness of the layers, is omitted. See Fig. 12 for the geometry and elastic parameters used in the following discussion. The model parameters selected represent

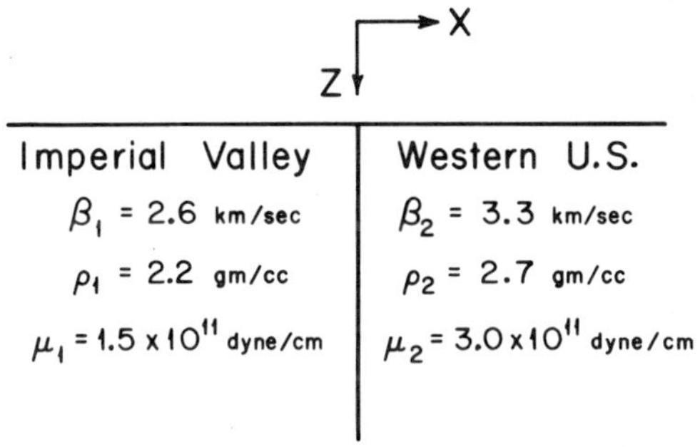

X

Z

Imperial Valley	Western U.S.
β_1 = 2.6 km/sec	β_2 = 3.3 km/sec
ρ_1 = 2.2 gm/cc	ρ_2 = 2.7 gm/cc
$\mu_1 = 1.5 \times 10^{11}$ dyne/cm	$\mu_2 = 3.0 \times 10^{11}$ dyne/cm

Figure 12
Model of boundary between Imperial Valley trough and the Western United States used to estimate Rayleigh wave transmission and reflection coefficients.

an interface between an averaged Imperial Valley structure and an averaged western United States structure. We require that the following conditions hold at the interface

$$\begin{aligned} &\int_0^\infty (\sigma_{xz}^1 - \sigma_{xz}^2)\cdot U_z^1 \, dz + \int_0^\infty (\sigma_{xx}^1 - \sigma_{xx}^2)\cdot U_x^1 \, dz = 0 \\ &\int_0^\infty (U_z^1 - U_z^2)\cdot \sigma_{xz}^1 \, dz + \int_0^\infty (U_x^1 - U_x^2)\cdot \sigma_{xx}^1 \, dz = 0. \end{aligned} \tag{5}$$

Here U_x and U_z are the displacements in the x and z directions, respectively, and the superscripts refer to the different media. Equation (5) simply requires continuity of energy density across the interface and yields results similar to those of McGarr and Alsop (1967). Applying Equation (5) to the model in Fig. 12, we obtain a reflection coefficient of 0.106 and a transmission coefficient of 0.917 for fundamental mode Rayleigh waves propagating out of the Imperial Valley. The transmission coefficient would be more closely equal to unity for a gradational contact. Therefore, there is less than a 10% decrease in the long wavelength amplitudes across the transition (long wavelengths here refer to wavelengths which are large compared to the thickness of the valley's low rigidity surface layers). The moments in this study are based on ten second surface waves which have a wavelength of about 25 km, and thus can be considered long by the above definition. Thus, we conclude that the boundary of the Imperial Valley has little effect on our estimates of moment, even though it may severely affect the shorter period surface wave energy ($T < 5$ sec).

Comparison of body wave and surface wave estimates of moment

Table 2 compares the moment estimates for events A and B obtained by three different methods: (1) from the synthetic surface waves as discussed above; (2) from the empirical relationship of WYSS and BRUNE (1968), log $M_0 = M_s + 19.2$ ($3 < M_L < 6$); and (3) from the apparently flat part of the locally recorded S-wave spectrum. Moments based on the surface wave synthetics are 4 to 6 times larger than those based on the local S-wave spectral estimates. The S-wave moments given in Table 2 have been corrected for the sediment amplification factor discussed earlier. Even if this correction were not made, the moments based on the synthetic surface waves would still be a factor of 1.5 to 2.5 times larger.

Table 2

Event	M_L	Station	Moment From Synthetics	Moment From M_s	Moment From S-wave Spectrum
A(0701Z Jan 25)	3.7	GSC	9.8×10^{21}	8.3×10^{21}	1.7×10^{21}
		PAS	1.1×10^{22}	6.0×10^{21}	
B(1431Z Jan 25)	4.3	GSC	2.3×10^{22}	1.5×10^{22}	5.0×10^{21}
		PAS	2.2×10^{22}	2.6×10^{22}	

The relationship log $M_0 = M_s + 19.2$ was used to obtain additional rough estimates of surface wave moment for five smaller swarm events for which no fault plane solution was available. Surface wave magnitudes were calculated using the formula of MARSHALL and BASHAM (1972). This formulation enables calculation of M_s at distances under 20 degrees and at periods other than 20 sec, in contrast to the Gutenburg definition (RICHTER 1958). Moments based on the above moment-magnitude relationship follow the same trend as those determined from the synthetic surface waves, being several times larger than the S-wave moments in all cases except one.

The above observations imply increasing spectral amplitudes between 1 and 10 sec and suggest the existence of two corner frequencies. The higher frequency corner is well resolved by the strong motion and digital data. However, the inferred corner frequency is at periods greater than 2 sec, where the spectral character is not resolvable from the available data.

Several mechanisms may be suggested to explain the above spectral shape. After-slip, or fore-slip with a relatively long time constant, as postulated by KANAMORI (1973) can increase the amplitude of the longer period energy. Similarly, after an initial high stress drop over a small source region, Brawley swarm earthquakes may continue to rupture with a characteristic time of 10 sec. The Imperial Valley's low rigidity surface layers would require less energy to rupture, and thus may encourage upward rupture. A partial stress drop model (BRUNE 1970) is another possible mechanism. In this model asperities or some other aspects of the fault prevent the slip from going to the equilibrium limit for uniform stress drop over the entire fault surface. The character of Brawley earthquake spectra may also be interpreted in terms of the long thin rectangular fault model of HASKELL (1964) and SAVAGE (1972). Here the two corner frequencies are related to the long and short dimensions of the fault with an intermediate spectral slope of ω^{-1}. Unfortunately, the present data cannot distinguish between the several models mentioned above. A more dense array of stations is needed to observe energy focussing, and other effects on the radiation pattern.

Seismic energy and apparent stress

The apparent stress $\eta\bar{\sigma}$ as defined by WYSS (1970) is given by

$$\eta\bar{\sigma} = \mu E_s/M_0. \tag{6}$$

Here E_s is the radiated seismic energy, $\bar{\sigma} = \frac{1}{2}(\sigma_1 + \sigma_2)$ the average shear stress, and η the seismic efficiency. To calculate the apparent stress, it is first necessary to determine the seismic energy. Several energy-local magnitude relationships have been proposed over the years, and it is one of the most uncertain relationships in seismology. Two formulations are considered in this study: the relation given by RICHTER (1958) as a correction to GUTENBURG and RICHTER (1956)

$$\log E_s = 9.9 + 1.9\, M_L - 0.024\, M_L^2 \tag{7}$$

and the relation of THATCHER and HANKS (1973)

$$\log E_s = 2.0\, M_L + 8.1 \tag{8}$$

For magnitudes greater than 6.0, both expressions yield approximately the same results. However, for smaller magnitudes, the curves diverge. The GUTENBERG–RICHTER (1956) relation gives the higher energy for a given magnitude. The two expressions above are plotted for comparison in Fig. 13, along with the energy-local magnitude data from the Brawley swarm. Local magnitudes in Fig. 13 are those of CARL JOHNSON (personal communication, 1976), as determined from Pasadena Wood–Anderson records.

Seismic energies for the Brawley events were determined in the same way as in the THATCHER and HANKS (1973) study, as follows. The Thatcher and Hanks relation was obtained by integrating spectra of many southern California earthquakes and was offered as a better fit to southern California data than the Gutenberg–Richter relation. The S-wave radiated energy, E_s, can be expressed in terms of the spectral amplitude, $\Omega(\omega)$, as

$$E_s = (\tfrac{1}{2}\pi)(I_s R^2\, \mu/\beta) \int_{-\infty}^{\infty} |\Omega(\omega)\cdot\omega|^2\, d\omega \tag{9}$$

(HANKS and THATCHER 1972) where ω is the angular frequency and I_s is an S-wave radiation pattern term. If the spectrum is assumed to have an ω^{-2} high frequency fall-off and a flat low frequency trend then the above integral reduces to

$$E_s = [1/(0.85)^2](128\pi^3/15)\mu R^2\Omega_0^2 f_c^3/\beta. \quad (10)$$

Here we have used the expression for moment given in (2). This expression for E_s is a minimum estimate since it does not include the P-wave energy. From Fig. 13 we see that the Gutenberg–Richter energy-magnitude relation gives the better fit to the larger ($M_L > 3.0$) Brawley earthquakes. Energies for the smaller events ($M_L < 3.0$)

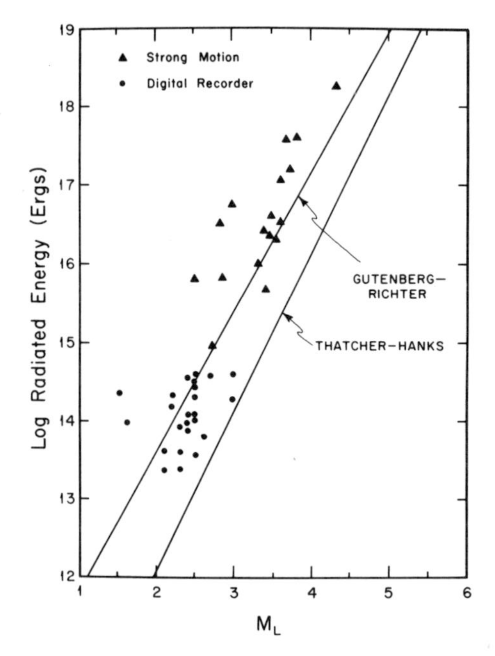

Figure 13
Log of the radiated seismic energy versus local magnitude. Radiated energies have been obtained by spectrum integration where ω^{-2} high-frequency spectrum fall off is assumed. Two commonly used energy versus magnitude relationships are plotted for comparison.

are still underestimated by the Thatcher and Hanks relation. Underestimation of energies by the Thatcher and Hanks relation implies that the corner frequencies found in this study are consistently higher than the average corner frequencies for their study for a given magnitude. Due to the cubic dependence on corner frequency, a small underestimate in corner frequency results in a large underestimation of seismic energy. Errors in the choice of Q are believed to be less critical in the present study because of the short ray paths.

Apparent stress is plotted against stress drop (Fig. 14) for Brawley earthquakes using three different definitions of seismic energy: Gutenberg–Richter, Thatcher–

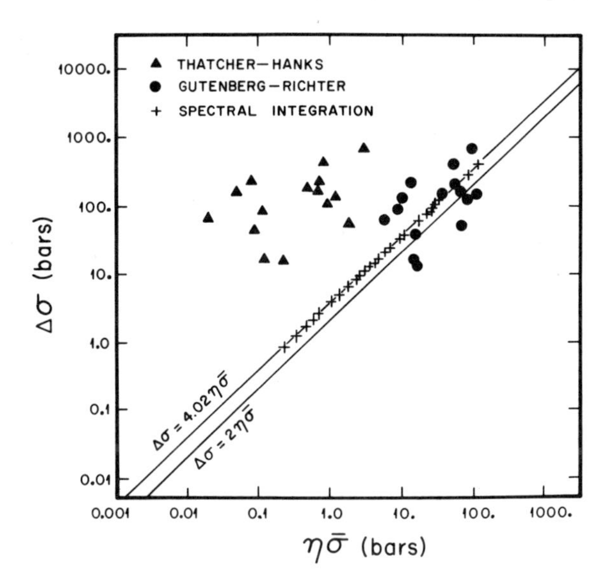

Figure 14
Plot of calculated values of stress drop ($\Delta\sigma$) versus apparent stress ($\eta\bar{\sigma}$) for three different methods of determining seismic energy. The curve $\Delta\sigma = 4.02\eta\bar{\sigma}$ assumes ω^{-2} high-frequency spectrum fall off and P-wave energy 1/3 the S-wave energy.

Hanks, and spectral integration. Dramatically different results are obtained for each of the three methods, illustrating the dangers of arbitrarily applying a particular energy-magnitude relationship to a given data set. The Thatcher-Hanks expression gives apparent stresses 2 to 3 orders of magnitude less than the stress drops, whereas for the Gutenberg-Richter formulation, they are only one order of magnitude less.

If the spectral integration method is used, (10) above, all the data for S-wave energy plot along the curve

$$\Delta\sigma = 5.37\,\eta\bar{\sigma}. \tag{11}$$

This is a direct result of Equations (2) through (4), (6), and (10). As stated above, (10) is a minimum estimate of seismic energy, since it does not account for the P-wave energy. Examination of a number of 3-component strong motion records suggest that the P-wave energy is typically $\frac{1}{3}$ the S-wave energy. The total energy is then approximately $1\frac{1}{3}$ times that given by (10), and the ratio in (11) between stress drop and apparent stress becomes 4.02, shown by the upper line in Fig. 14.

Regional variations in apparent stress in the southwestern United States

WYSS and BRUNE (1971) used the concept of apparent stress to study regional variations in spectra of earthquakes in the Southwestern United States. In that study apparent stress was estimated by determining moment from surface waves and energy from local magnitude M_L. The energy-magnitude relationship assumed was that used by ALLEN *et al.* (1965) ($\log E = 11.8 + 1.5\, M_L$) although that formula was not strictly applicable for use with local magnitudes. Thus, the results for apparent stress were useful for comparing relative spectral differences from region to region, but not for accurate determination of the true apparent stresses. When more recent energy vs. magnitude relations are used the apparent stresses for the earthquakes studied by Wyss and Brune are reduced considerably (by a factor of about 3 at $M_L = 5$, 5 at $M_L = 4$, and 8 at $M_L = 3$ for the THATCHER and HANKS 1973 relationship).

The revised estimates of apparent stress are less than 100 bars for all events of magnitude greater than 4 (events probably have to be this large to assure adequate recording of surface waves at one or more stations). In a few cases events with relatively high apparent stress in the Wyss and Brune study were found to have considerably lower stresses when additional long period data were examined, obviously a result of variations in radiation pattern. From the above considerations we conclude that there are no reliable estimates of true apparent stress exceeding about 100 bars. This is consistent with the observation that stress drops are generally less than about two-hundred bars.

Seismic discrimination

Body wave magnitudes were calculated for the largest of the Brawley swarm events using Canadian stations at distances generally greater than 20° (BASHAM, personal communication, 1976). Results are plotted in Fig. 15 along with other

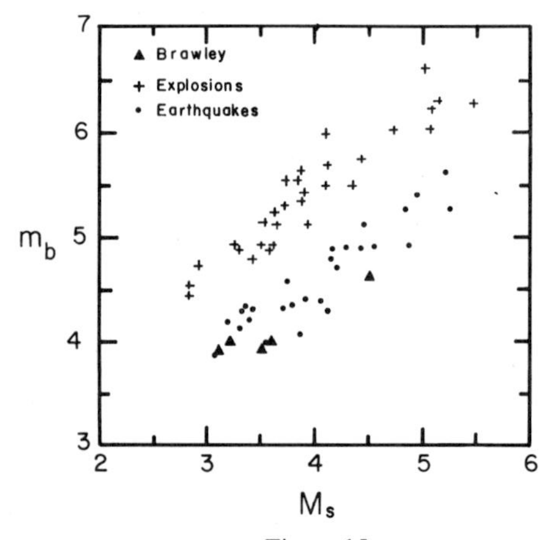

Figure 15

$M_s : m_b$ plot for the largest earthquakes of the Brawley swarm. Swarm events discriminate down to at least as low as $m_b = 4.0$. Additional values for other earthquakes and explosions after MARSHALL and BASHAM (1972) are shown for comparison.

earthquakes and explosions from MARSHALL and BASHAM (1972) for comparison. M_s values are based on US WWSSN readings (applying the method of MARSHALL and BASHAM 1972) and represent averages of at least two stations in all cases. m_b values represent averages of up to six determinations using the stations in Fig. 2. Events investigated discriminate down to an m_b value of 4.0. Accurate discrimination checks for smaller earthquakes is not possible due to small trace amplitudes.

Summary and discussion

The January 1975 Brawley earthquake swarm is characterized by a wide range in estimated stress drops, from 1 to 636 bars, over a magnitude range from 1.0 to 4.3. A trend of higher stress drop with greater depth has been observed, although there is considerable scatter in the data. Increasing stress drop with depth is consistent with increasing compaction of the sediments. Rocks near the bottom of the sedimentary column at a depth of 6 km are most likely highly metamorphosed and near granite in rigidity and failure characteristics, thus high stress drops are not unreasonable, even though temperature gradients are high and variable in the Imperial Valley. The high temperature gradients may explain why earthquakes do not occur at depths greater than about 8 km (ELDERS *et al.* 1972).

It is estimated that the low frequency spectral amplitude of signals recorded at stations on the Imperial Valley sedimentary column should be corrected by a factor of about 2.5 to estimate the source moment. With this correction, estimates of moment from 10 sec surface waves from distant stations range up to a factor of six times greater than the moment estimated from the locally recorded body wave spectrum at a period of 1 sec. This can be interpreted in terms of after-slip or fore-slip with a relatively long time constant, an expanding rupture, partial stress drop, or the Haskell long-thin fault model.

The explanation of the discrepancy between moments estimated from distant surface waves and moments estimated from near-source *S*-waves is an important problem in seismology because of its bearing on earthquake mechanism. Although we have tried to account for the structural effects of the Imperial Valley, there is considerable uncertainty in many of the parameters, such as Q. It is therefore possible, but we feel unlikely, that the discrepancy in surface wave and body wave estimates of moment may be due to the incomplete knowledge of the crustal structure and its effects. The sediment amplification factor used in this paper obviously contributes to the difference in the surface wave and body wave estimates of moment, but it appears to be a necessary consideration. A much more densely instrumented study is needed to reduce the uncertainties and provide a well substantiated physical model for the spectral shape of earthquakes.

Acknowledgements

The authors benefited greatly from numerous discussions with Gerald A. Frazier. Jerry King and William Prothero helped in the field work. The instruction in the operation and use of the digital field recorders by William Prothero is gratefully acknowledged. We wish to thank Thomas Hanks for making the strong motion data available, Carl Johnson and David Hadley for supplying unpublished data, and David Harkrider for allowing us to use his synthetic surface wave programs. This research is partially supported by NSF Grant #DES74-03188 and NASA Grant #NGR 05-009-246, and is a contribution of Scripps Institution of Oceanography of the University of California at San Diego.

References

ALEWINE, R. W. (1974), *Application of linear inversion theory toward the estimation of seismic source parameters*, Ph.D. Thesis, Calif. Inst. of Tech.

ALLEN, C. R., ST. AMAND, P., RICHTER, C. F. and NORDQUIST, J. M. (1965), *Relationship between seismicity and geologic structure in the Southern California region*, Bull. Seismo. Soc. Amer. *55*, 753–797.

BIEHLER, S., KOVACH, R. L. and ALLEN, C. R. (1964), *Geophysical framework of northern end of Gulf of California structural province*, AAPG Memoir No. 3, 126–143.

BRUNE, J. N. and ALLEN, C. R. (1967), *A low-stress drop low-magnitude earthquake with surface faulting: The Imperial, California earthquake of March 4, 1966*, Bull. Seismo. Soc. Amer. *57*, 501–514.

BRUNE, J. N. (1970), *Tectonic stress and the spectra of seismic shear waves from earthquakes*, J. Geophys. Res. *75*, 4997–5009.

BRUNE, J. N. (1971), (Correction) *Tectonic stress and the spectra of seismic shear waves from earthquakes*, J. Geophys. Res. *76*, 5002.

DIBBLEE, T. W. (1954), *Geology of the Imperial Valley region, California* in *Geology of Southern California.* Calif. Dept. Nat. Res., Div. Mines Bull. *170*, 21–28.

ELDERS, W. A., REX, R. W., MEIDAV, T., ROBINSON, P. T. and BIEHLER, S. (1972), *Crustal spreading in Southern California*, Science *178*, 15–24.

GUTENBERG, B. and RICHTER, C. F. (1956). *Earthquake magnitude, intensity, energy, and acceleration, 2*, Bull. Seismo. Soc. Amer. *46*, 105–145.

HANKS, T. C. and THATCHER, W. (1972), *A graphical representation of seismic source parameters*, J. Geophys. Res. *77*, 4393–4405.

HARKRIDER, D. G. (1964), *Surface waves in multilayered elastic media. Part I. Rayleigh and Love waves from buried sources in a multilayered elastic halfspace*, Bull. Seismo. Soc. Amer. *54*, 627–679.

HARKRIDER, D. G. (1970), *Surface waves in multilayered elastic media. Part II. Higher mode spectra and spectral ratios from point sources in plane layered earth models*, Bull. Seismo. Soc. Amer. *60*, 1937–1987.

HASKELL, N. (1964), *Total energy and energy spectral density of elastic wave radiation from propagating faults*, Bull. Seismo. Soc. Amer. *54*, 1811–1842.

JOHNSON, C. E. and HADLEY, D. M. (1976), *Tectonic implications of the Brawley earthquake swarm, Imperial Valley, California, January 1975*, Bull. Seismo. Soc. Amer. *66*, 1133–1144.

JOHNSON, D. A. and Hanks, T. C. (1976), *Strong-motion earthquake accelerograms at Brawley, California: January 25, 1975*, Bull. Seismo. Soc. Amer. *66*, 1155–1158.

KANAMORI, H. (1973), *Mode of strain release associated with major earthquakes in Japan*, Ann. Rev. Earth Planet. Sci. *1*, 213–239.

MARSHALL, P. D. and BALSHAM, P. W. (1972), *Discrimination between earthquakes and underground explosions employing an improved M_s scale*, Geophys. J. R. astr. Soc. *28*, 431–458.

McGarr, A. and Alsop, L. E. (1967), *Transmission and reflection of Rayleigh waves at vertical boundaries*, J. Geophys. Res. *72*, 2169–2180.

Prothero, W. A. (1976), *A portable digital seismic recorder with event recording capacity*, Bull. Seismo. Soc. Amer. *66*, 979–985.

Richter, C. F., *Elementary Seismology* (W. H. Freeman, San Francisco, California 1958), p. 343.

Savage, J. C. (1972), *Relation of corner frequency to fault dimension*, J. Geophys. Res. *77*, 3788–3795.

Thatcher, W. and Brune, J. N. (1971). *Seismic study of an oceanic ridge earthquake swarm in the Gulf of California*, Geophys. J. R. astr. Soc. *22*, 473–489.

Thatcher, W. and Hanks, T. C. (1973), *Source parameters of Southern California earthquakes*, J. Geophys. Res. *78*, 8547–8576.

Trifunac, M. D. and Lee, V. (1973), *Routine computer processing of strong-motion accelerograms*, Earthquake Eng. Res. Rept., Calif. Inst. of Tech.

Trifunac, M. D., Udwadia, F. E. and Brady, A. G. (1973), *Analysis of errors in digitized strong-motion accelerograms*, Bull. Seismo. Soc. Amer. *63*, 157–187.

Wyss, M. (1970), *Stress estimates for South American shallow and deep earthquakes*, J. Geophys. Res. *75*, 1529–1544.

Wyss, M. and Brune, J. N. (1968), *Seismic moment, stress, and source dimensions for earthquakes in the California-Nevada region*, J. Geophys. Res. *73*, 4681–4694.

Wyss, M. and Brune, J. N. (1971), *Regional variations of source properties in Southern California estimated from the ratio of short to long-period amplitudes*, Bull. Seismo. Soc. Amer. *61*, 1153–1167.

(Received 14th February 1977)

Pageoph, Vol. 115 (1977), Birkhäuser Verlag, Basel

Anomalous Seismicity prior to Rock Bursts: Implications for Earthquake Prediction

By B. T. BRADY[1])

Abstract – Anomalous seismicity changes (increase followed by a decrease) were recorded prior to three moderate rock bursts in the Star mine, Burke, Idaho. In each case, based upon the anomalous seismicity behavior, miners were evacuated or were prohibited from entering active mine stopes that were located in the immediate vicinity of the seismicity buildup prior to the bursts. Analyses of pre- and post-seismic activity are interpreted in terms of, and shown to be consistent with, the inclusion theory of failure. Implications of these observational results for the problem of rock bursts and earthquake prediction are discussed.

Key words: Seismicity; Earthquake prediction; Stress *in-situ*.

Introduction

Evidence is presented here that seismic precursor effects, such as anomalous seismicity behavior, were observed prior to rock bursts that occurred in a deep (~2.3 km) silver mine in northern Idaho. This behavior appears to be similar to that reported to precede some earthquakes and consists of a dramatic increase of seismicity in the hypocentral region of the impending burst, followed by a distinct decrease prior to the occurrence of the burst [1, 3, 5]. Seismicity anomaly information for these rock bursts and their implications for earthquake prediction are discussed.

Rock bursts at the Star mine

The Star mine, Burke, Idaho, is the deepest lead–zinc mine in the world and is located in the Coeur d'Alene mining district of northern Idaho. The Star mine occupies the westerly portion of the Star–Morning ore body, and both the Star and Morning mines are presently operated by the Hecla Mining Co. as the Star Unit.

The Star–Morning ore shoot is in a nearly vertical shear zone striking N 79° W (Fig. 2) and has a maximum strike length of approximately 1.3 km and an average stoping width of 3 meters. Wall rock is almost entirely dry, brittle, Precambrian Revett Quartzite. The principal mining method is horizontal-slice cut-and-fill stoping. In the lower levels of this mine, where the walls are subject to high stress, the ore body is developed by lateral drifts of 60 meters' average vertical separation in one

---

[1]) Physicist, U.S. Department of the Interior, BuMines, Denver Mining Research Center, Denver, USA.

of the walls with crosscuts through the ore. A blind stoping method is used with cribbed raises carried upward from the crosscuts in the ore zone (for details on the mining method and terminology, refer to [6]).

Detailed microseismic monitoring of potential rock burst zones in the Star mine was begun in 1975. The Star system consists of 21 geophones and associated peripheral computer equipment (see [7] for system details). The geophones are located on levels 6900, 7100, 7300, 7500, and 7700 (numbers refer to distances in feet measured from the surface entrance to the Star mine, 1 foot = 0.308 meter). The Star microseismic system covers approximately 490 meters and 215 meters parallel and normal, respectively, to the strike of the ore body. The minimum distance between any two geophones in the Star mine in approximately 50 meters. Complete details as to geophone locations within the Star mine are available elsewhere [7, Fig. 3].

Once a seismic event occurs and is picked up by a geophone, the signal is amplified at the geophone location, filtered, and transmitted to a voltage monitor. If the signal exceeds a preset threshold level of 0.5 volt peak-to-peak, a 15-volt pulse is transmitted by the voltage monitor to the computer and the event is registered. At this time, a 100 ms (1 ms = 1 millisecond) time window is opened. If four additional geophones are triggered by the event, the event is registered and located.

The average V_p velocity between geophones is determined by detonating one stick of 60% dynamite in known locations and observing the travel time to various geophones. Experimental results suggest an average velocity of 5.5 km/sec [7]. These explosions are consistently located to within 6 meters of the known source. Thus, location accuracy of seismic events that occur within the Star system is probably within a spherical region of radius less than 10 meters surrounding the source.

Unfortunately, information relating to magnitudes and fault plane solutions for the seismic events reported in this article is not available. However, surface seismographs located within the Coeur d'Alene mining district do permit a relative determination of the relative strength of the seismic event. For example, a typical rock burst will be recorded on the surface seismographs, while the individual seismic events that *precede* and *follow* (aftershocks in this article) the burst are generally too low in strength to be registered.

The seismicity behaviors of three rock bursts are examined in detail in this article. For sake of brevity, I have combined both the observational data of each burst and its interpretation in terms of the inclusion theory of failure in the analysis of the burst. The distinction between observation and interpretation is made clear throughout the text. In addition, it may be useful for the reader to be aware that the seismicity, as indicated by the number of events per day, in the Star mine shows that seismicity prior to rock bursts exhibits a *clear* increase (approximately 10 events per day [7, Fig. 9]) over a normal background level of approximately 8 events per day. For example, in the two months that preceded the rock bursts discussed in this article, the average number of events that occurred in a quarter day period (6 hours) was approximately two (2) with a standard deviation of (±) two (2). The 'background' seismic events occurred over a broad region (usually a region whose dimensions are at least a factor of five (5) over an average dimension of the aftershock region produced by the rock burst). In addition, there is no tendency for these 'background' events to cluster spacially. The increase in seismicity prior to rock bursts reported in this article exhibit both a spacial as well as a temporal clustering that are statistically significant.

The 3 September 1975 rock burst

A moderate burst ($M_L < 1.5$) occurred on the 7500 level of the Star mine at 10:09:04 a.m. (18:09:04, UTC) on 3 September 1975. The burst was preceded by a dramatic increase of seismic activity in a localized region, which was followed by a distinct decrease prior to the burst as shown in Fig. 1. *Miners were evacuated from the stope (located in the immediate area of this seismic buildup) at this time.* In Fig. 1,

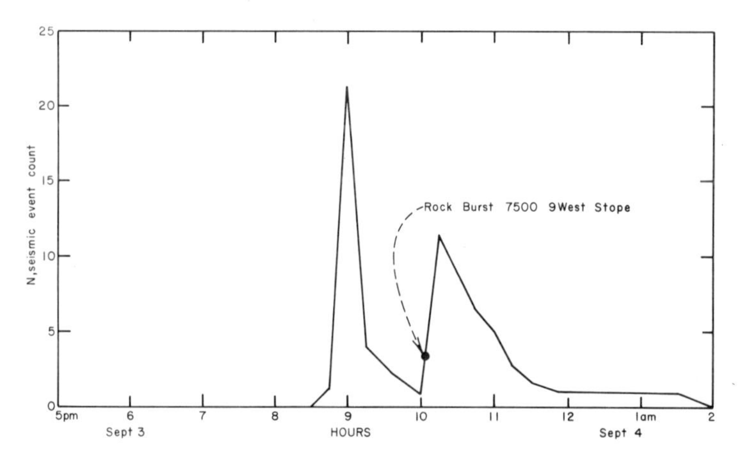

Figure 1
Seismic count (N) against time (hours) for 3 September rock burst. Seismic event count, N, refers to events that form the inferred primary inclusion zone of the impending burst.

number of seismic events refers to the number of events within a 15 minute time window. Table 1 lists seismic event numbers and their time of occurrence prior to the burst. Figures 2 and 3 illustrates the locations of the seismic events projected onto the horizontal plane (Fig. 2) and the projections of the aftershocks and their relationship to the mainshock onto the vertical plane along sections A–A' (Fig. 3a) and B–B' (Fig. 3b).

Four results are noteworthy from these data. *First*, the rapid increase of seismic activity prior to the burst was primarily associated with events 1 through 14 (Table 1, circles in Fig. 2), outlining a zone of early intensified activity (the shaded, circular zone shown in Fig. 2). The relative area, A_i, of this zone that apparently contains and is delineated by these events (marked by an asterisk in Table 1, circles in Fig. 2) is approximately 7.86×10^5 cm^2, and the total time, τ_d, required for the formation of this circularly shaped region was 188 seconds. *Second*, the seismic events that followed (squares, Figs. 2 and 3) the formation of this zone were concentrated outside this circular zone and were primarily located near the outer boundaries of what was to be

Table 1
Seismic event number and time prior to 3 September burst

Event number	Time (a.m.)	Event number	Time (a.m.)
1*	9:00:32	15	9:05:53
2	9:00:38	16	9:06:09
3*	9:00:48	17	9:08:46
4*	9:00:53	18	9:10:44
5	9:00:55	19	9:22:27
6*	9:00:57	20	9:24:32
7	9:01:11	21	9:30:40
8*	9:01:38	22	9:44:52
9*	9:01:39	23	10:01:03
10	9:02:29	24	10:01:58
11	9:02:32	25	10:02:08
12*	9:02:40	26	10:02:24
13*	9:03:00	27	10:02:42
14*	9:03:38	28	10:05:10

Burst was event number 29 and occurred at 10:09:04 a.m.
* Events that are taken to have formed the primary inclusion zone.

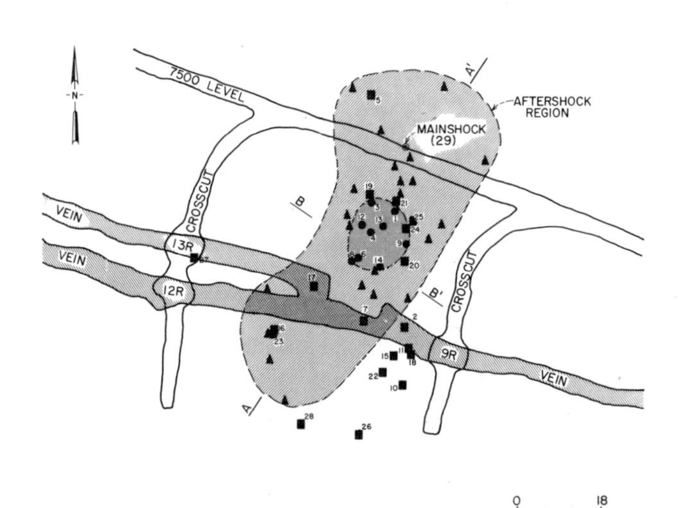

Figure 2
Plan view of 7500 level of the Star Mine including aftershock region, 'primary inclusion zone', and burst epicenter of 3 September 1975, rock burst. Circles refer to events that form the inferred primary inclusion zone of the impending burst; squares to pre-burst events not involved in the formation of the primary inclusion zone; triangles to events following the burst.

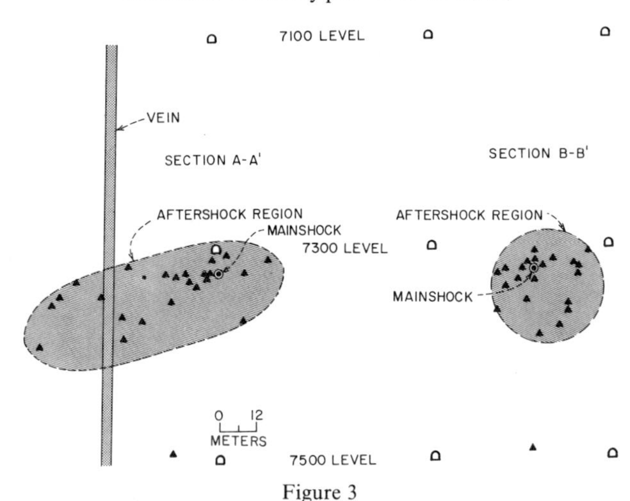

Figure 3

Vertical section of aftershock region of 3 September 1975, rock burst sections *A–A′* (a) and *B–B′* (c). Symbols (circles, squares, triangles) have the same meaning as in Fig. 2.

the aftershock region, or just outside the initially formed, circular zone (events 19–21, 24, and 25). Of particular interest, note that the events that followed the formation the primary inclusion zone were concentrated in a direction that was the eventual rupture propagation direction. Activity then abruptly diminished after event 25 for about 6.5 minutes until the burst occurred at 10:09:04 a.m. (18:09:04, UTC), and was followed by 22 aftershocks (triangles, Figs. 1a, b, and c) that defined an elliptical zone of area, A, equal to approximately 1.74×10^7 cm^2. The total time to the burst, measured from the initiation of seismic activity *assumed* to signal the growth (event 1, Table 1) of the primary inclusion zone, was 68.5 minutes. *Third*, the cross sections of the aftershock zone reveal an ellipsoidal zone (Figs. 3a and 3b) whose major axis is parallel to the eventual rupture propagation direction. In particular, note the circularly shaped cross section of this ellipsoid (Fig. 3b). This geometry is remarkably similar to that of the aftershock zone generally predicted by the inclusion theory when the far-field least and intermediate principal stresses are equal, an observation consistent with *in situ* stress measurements in this mine (Table 2) ([3, 4], see reference [4] Fig. 1). This figure suggest a 'volumetric' failure mechanism of this burst. However, the location accuracy of these events (± 10 meters) signifies that a 'planar' (fault) failure mechanism cannot be ruled out.

Table 2

In situ *stress distribution at east lateral drift 7300 level – Star mine*

Stress	Magnitude (kb)	Bearing
Major horizontal (σ_1)	0.66	N 14° W
Minor horizontal (σ_3)	0.42	N 76° E
Vertical (σ_2)	0.43	—

Fourth, the burst hypocenter was located approximately 8 meters from the predicted (on the basis of the inclusion theory of failure [3]) location on the boundary of the primary inclusion zone. This apparent displacement may or may not be real, as it falls within the reliability of the location accuracy of the microseismic system. However, this displacement is consistent with the framework of the inclusion theory in that, as the focal volume of the impending burst is predicted to become elastically stiffer than its immediate surroundings, the seismic velocities will increase [3]. This predicted increase would result in a migration of the actual burst hypocenter away from the boundary of the primary inclusion zone, as is shown in Fig. 2. To account for a relative displacement of this magnitude ($\simeq 8$ meters) of the burst hypocenter, a v_p increase of approximately 8% would be required in the area that was to become the aftershock region.

Theoretical and experimental studies have shown that the predicted functional relationships of the time required to form the primary inclusion zone (τ_{do}) and the time required for the failure to occur (τ_0) to the areas A_i and A are [3]:

$$\begin{aligned} \tau_{do} &= 2.43 \times 10^{-4} A_i \\ \tau_0 &= 2.43 \times 10^{-4} A, \end{aligned} \tag{1}$$

where times and areas are measured in seconds and square centimeters, respectively. Studies by the United States Bureau of Mines in the Galena mine, Wallace, Idaho, have also shown that a lower limit to the ratio τ_0/τ_{do} is 12.5 in hard, brittle rock [3]. Thus, once τ_d is known, a minimum predicted time to the burst is possible. Based upon the above observational data, the minimum predicted time to the 3 September burst is 40 minutes, where the observed values of τ_d ($=188$ s) is used for τ_{do}. This value compares quite favorably with the observed value of 68.5 minutes. However, based upon the observed areas, A_i and A, the 'predicted' times for τ_{do} and τ_0 from equation (1) are 191 seconds and 70.3 minutes, respectively, in excellent agreement with observation. Thus, based on the observed seismicity increase that occurred within a highly localized region followed by a decrease, this burst was theoretically predictable, and the analysis further admits a more realistic estimate of the ratio τ/τ_d ($=\tau/\tau_{do} = 21.8$) required for accurate prediction time of future failures in this mine, assuming that the failures occur under comparable conditions.

This rock burst was *not* detected by the Newport, Wash., seismic station, operated by the United States Geological Survey. The Newport station, which is located approximately 150 km from the Star mine, is capable of detecting events of magnitude greater than $M1.5$ in the Coeur d'Alene district (KERRY, USGS, personal communication 1976). The functional relationship between aftershock area, A, and magnitude, M, can be estimated by using the Utsu relationship, $\log_{10} A = M + 6.3$, where A is in square centimeters [3]. Substituting the observed area of 1.74×10^7 cm^2 into this equation gives an estimated magnitude of $M0.9$ for the 3 September burst, well below the threshold value of $M1.5$.

The 24 October 1975 rock burst

A rock burst whose strength was considerably larger than the 3 September burst, but not detected at the Newport, Wash., seismic station, occurred in nearly the same location as the latter event. The burst was also preceded by a dramatic increase in seismic activity, which was followed by a distinct decrease prior to the burst, as shown in Fig. 4. *Miners were evacuated from the stope region at this time.*

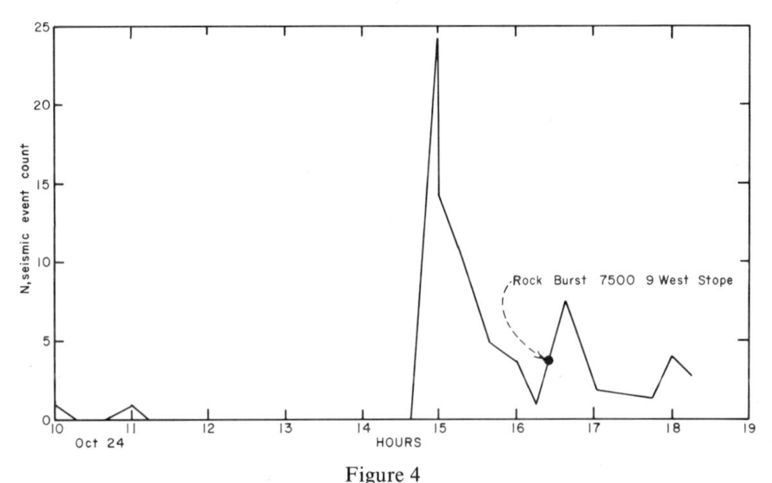

Figure 4
Seismic count (N) against time (hours) for 24 October, 1975, rock burst.

The seismicity increase occurred in a highly localized region shown in Fig. 5. The time duration, τ_d, of this increase was equal to 434 seconds, and mapped out an elliptical zone of area, A_i, equal to approximately 1.78×10^6 cm^2. The aftershock area, A, is equal to approximately 3.55×10^7 cm^2. Thus, the observed ratio τ_0/τ_{do} is equal to approximately 20.0, in close agreement with the value (21.8) obtained from the 3 September burst, the estimated time to this burst, τ_0, is 21.8 τ_0, or 9460 sec (=158 minutes). This value compares with the observed time, τ, of approximately 5239 sec (=87 minutes), a factor of two lower than the predicted value of 158 minutes.

Analysis of the seismic data for this burst reveals two interesting observations. *First*, while the seismicity behavior following the formation of the primary inclusion zone of this burst closely paralleled the behavior observed for the 3 September burst, there was a large seismic event (see solid arrow, Fig. 5) near the hypocenter of the burst that occurred approximately 6 minutes prior to the burst. This event was of sufficient strength to be recorded on the seismograph located on the surface of the mine and may possibly have led to a premature triggering of the mainshock. This behavior is similar to large foreshocks in the hypocentral regions of some large earthquakes just prior ($\sim$few hours) to the mainshock. *Second*, the aftershock data suggest a general lack of concentration. This observation can be interpreted in terms of the inclusion theory as indicating that the focal region of this burst may have had

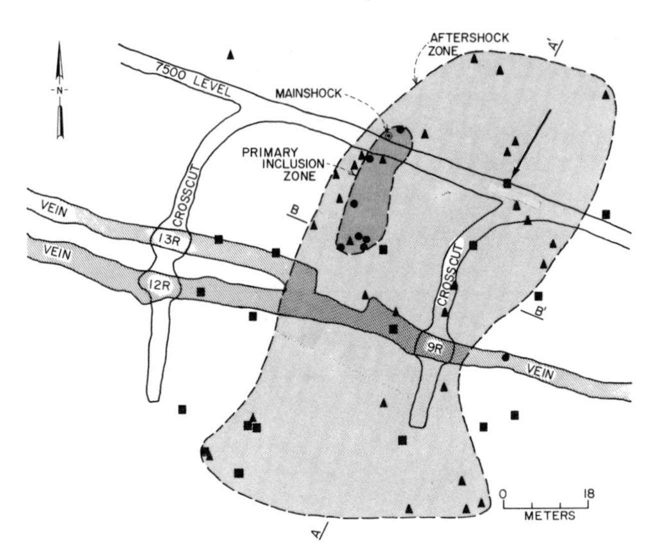

Figure 5
Plan view of 7500 level of the Star mine including aftershock region 'primary inclusion zone', and burst epicenter of 24 October 1975, rock burst. Symbols (circles, squares, triangles) have the same meaning as in Fig. 2. Note the location of the arrow. This event occurred approximately 7 minutes prior to the mainshock and radiated sufficient energy to trigger the surface seismograph.

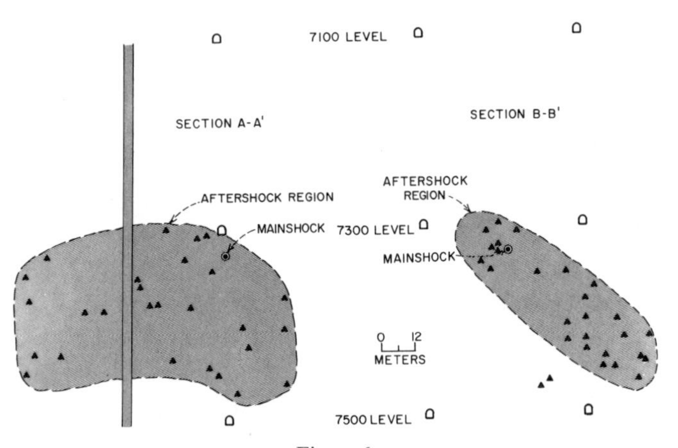

Figure 6
Vertical section of aftershock region of 24 October 1975, rock burst along sections A–A' (a) and B–B' (b). Symbols (circles, squares, triangles) have the same meaning as in Fig. 2.

enough time required for the storage of strain energy. The reader will recall that precursor time in the inclusion theory denotes the time interval during which the focal region of an impending failure stores strain energy. Similar behavior, that is, possible triggering of failures by large foreshocks, has been observed prior to some major shallow earthquakes [3].

The reader may, at this time, be somewhat puzzled by the observation that the numbers of foreshocks and aftershocks for both the 3 September and 24 October bursts appear to be comparable in number. It is quite likely, however, that the numbers of aftershocks are greater than reported here as the computer is saturated during the first several minutes or so following the burst. This type of behavior is also observed with bursts that occur in the Galena mine.

The 8 October 1975 rock burst sequence

A sequence of rock bursts, who strengths were considerably less than the 3 September and 24 October events, but nevertheless large enough to be registered on the seismograph located on the surface of the Star mine, occurred on the 7500 level of the Star mine on 8 October 1975. Figure 7 illustrates the seismic event count

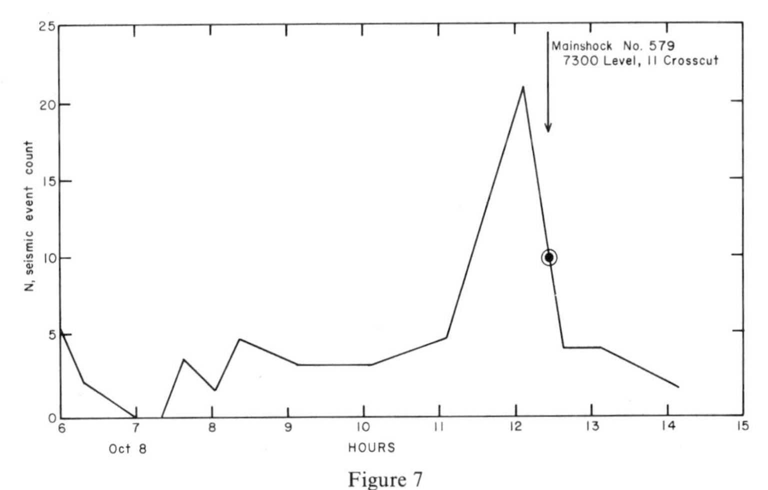

Figure 7
Seismic count (N) against time (hours) for 8 October 1975, rock burst sequence.

against time during 8 October. *Men were prohibited from entering 7500–9 west stope on the basis of this apparent seismicity increase.* A small, damaging rock burst occurred at 12:53:32 (#579, Table 3) shortly after the men were prohibited from entering the stope.

There are features of this burst that are readily distinguishable from the 3 September and 24 October bursts, and as such, demonstrate that a seismicity increase followed by a decrease is not a necessary and sufficient condition for failure to occur, although in this situation, a small burst actually occurred. For example, the seismicity behavior is distinctly different, as is apparent from examination of Figs. 1, 4, and 7. In addition, while increases in Fig. 1 and 4 are both interpreted as being associated with the formation of primary inclusion zones that led to the bursts, the seismicity increase

Table 3

Data relating to 8 October 1975, rockburst sequence

Burst number	Time	Aftershock area (A) (cm^2)	Precursor time (τ_0) calculated (min)	Primary inclusion formation time (τ_{do}) calculated (min)
522	6:08:09	7.86×10^6	31	1.6
562	11:41:56	3.75×10^6	15.2	0.8
569	12:40:08	5.77×10^6	23	1.2
579*	12:53:32	3.47×10^6	14	0.7
589	13:16:38	8.47×10^5	3.4	0.2

* Damaging burst (note inferred rupture direction toward mine workings).

associated with the 8 October burst was apparently due to aftershocks from events 562 and 569, and not to the formation of a primary inclusion zone (Fig. 8). The apparent decrease was associated with the 'normal' hyperbolic aftershock behavior of events 562 and 569.

In the 8 October burst sequence, *no* seismic events were detected in the hypocentral regions prior to the occurrence of the damaging burst (event 579). The low intensities of events 522, 562, 569, 579, and 589, listed in Table 3, are inferred both by observation of mine damage and their recording on the surface seismograph. In addition, the

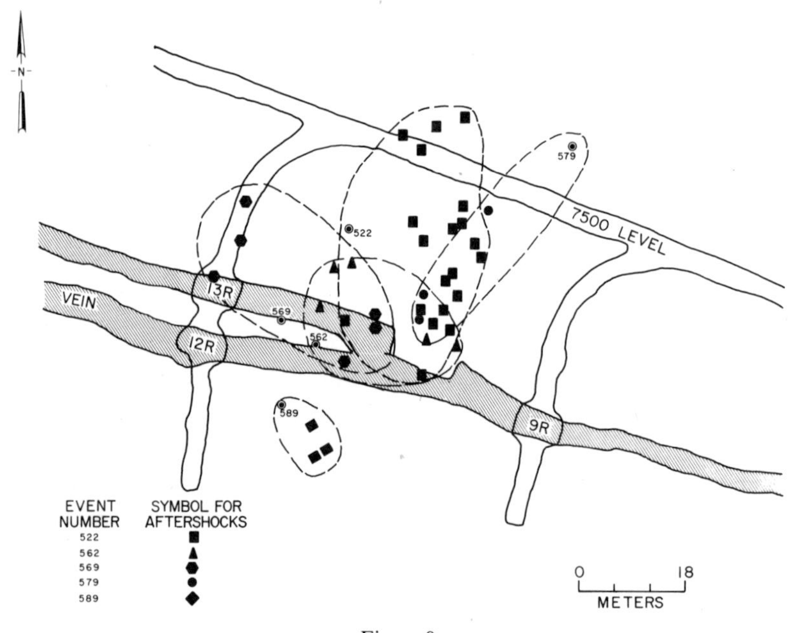

Figure 8

Plan view of 7500 level of the Star mine illustrating aftershock regions of the four low strength seismic events that occurred 8 October 1975.

low strengths of these same events suggest that the (inferred) seismic events that may have produced the primary inclusion zones of events listed in Table 3 may not have had sufficient energy to trigger the required five geophones so as to be recorded on the mine computer. Note also that the calculated formation times (τ_0) listed in Table 3 for these bursts are small ($\sim$several minutes), which adds support to this conjecture. Each of these bursts, listed in Table 3, was recorded on the seismograph located on the surface of the mine. The relative amplitudes of their seismic signatures were found to correspond with the observed aftershock areas of each burst, that is, the larger the aftershock area, the larger the peak-to-peak displacement recorded on the seismograph for the event. Event 579 was observed to have the largest peak-to-peak displacement.

Lastly, it is of interest to note that event 569 may have been instrumental in triggering the damaging burst 579. The calculated time, τ_0, for 579 would have been 14.1 minutes. The observed time interval between 569 and 579 was 13.5 minutes.

Discussion

Based upon the three cases of anomalous seismicity behavior discussed in this article, miners were either evacuated (3 September and 24 October) or prohibited (8 October) from entering regions where moderate rock bursts occurred. Interpretation of the seismicity that preceded each burst shows that the increased seismicity of the 3 September and 24 October bursts was associated with the development of zones of apparent concentrated dilatancy, termed primary inclusion zones, in the hypocentral region of the impending bursts. The seismicity increase and its associated decrease prior to the damaging but small 8 October burst were due to aftershocks of the preceding small bursts, and not the formation of a primary inclusion zone (PIZ) as was the case in the 3 September and 24 October bursts. Data to correctly analyze the 8 October damaging burst (#579, Table 3) were not available because of its apparent very low magnitude, although there is no way to test this conjecture. The 8 October burst sequence (Table 3) does, however, point out the difficulties in predicting small bursts in mines, as such bursts can inflict severe damage to miners who may be located in the immediate vicinity of these bursts.

Experimental data presented in this article and elsewhere [1–5] support the hypothesis that failure of rock materials satisfies a scale-invariant process. Because of this property, these data indicate that reliable prediction of impending failure is a distinct possibility. The time intervals over which the anomalies develop on the intermediate scale are on the order of a few minutes or hours. Thus, prediction of failures on the intermediate scale, such as mine failures like rock bursts, coal bumps, gas outbursts, roof failures, and slope failures, should be possible.

A word of caution on using seismicity anomalies as predictors of impending failure is in order. In the inclusion theory, the seismicity anomaly, that is, seismicity increase followed by a decrease, which is associated with the formation of a primary inclusion zone, is both a necessary and sufficient condition for the occurrence of failure. However, the converse is not true. A seismicity increase followed by a decrease is not a sufficient condition for an impending failure. Seismicity behavior, in order to be utilized as a predictor of failure, must satisfy certain requirements. First, the siesmicity must be associated with the formation of a primary inclusion zone (PIZ). Once this zone has formed, the system *must* evolve to failure. Second, the seismic events that form the PIZ must be characterized by anomalously long rupture lengths because these seismic events are forming within a stress field whose least principal stress is either zero or tensile [3]. Thus, these events are predicted to exhibit spectra shifted toward low frequencies in comparison with the events that occur in the surrounding medium. Consequently, the energy that would normally be dissipated in frictional processes is now available to power crack growth. This type of analysis is required to determine absolutely that a PIZ of an impending burst has formed. This aspect of the prediction problem is now under active investigation in the Coeur d'Alene mining district.

Implications for earthquake prediction

Figure 9 illustrates the predicted precursor time, τ_0, in seconds against 'effective length', L, in centimeters of selected failures, including a rock burst in the Galena mine [located in the same mining district as the Star mine] [2], a coal mine roof fall in Pennsylvania [2], laboratory size rock failures [4], and several earthquakes. Precursor time, τ_0, refers to the time interval between the initiation of the process that leads to the occurrence of the failure and the actual occurrence of the rupture. This process, once begun, proceeds independently of any changes in the far-field boundary conditions and any changes in material properties within the PIZ and its associated focal region, such as might occur by creep related processes [3]. These changes can, however, produce changes in the time required for storage of strain energy in the focal region of the PIZ. The 'effective length' denotes an *average* linear dimension of the aftershock region that is generated by the failure. The 3 September and 24 October Star events are also shown in Fig. 9. The average 'effective lengths' for these events were calculated by approximating the observed elliptical shaped aftershock areas as circular zones of average diameters of 45 meters and 55 meters, respectively.

Two important observations can be drawn from the data listed in Fig. 9. First, failure of a wide variety of rock materials satisfies a scale invariant process. By scale invariance is meant that the physical processes that lead to failure are independent of scale. Thus, an understanding of the physical processes which lead to failure on the small (laboratory) scale admits an understanding of physical processes leading to failure on the intermediate (mine) scale and the large scale, such as earthquakes. Second, since the relationship between τ_0 and A is approximately linear, and since observational data suggest that earthquakes occur repeatedly in the same region, apparently on preexisting fault planes, it does not appear to be important to the physical processes which lead irrevocably to failure whether failure occurs in fresh, unbroken material (such as laboratory failures and rock bursts listed in Fig. 9) or along preexisting fault zones, as clearly most of the earthquakes in Fig. 9 have occurred. This result probably suggests that 'old' fault zones probably heal with time, due perhaps, to high temperatures and deposition of minerals by solutions along the fault zone, prior to the occurrence of the next shock.

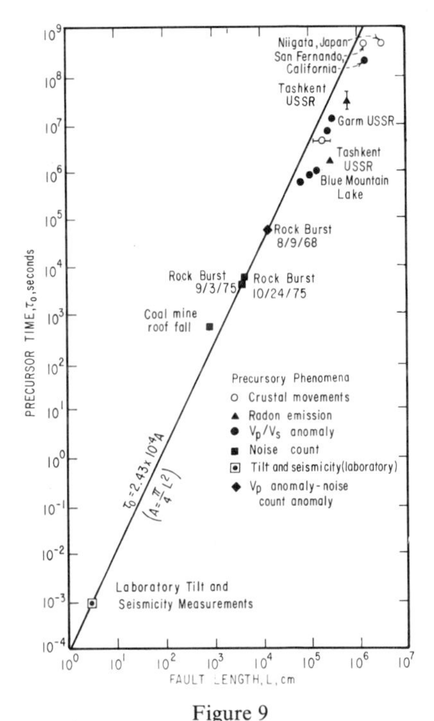

Figure 9

Precursor time – 'fault' length relationships for selected earthquakes, mine failures, and laboratory failures.

Other, perhaps less important, aspects of the data listed in Fig. 9 warrant attention: (1) The relationship between τ_0 and A, shown in Fig. 9, is only valid when changes are not occurring in the far-field boundary conditions during the time interval τ_0. Obviously, this condition can be relaxed, as evidenced by the presence of large foreshocks, such as occur prior to large earthquakes and rock bursts (for example, the 24 October, 1975, Star burst). Foreshocks tend to increase load within the potential failure zone, and consequently, can serve to 'prematurely' trigger the impending failure. (2) Changes (increases) in the far-field boundary strains due to plate motions can shorten the theoretically determined precursor time. These changes are consistent with the inclusion theory as the precursor time is inversely proportional to loading rate changes that occur during the preparation time of the impending shock. Thus, there is *no* compelling reason to presume that the functional relationship specified by equation 1 (Fig. 9) will hold for all classes of failures. In addition, the reader will note the tendency for precursor times to drop off at increasing sizes of impending failure, probably the result of changes in the far-field stresses due to plate motions occurring during the required preparation time of the impending shock. Note also that the predicted and observed precursor time of approximately 1 ms prior to failure of rock on the laboratory scale should be considered as representing an upper limiting value. For example, shock loading tests, entailing large increases occurring over a short time interval in a material within which a PIZ is nucleated, can decrease the predicted precursor time (τ_0) by several orders of magnitude, the exact value depending on how rapidly the far-field strains are changing.

The ratio τ_0/τ_{do} $(=A/A_i)$ was shown earlier to be approximately 21.0 for rock bursts that were known to develop in a rock mass characterized by very low intrinsic porosity. (Intrinsic porosity, ϕ_i, in the context used here, refers to the porosity existing within material prior to the application of applied loads.) Excluded from the magnitude of ϕ_i is any rehealing, due for example, to pressure solution or cementation from pore fluids during the time interval between when the load is applied and the occurrence of failure.

It is shown in the Appendix that the influence of intrinsic porosity on the predicted τ_0 versus A relationship is an important factor in calculating realistic accurate prediction times to failure. The function relationship of τ_0/τ_{do} to ϕ_i is shown to be

$$\tau_0 \simeq \tau_{do} + \tau_{do} \frac{K(\tau_0)}{n} \left(\frac{1 - \phi_0}{1 - \phi} \right), \tag{2}$$

where $\phi_0 = \phi_i + \phi_f + \phi_p$
$\phi = \phi_i$
$n = C_0/T_0$
and ϕ_i = intrinsic porosity of the material,
ϕ_f = fracture porosity prior to the formation of the PIZ,
ϕ_p = fracture porosity within the PIZ due to the seismic events that occurred to form the PIZ,
C_0/T_0 = ratio of uniaxial compressive strength to the tensile strength of the material.

In equation (2), $K(\tau_0)$ is the stress concentration factor existing at the boundary of the PIZ and its associated focal region at the instant of failure.

Two useful results are a direct consequence of equation (2). First, the inclusion theory provides a simple explanation of the transition in material behavior from violent brittle failure occurring in fresh, unbroken materials with no intrinsic porosity ($\phi_i = 0$) to 'quiet' aseismic creep observed to occur along well-fractured fault zones whose intrinsic porosities are high and equal to $\phi_i \simeq 1 - (\phi_f + \phi_p)$. Note that in the latter case, $A \simeq A_i$, that is, the effective tensile strength of the fault zone (gouge) is approximately zero. Second, the ratio (α) of focal region area (A) to the primary inclusion zone area (A_i), A/A_i, is not an invariant. Its value can range from high ($\simeq 21.8$) for fresh failures in tight, low porosity rock to low ($\simeq 1.0$) along highly fractured fault zones exhibiting aseismic or steady-state creep. A determination of α in given earthquake zones is critical in order to permit accurate predictions of impending earthquakes.

In conclusion, one of the more important findings of this study is that failures on the intermediate (mine) scale and small (laboratory) scale exhibit characteristics consistent with failures on the large scale, such as earthquakes. Thus, failure characteristics of rock bursts, such as their associated foreshocks and aftershocks, should produce results compatible with known observational characteristics of earthquakes. The study of rock bursts, to date, supports this hypothesis, although much more thorough and detailed investigations are required. The Bureau of Mines experience in rock burst prediction, as well as that of the Hecla Mining Company, suggests that reliable and accurate prediction of rock bursts will be difficult. For instance, small bursts (like the 8 October 1975 sequence reported earlier) that have occurred in the Star mine apparently exhibit no seismic anomalies. However, the energy content of these bursts is probably low so as not to register on the existing seismic network. Similarly, large bursts ($M_L \geq 2.0$), which are often located several tens to hundreds of meters from the mine workings, have occurred with no observable seismicity anomaly. This class of bursts has been observed in both the Star and Galena mines, both of which have active seismic networks. Our preliminary studies suggest these bursts occur along old fault zones that are apparently activated by the mining process. A working hypothesis at this time is that these bursts occur along existing zones that are 'hung up' along small asperities or lock points. Failure of the asperities, showing perhaps the classic seismic anomaly discussed earlier but too small to register on the existing network, causes releases of strain energy that had been stored along the fault zone by nearby mining operations. *Thus, if the characteristics of rock bursts carry over to large-scale failure phenomena, and preliminary data suggest this to be the case*, then it is conceivable that a low-magnitude earthquake, predicted to occur along an active fault zone, could release stored strain energy outside the locked zone and produce an energy release equivalent to a *much larger* shock. This situation clearly requires further study.

Acknowledgments

The Hecla Mining Co., Wallace, Idaho, generously provided the data used in this report. In particular, special thanks are due to Jon Langstaff, Senior Rock Mechanics Engineer, Hecla Mining Co., for discussions of this data. The opinions expressed in this report represent those of the author and not necessarily those of the Hecla Mining Co. W. Spence, R. Martin, and in particular, M. Wyss offered valuable comments on early drafts of this manuscript.

APPENDIX

Effect of intrinsic porosity on the precursor time-aftershock area relationship

Intrinsic porosity, ϕ_i, will be defined as the porosity existing within a material prior to the application of applied loads. Excluded from the value of ϕ_i is any rehealing due, for example, to pressure solution or cementation resulting from pore fluids, that may occur during the time interval between load application and the occurrence of failure.

Let σ_t and σ_c denote the local values of the tensile and compressive stresses, in a direction normal to the eventual rupture propagation direction, existing within the evolving PIZ and its associated focal region at a time, t, prior to the mainshock. Force equilibrium in the material requires

$$\int \sigma_t \, dA_i = \int \sigma_c \, dA, \quad 0 \leq t \leq \tau_0 \tag{A1}$$

where the integration is over the *solid* cross sectional areas. It is assumed, for calculational simplicity, that no changes occur in the far-field boundary conditions during the preparation time required for the failure. For ease of calculation, assume the stresses in equation A1 are uniform throughout their respective areas of integration. Thus, to a first-order approximation

$$\bar{\sigma}_t A_i^* \simeq \bar{\sigma}_c A^*,$$

where

$$\bar{\sigma}_t = \frac{1}{A_i^*} \int \sigma_t \, dA_i \quad , \quad \bar{\sigma}_c = \frac{1}{A_i^*} \int \sigma_c \, dA \tag{A3}$$

and A_i^* and A^* represent the effective solid cross sectional areas of the PIZ and its associated focal region. Define

ϕ_i = intrinsic porosity of the material,
ϕ_f = fracture porosity existing in the material prior to the formation of the PIZ,
ϕ_p = fracture porosity induced within the PIZ during the formation of the PIZ.

Equations 1, A1, and A2 can then be combined to give the total time required for the failure preparation process $\{[A_i^* = A_i(1 - \phi_0), A^* = A(1 - \phi(t)]\}$

$$\tau \simeq \tau_{do} + \frac{\bar{\sigma}_c \tau_{do}}{\bar{\sigma}_t \phi_p} \frac{1 - \phi_0}{1 - \phi(t)}, \tag{A4}$$

where $\phi_0 = \phi_i + \phi_f + \phi_p$, $\phi = \phi_f(t)$, and $\tau/\tau_{do} = A/A_i$. Note that the fracture porosity, ϕ_f, existing within the focal region of the evolving PIZ is time dependent, since cracks within this region will close in response to an increasing level in $\bar{\sigma}_c$ as the system evolves to failure.

In the inclusion theory, a long, narrow macrocrack forms within the primary inclusion zone at the instant of failure. Let the value of $\bar{\sigma}_t$ just prior to crack coalescence be denoted by T_0, where T_0 denotes the uniaxial tensile strength of the material. Let $C_0 = K(\tau_0)\bar{\sigma}_c$, where C_0 is the uniaxial compressive strength of the material at the boundary between the PIZ and its focal region, and $K(\tau_0)$ is a stress concentration factor denoting the ratio of C_0 to the average compressive stress existing in the focal region of the impending failure at the instant of failure. Equation A4 becomes

$$\tau_0 \simeq \tau_{do} + \tau_{do} \frac{K(\tau_0)}{n\phi_p} \frac{1 - \phi_0}{1 - \phi_i},$$

where $\phi(t)$, at time $t = \tau_0$, is assumed to ϕ_i, i.e., $\phi_i(\tau_0) \simeq 0$ throughout the focal region, and $n = C_0/T_0$. The relationship between A and A_i can then be written

$$\alpha = \frac{A}{A_i} \simeq 1 + \frac{K(\tau_0)}{n\phi_p} \frac{1 - \phi_0}{1 - \phi_i}. \tag{A6}$$

Equations A5 and A6 show that both precursor time, τ_0, and aftershock area, A, decrease as the intrinsic porosity, ϕ_i, of the material increases. For example, τ_0 and A attain their maximum possible values when $\phi_i = 0$ and their lowest values, τ_{do} and A_i, respectively, in the limit as ϕ_i approaches $1 - (\phi_f + \phi_p)$. The effective tensile strength of the material is identically zero under these conditions. No seismicity anomaly will be observed on any scale of failure, in the unlikely situation that ϕ_i remains scale invariant. The system enters a state of aseismic or steady-state creep at this time.

The rock burst data reported in the test provide a reasonable upper limit for α ($\simeq 21$) in equation A6. In this situation

$$K(\tau_0) \simeq \left[\frac{20n}{1 - (\phi_f + \phi_p)}\right]^{-1} \tag{A7}$$

The theoretical ratio of the compressive strength, that is, the value of the applied compressive stress required to initiate crack growth from a critically oriented flaw, to the tensile strength, n, is provided by the Griffith model of failure. The result is $n = 8.0$. Thus, $K(\tau_0 \simeq 0.006[1 - (\phi_f + \phi_p)]$, or simply, the stress concentration factor at the PIZ–focal region boundary is dependent only on the total fracture porosity existing within the PIZ at a time just prior to crack coalescence in this zone. A reasonable value for $(\phi_f + \phi_p)$ is 0.5, giving $K(\tau_0) \simeq 0.003$. To a first-order approximation, the quantities $K(\tau_0)$, n, and $(\phi_f + \phi_p)$ are scale invariant.

References

[1] Aggarwal, Y. P., Sykes, L. R., Armbruster, J. and Sbar, M. L. (1973), *Premonitory changes in seismic velocities and prediction of earthquakes*, Nature *241*, 5385.

[2] Brady, B. T. (1974), *Seismic precursors before rock failures in mines*, Nature *252*, 5484.

[3] BRADY, B. T. (1966), *Theory of earthquakes, IV. General implications for earthquake prediction*, Pure appl. Geophys. (in press).

[4] BRADY, B. T. (1977), *An investigation of the scale invariant properties of failure*, Int. J. Rock Mech. Mining Sci. (in press).

[5] NERSESOV, I. L., LUKK, A. A., PONOMAREV, V. S., RAUTAIN, T. G., RULEV, B. G., SEMENOV, A. N. and SIMBRIEVA, I. G., *Possibilities of Earthquake Prediction Exemplified by the Garm Area of the Tadzhik, S.S.R., in Earthquake Precursors* (ed. M. A. Sadovskii, I. L. Nersesov and L. G. Latynina) (U.S.S.R. Academy of Sciences, Moscow 1975), pp. 79–99.

[6] MCWILLIAMS, J. R. and ERICKSON, E. G., *Methods and Costs of Shaft Sinking in the Coeur d'Alene District, Shoshone County, Idaho*, (BuMines IC 7961, 1960).

[7] LANGSTAFF, J., *Hecla's seismic detection system*, Proceedings of the 17th Rock Mechanics Symposium, Snowbird, Utah, 1976.

(Received 2nd November 1976)

Pageoph, Vol. 115 (1977), Birkhäuser Verlag, Basel

Seismicity Gap near Oaxaca, Southern Mexico as a Probable Precursor to a Large Earthquake

By MASAKAZU OHTAKE[1]), TOSIMATU MATUMOTO[2]) and GARY V. LATHAM[2])

Summary – An area of significant seismic quiescence is found near Oaxaca, southern Mexico. The anomalous area may be the site of a future large earthquake as many cases so far reported were. This conjecture is justified by study of past seismicity changes in the Oaxaca region. An interval of reduced seismicity, followed by a renewal of activity, preceded both the recent large events of 1965 and 1968. Those past earthquakes have ruptured the eastern and western portions of the present seismicity gap, respectively, so that the central part remaining is considered to be of the highest risk of the pending earthquake.

The most probable estimates are: $7\frac{1}{2} \pm \frac{1}{4}$ for the magnitude and $\varphi = 16.5° \pm 0.5°$N, $\lambda = 96.5° \pm 0.5$W for the epicenter location. A firm prediction of the occurrence time is not attempted. However, a resumption of seismic activity in the Oaxaca region may precede a main shock.

Key words: Seismicity; Earthquake prediction.

Introduction

From studies of the seismicity of the northwestern Circum-Pacific seismic belt, FEDOTOV (1965) and MOGI (1968a) found that 'gaps' in activity have been successively filled within several tens of years by a series of great earthquakes without significant overlap of their rupture zones. More detailed investigations by UTSU (1968), MOGI (1968b, 1969) and NAGUMO (1973) revealed that some of the greatest earthquakes ($M \geq 8$) which occurred in and near Japan were preceded by seismically quiescent periods of several to a few tens of years. Similar phenomena have been reported also from other seismic zones including Alaska and the Aleutians (SYKES, 1971), South America (KELLEHER, 1972), and major portions of the Circum-Pacific belt (KELLEHER *et al.*, 1973; KELLEHER, 1972; KELLEHER and SAVINO, 1975). The search for preseismic reduction in regional seismicity has been successfully extended to smaller earthquakes of magnitude 6 to 7 (*e.g.*, BOROVIK *et al.*, 1971; OHTAKE, 1976).

As a result of these studies, the recognition of seismicity gaps in active seismic zones is considered to be a promising method for prediction of major earthquakes. This notion was suggested for California earthquakes in an earlier study by ALLEN

[1]) International Institute of Seismology and Earthquake Engineering, Hyakunin-cho 3–28–8, Shinjuku-ku, Tokyo, Japan. On leave from the Marine Science Institute, University of Texas, USA.

[2]) Marine Science Institute, Geophysics Laboratory, University of Texas, 700 The Strand, Galveston, Texas 77550, USA.

et al. (1965). Some investigators, on the other hand, have emphasized that strong earthquakes are frequently preceded by a marked increase in local seismicity (*e.g.*, FEDOTOV, 1967; SADOVSKY *et al.*, 1972; SUYEHIRO and SEKIYA, 1972; WESSON and ELLSWORTH, 1973). KELLEHER and SAVINO (1975), however, revealed that prior seismic activity, if any, is generally confined to the vicinity of the epicenters of pending large earthquakes, and extensive portions of the future rupture zones commonly remain aseismic until the time of the main shocks.

SCHOLZ *et al.* (1973) ascribed both the quiescence and resumption of seismic activity, together with other premonitory phenomena, to pre-seismic dilatancy of the crust and consequent fluid diffusion. According to this hypothesis, seismic quiescence is attributed to the increased strength of a medium as pore pressure drops in the dilatancy-hardening stage; while the renewal of seismic activity occurs with the recovery of pore pressure and consequent weakening of the medium under shear stress. Further investigations of the nature of seismicity gaps can be expected to break new ground, not only for phenomenological earthquake prediction, but in the physical understanding of earthquake occurrence.

The present paper reports a significant decrease in shallow seismicity near Oaxaca,

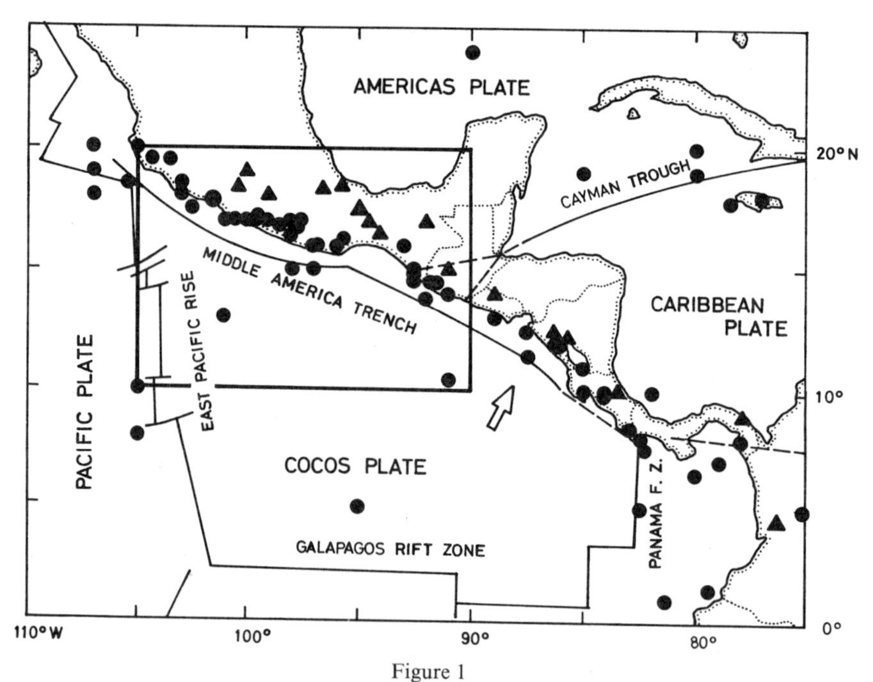

Figure 1

Seismicity and tectonic setting of the Mexico-Middle American region. Small circles and triangles are epicenters of major shallow ($H \leq 65$ km) and deep ($H > 65$ km) focus earthquakes, respectively. The seismic data are taken from the table compiled by DUDA (1965) for the period 1897–1964 ($M_L \geq 7$), and PDE for 1965–1975 ($Ms \geq 7$ or $mb \geq 6.5$). Plate boundaries are delineated after MOLNAR and SYKES (1969). The region of the present study is shown by a rectangle covering southern Mexico and western Guatemala.

southern Mexico, in recent years. Based on this finding, we attempt to predict some of the characteristics of future earthquakes which may take place in this region.

2. *Tectonic setting*

Figure 1 shows the epicenters of major earthquakes which have occurred in the Mexico-Central America region since 1897. The tectonic elements of MOLNAR and SYKES (1969) are also given in this figure. An active seismic belt follows the Pacific margin of Mexico and Central America, paralleling the Middle America Trench. This activity is associated with the subduction of the Cocos plate under the Americas and the Caribbean plates. The pattern of activity is typical of island arcs, and the deep seismic zone is well delineated down to a depth of 250 km (MOLNAR and SYKES, 1969; STOIBER and CARR, 1973). The greatest earthquakes of this region have occurred along the Pacific coast, and were mostly of the low-angle underthrusting type in accordance with the subduction model (MOLNAR and SYKES, 1969). The region of our present study includes the most active part of the seismic belt as outlined by a rectangle in Fig. 1.

3. *Seismicity gap*

Figure 2 compares the shallow seismicity ($H < 60$ km) of the southern Mexico-Guatemala region for two successive time intervals: (a) June 1971–May 1973, and (b) June 1973–May 1975. The data are taken from the *Preliminary Determination of Epicenters* (PDE) published by the National Earthquake Information Service (NEIS) of the United States Geological Survey.

For period (b), an area bounded by 95.5°W and 98°W, as outlined by a rectangle, is almost completely free of shallow earthquakes of sufficient magnitude to be reported by PDE ($M \cong 4$ or larger). Assuming that epicenters are distributed randomly throughout the active belt, the probability that this gap could occur by chance in a given 2-year period is about 1 in 36,000 (See Appendix). An additional 40 to 50 earthquakes have been located by the NEIS in the seismic zone of Fig. 2 between 1 June and 31 December, 1975 (latest data available to us). None of these occurred within the quiescent zone, so that the assertion of nonrandomness is further strengthened.

The seismic history of the anomalous area is illustrated in Fig. 3 by a magnitude versus time diagram for shallow earthquakes ($H < 60$ km) which occurred in the present seismicity gap since 1963. This figure demonstrates a clear reduction in earthquake occurrence, starting in mid-1973, followed by complete quiescence for at least 2.5 years. Such a prolonged period of quiescence is clearly unusual for this region.

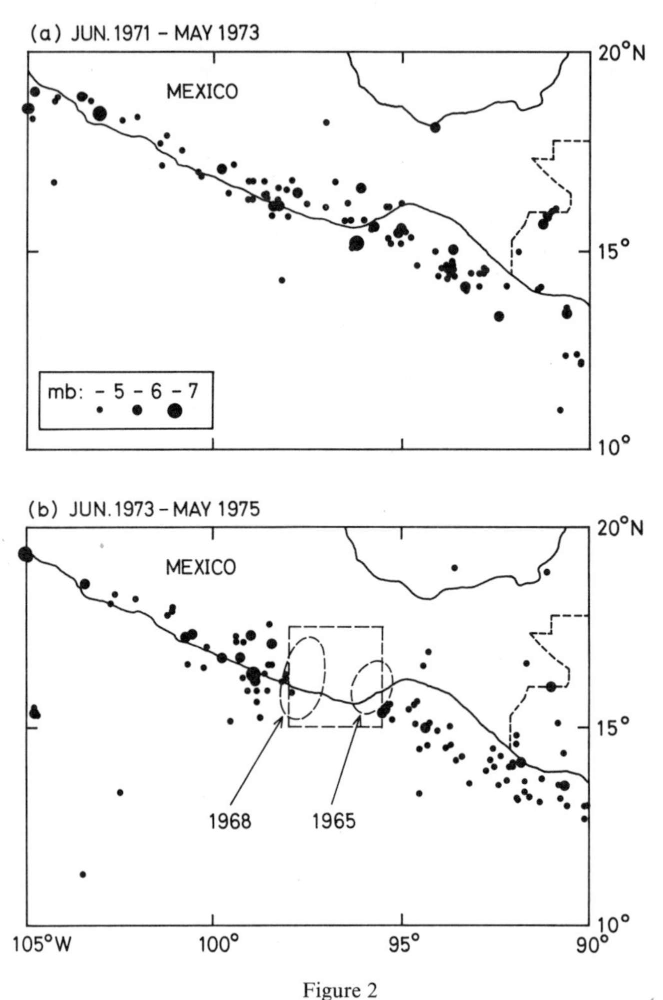

Figure 2
Epicentral distribution of shallow earthquakes (H < 60 km) which occurred in the southern Mexico-Guatemala region during the period of (a) June 1971–May 1973, and (b) June 1973–May 1975. The area of seismic quiescence during the latter period is outlined by a square. Two ellipses show aftershock zones of the Oaxaca earthquakes of 1965 and 1968, respectively.

4. Oaxaca earthquakes of 1965 and 1968

Two large earthquakes have occurred in the Oaxaca region during the past decade:

Aug. 23, 1965 $\varphi = 16.33°\mathrm{N}$, $\lambda = 95.80°\mathrm{W}$, $H = 29$ km, $Ms = 7\frac{1}{2} - 7\frac{3}{4}$

Aug. 2, 1968 $\varphi = 16.56°\mathrm{N}$, $\lambda = 97.79°\mathrm{W}$, $H = 36$ km, $Ms = 7.5$

(after the *Bulletin of the International Seismological Centre*). The approximate

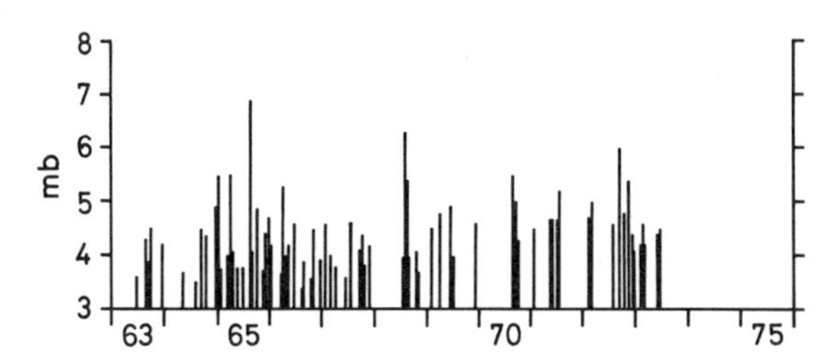

Figure 3
Magnitude versus time diagram of shallow earthquakes ($H < 60$ km) which took place within the square in Fig. 2(b) since 1963. Each event reported by PDE is plotted by a vertical segment of which length corresponds to its body wave magnitude.

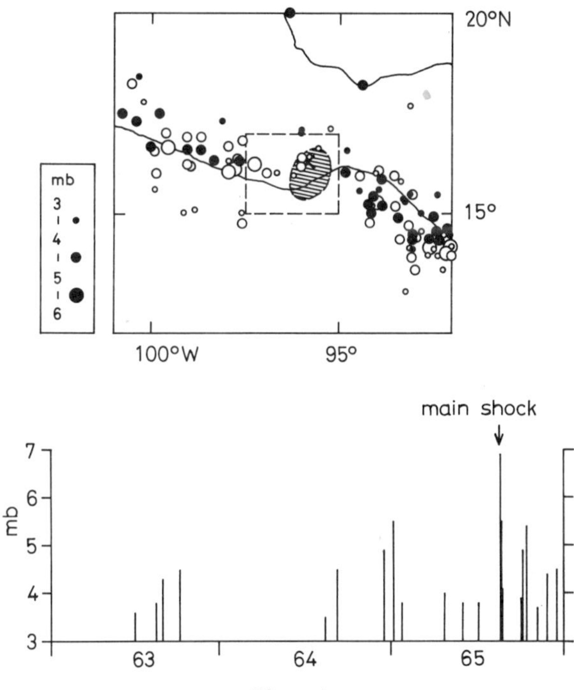

Figure 4
Change in regional seismicity prior to the Oaxaca earthquake of August 23, 1965. Upper: Solid and open circles indicate shallow earthquakes ($H < 60$ km) for the periods of seismic quiescence and resumption preceding the main shock, respectively. The shaded area is the aftershock zone of the main shock (cross mark). Lower: Magnitude versus time diagram for the earthquakes which took place inside of the rectangular area shown in the upper figure.

aftershock zones of these events are plotted in Fig. 2(b) by ellipses. Both of these earthquakes were preceded by anomalous changes in local seismicity. For the former case, local seismicity exhibited a quiescent stage from late 1963 to mid 1964, followed by a renewal of activity prior to the main shock (Fig. 4). For the latter case, preseismic quiescence was more distinct while resumption of activity was weak (Fig. 5).

In this paper, we refer to the period of quiescence as the α stage, and to the period of renewed activity as the β stage. The time interval between the onset of the α stage and occurrence of main shock was 1.5 to 2 years for the Oaxaca earthquakes. It is not clear whether the activities of the β stage were confined in the vicinity of the epicenters of the main shocks or not. Such a detailed analysis seems to require more reliable location of the epicenters, and also more complete coverage of small earthquakes.

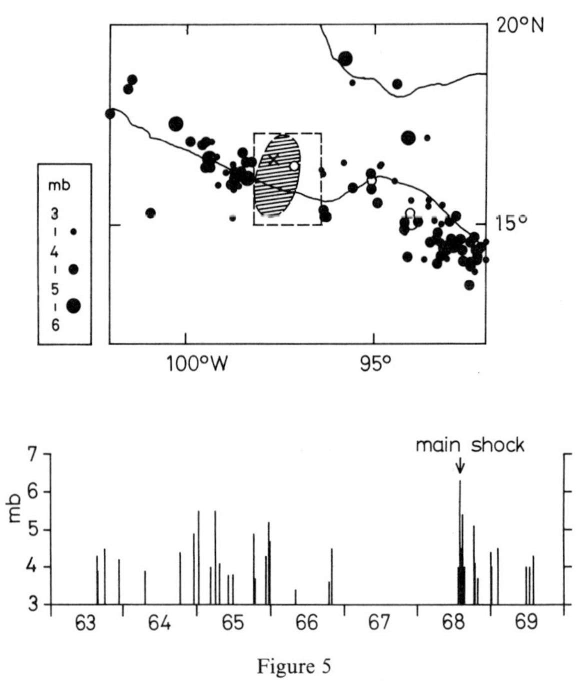

Figure 5
Change in regional seismicity prior to the Oaxaca earthquake of August 2, 1968. See Fig. 4 for explanation.

It is interesting to note that aftershocks did not completely fill the seismicity gap suggesting the smaller dimension of the rupture zone compared with the seismically anomalous area.

The focal mechanisms of the earthquakes were typical underthrust with shallow dip angle. The azimuth and plunge of the slip vector for the 1965 earthquake is reported by MOLNAR and SYKES (1969) as ϕ = N36°E and δ = 14°, respectively. Fig. 6 shows distribution of *P* initial motions of the 1968 event based on readings from microfilm records of the World-Wide Standard Station Network. The slip corresponds to a low angle thrust with ϕ = N49°E and δ = 6°.

Both of the main shocks were located near the northern extremes of their aftershock zones. This suggests that the rupture started at depth beneath the landward (NE) end of the dislocation, and unilaterally propagated towards shallower depth along the subducted plate interface. This pattern of rupture propagation is frequently found for large, thrust-type earthquakes as discussed by KELLEHER *et al.* (1973).

Summarizing the above discussions, the Oaxaca earthquakes of 1965 and 1968 are characterized by marked similarities in magnitude, space-time patterns of precursory seismicity change, focal mechanisms, and rupture patterns. These systematics may be useful in predicting the characteristics of future earthquakes in this region.

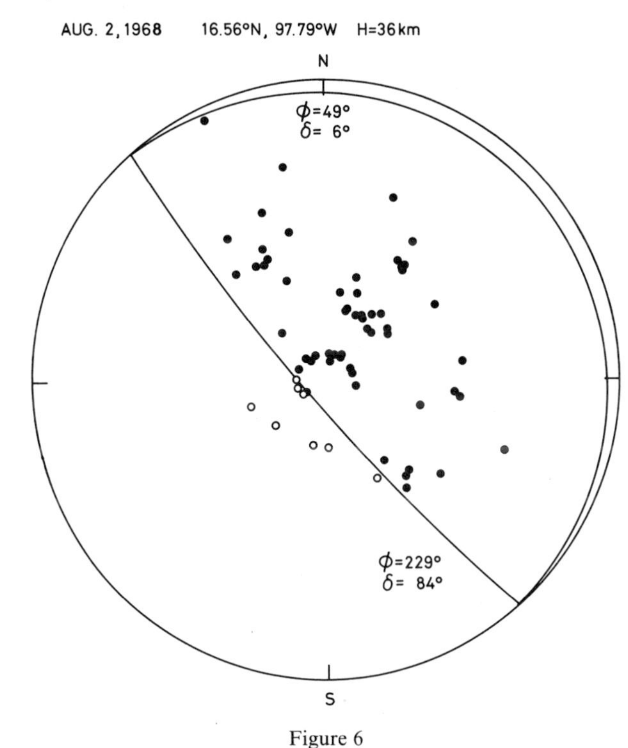

Figure 6
Equal area projection (lower hemisphere) of seismic radiation pattern for the Oaxaca earthquake of August 2, 1968. Solid and open circles indicate compressional and dilatational initial motions of P wave, respectively. This radiation pattern can be interpreted as representing underthrusting on a fault plane dipping at a shallow angle to the northeast.

5. *The impending earthquake*

The arguments presented above suggest to us that the relative seismic quiescence near Oaxaca may be signaling the eventual occurrence of a large earthquake in this region. We try to predict the most probable characteristics of the anticipated earthquake in the following.

Magnitude

The size of the present seismicity gap is roughly 7×10^4 km^2 in area, and 300 km in linear dimension, respectively. UTSU and SEKI (1955) and UTSU (1961) found statistical relations between the magnitude of a main shock M and area A or linear dimension l of its aftershock zone:

$$\log A = 1.02\, M - 4.0 \tag{1}$$

$$\log l = 0.5\, M - 1.8 \tag{2}$$

Direct application of those formulae results in magnitude $8.6 \sim 8.7$ for the future earthquake if we assume that the aftershock zone will completely cover the present zone of quiescence. This seems unlikely, however, based upon the data presented above which suggests that the size of the aftershock zone is normally smaller than the preceding seismicity gap, at least in the Oaxaca region. Moreover, strain energy has been released in the eastern and western parts of the seismicity gap by the recent earthquakes of 1965 and 1968 (see Fig. 2). Thus, we would expect that the central area between those aftershock zones is the most likely site of the next large earthquake in this region. The separation between those two previous aftershock zones is roughly comparable with their dimensions. Therefore, the magnitude of the predicted earthquake may also be comparable to the previous events, that is, $7\frac{1}{2} \sim 7\frac{3}{4}$.

According to ANDERSON and WHITCOMB (1975), the linear dimension, L, of an area showing anomalous behaviour prior to an earthquake is empirically related to l by a simple formula,

$$\log l/L^2 = -3. \tag{3}$$

Combining (3) with (2), magnitude 7.3 is predicted for $L^2 = 7 \times 10^4$ km^2. In conclusion, $7\frac{1}{2} \pm 1/4$ is the best estimate we can make for the magnitude.

Location

As stated above, it appears likely that the aftershock zone of a future earthquake will occupy the central part of the present seismicity gap. The earthquake will probably be of the thrust type. The rupture is likely to begin in the northern portion of the zone and propagate southward, judging from the past instances. If this is the case, the epicenter of the main shock should be in the region, $\varphi = 16.5° \pm 0.5°$N and $\lambda = 96.5° \pm 0.5°$W.

Occurrence time

More than three years have passed since the onset of seismic quiescence in the Oaxaca region. This is substantially longer than the precursor times of the previous Oaxaca earthquakes. As yet, we have no reliable basis for predicting the time of occurrence. The precursor times observed for the 1965 and 1968 Oaxaca earthquakes (magnitude $7\frac{1}{2} \sim 7\frac{3}{4}$) were much shorter than the 13 to 31 years expected from the magnitude-precursor time relations by WHITCOMB *et al.* (1973), SCHOLZ *et al.* (1973), and RIKITAKE (1976). This suggests that precursory time relations may vary greatly between various tectonic provinces. It may also indicate that some completely different phenomenon from the previous studies is being observed. In any case, previous *standard* magnitude-precursor time relations do not seem to be applicable in the present examples. However, a resumption of seismic activity corresponding to the β stage of activity, may be expected prior to the future, large earthquake as in the cases of 1965 and 1968.

6. *Discussions and conclusions*

Since June 1973, the frequency of shallow earthquakes in the area of Oaxaca, southern Mexico, has been unusually low, This area experienced two major earthquakes in the past decade: an event of magnitude 7.5 in 1965, and one of magnitude $7\frac{1}{2}$–$7\frac{3}{4}$ in 1968. These were both preceded by intervals of quiescence (α stage) and following resumption (β stage) of local seismicity in advance of the main shocks.

Previous case studies suggest that the β stage is not a universal phenomenon as is the α stage. Although it may partly be attributed to insufficient detection capability, there are some cases which definitely contradict our finding. KELLEHER and SAVINO (1975), for instance, reported that the prior seismic activity was not detected with reasonable coverage of small earthquakes for the Sitka earthquake of 1972, $Ms = 7.6$ and others. However, the contrast is no wonder because the pattern of earthquake sequences greatly depends on regional characteristics of the lithosphere. Based on laboratory experiments of rock fracture, MOGI (1963) revealed that the main rupture of nonuniform medium tends to accompany distinct activity of foreshocks, which does not appear for a uniform medium. Structural irregularity of the medium and consequent local concentration of stress is considered to be the basic precondition for increase in seismicity prior to the main shocks.

Therefore, the β stage is expected to appear only for regions of nonuniform structure. Really, the region of our present concern exhibits various indications of nonuniform, block-like structure. It is the conclusion of STOIBER and CARR (1973) that the underthrust lithosphere of the Mexico-Central America region is broken along tear faults into 100 to 300 km long segments. In such nonuniform regions, the weakest portions of the lithosphere will sensitively respond to the tectonic force building up so that seismicity change is expected to be a useful measure for predicting large earthquakes.

Based upon the significant scale of the current seismicity gap, both in time and space, together with the previous pattern of earthquake occurrence, the probability of a future, large earthquake in this area is considered to be high. The focal mechanism of the impending earthquake is expected to be a low angle underthrust associated with subduction of the Cocos plate under the Mexico-Middle America arc. The magnitude and epicentral coordinates are estimated to be $Ms = 7\frac{1}{2} \pm \frac{1}{4}$, and $\varphi = 16.5° \pm 0.5°$N, $\lambda = 96{\cdot}5° \pm 0.5°$W. No reliable estimate of occurrence time can be given.

While the evidence presented here is certainly not conclusive, we believe that it is strong enough to justify the initiation of a program of measurements in the Oaxaca region. If possible, the program should include monitoring of microearthquake activity, seismic wave velocity, and crustal deformation. Among these, systematic monitoring of microearthquakes may be the most effective and simplest method since a renewal of activity (β stage) preceding the main event is expected. Detection of a significant increase in seismic activity in the region would not pinpoint the time-of-occurence, but it would serve as a warning that the time remaining before the earthquake might be quite short.

Acknowledgements

We gratefully acknowledge the assistance of Dr. Yoshio Nakamura and Dr. James Dorman for critical review of the manuscript and helpful advice. This work was partially supported by the Harry Oscar Wood Fund, Marine Science Institute, Geophysics Laboratory, Contribution No. 199.

References

ALLEN, C., AMAND, P. ST., RICHTER, C. and NORDQUIST, J. (1965), *Relationship between seismicity and geologic structure in the southern California region,* Bull. Seism. Soc. Amer. *55*, 753–797.

ANDERSON, D. and WHITCOMB, J. (1975), *Time-dependent seismology,* J. Geophys. Res. *80*, 1497–1503.

BOROVIK, N., MISHARINA, L. and TRESKOV, A. (1971), *On the possibility of strong earthquakes in Pribayakalia in the future,* Izv. Acad. Sci. USSR, Phys. Solid Earth, Engl. Transl. 13–16.

DUDA, S. (1965), *Secular seismic energy release in the circum-Pacific belt,* Tectonophys. *2*, 409–452.

FEDOTOV, S. (1965), *Regularities of the distribution of strong earthquakes in Kamchatka, the Kuril Islands, and northeastern Japan,* Trudy Inst. Fiz. Zemli, *36* (203), 66–93 (in Russian).

FEDOTOV, S. (1967), *Long-range seismic forecasting for the Kurile-Kamchatka zone,* Transactions of the Meetings of the Far East Earth Science Division, Akademii Nauk SSSR, Moscow, 1–16 (in Russian).

HOGG, R. and CRAIG, A., *Introduction to Mathematical Statistics,* (MacMillan, New York, 1959), 152–160.

KELLEHER, J. (1972), *Rupture zones of large South American earthquakes and some predictions,* J. Geophys. Res. *77*, 2087–2103.

KELLEHER, J. and SAVINO, J. (1975), *Distribution of seismicity before large strike slip and thrust-type earthquakes,* J. Geophys. Res. *80*, 260–271.

KELLEHER, J., SYKES, L. and OLIVER, J. (1973), *Possible criteria for predicting earthquake locations and their application to major plate boundaries of the Pacific and the Caribbean,* J. Geophys. Res. *78*, 2547–2585.

MOGI, K. (1963), *Some discussions on aftershocks, foreshocks and earthquake swarms —— the fracture of a semi-infinite body caused by an inner stress origin and its relation to the earthquake phenomena,* Bull. Earthq. Res. Inst., Univ. Tokyo, *41*, 615–658.

MOGI, K. (1968a), *Sequential occurrences of recent great earthquakes,* J. Phys. Earth, *16*, 30–36.

MOGI, K. (1968b), *Some features of recent seismic activity in and near Japan* (*1*), Bull. Earthq. Res. Inst., Univ. Tokyo, *46*, 1225–1236.

MOGI, K. (1969), *Some features of recent seismic activity in and near Japan* (*2*), *Activity before and after great earthquakes,* Bull. Earthq. Res. Inst., Univ. Tokyo, *47*, 395–417.

MOLNAR, P. and SYKES, L. (1969), *Tectonics of the Caribbean and Middle America regions from focal mechanisms and seismicity,* Geol. Soc. Amer. Bull. *80*, 1639–1684.

NAGUMO, S. (1973), *Activation mode of great submarine earthquakes along the Japanese Islands,* Publications for the 50th Anniversary of the Great Kanto Earthquake, 1923, Earthq. Res. Inst., Univ. Tokyo, 273–291.

OHTAKE, M. (1976), *Search for precursors of the 1974 Izu-Hanto-oki earthquake, Japan*, Pure Appl. Geophys. *114* (in press).

RIKITAKE, T., *Earthquake Prediction*, (Elsevier, Amsterdam, 1976), pp. 357.

SADOVSKY, M., NERESOV, I., NIGMATULLAEV, S., LATYNINA, L., LUKK, A., SEMENOV, A., SIMBIREVA, I. and ULOMOV, V. (1972), *The processes preceding strong earthquakes in some regions of middle Asia*, Tectonophys. *14*, 295–307.

SCHOLZ, C., SYKES, L. and AGGARWAL, Y. (1973), *Earthquake prediction: a physical basis*, Science, *181*, 803–810.

STOIBER, R. and CARR, M. (1973), *Quaternary volcanic and tectonic segmentation of Central America*, Bull. Volc. *37*, 304–325.

SUYEHIRO, S. and SEKIYA, H. (1972), *Foreshocks and earthquake prediction*, Tectonophys. *14*, 219–225.

SYKES, L. (1971), *Aftershock zones of great earthquakes, seismicity gaps, and earthquake prediction for Alaska and the Aleutians*, J. Geophys. Res. *76*, 8021–9041.

UTSU, T. (1961), *A statistical study on the occurrence of aftershocks*, Geophys. Mag. *30*, 521–605.

UTSU, T. (1968), *Seismic activity in Hokkaido and its vicinity*, Geophys. Bull. Hokkaido Univ. *20*, 51–75 (in Japanese).

UTSU, T. and SEKI, A. (1955), *A relation between the area of after shock region and the energy of main-shock*, Zisin, J. Seism. Soc. Japan, Ser. 2, *7*, 233–240 (in Japanese).

WESSON, R. and ELLSWORTH, W. (1973), *Seismicity preceding moderate earthquakes in California*, J. Geophys. Res. *78*, 8527–8546.

WHITCOMB, J., GARMANY, J. and ANDERSON, D. (1973), *Earthquake prediction: variation of seismic velocities before the San Fernando earthquake*, Science, *180*, 632–635.

(Received 20th September 1976)

Pageoph, Vol. 115 (1977), Birkhäuser Verlag, Basel

Kinetic Shear Resistance, Fluid Pressures and Radiation Efficiency During Seismic Faulting

By Richard H. Sibson

Summary – During earthquake faulting, radiation efficiency and the degree of stress relief are critically dependent on the kinetic shear resistance. This is often assumed to stay constant during slip, but geological evidence suggests that for moderate or large shallow earthquakes it may decrease dramatically to near-zero values once slip is initiated, either by melt formation or by transient increases in fluid pressure on the fault plane. The latter, probably more common process may arise partly through an interaction between temperature and water pressure, and partly through dilatancy recovery as shear stress is relieved. If the fault remains undrained, stress relief should be absolute with seismic efficiency reaching high values, so that stress drops give a measure of the level of tectonic shear stress in fault zones. Supporting evidence comes from the observation that apparent stress is generally about half the stress drop.

Key words: Seismic faulting; Pore fluid pressure; Radiation efficiency.

1. *Introduction*

Absolute values of tectonic shear stress, τ, in fault zones are largely unknown, though from heat flow studies an upper limit of a few hundred bars is considered likely for the San Andreas Fault (Brune *et al.* 1969). Minimum values for shear stress are provided by estimates of stress drop,

$$\Delta\tau = \tau_1 - \tau_2 \tag{1}$$

and accelerating stress,

$$\tau_{ac} = \tau_1 - \tau_{fk} \tag{2}$$

for shallow earthquakes; τ_1 and τ_2 being respectively the initial and final values of shear stress on the fault and τ_{fk} the kinetic shear resistance. (Note that *accelerating stress* is synonymous with the term *effective stress* used by Brune (1970) and others, which is avoided here because of possible confusion with the effective stress principle applied to fluid-saturated rocks.) Large earthquakes ($M_s \geqslant 6$) typically have stress drops of about 100 bars in intraplate regions, or about 30 bars when they occur in long-established fault zones at plate boundaries (Kanamori and Anderson 1975). On the basis of observed ground accelerations and velocities, Brune (1970) finds that τ_{ac} is of the order of 100 bars or less. However, if shear resistance remains high

[1]) Dept. of Geology, Imperial College, London SW7 2BP, England.

throughout slip and only a small fraction of the shear stress is relieved, as in laboratory studies of frictional stick-slip, *kilobar* shear stresses may still occur in natural fault zones.

This paper explores the possibility that moderate to large earthquakes at shallow depths incorporate *feedback* mechanisms whereby kinetic shear resistance is drastically reduced once slip is initiated, so that substantial stress relief occurs and a high radiation efficiency is achieved.

2. *Models for frictional stick-slip*

Energy budget

During shallow earthquake faulting, the energy released (E) is partitioned in the manner,

$$E = E_s + E_f + E_c + E_g \tag{3}$$

where E_s is the seismic wave energy, E_f is the work dissipated in frictional traction over the fault surface, E_c is the fracture energy expended at the crack-tip as the rupture propagates and E_g is the work done against gravity. Estimates of E_c are uncertain; HUSSEINI *et al.* (1975) suggest a range of 10^3–10^7 erg cm^{-2} for frictional rupture on existing faults, and for the time being this factor is neglected. To simplify matters further, we consider the case of strike-slip faulting so that E_g also may be disregarded. Under these conditions, the seismic radiation efficiency may be expressed as,

$$\eta = E_s/E = \frac{E - E_f}{E} \tag{4}$$

Source parameters

In the simple models for frictional stick-slip which are generally considered (see DIETERICH 1974), slip starts when the static frictional resistance on the fault, τ_{fs} is overcome so that,

$$\tau_1 = \tau_{fs} = \mu_s \cdot \sigma_n' = \mu_s(\sigma_n - P) \tag{5}$$

where μ_s is the static coefficient of friction (typically about 0.75), σ_n is the normal stress across the fault and P is the fluid pressure. Once slip has begun, motion is opposed by the kinetic shear resistance,

$$\tau_{fk} = \mu_k \cdot \sigma_n' = \mu_k(\sigma_n - P) \tag{6}$$

where $\mu_k (< \mu_s)$ is the kinetic coefficient of friction (Fig. 1a). Initially the stress available for accelerating fault motion is $(\tau_1 - \tau_{fk})$, but linear relief of tectonic shear stress takes place with displacement (probably at around 10^2–10^3 bars m^{-1}), so that deceleration occurs when τ drops below τ_{fk}.

Thus from the fault models of OROWAN (1960) and SAVAGE and WOOD (1971), if rupture propagation effects are neglected and an average slip, $\bar{u}$, takes place over a shear dislocation of area, A, on a strike-slip fault, the total energy released is,

$$E = \tfrac{1}{2}(\tau_1 + \tau_2)\cdot\bar{u}\cdot A = \bar{u}\cdot A\cdot\bar{\tau} \tag{7}$$

while the frictional work is,

$$E_f = \bar{u}\cdot A\cdot\bar{\tau}_{fk}, \tag{8}$$

where $\bar{\tau}_{fk}$ is the average kinetic shear resistance during slip. Then from equation 4 the efficiency is,

$$\eta = \frac{\bar{\tau} - \bar{\tau}_{fk}}{\bar{\tau}} \tag{9}$$

and the apparent stress (which can be estimated from the ratio of seismic energy to moment (AKI 1966)) is,

$$\eta\cdot\bar{\tau} = \bar{\tau} - \bar{\tau}_{fk} = \frac{\Delta\tau}{2} + (\tau_2 - \bar{\tau}_{fk}). \tag{10}$$

Following SCHOLZ *et al.* (1972), equation (10) can be rewritten in the form,

$$\Delta\tau = 2\left\{\tau_1 - \frac{\bar{\tau}_{fk}}{1 - \eta}\right\} \tag{11}$$

illustrating the critical interdependence of stress drop, efficiency and kinetic shear resistance. From Fig. 1 and equations (9) and (11) it is clear that the lower the value of $\bar{\tau}_{fk}$ with respect to τ_1, the greater is the radiation efficiency and the degree of stress relief. Though direct evidence is lacking, τ_{fk} is generally assumed to stay roughly constant during experimental stick-slip (Fig. 1a), and the ratio μ_s/μ_k (which corresponds to the ratio τ_1/τ_{fk} for the same effective normal stress) usually has values of

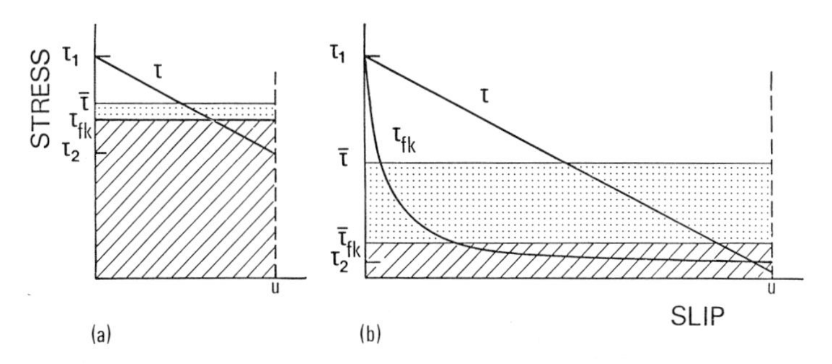

Figure 1

Stress-displacement models for stick-slip; (a) τ_{fk} stays constant, (b) τ_{fk} decreases during slip. Dotted area is a measure of E_s, diagonal hachuring represents E_f.

around 1.1 to 1.4 (DIETERICH 1974). Taking an average value of 1.25 and a typical fractional stress drop ($\Delta\tau/\tau_1$) of 0.1, the radiation efficiency may be estimated from equation (11) to be about 16%. However, it will be shown that when natural earthquake faulting involves displacements of more than a few centimetres (i.e., $M_s > 4$, say) at depths in excess of a kilometre or so, it is most unlikely that τ_{fk} stays constant throughout slip. Rather, it may be expected to drop quickly to low values as in Fig. 1b, so that the fractional stress drop is high and much of the released energy is expended in seismic radiation.

Savage–Wood inequality

On the assumption that fault motion is brought to a halt solely by the kinetic shear resistance so that $\tau_2 < \bar{\tau}_{fk}$, SAVAGE and WOOD (1971) used equation (10) to derive the inequality,

$$\eta \cdot \bar{\tau} < \frac{\Delta\tau}{2}. \tag{12}$$

KANAMORI and ANDERSON (1975) find, however, that for moderate to large earthquakes $\eta \cdot \bar{\tau} \sim \Delta\tau/2$ which suggests that $\tau_2 \sim \bar{\tau}_{fk}$. This situation is likely to arise for the model in Fig. 1b, when both τ_2 and $\bar{\tau}_{fk}$ have values approaching zero.

3. *Role of aqueous fluids in determining feedback mechanism*

Two alternative mechanisms are envisaged whereby τ_{fk} is sharply reduced as seismic slip progresses. The first involves lubrication of the fault plane by formation of a frictionally generated melt; this is thought to be possible only under dry conditions in the *absence* of an intergranular fluid (SIBSON 1973). The second mechanism depends on the *presence* of an aqueous intergranular fluid, and involves the generation of a transient increase in fluid pressure on the fault during slip. In accordance with equation (6) this serves to lower the effective normal stress across the fault, thereby decreasing the kinetic shear resistance.

One may infer that an aqueous fluid phase is generally present during faulting from the widespread occurrence of hydrothermal veining and alteration products in the brittle regions of ancient fault zones, and the retrograde metamorphic assemblages found in deep-level shear zones (BEACH 1976, SIBSON 1977). Faulting is likely to occur under dry conditions only in more or less intact crystalline rocks of low porosity. This is borne out by the general scarcity of fault-generated pseudotachylyte melt-rock, which is developed only in host rocks of this sort (HIGGINS 1971, SIBSON 1975).

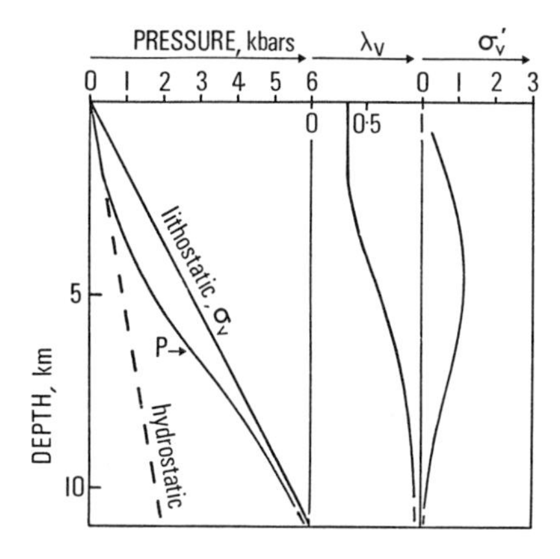

Figure 2
Schematic representation of fluid pressure variation within a major fault zone.

Fluid pressures in fault zones

According to the principle of effective stress (HUBBERT and RUBEY 1959), the effective vertical pressure at a depth, z, in the crust may be written,

$$\sigma_v' = \sigma_v - P = \rho g z(1 - \lambda_v) \tag{13}$$

where the pore–fluid factor,

$$\lambda_v = \frac{P}{\rho g z} \tag{14}$$

ρ being the average crustal density and g the acceleration due to gravity. At high crustal levels, where pore space is connected through to the surface, λ_v has a value of 0.3 to 0.4, corresponding to the ratio of water to rock densities, but evidence from deep oil wells suggests that λ_v may approach unity at depth as a result of increasing sediment compaction. PRICE (1975) argues that tectonic strain-rates, probably several orders of magnitude greater than those associated with sediment compaction should also serve to increase fluid pressure, so high λ_v values may be expected in the vicinity of active fault zones. BERRY (1973) has cited evidence for the existence of such high pressures in the vicinity of the San Andreas Fault. Further, if *kilobar* stresses are not to occur in natural fault zones, and the frictional properties of faults bear any resemblance to those measured in the laboratory, high fluid pressures are needed to lower the frictional constraints on faulting (SIBSON 1974, STESKY *et al.* 1974). In fact, the frequent occurrence of extensional veining in ancient fault zones suggests that fluid pressure locally exceeded the least principal compressive stress to an extent that hydraulic fracturing took place (SECOR 1969). Certainly it is commonly assumed that $\lambda_v = 1$ under greenschist grades of metamorphism and above, when mineral grains can deform plastically to decrease void space (TURNER 1968, p. 60). Under normal geothermal gradients of 20–30°C km^{-1}, such conditions should exist at depths in excess of 10–15 km in major fault zones.

Thus the maximum effective vertical stress at any depth in established fault zones is probably about 0.65 times the lithostatic load, and may well be much less, reaching a peak value somewhere in the depth range 0–15 km (Fig. 2).

Stress relationships

In a compressive stress field with principal stresses $\sigma_1 > \sigma_2 > \sigma_3$, strike-slip faulting will take place with $\sigma_v = \sigma_2$ (ANDERSON 1951). As both σ_2 and the normal stress across the fault, σ_n, must lie between σ_1 and σ_3, for the purposes of this discussion we consider the effective stresses, $\sigma'_2 = \sigma'_v = \sigma'_n$.

4. *Friction-melting*

The possibility of a melt forming on a fault plane, as the result of frictional heat generated during seismic slip, has been considered by JEFFREYS (1942), ANDERSON (1951), AMBRASEYS (1969), PRICE (1970), and especially McKENZIE and BRUNE (1972), who point out that friction-melting should lead to particularly high efficiency earthquakes. All these analyses are of *dry* frictional sliding and neglect the possible effects of an intergranular fluid, the presence of which inhibits large increases in temperature (see below). However, in the rather rare circumstances of dry faulting, the general conclusion is that temperature rises sufficient to bring about melting ($\Delta T \sim 1000$°C) are to be expected for displacements of more than a few centimetres at seismic slip-rates of 10–100 cm s^{-1}, provided τ_{fk} is of the order of 100 bars or more. Thus for our model of a strike-slip fault, friction-melting should occur under dry conditions (i.e., $\lambda_v = 0$) for moderate to large earthquakes at depths greater than about 1 km, assuming $\mu_k = 0.5$ (see also Fig. 6).

From an analysis of ancient faults containing pseudotachylyte, SIBSON (1975) deduced the empirical equation,

$$\tau_{fk} = 1.13 \times 10^9 \cdot u^{-1/2} = 5.4 \times 10^7 \cdot a^{-1} \qquad \text{c.g.s.} \tag{15}$$

where u is the displacement and a is the average thickness of the melt layer. This relationship has the implication that once a melt forms, the kinetic shear resistance drops extremely quickly from an initially rather high value (>1.6 kbar) to a figure possibly representing the viscous resistance of the thickening layer of melt. Because there is a strong tendency for melt to be ejected from the generating surface (Fig. 3), it is important to note that equation (15) holds only while the fault plane remains undrained.

Figure 3
Pseudotachylyte melt-rock in an injection-vein complex leading off a generating fault plane (arrowed). Host rock is quartzo-feldspathic Lewisian gneiss. From the zone of the Outer Hebrides Thrust, Isle of Barra, Scotland.

5. *Transient fluid pressures*

In the more usual situation where an intergranular fluid is present during seismic slip, some combination of the two processes described below should lead to a transient increase in fluid pressure, ΔP, about the fault plane. From equation (6), by taking differentials,

$$\Delta\tau_{fk} = -\mu_k \cdot \Delta P \tag{16}$$

so that the rise in fluid pressure is accompanied by a drop in shear resistance.

Dilatancy recovery

In the dilatancy/fluid-diffusion model for shallow earthquakes (NUR 1972, SCHOLZ *et al.* 1973, WHITCOMB *et al.* 1973), seismic failure occurs when the crust around a fault zone is in a fluid-saturated dilatant state. In theory, the degree of dilatancy ($\Delta V/V$) increases with the level of tectonic shear stress and dilatancy recovery should therefore accompany slip (SCHOLZ and KRANZ 1974), but because little is known of the micromechanics involved, the constitutive law is not well established. NUR (1975) suggests that for one-dimensional behaviour it may be approximated by a power law of general form,

$$(\Delta V/V) \propto \tau^n \tag{17}$$

where $n = \frac{1}{2}$ for 'sand-pile' dilatancy (stress sensitive at low ambient stress levels), $n = 1$ for 'joint' dilatancy (constant stress sensitivity), and $n = 2$ for 'microcrack' dilatancy (stress sensitive at high stress levels). At high crustal levels (<10 km, say), whether regional dilation occurs or not, one may expect elements of 'joint' dilatancy in secondary fracture systems and 'sand-pile' dilatancy in granular cataclastic detritus *inside* fault zones themselves. 'Microcrack' dilatancy may also play a role if the shear stress is sufficiently high.

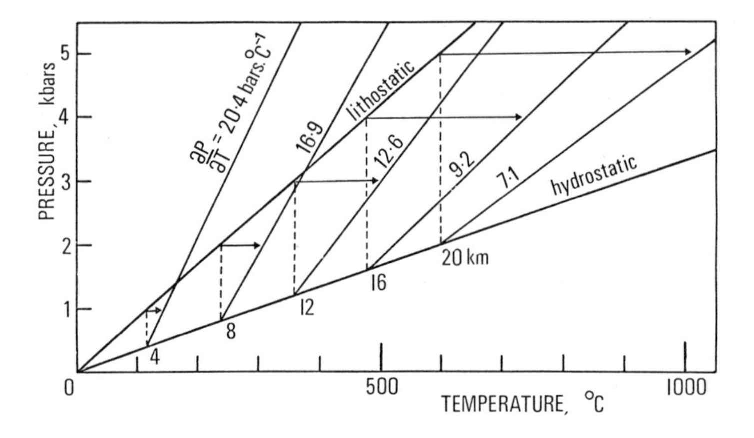

Figure 4
P–T gradients at constant volume for water initially under hydrostatic pressure at various depths along a geothermal gradient of 30°C km^{-1}.

Thus whatever the initial value of shear stress, the stress relief accompanying seismic slip must lead to immediate dilatancy recovery adjacent to the fault plane, and a concomitant increase in fluid pressure. However, as the constants in the constitutive law are not known, it is not yet possible to make meaningful estimates of the magnitude of the effect.

Temperature–fluid pressure interaction

SIBSON (1973) showed that provided void space does not increase greatly, the presence of an aqueous intergranular fluid prevents a large rise in temperature during seismic slip, because the first increments of frictionally generated heat are sufficient to cause a marked increase in fluid pressure on the fault plane. This lowers kinetic shear resistance in accordance with equation (16), and thus the further capacity of the movement to generate heat. It has been suggested above, though, that a *decrease* in void space accompanies slip, so that calculations of the increase in fluid pressure for a given temperature rise, assuming constant volume conditions (as below), are probably well on the conservative side.

For the strike-slip fault model we consider the extreme case with water initially at hydrostatic pressure only, along a geothermal gradient of 30°C km^{-1} (Fig. 4). Under constant volume conditions $\partial P/\partial T$ is very close to linear and may be plotted for various depths using the tables of BURNHAM, HOLLOWAY and DAVIS (1969). If at any depth fluid pressure is raised to equal the lithostatic load, the effective normal stress to the fault and thus τ_{fk} drop to zero. The requisite rise in temperature is plotted against depth in Fig. 5. As the feedback mechanism involves interdependent and non-steady-state sliding velocities, heat flow and fluid flow, full analysis would be difficult, but the following arguments indicate its potential for lowering kinetic shear resistance.

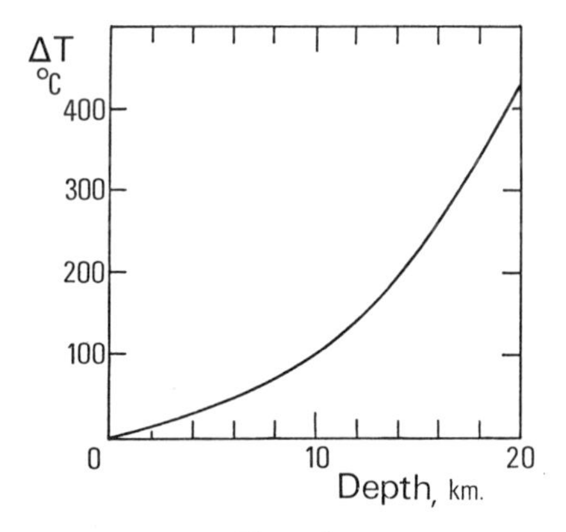

Figure 5
Temperature increase, ΔT, needed to raise fluid pressure from hydrostatic to lithostatic values versus depth, for water held at constant volume.

Moderate or large earthquake fault displacements take place in a few seconds or tens of seconds with maximum slip-rates ranging up to perhaps 100 cm s^{-1} or so (BRUNE 1970). We investigate the effect of constant power dissipation on the fault surface during an initial increment, Δt, of the total time required for a seismic displacement. The power dissipated per unit area of the fault plane is,

$$\dot{Q} = \tau_{fk} \cdot v \tag{18}$$

where v is the slip-rate, and the resulting rise in temperature, ΔT, can be calculated using the equation for constant heat flux on a plane in an infinite medium (CARSLAW and JAEGER 1959, p. 76). If values of 10^{-2} erg cm^{-2} and 2×10^5 erg cm^{-1} °C^{-1} are adopted for thermal diffusivity and conductivity respectively, we can write,

$$\Delta T = 2.82 \times 10^{-7} \cdot \tau_{fk} \cdot v \cdot \sqrt{\Delta t} \quad \text{c.g.s.} \tag{19}$$

Using this relationship, lines of equal temperature rise (and power dissipation) may be plotted on a graph of τ_{fk} versus v for a given time increment, in this case lasting one second (Fig. 6). Thus a temperature rise of say 40°C (nearly sufficient to reduce τ_{fk} to zero at 5 km depth for the conditions in Fig. 4) could be accomplished in one second if, for example, a linear decrease in τ_{fk} from 142 to 1.42 bars is accompanied by an increase in slip-rate from 1 to 100 cm s^{-1} (corresponding to a seismic acceleration of $\sim$0.1 g).

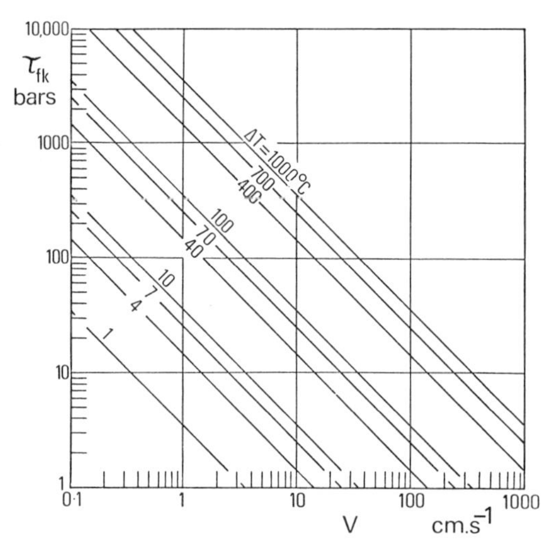

Figure 6
Lines of equal temperature rise, ΔT, on a plot of kinetic shear resistance, τ_{fk}, versus slip-rate, v, for an increment of constant power dissipation lasting 1 s.

In a more general analysis, we can consider *any* initial value of kinetic shear resistance, τ^*_{fk}, and the constant fractional value thereof, $q\tau^*_{fk}$, which has to act at constant sliding velocity to lower τ^*_{fk} to $q\tau^*_{fk}$ during a short time interval, Δt. Clearly, the parameter q provides a *minimum* estimate of the factor by which τ_{fk} is reduced during the slip increment. By putting $\Delta\tau = \tau^*_{fk}(1 - q)$ in equation (16) and $\tau_{fk} = q\tau^*_{fk}$ in equation (19), the expressions may be combined to give,

$$q = \left\{1 + 2.82 \times 10^{-7} \cdot \mu_k \cdot v \cdot \sqrt{\Delta t} \cdot \left(\frac{\partial P}{\partial T}\right)\right\}^{-1} \quad \text{c.g.s.} \tag{20}$$

which is independent of τ_{fk}. In Fig. 7, values of q are plotted against the lower range of seismic slip-rates for initial slip increments lasting 0.1 and 1 s, with $\mu_k = 0.5$ and $\partial P/\partial T$ having values of 15 and 1.5 bar °C^{-1}. The first of these gradients corresponds to water held at constant volume without leakage at a depth of 10 km. For the same depth the lower gradient could result either from an increase in void space, which has been shown to be unlikely, or from mass leakage of fluid by about 16% in the first 100°C of temperature rise.

It is apparent that q must underestimate the true reduction factor by a proportion which increases as q decreases. With this in mind, together with the likelihood that slip-rate is accelerating at 100 cm s^{-2} or more during the early stages of a seismic displacement (HANKS and JOHNSON 1976), it is evident from Fig. 7 that the interaction between temperature and fluid pressure cannot be neglected even if a considerable loss of fluid occurs from the vicinity of the fault plane.

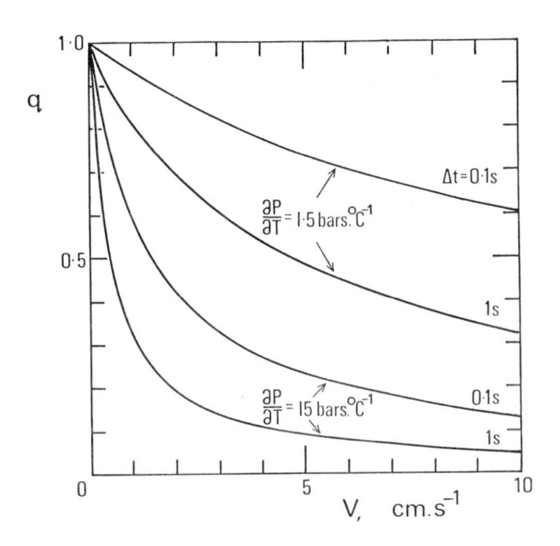

Figure 7
'q' reduction factor versus slip-rate for $\partial P/\partial T = 15$ and 1.5 bars °C^{-1} and initial slip increments lasting 0.1 and 1 s.

6. *Fault drainage and the degree of stress relief*

Both friction–melting and the temperature–fluid pressure interaction will lead to fluid pressure gradients normal to and immediately adjacent to the fault. Thus the effectiveness of both these mechanisms for reducing τ_{fk} depends on the extent to which these gradients are maintained and the fault stays undrained during slip. Over the short time intervals involved, through-rock drainage will not, of course, occur for a viscous rock melt, but can be significant for aqueous fluids in permeable rocks like sandstones. Inside fault zones, however, fine cataclastic gouge with a high content of clay minerals is invariably present at supra-metamorphic levels. Such material may be expected to have properties similar to compacted argillaceous sediments which, while retaining comparatively high porosities (say, 5%), possess very low permeabilities ($<10^{-6}$ darcies) approaching those of crystalline rocks (PRICE 1975). For permeabilities of this order, fluid loss from the fault plane during short time increments will be negligible. Note also that increases in fluid pressure which result from dilatancy recovery during slip do not depend on the creation of a steep pressure gradient normal to the fault, and are therefore unaffected by permeability characteristics.

However, it should be borne in mind that rock permeabilities increase markedly as fluid pressure rises to decrease the effective confining pressure (BRACE *et al.* 1968). Also if fluid pressures rise above the least principal compressive stress, existing fractures may be forced open when suitably oriented, allowing rapid drainage of the fault plane to take place by channel flow. New extension fractures may also be created by hydraulic fracturing (or magmatic fracturing in the case of friction-melting – see Fig. 3).

Thus although kinetic shear resistance may be drastically lowered by the cited feedback mechanisms, it is doubtful whether complete reduction to zero values can ever be achieved.

Peak acceleration data

HANKS and JOHNSON (1976) have interpreted high frequency (>5 Hz), large amplitude (>0.5 g) acceleration pulses as resulting from localised and isolated bursts of intense faulting superimposed on the gross motion of earthquake fault displacements. Such behaviour might well be caused by spasmodic transitions from low to high efficiency faulting induced by one of the feedback mechanisms. Each episode of acceleration could result from localised *lift-off* and reduction of shear resistance, followed by sudden drainage of the fault plane.

7. *Conclusions*

An aqueous intergranular fluid is generally present in established fault zones at pressures at least as great as hydrostatic. However, irrespective of whether faulting occurs under wet or dry conditions, it has been shown that feedback mechanisms should bring about a reduction in kinetic shear resistance once slip exceeds a few centimetres at depths greater than perhaps a kilometre. In a dry environment the mechanism involves friction-melting, while under wet conditions transient increases in fluid pressure arise through some combination of local dilatancy recovery and the temperature–fluid pressure interaction. The effectiveness of both feedback mechanisms is critically dependent on the extent to which the fault remains undrained during slip.

If little drainage occurs, τ_{fk} will be lowered to near-zero values and as a consequence, both radiation efficiency and the degree of stress relief will be high. Thus stress drops should be an approximate measure of the initial tectonic shear stress, which satisfies the observation that in general the Savage–Wood inequality is not obeyed, and apparent stress is about half the stress drop. However, these conclusions are based on the assumption that the fracture energy expended to propagate the earthquake fault, E_c, can be neglected. As E_c will increase in direct proportion to the earthquake fault dimension, it may be this factor which is responsible for arresting rupture propagation if tractional forces over the fault surface are nullified by one or other of the feedback mechanisms.

If on the other hand, the feedback mechanisms are only partially operative during slip, they could be responsible for the high frequency pulses of acceleration which are sometimes observed, but the overall stress drop would not then be a measure of tectonic shear stress.

Acknowledgements

I thank Dr. E. H. Rutter for reading the manuscript. This work forms part of a project supported by the USGS National Earthquake Hazards Reduction Programme, Contract No. 14-08-0001-G-377.

References

AKI, K. (1966), *Generation and propagation of G waves from the Niigata earthquake of June 16, 1964, 2, Estimation of earthquake moment, released energy, and stress–strain drop from G-wave spectrum*, Bull. Earthquake Res. Inst., Tokyo Univ. *44*, 73–88.

AMBRASEYS, N. N. (1969), *Maximum intensity of ground movements caused by faulting*, Proc. 4th World Conf. Earthquake Engineering, Santiago *1*, 154–171.

ANDERSON, E. M., *The Dynamics of Faulting*, 2nd edn. (Oliver & Boyd, Edinburgh 1951, 206 pp.

BEACH, A. (1976), *The interrelations of fluid transport, deformation, geochemistry, and heat flow in early Proterozoic shear zones in the Lewisian complex*, Phil. Trans. R. Soc. Lond. *A280*, 569–604.

BERRY, F. A. F. (1973), *High fluid potentials in the California Coast Ranges and their tectonic significance*, Bull. Am. Assoc. Petroleum Geologists *57*, 1219–1249.

BRACE, W. F., WALSH, J. B. and FRANGOS, W. T. (1968), *Permeability of granite under high pressure*, J. Geophys. Res. *73*, 2225–2236.

BRUNE, J. N. (1970), *Tectonic stress and the spectra of seismic shear waves from earthquakes*, J. Geophys. Res. *75*, 4997–5009.

BRUNE, J. N., HENYEY, T. L. and ROY, R. F. (1969), *Heat flow, stress and rate of slip along the San Andreas Fault, California*, J. Geophys. Res. *74*, 3821–3827.

BURNHAM, C. W., HOLLOWAY, J. R. and DAVIS, N. F. (1969), *Thermodynamic properties of water to 1000°C and 10,000 bars*, Geol. Soc. Am. Spec. Paper, 132.

CARSLAW, H. S. and JAEGER, J. L., *Conduction of Heat in Solids*, 2nd edn. (Clarendon Press, Oxford 1959), 510 pp.

DIETERICH, J. H., *Earthquake mechanisms and modelling*, in *Annual Review of Earth and Planetary Science*, Vol. 2 (ed. F. A. Donath) (Annual Reviews Inc., Palo Alto, Calif., 1974), 275–301.

HANKS, T. C. and JOHNSON, D. A. (1976), *Geophysical assessment of peak accelerations*, Bull. Seism. Soc. Am. *66*, 959–968.

HIGGINS, M. W. (1971), *Cataclastic Rocks*, U.S. Geol. Survey Prof. Paper, 687.

HUBBERT, M. K. and RUBEY, W. W. (1959), *Role of fluid pressure in mechanics of overthrust faulting*, Bull. Geol. Soc. Am. *70*, 115–206.

HUSSEINI, M. I., JOVANOVICH, D. B., RANDALL, M. J. and FREUND, L. B. (1975), *The fracture energy of earthquakes*, Geophys. J. R. astr. Soc. *43*, 367–385.

KANAMORI, H. and ANDERSON, D. L. (1975), *Theoretical basis of some empirical relations in seismology*, Bull. Seism. Soc. Am. *65*, 1073–1095.

JEFFREYS, H. (1942), *On the mechanics of faulting*, Geol. Mag. *79*, 291–295.

MCKENZIE, D. and BRUNE, J. N. (1972), *Melting on fault planes during large earthquakes*, Geophys. J. R. astr. Soc. *29*, 65–78.

NUR, A. (1972), *Dilatancy, pore fluids and premonitory variations of t_s/t_p travel times*, Bull. Seism. Soc. Am. *62*, 1217–1222.

NUR, A. (1975), *A note on the constitutive law for dilatancy*, Pure appl. Geophys. *113*, 197–206.

OROWAN, E. (1960), *Mechanism of seismic faulting*, in *Rock Deformation*, Geol. Soc. Am. Mem. *79*, 323–345.

PRICE, N. J. (1970), *Laws of rock behaviour in the earth's crust* in *Rock mechanics – theory and practice*', Proc. 11th Symp. Rock Mech., Berkeley California (ed. W. H. Somerton) (Am. Inst. Min. Metall. & Petrol. Eng., New York), 1–23.

PRICE, N. J. (1975), *Fluids in the crust of the earth*, Sci. Progr. *62*, 59–87.

SAVAGE, J. C. and WOOD, M. D. (1971), *The relation between apparent stress and stress drop*, Bull. Seism. Soc. Am. *61*, 1381–1388.

SCHOLZ, C. H. and KRANZ, R. (1974), *Notes on dilatancy recovery*, J. Geophys. Res. *79*, 2132–2135.

SCHOLZ, C. H., MOLNAR, P. and JOHNSON, T. (1972), *Detailed studies of frictional sliding of granite and implications for the earthquake mechanism*, J. Geophys. Res. *77*, 6392–6406.

SCHOLZ, C. H., SYKES, L. R. and AGGARWAL, Y. P. (1973), *Earthquake prediction – a physical basis*, Science *181*, 803–810

SECOR, D. T. (1969), *Mechanics of natural extension fracturing at depth in the earth's crust*, Can. Geol. Survey, Paper 68-52, 3–48.

SIBSON, R. H. (1973), *Interactions between temperature and fluid pressure during earthquake faulting – a mechanism for partial or total stress relief*, Nature Phys. Sci. *243*, 66–68.

SIBSON, R. H. (1974), *Frictional constraints on thrust, wrench and normal faults*, Nature *249*, 542–544.

SIBSON, R. H. (1975), *Generation of pseudotachylyte by ancient seismic faulting*, Geophys. J. R. astr. Soc. *43*, 775–794.

SIBSON, R. H. (1977), *Fault rocks and fault mechanisms*, J. Geol. Soc. Lond. (in press).

STESKY, R. M., BRACE, W. F., RILEY, D. K. and ROBIN, P. Y. F. (1974), *Friction in faulted rock at high temperature and pressure*, Tectonophys. *23*, 177–203.

TURNER, F. J., *Metamorphic Petrology* (McGraw-Hill, New York 1968), 403 pp.

WHITCOMB, J. H., GARMANY, J. D. and ANDERSON, D. L. (1973), *Earthquake prediction: variation of seismic velocities before the San Fernando earthquake*, Science *180*, 632–641.

(Received 20th September 1976)

Pageoph, Vol. 115 (1977), Birkhäuser Verlag, Basel

A Viscoelastic Relaxation Model for Post-Seismic Deformation from the San Francisco Earthquake of 1906

By JOHN B. RUNDLE and DAVID D. JACKSON[1])

Abstract – We present a realistic model of the San Andreas fault zone. We propose that aseismic ground displacement is a sum of visco-elastic relaxation following large earthquakes, transient fault slip, steady fault slip and a large-scale relative plate motion. We used the model to explain the aseismic ground displacements observed after the San Francisco earthquake of 1906.

The data do not resolve the question of which is the dominant mechanism, but viscoelastic relaxation can contribute a significant fraction of the displacement if the elastic plate thickness is 50 km or less. If the relative plate motion is taken to be 5.5 cm/yr, as found from plate rotation pole studies, then the zone of significant shearing in the mantle is probably at least 100 km thick beneath California.

Key words: Viscoelastic relaxation; Earthquake San Francisco 1906; Stress in lithosphere.

Examination of horizontal geodetic triangulation and trilateration data has led to the conclusion that relative motion across the San Andreas fault occurs in both the spasmodic fashion characteristic of an earthquake and in a non-catastrophic manner extending over weeks, months or years (SAVAGE and BURFORD, 1973; THATCHER, 1975a; THATCHER, 1975b). To explain the data, standard techniques in the mathematical theory of elasticity have been used to construct models of the fault in which the coseismic ground displacements originate from a dislocation in an elastic half-space (KNOPOFF, 1958; CHINNERY, 1963; PRESS, 1965). Models describing aseismic strain accumulation across the fault are presently based upon stable sliding on a vertical plane below a shallow locked region in either an elastic half-space (SAVAGE and BURFORD, 1973) or in an elastic plate (TURCOTTE and SPENCE, 1974).

The purpose of the work reported here was to develop a more realistic model of the San Andreas fault. Many of the models previously mentioned assumed the earth to be homogeneous in its elastic properties. Although this may be a good assumption for the uppermost 20 km of the San Andreas, it appears that a transition to some kind of anelasticity (BRACE and BYERLEE, 1970) occurs at greater depths. While one of the models (TURCOTTE and SPENCE, 1974) described above allows for such a transition, it does so only for the steady strain accumulation phase of the

[1]) Department of Geophysics and Space Physics, University of California, Los Angeles, California 90024 USA.

earthquake cycle. Transient effects due to the interaction of the earthquake stresses with the anelastic zone are ignored.

We propose to describe the San Andreas fault as a fracture in an elastic layer overlying a Maxwell viscoelastic half-space (Fig. 1).

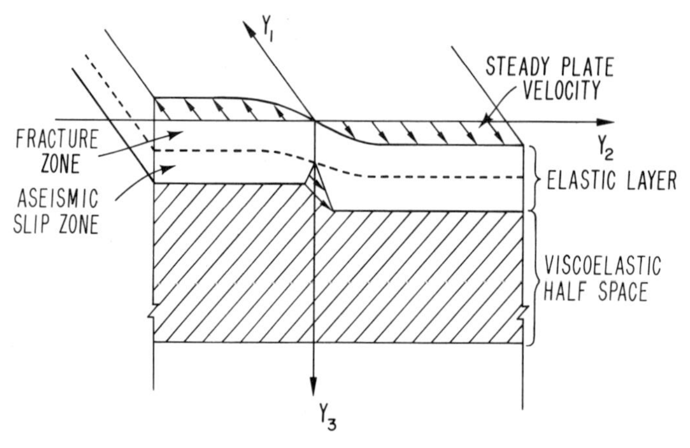

Figure 1
Proposed model for the San Andreas fault. An elastic surface layer, divided into an upper 'brittle' seismic zone and a lower 'aseismic slip' zone overlies a viscoelastic half-space. The forces driving the elastic layer produce a steady relative plate velocity at the free surface which approaches a constant value far from the fault.

This model is similar to that used by NUR and MAVKO (1974) to explain vertical displacements following thrust faulting in Japan, and to those of SMITH (1974) and BARKER (1975). Episodic and steady aseismic slip are allowed to exist in the lower portion of the elastic layer, and we admit the existence of surface displacements due to a large-scale relative plate driving force (RUNDLE and JACKSON, 1976a). We can therefore explain the long-term aseismic surface displacements observed across the fault by a sum of transient viscoelastic relaxation following nearby earthquakes, aseismic episodic fault slip, aseismic steady fault slip and the steady plate driving force.

The surface displacements due to viscoelastic relaxation are computed analytically by means of an approximation to the Green's function (RUNDLE and JACKSON, 1976b) for a static double-couple nucleus of strain in an elastic layer over a Maxwell viscoelastic half-space. The approximate Green's function, which was obtained by superposing image sources, differs from the exact Green's function in that the displacements and normal tractions are only approximately continuous at the boundary between layer and half-space. The superposition method leads to the exact Green's function for a simpler problem, that of a single-couple line source in an elastic layer over an elastic half-space (RYBICKI, 1971). A test of the surface displacements from a rectangular surface fault computed using both an analytic integration of the approximate Green's function and a numerical integration of the exact Green's function

(BEN-MENAHEM and SINGH, 1968) indicates that use of the approximate Green's function introduces errors of only a few per cent. The advantage of using the approximate Green's function is that all further mathematical operations can be done analytically, greatly facilitating the inverse problem.

Viscous properties are introduced into the half-space using the correspondence principle (BIOT, 1954; LEE, 1955), which assumes linear viscoelasticity. Although the rheology of the anelastic region is unknown, the assumption of a linear flow law is probably good to a first approximation (MCKENZIE, 1967; MCKENZIE, 1968; POST and GRIGGS, 1973; STOCKER and ASHBY, 1973; FROIDEVAUX and SCHUBERT, 1975; PARMENTIER, TURCOTTE and TORRANCE, 1976). Alternatively, it may be that at relatively shallow depths, stress relaxation occurs by hydrogeologic flow. Darcey's law, which governs the transient flow of water in porous media, is a linear diffusion equation, and thus to a first order of approximation this stress relaxation process is linear. In any case, we feel that the inclusion of both viscous and elastic properties in the half-space is necessary if we are to realistically describe ground displacements from times shortly after an earthquake to times long after the event. Use of linear viscoelasticity by way of the correspondence principle is the most convenient way to produce such a realistic model of an earthquake.

We have chosen the mechanical properties of the half-space to be those of a Maxwell viscoelastic solid whose elastic constants are the same as those of the elastic layer. A Maxwell element is shown in Fig. 2. A choice of Maxwell mechanical properties is motivated by the observation that for times short compared to a day, the earth behaves elastically, while for much longer times, it appears to deform by 'flowing' under applied stress (MCCONNELL, 1965; CRITTENDEN, 1968; PELTIER, 1974). This dual behavior is similar to that exhibited by the Maxwell element.

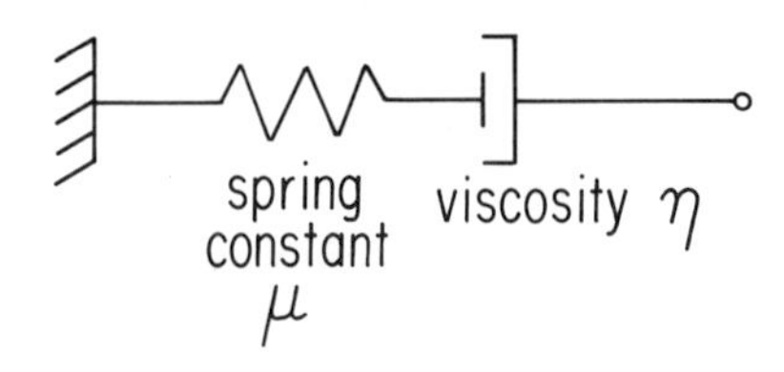

Figure 2
A Maxwell element.

We used THATCHER's (1974, 1975a, 1975b) fault parameters for the 1906 San Francisco earthquake in our model. In particular, the depth of faulting was assumed to be 10 km for the section from San Juan Bautista to Point Arena. With this value of fault depth, the elastic layer thickness can be no greater than about 50 km if viscoelastic relaxation is to play a significant role in the post-seismic ground displacements.

We have also assumed that aseismic fault slip occurs on a strip of the fault plane extending from the bottom of the fractured zone, 10 km, to the bottom of the elastic layer. Episodic slip was taken to be greatest at 10 km depth and to decrease linearly to zero at the bottom of the elastic layer, while steady state slip was taken to be greatest at the bottom of the elastic layer and to decrease linearly to zero at 10 km depth. The surface displacements due to both kinds of slip are computed using RYBICKI's (1971) Green's function for a single-couple line source in an elastic layer over an elastic half-space. For episodic fault slip, Maxwell viscoelastic properties are again introduced into the half-space by means of the correspondence principle, while for steady state fault slip the rigidity of the half-space is set equal to zero.

Since we wish to eventually fit our model to the available geodetic data, we must allow for all possible displacement generating mechanisms. One of these is the gross relative plate motion produced by the plate driving forces in the absence of faulting, aseismic fault slip, and transient viscoelastic relaxation. No matter what the origin of the driving forces, we shall assume that the surface motions they induce across the fault can be represented by

$$u_0 = \frac{2V_0\,\Delta t}{\pi} \tan^{-1}(y_2/y_0) \tag{1}$$

Here y_2 is the horizontal coordinate normal to the fault strike with coordinate origin at the fault, y_0 is a characteristic distance scale, V_0 is one-half the relative plate velocity with respect to points far from the fault and Δt is the time interval of interest. Equation (1) provides the physically reasonable result that the steady driving forces cannot by themselves produce surface motion near the fault except indirectly by some form of faulting, aseismic slip or transient viscoelastic relaxation. Far away from the fault however, we assume that the driving forces will produce the relative displacement $|2V_0\,\Delta t|$.

If in fact the driving forces are convective in origin, we should physically expect them to be near zero at small distances on either side of the fault zone and increase to some constant value at much larger distances. The horizontal distance scale y_0 would then have the interpretation of a horizontal convective wavelength. If the driving forces arise instead from trench or ridge forces, the physical meaning of y_0 is somewhat unclear.

After constructing a model for the fault zone, we adjust the unknowns to best fit the available data. The parameters in the model are H, the elastic layer thickness; τ, a characteristic relaxation time constant for the half-space; V_0, the plate velocity; and y_0, the horizontal distance scale for the velocity. It should be noted that $2V_0$ has also been used for the relative steady fault slip velocity at the boundary between layer and half-space. τ is the quantity η/μ_0, where μ_0 is the rigidity for both layer and half-space and η is the effective viscosity of the half-space. The times of onset and ending of episodic fault slip, and the constant slip velocity at 10 km depth were determined independently by THATCHER (1974, 1975b) and are treated as known. The data were first used to fix H, τ, y_0 and V_0 for models without episodic slip. Later, the full model with all ground displacement mechanisms operating and with the episodic slip parameters chosen on the basis of THATCHER's (1975b) work was employed in fitting the data.

A realistic fault model should satisfy the constraint that the net displacement from all sources, averaged over geologic time and over a relatively narrow fault zone, should be independent of depth for shallow depths. That is, the plate behaves elastically except near a fault, or near its lower boundary. We do not impose such a constraint, because we do not believe that the time period studied here, less than one century, qualifies as 'geologic time'.

Throughout most of California, earthquakes on the San Andreas fault occur at shallow depths of less than about 20 km (BRACE and BYERLEE, 1970). At greater depths, either stable aseismic slip or the existence of a viscoelastic medium can explain the observed lack of earthquakes. The model we propose can represent both of these possibilities. In the data inversion, we found models in which the elastic layer thickness ranged from 16 to 50 km. For small thicknesses, the elastic layer can be identified with the topmost brittle region of the crust and the half-space with a lower crustal zone in which the conditions of temperature and pressure allow creep to occur. For larger thicknesses, we identify the elastic layer with a 'mechanical' lithosphere and the half-space with a 'mechanical' asthenosphere, where the distinction is based solely on the relative importance of elastic and viscous mechanical properties of the earth. We do not expect that the lithosphere so defined will necessarily be identical to that inferred from gravity, heat flow, crustal rebound, seismic velocities, or seismic attenuation (MCCONNELL, 1965; CRITTENDEN, 1968; JACKSON, 1971a; JACKSON, 1971b; PELTIER, 1974; CROUGH, 1975; LEEDS, 1975) as the physical properties to which these methods are sensitive are to some extent independent.

The data for the inversions are made up of a set of post-seismic angle changes obtained from a series of geodetic surveys conducted in northern California over the years 1907–1947. Forty angle changes from survey nets in Point Arena over the period 1907–1930 (Fig. 3) and 28 angle changes from observations in the San Francisco Bay area over the period 1907–1947 (Fig. 4) were used to fix the values of the model parameters. The data are described in considerable detail by THATCHER (1974, 1975a, 1975b), and the method of inversion is taken from JACKSON (1972, 1973, 1974).

We found that our viscoelastic model can fit the data well (see also BARKER, 1976). This is a new result, and it implies that our anelastic model of the fault zone is at least as representative of the San Andreas as any of the previous elastic half-space or elastic plate models. However, it was apparent that many different combinations of viscoelastic displacement mechanisms can explain the data. The values of H found range from 16 to 50 km; of τ from 3.8 to 1000 yr; of y_0 from 5 to 264 km; and of V_0 from -0.08 to -2.75 cm/yr.

The model nonuniqueness is due principally to the poor quality and small number of data that are available. From an analysis of the model parameter covariance matrix (RUNDLE and JACKSON, 1976a) it can be shown that on the order of 10^5

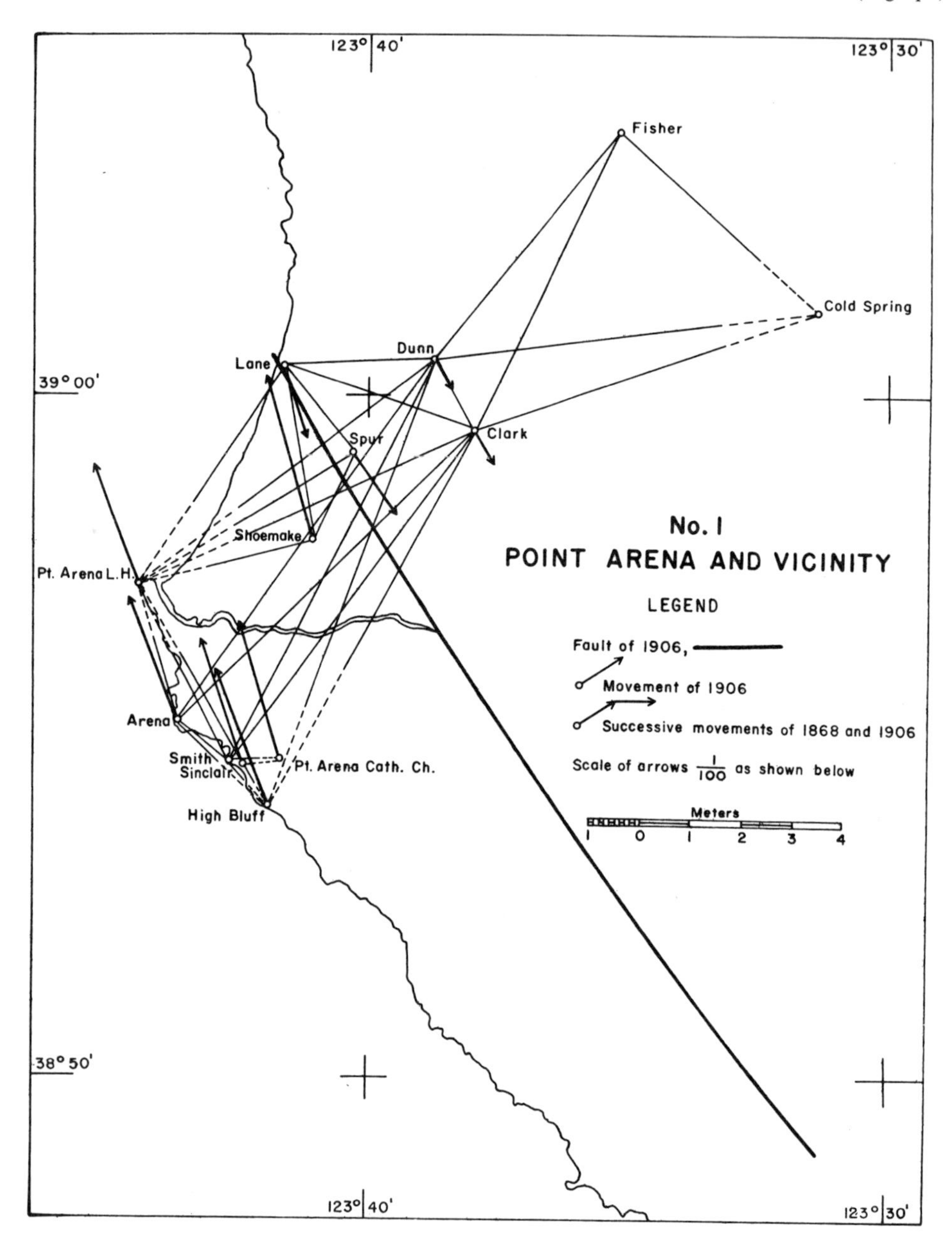

Figure 3
Map of the Point Arena survey net. Figure is taken from HAYFORD and BALDWIN (1907).

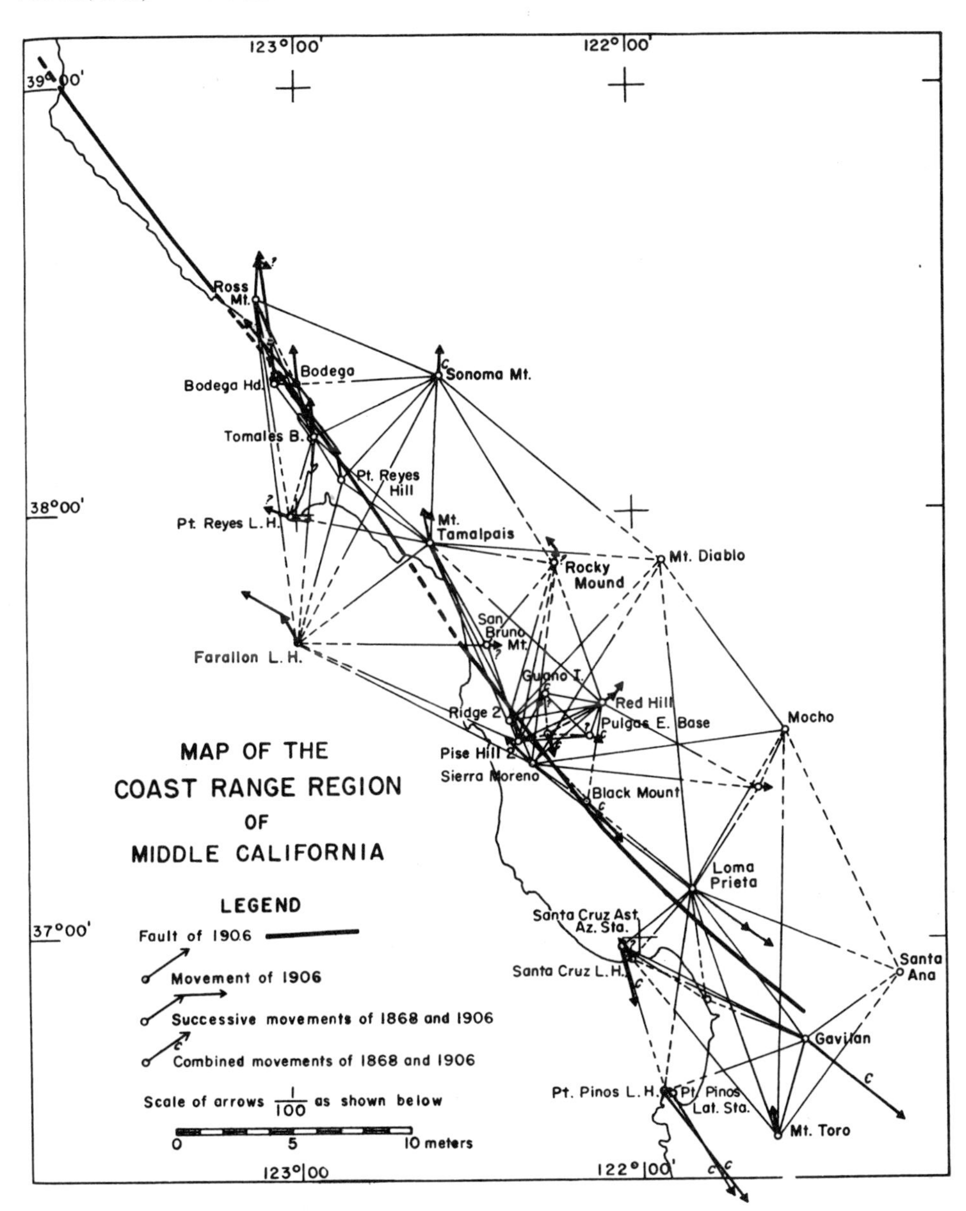

Figure 4
Map of the San Francisco survey net. Figure is taken from HAYFORD and BALDWIN (1907).

statistically independent data are needed to uniquely fix the value of H to within 3 km, of τ to within 2 yr, of y_0 to within 5 km and of V_0 to within 1 cm/yr.

The indeterminancy of the model parameters can be portrayed graphically by using an r.m.s. residual map. As explained in JACKSON (1972), the residual measures the fit between the model predictions and the data. A model is usually judged to be a good fit if the residual is less than 1.0 (standard deviation) and the best fitting model has minimum r.m.s. residual.

In Fig. 5 we show a residual map with H and τ fixed at 10.1 km and 993.3 years. For this value of τ, viscoelasticity has been effectively removed from the model. V_0 and y_0 are allowed to vary and contours of equal residual have been plotted. Neither episodic nor steady fault slip is allowed. We now fix the relative plate velocity, $2V_0$, at the value for northern California found from plate rotation pole studies (MINSTER *et al.*, 1974), -5.5 cm/yr. If we require that a model residual be less than or equal to 1.0, we find that y_0 can vary from less than 50 km to 280 km with the preferred value being 100 km.

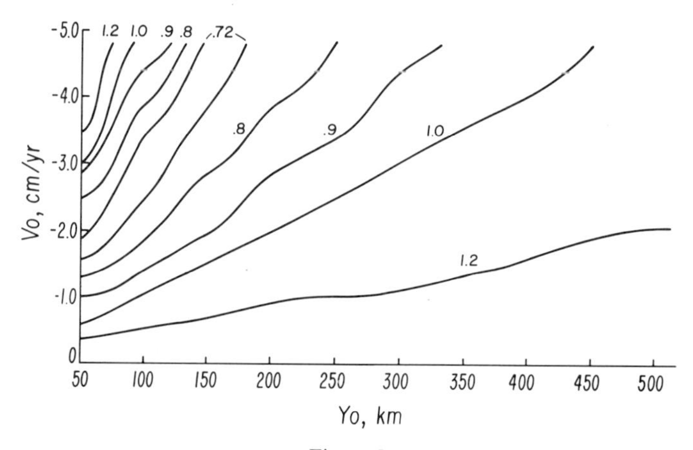

Figure 5
r.m.s. residuals of Bay area data with H and τ fixed at 10.1 km and 993 yr.

Similarly, Fig. 6 is a residual map whose values of H and τ have been fixed at 20.0 km and 5.0 yr. Again the models contain no aseismic fault slip. In this second case a typical constant residual contour has been displaced to the right and down from the same contour in Fig. 5. Because viscoelasticity has been included in the models of Fig. 6, not as much relative plate motion near the fault is needed to account for the observed angle changes as in Fig. 5. Hence y_0 tends to be larger than in Fig. 3 for the same value of V_0. If we again fix $2V_0$ at -5.5 cm/yr, y_0 has only a lower bound of 65 km, an increase over the previous lower bound. No upper bound is apparent and the preferred value of y_0 has also increased to 150 km. Although we can accept values for y_0 of less than 50 km, the preferred y_0 is 100 km.

If the plates are driven by stresses induced by fluid motions, a preferred minimum value of 100 km for y_0 is significant. Physically, the horizontal variation should have a wavelength which is of the same order of magnitude as that for the vertical variation. Thus we conclude that the vertical variations in the fluid motions occupy a zone whose thickness is greater than 100 km.

A final implication of our model is that the stress contained in the viscoelastic region due to a brittle fracture in the elastic layer will be transferred to the layer as the stress relaxation proceeds. Numerical experiments indicate that for a visco-elastic model of the 1906 San Francisco earthquake whose elastic layer thickness is 20 km, the total surface stress regained by the elastic layer in the vicinity of the fault is about 20% of the stress released by the fracture event. For a model whose elastic layer thickness is 50 km, the stress recovery is only about 5%. Thus in our model an earthquake tends to transfer stress from the elastic layer to the visco-elastic zone, and stress relaxation of the viscoelastic region transfers stress back to the layer. This behavior is similar to that exhibited by the model in SAVAGE (1975).

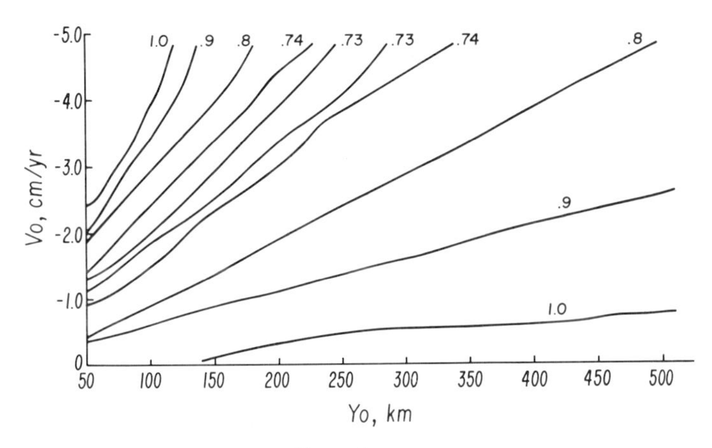

Figure 6
r.m.s. residuals of Bay area data with H and τ fixed at 20 km and 5.0 yr.

The model we have outlined here contains brittle fracture, viscoelastic relaxation and both steady and transient fault slip. We feel that all of these effects must be included if the San Andreas fault zone is to be realistically represented. Our difficulty in finding a unique model stems from both the many ways in which aseismic surface motion can occur and the poor quality of the data. We found that a unique model in which only one parameter is free can be found, but this uniqueness is obtained at the expense of knowledge about the other model parameters.

REFERENCES

BARKER, T. (1976), *Quasi-static motions near the San Andreas fault zone*, Geophys. J. R. astr. Soc. *45*, 689–706.

BEN-MENAHEM, A. and SINGH, S. (1968), *Multipolar elastic fields in a layer half space*, Bull. seism. Soc. Am. *58*, 1519–1572.

BIOT, M. (1954), *Theory of stress–strain relation in anisotropic viscoelasticity and relaxation phenomena*, J. appl. Phys. *25*, 1385–1391.

BRACE, W. and BYERLEE, J. (1970), *California earthquakes: Why only shallow focus?* Science, *168*, 1573–1575.

CHINNERY, M. (1963), *The stress changes that accompany strike-slip faulting*, Bull. seism. Soc. Am. *53*, 921–932.

CRITTENDEN, M. (1968), *Effective viscosity of the earth derived from isostatic loading of Pleistocene Lake Bonneville*, J. geophys. Res. *68*, 5517–5530.

CROUGH, S. (1975), *Thermal model of oceanic lithosphere*, Nature, *256*, 388–390.

FROIDEVAUX, C. and SCHUBERT, G. (1975), *Plate motion and structure of the continental asthenosphere: A realistic model of the upper mantle*, J. geophys. Res. *80*, 2553–2564.

HAYFORD, J. and BALDWIN, A. (1907), *The earth movements in the California earthquake of 1906*, in *Reports on Geodetic Measurements of Crustal Movements* (ed. B. K. Meade), U.S.G.P.O. Publ. No. 0317-00167, Rockville, Md., 1973.

JACKSON, D. (1971a), *The attenuation of Love waves and toroidal oscillations of the earth*, Geophys. J. R. astr. Soc. *25*, 25–34.

JACKSON, D. (1971b), *Attenuation of seismic waves by grain boundary relaxation*, Proc. Nat. Acad. Sci. *68*, 1577–1579.

JACKSON, D. (1972), *Interpretation of inaccurate, insufficient and inconsistent data*, Geophys. J. R. astr. Soc. **28**, 97–109.

JACKSON, D. (1973), *Marginal solutions to quasi-linear inverse problems in geophysics: The edgehog method*, Geophys. J. R. astr. Soc. *35*, 121–136.

JACKSON, D. (1974), *Tradeoffs and simultaneous optimizations in inverse problems*, unpublished manuscript.

KNOPOFF, L. (1958), *Energy release in earthquakes*, Geophys. J. R. astr. Soc. *1*, 44–52.

LEE, E. (1955), *Stress analysis in visco-elastic bodies*, Quart. J. appl. Math. *13*, 183–190.

LEEDS, A. (1975), *Lithospheric thickness in the Western Pacific*, Phys. Earth Planetary Interiors, *11*, 61–64.

MCCONNELL, R. (1965), *Isostatic adjustment in a layered earth*, J. geophys. Res. *70*, 5171–5188.

MCKENZIE, D. (1967), *The viscosity of the mantle*, Geophys. J. R. astr. Soc. *14*, 297–305.

MCKENZIE, D., *The geophysical importance of high temperature creep*, in *The History of the Earth's Crust* (ed. R. Phinney) (Princeton Univ. Press., Princeton 1968).

MINSTER, J., JORDAN, T., MOLNAR, P. and HAYNES, E. (1974), *Numerical modelling of instantaneous plate tectonics*, Geophys. J. R. astr. Soc. *36*, 541–576.

NUR, A. and MAVKO, G. (1974), *Postseismic viscoelastic rebound*, Science, *183*, 204–206.

PARMENTIER, E., TURCOTTE, D. and TORRANCE, K. (1976), *Studies of finite amplitude non-Newtonian thermal convection with application to convection in the earth's mantle*, J. geophys. Res. *11*, 1839–1846.

PELTIER, W. R. (1974), *The impulse response of a Maxwell earth*, Rev. Geophys. space Phys. *12*, 649–699.

POST, R. and GRIGGS, D. (1973), *The earth's mantle: Evidence of non-Newtonian flow*, Science, *181*, 1242–1244.

PRESS, F. (1965), *Displacement, strains and tilts at teleseismic distances*, J. geophys. Res. *70*, 2395–2412.

RUNDLE, J. and JACKSON, D. (1976a), *A three-dimensional viscoelastic model of a strike-slip fault*, Geophys. J. R. astr. Soc. in press.

RUNDLE, J. and JACKSON, D. (1976b), *A realistic model of the San Andreas fault zone applied to the San Francisco earthquake of 1906*, submitted to Geophys. J. R. astr. Soc.

RYBICKI, K. (1971), *The elastic residual field of a very long strike-slip fault in the presence of a discontinuity*, Bull. seism. Soc. Am. *61*, 79–92.

SAVAGE, J. and BURFORD, R. (1973), *Geodetic determination of relative plate motion in central California*, J. geophys. Res. *78*, 832–845.

SAVAGE, J. (1975), *Comment on 'An analysis of strain accumulation on a strike slip fault' by D. L. Turcotte and D. A. Spence*, J. geophys. Res. **80**, 4111–4114.

SMITH, A. T. (1974), *Time dependent strain accumulation and release at island arcs*, Trans. Am. Geophys. Union, *55*, 427, 1974.

STOCKER, R. and ASHBY, M. (1973), *On the rheology of the upper mantle*, Rev. Geophys. space Phys. *11*, 392–426.

THATCHER, W. (1974), *Strain release mechanism of the 1906 San Francisco earthquake*, Science *184*, 1283–1285.

THATCHER, W. (1975a), *Strain accumulation and release mechanism of the 1906 San Francisco earthquake*, J. geophys. Res. *80*, 4862–4872.

THATCHER, W. (1975b), *Strain accumulation on the northern San Andreas fault zone since 1906*, J. geophys. Res. *80*, 4873–4880.

TURCOTTE, D. and SPENCE, D. (1974), *An analysis of strain accumulation on a strike slip fault*, J. geophys. Res. *79*, 4407–4412.

(Received 17th August 1976)

Pageoph, Vol. 115 (1977), Birkhäuser Verlag, Basel

Stress Accumulation and Release on the San Andreas Fault[1])

By D. L. Turcotte[2])

Summary – The San Andreas fault can be divided into locked and free sections. On the locked sections accumulated slip is released in great earthquakes. On the free sections slip is occurring continuously either aseismically or during smaller earthquakes. Stress drops during earthquakes can be estimated from the ratio of short to long period amplitudes and from surface strain. Surface heat flow may provide an upper bound on the absolute stress. The failure or yield stress must reach a maximum at some depth on the fault. This maximum may occur in the near-surface brittle zone or deeper in the plastic zone of the fault. The historic distribution of seismic activity provides information on the stress level. The accumulation of strain and stress on the fault can be predicted using elastic theory. It is necessary, however, to include the viscous coupling of the lithosphere to the asthenosphere in order to fully model the problem.

Key words: Stress in lithosphere; San Andreas fault; Viscoelastic relaxation.

1. Introduction

The San Andreas fault stretches almost the entire length of the state of California as shown in Fig. 1. Great earthquakes are known to occur on the San Andreas, the last on April 18, 1906. In terms of plate tectonics the San Andreas is believed to accommodate lateral sliding between the Pacific and North American plates. An understanding of the time dependence of relative motion across the fault is essential to understanding the behavior of the fault.

On some sections of the fault there is little seismic activity, the fault appears to be locked and strain is being accumulated. On other sections of the fault, small earthquakes and aseismic creep relieve at least a fraction of the accumulating strain. On the locked sections of the fault the accumulated strain appears to be relieved in great earthquakes. There is no evidence that great earthquakes occur on the sections where the strain is being continually relieved.

It is appropriate to divide the fault into four sections: these are the northern locked section, the central free section, the southern locked section, and the southern free section. A coordinate system is introduced to measure distance along the fault; its origin is at the center of the central free section as shown in Fig. 1. The central free section extends 280 km from near Cholame to Redwood City just south of San

---

[1]) Contribution No. 588 of the Department of Geological Sciences, Cornell University.

[2]) Department of Geological Sciences, Cornell University, Ithaca, New York 14853, USA.

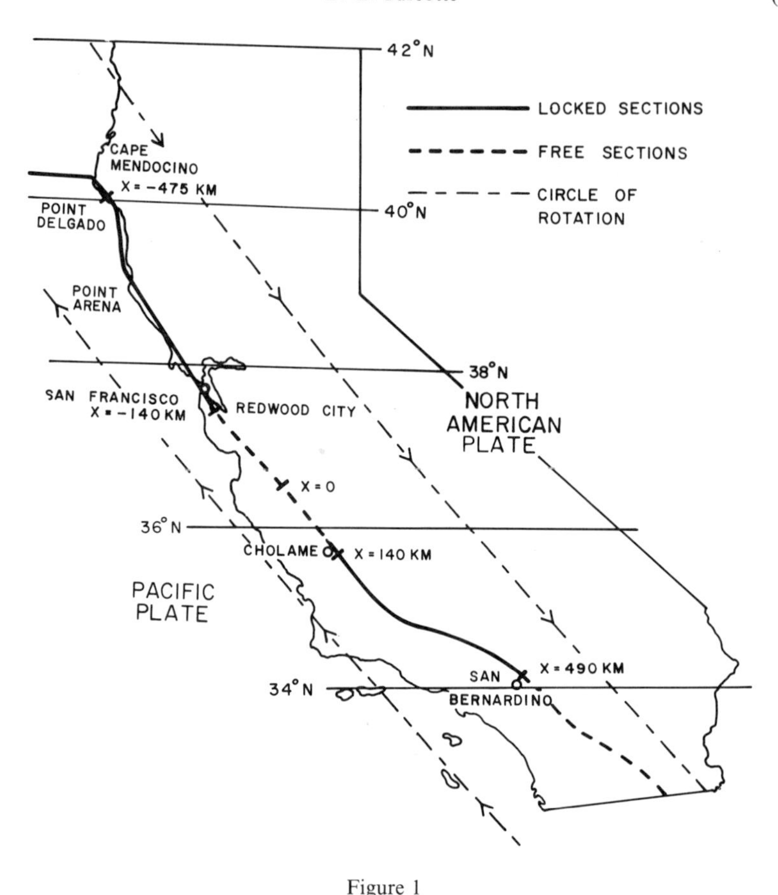

Figure 1
The locked and free sections of the San Andreas fault are shown. Distances are measured from the center of the central free section. Also shown are two circles of rotation for the motion of the Pacific plate relative to the North American plate about the pole of rotation given by MINSTER *et al.* (1974).

Francisco. Fault creep and small earthquakes occur in this section of the fault continuously.

The northern locked section extends about 330 km from Redwood City to Cape Mendocino. A fault break was reported for the entire length of this section in the 1906 earthquake and the measured surface displacements are given in Fig. 2. There has been no reported fault creep or earthquake on this section of the fault since 1906. Clearly this section is locked and is accumulating slip. The southern locked section extends some 350 km from Cholame to near San Bernardino. There are reports of a fault break along the entire length of this section in 1857. Since then, no fault creep or earthquakes have been reported on this section. Although the magnitude 7.7 Kern County earthquake of 1952 and the magnitude 6.4 San Fernando earthquake in 1971 occurred near this section, it is reasonable to conclude that this section of the San Andreas is locked and that slip has accumulated since 1857.

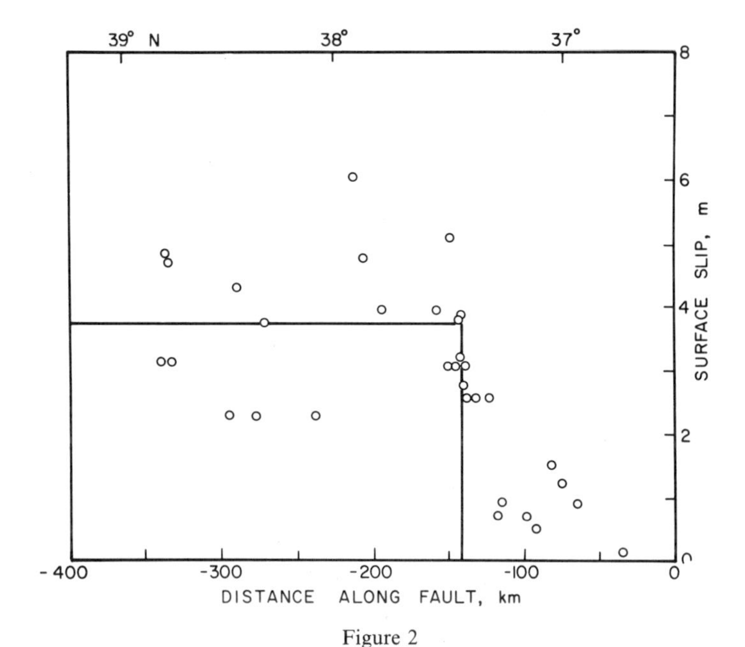

Figure 2
Surface displacements on the San Andreas fault during the April 18, 1906 earthquake (after THATCHER, 1975a). The solid line is the predicted displacement on the northern locked section assuming that 5.5 cm/yr accumulated since the earthquake of 1838.

South of San Bernardino the San Andreas fault splays off into a series of faults. Small and moderate earthquakes and fault creep occur on a number of these and it appears that this section of the fault is not locked.

2. *Tectonic significance of the San Andreas fault*

There is no question that the San Andreas fault represents a major boundary between the Pacific and North American plates. If the fault accommodates the relative motion between the two plates then it must lie on a small circle about the pole of rotation that defines the relative motion of the plates. Taking the pole of rotation (50.9°N, −66.3°E) given by MINSTER *et al.* (1974), two circles of rotation are compared with the trend of the fault in Fig. 1. The good agreement between the circles of rotation and the general trend of the San Andreas fault is evidence that the fault is the primary boundary between the Pacific and North American plates.

WOODFORD and MCINTYRE (1976) have obtained a best fit for a small circle to the San Andreas fault and have found that the resulting pole of rotation deviates considerably from the pole position given by MINSTER *et al.* (1974). This approach, however, fails to take into account the tectonic significance of the bends in the San Andreas fault. As shown in Fig. 1, the straight sections of the fault are in good agreement with the pole of rotation given by MINSTER *et al.* (1974). The convergence of the plates at the bend in the fault north of Los Angeles requires thrusting and this has resulted in the formation of the Transverse Ranges.

Studies utilizing marine magnetic anomalies have shown that the relative velocity between the Pacific and North American plates is 5.5 cm/yr (ATWATER and MOLNAR, 1973). This velocity is an average over several million years. The most recent data point is the last reversal of the earth's magnetic field about 700,000 years ago. If it is hypothesized that this velocity is representative of the motion between the plate on a time scale of hundreds of years and if it is further hypothesized that a large fraction of this velocity occurs across the San Andreas fault, several conclusions and predictions can be made.

A major earthquake occurred in northern California during late June 1838. LAUDERBACK (1947) has carried out a study of this earthquake using the available sources of information – diaries, books, etc. He concluded: 'This study has convinced me that the earthquake of June, 1838, was the next preceding earthquake corresponding to the earthquake of April, 1906, involving a major fault-trace phenomenon along the central (and perhaps northern) Coast Range course of the San Andreas fault.'

If the 1838 earthquake relieved the accumulated slip on the northern section of the San Andreas fault, assuming a relative velocity across the fault of 5.5 cm/yr, the accumulated slip in 1906 would have been 3.74 m. This predicted slip is compared with the measured surface slip during the 1906 earthquake (THATCHER, 1975a) in Fig. 2. Although there is considerable scatter, reasonable agreement is obtained.

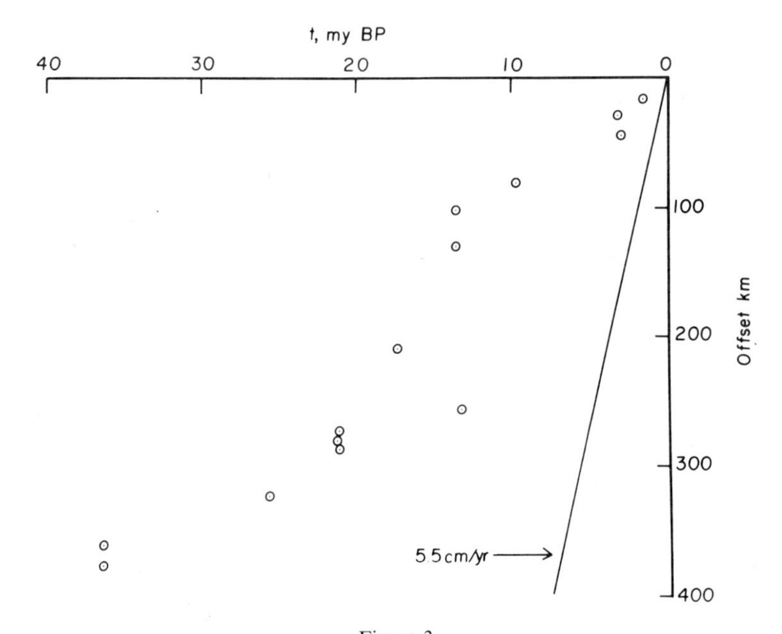

Figure 3
Dated geological offsets on the San Andreas fault after GRANTZ and DICKINSON (1968).

This approach can also be used to predict the accumulated slip on the San Andreas fault at the present time. On the northern section, 2.8 m of slip has accumulated since 1906. The southern section of the San Andreas fault between Cholame and San Bernardino is also apparently a locked section. A great earthquake with large surface displacement was recorded on this section on January 9, 1857 (WOOD, 1955). Since that time 6.5 m of slip has accumulated.

Although considerable evidence can be presented supporting the hypothesis that 5.5 cm of slip accumulated on the locked sections of the San Andreas fault each year, evidence to the contrary is also available. Geological evidence for offsets on the San Andreas fault has been summarized by GRANTZ and DICKINSON (1968). Their results are compared with the 5.5 cm/yr in Fig. 3. It is seen that the geological evidence favors a considerably smaller slip velocity on the fault.

Direct measurements of the creep velocity on the central section of the San Andreas fault have been carried out. These results have been summarized by SAVAGE and BURFORD (1971) and the mean of the observations is 2.28 cm/yr. The change in astronomical azimuth across the San Andreas fault between Mount Toro and Santa Ana just south of Hollister has been measured since 1886 (Savage, 1976). The base line is 53.8 km and it is nearly perpendicular to the fault. The change in shear strain, $\partial u/\partial y$, between observations is given in Fig. 4. If it is assumed that the two observation points are on rigid translating plates, the mean velocity across the fault is 3.3 cm/yr.

Certainly a significant fraction of the relative motion between the Pacific and North American plates takes place on the San Andreas fault. How large this fraction is remains unresolved.

3. *Stress level and fault rheology*

An understanding of the rheology of the rock in and adjacent to the San Andreas fault is essential to a full understanding of the fault. Certainly the fault must penetrate the full lithosphere. However, the nature of the fault at depth is likely to be entirely different from the nature of the fault at the surface. At low temperatures and low lithostatic pressures, rock behaves in an elastic-brittle manner. The occurrence of earthquakes to depths of about 15 km indicates a brittle-like behavior. However, there may be little resemblance between the rheological behavior of homogeneous rock and the behavior of a relatively wide fault zone made up of fault gauge (broken and fractured rock). The stress drop during a great earthquake is some fraction of the stress level that can be transmitted across the fault gauge before failure occurs.

Based on studies of seismograms, Brune and Allen (1967) suggested an average stress drop of 70 bars in the 1906 earthquake. The surface stress drop during the 1906 earthquake can be estimated directly from surface displacements. A least square fit to observed displacements has been given by Knopoff (1958)

$$u = 0.72 \times 10^{-3}[(y^2 + \{3.2 \times 10^5\}^2)^{1/2} - |y|] \tag{1}$$

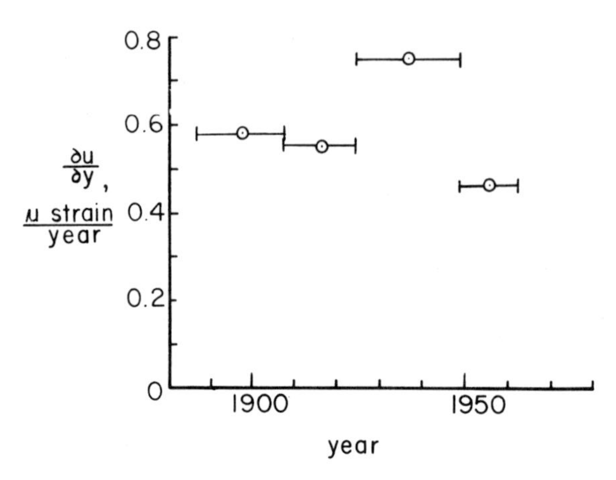

Figure 4
Strain accumulation (relative motion) between Mount Toro and Santa Ana (separation 53.8 km) after SAVAGE (1976). Horizontal bars represent intervals between measurements of astronomical azimuth.

with the displacement u and the distance from the fault y in cm. Taking the shear modulus in the crust $G = 3 \times 10^{11}$ dyne/cm^2 and

$$\sigma_{xy} = G \frac{\partial u}{\partial y} \tag{2}$$

we find that the surface stress drop $\sigma_{xy} = 216$ bars.

Another source of information on the stress level on the San Andreas fault is the heat flow. Heat production on the fault per unit area q due to friction is related to the mean stress on the fault $\bar{\sigma}_{xy}$ and the velocity across the fault u by

$$q = \bar{\sigma}_{xy} u \tag{3}$$

Detailed calculations of the surface heat flow resulting from frictional heating on a strike slip fault have been given by HENYEY and WASSERBURG (1971). The surface heat flow is elevated by approximately the frictional heating over a width equal to the depth over which the heating takes place.

Both HENYEY and WASSERBURG (1971) and LACHENBRUCH and SASS (1973) have concluded that there is no measurable heat flow anomaly in the immediate vicinity of the fault although there is a broader heat flow anomaly associated with the fault which has a width of about 100 km and a magnitude of 0.5–1.0 h.f.u.

Taking the maximum heat flow anomaly which would escape detection in the immediate vicinity of the fault to be $q = 0.5$ h.f.u. and $u = 5.5$ cm/yr, the maximum mean stress on the fault from (3) is about 100 bars. It should be noted that the transport of heat by ground water circulation has not been considered. Since the fault gauge is quite porous convective cooling could be important in the transport of frictional heat. Therefore, this upper limit for the stress on the fault must be used with caution.

On the deeper portion of the fault displacements must occur by plastic deformation. The behavior of the fault is critically dependent on the failure and yield stress on the fault as a function of depth. Is the deeper plastic zone stronger than the near-surface brittle zone? The failure stress is probably related to a coefficient of friction. Since the hydrostatic pressure increases with depth, it is expected that the failure stress will increase with depth. However, the temperature also increases with depth. And with increasing temperature the yield stress decreases.

4. *Seismicity*

The record of seismicity on and adjacent to the San Andreas fault should also provide information on the state of stress. High levels of seismic activity should be indicative of high stress levels.

The felt intensities of earthquakes associated with the San Andreas fault system are given in Fig. 5 as a function of latitude for the time period from 1880 to 1972. All earthquakes of intensity 5 or larger which were centered within 100 km of the San Andreas fault are included. For the period between 1880 and 1928 the *Townley and Allen Catalog of California Earthquakes* was used. For the period between 1929 and 1972 the *Catalog of United States Earthquakes* published by the U.S. Coast and Geodetic Survey (since 1969 by the National Oceanic and Atmospheric Administration) was used. Results prior to 1880 were not used because of inadequate data. From 1880 to 1930 the Rossi-Forel scale of felt intensities was used. From 1931 to 1972 the modified Mercalli intensity scale of 1931 was used. The extent of the 1906 fault break and the division of the fault into sections are also shown in Fig. 5.

It is clear from Fig. 5 that there is some reporting bias in the data, particularly for the smaller events. This is to be expected since the great increase in the number of seismometers in California in the last 20 years has greatly improved the recording of earthquakes. However, despite the variation in the quality of the data there are some very clear trends illustrated. The most obvious is the reduction in the number of small and intermediate size earthquakes adjacent to the northern locked section following the 1906 earthquake. This reduction is strong evidence that the stress level in the elastic lithosphere adjacent to the northern locked section was significantly reduced in the 1906 earthquake. This reduction of stress is consistent with the hypothesis that the maximum stress is transmitted in the elastic-brittle lithosphere and that this stress is relieved during a great earthquake and its subsequent aftershock sequence. The alternative hypothesis that the maximum stress is transmitted beneath the brittle region and that it actually is increased after a great earthquake would allow a reduction of stress only in the immediate vicinity (a few tens of kilometers) of the fault. The stress and therefore the seismicity at distances of, say, 20–100 km from the fault would not be affected.

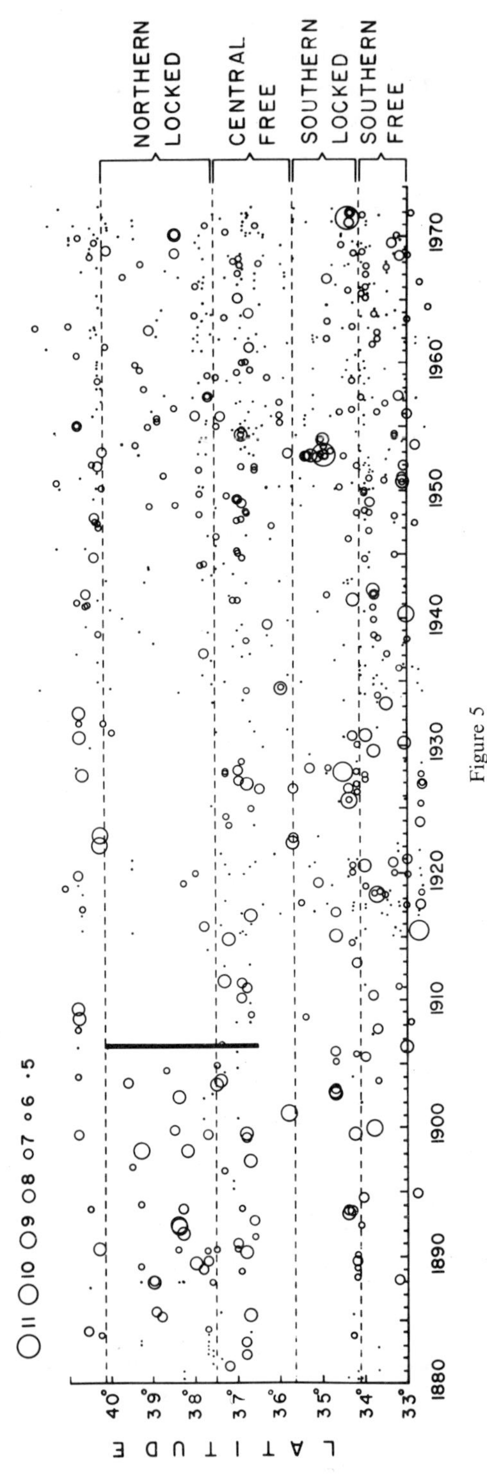

Figure 5

Felt intensities of earthquakes within 100 km of the San Andreas fault as a function of latitude from 1880 to 1972. The solid line shows the surface break of the 1906 earthquake.

Returning to Fig. 5, there appears to be a systematic increase in the number of intermediate size earthquakes adjacent to the southern locked section. This is probably due to an increase in the stress level on this section of the fault. However, it does not appear that this increase is sufficiently systematic to be of use in predicting when a great earthquake will occur on the southern locked section. As the strain and stress increase in the elastic plates adjacent to the southern locked section, stress concentrations occur causing earthquakes. This is particularly true in the vicinity of the bend in the San Andreas north of Los Angeles. Because of the bend, simple strike-slip motion on the San Andreas cannot completely relieve the accumulating strain. Deformation associated with the Transverse Ranges and their associated fault systems is required. The Kern County earthquake of 1952 and the San Fernando earthquake of 1971 are examples of this deformation.

It is also clear from Fig. 5 that there are a considerable number of earthquakes on the central and southern free sections of the fault during the entire period. This is consistent with the hypothesis that a significant fraction of the accumulating strain is being continuously relieved on these sections.

5. *Elastic models*

In order to understand the cyclical accumulation and release of stress and strain on the San Andreas fault it is necessary to hypothesize appropriate models. A two-dimensional model for the accumulation and release of slip on a strike-slip fault has been proposed by Turcotte and Spence (1974). In this model the upper part of the fault was assumed to be locked (zero displacement) and the lower part of the fault zone was assumed to be free sliding (zero shear stress). An analytical solution for the distribution of stress and strain was given. During a great earthquake the accumulated stress and strain was assumed to be relieved.

Implicit in this model is the assumption that the failure stress on the locked brittle zone is large compared with the yield stress on the deeper plastic zone. After a great earthquake the remaining stress on both the upper brittle zone and the lower plastic zone is small compared with the initial stress on the initially locked brittle zone before the earthquake. This assumption has been questioned by Savage (1975) who proposed a transfer of stress between the upper brittle zone and the lower plastic zone. No quantitative calculations of the expected consequences of this model have been given.

In order to explain the longitudinal behavior of the San Andreas fault a three-dimensional model for the accumulation and release of stress and strain was proposed by Spence and Turcotte (1976). This model is illustrated in Fig. 6. Two semi-infinite plates acted on by shearing forces were assumed to be sliding freely past each other except for two locked sections. The northern and southern locked sections were assumed to be locked to a depth *d*. At greater depths the fault was assumed to be free sliding as well as on other sections of the fault. Using the technique of matched

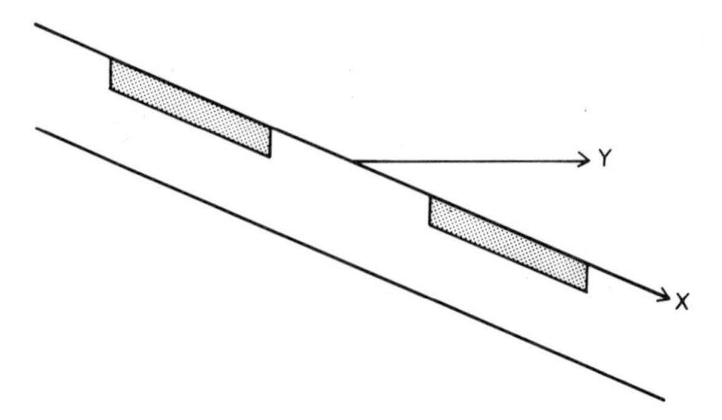

Figure 6
Illustration of the elastic model for strain accumulation on the San Andreas fault after SPENCE and TURCOTTE (1976). Two semi-infinite plates are sliding past each other except for the two shaded locked sections.

asymptotic expansions, analytic expressions were obtained for the three-dimensional distribution of strain and stress.

Assuming that a relative plate velocity of 5.5 cm/yr is applied 1000 km away from the fault, the surface strain accumulation in the region $x > 0$ and $y > 0$ is given in Fig. 7. This corresponds to the strain on the east side of the southern locked section. The strain accumulation is given in terms of the components

$$\gamma_1 = \frac{\partial u_x}{\partial x} - \frac{\partial u_y}{\partial y}, \qquad \gamma_2 = \frac{\partial u_x}{\partial y} + \frac{\partial u_y}{\partial x} \tag{4}$$

as introduced by Frank (1966). The coordinate system is centered on the center of the central free section as in Fig. 1. It should be noted that the analysis is for a straight fault so that the bend in the San Andreas north of Los Angeles is not considered.

Many measurements of strain adjacent to the San Andreas fault have been carried out (SAVAGE and BURFORD, 1970, 1973; THATCHER, 1975a, 1975b). Although in some cases the accumulation of strain between two points appears to increase linearly with time, the point to point correlation of strain accumulation reveals considerable scatter; see for example TURCOTTE and SPENCE (1974). This scatter can be attributed to regional variations in topography and elastic properties and to local faults. However, the magnitudes of the strain accumulation given in Fig. 7 are in reasonable agreement with the observations.

A primary conclusion from Fig. 7 is the large strain and stress accumulation at the tips of the locked sections of the fault. This result is consistent with the observations of WYSS and BRUNE (1971). Using the ratio of short- to long-period amplitudes to estimate apparent stress they concluded that the San Bernardino–San Gorgonio Mountain region is under higher stresses than other portions of the San Andreas fault. This area corresponds closely to the southern end of the southern locked section.

According to the elastic analysis the locked sections impede the lateral sliding between the plates. The plate interactions extend large distances from the locked

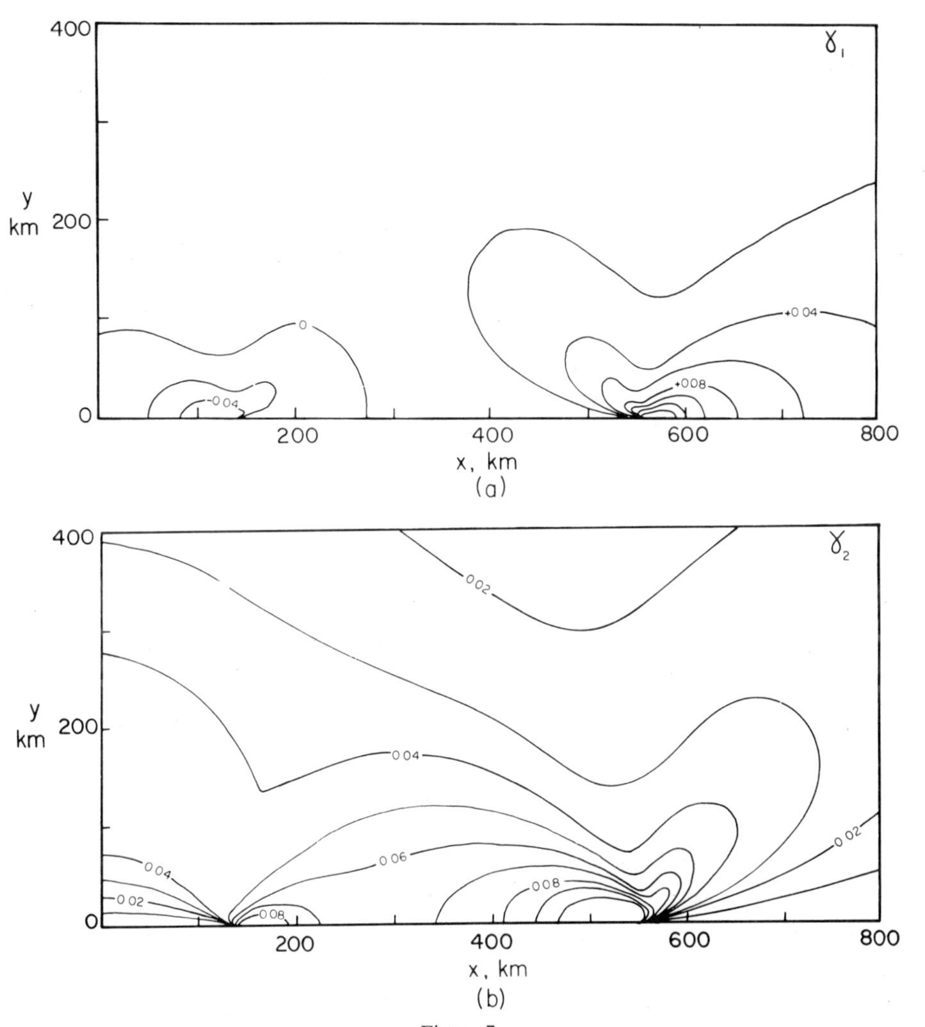

Figure 7
Surface strain accumulation in the region $x > 0$ and $y > 0$:

(a) $\gamma_1 = \partial u_x/\partial x - \partial u_y/\partial y$; (b) $\gamma_2 = \partial u_x/\partial y + \partial u_y/\partial x$.

Contours are in units of μ strain/yr. Results taken from SPENCE and TURCOTTE (1976).

sections. The analysis suggests that the locked sections reduce the creep on the central free section below the relative velocity between the plates. This is in agreement with the creep measurements given by SAVAGE and BURFORD (1971). The mean of their observations is 2.28 ± 1.37 cm/yr. According to the theory the accumulating slip would be relieved by accelerated creep after great earthquakes. It should be noted that the observations (SAVAGE, 1976) of the azimuth change across the San Andreas fault between Mount Toro and Santa Ana as illustrated in Fig. 4 do not show an acceleration of creep after the 1906 earthquake.

Certainly the elastic analysis can only be approximately valid. The viscous coupling with the asthenosphere will tend to damp out sudden seismic displacements and the propagation of stress will be by a diffusion mechanism rather than by a seismic wave. The distance y_s that stress will diffuse in a time t is given approximately by (BOTT and DEAN, 1973)

$$y_s = (Gbdt/\eta)^{1/2} \tag{5}$$

where d is the thickness of the lithosphere and b and η are the thickness and viscosity of the asthenosphere. Taking $G = 3 \times 10^{11}$ dyne/cm², $\eta = 4 \times 10^{20}$ dyne sec/cm², $d = 100$ km, $b = 300$ km, and $t = 100$ yr, the value of y_s from (5) is 270 km. The oscillatory motions on the fault would be expected to be damped out several hundred kilometers into the plate.

6. *Cyclical behavior*

If the locked sections of the San Andreas fault provide the primary resistance to the lateral sliding between the two plates, a cylindrical behavior is suggested. Between great earthquakes there is no relative displacement between the Pacific and North American plates on the northern and southern locked sections. Slip is accumulating through elastic strain in the plates. Slip occurs continuously on the central and southern free sections.

The accumulated slip on the locked sections is relieved in great earthquakes. All relative displacements between the surface plates on the locked sections must occur as slip during great earthquakes. Great earthquakes occur sequentially. When a great earthquake occurs on one locked section, the stress on this section is relieved. However, a fraction of the stress may be transferred to the other locked section. The central free section acts as a buffer zone between the two locked sections.

The cyclical behavior of the San Andreas fault since 1838 is illustrated in Fig. 8. For simplicity, equal units of slip are assumed between great earthquakes. Just prior to the 1838 earthquake on the northern locked section, two units of slip had accumulated on the northern locked section and one unit of slip (1Δ) had accumulated on the southern locked section. In the 1838 earthquake the two units of slip were relieved on the northern locked section and a new phase of slip accumulation began. Between 1838 and 1857 one additional unit of slip accumulated on the northern and southern locked sections. Two units of slip were relieved on the southern locked section during the 1857 earthquake. Between 1857 and 1906 an additional unit of slip accumulated on the locked sections. Two units of slip were relieved on the northern locked section during the 1906 earthquake. Since 1906 an additional unit of slip has accumulated on the locked sections. The next great earthquake on the San Andreas fault will relieve the two accumulated units of slip on the southern locked section.

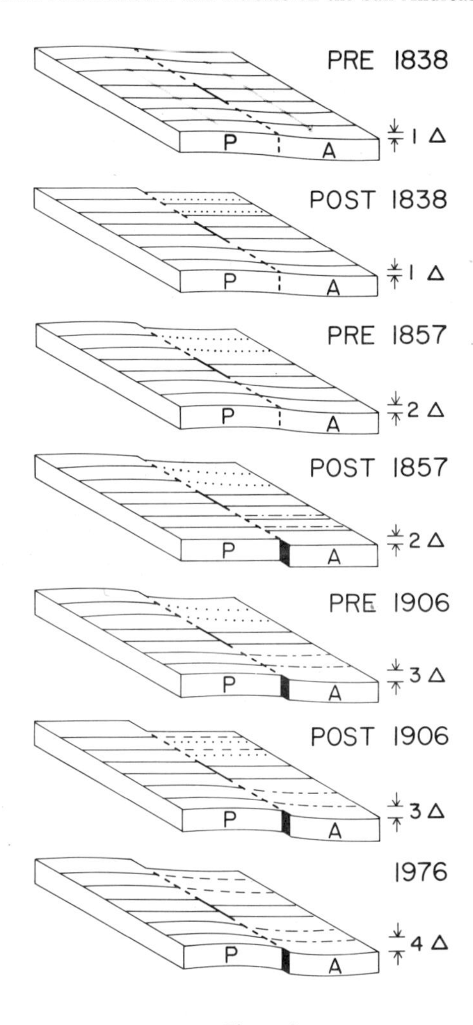

Figure 8
Illustration of the cyclical behavior of the San Andreas fault. The Pacific plate (*P*) is sliding laterally past the North American plate (*A*) in increments of Δ. Accumulated slip on the northern locked section was relieved in 1838 and 1906 and on the southern locked section in 1857.

7. *Conclusions*

There are several important questions that must be answered before a complete understanding of the San Andreas fault can be developed.

(1) What is the mean velocity of motion on the fault? Is it the relative plate velocity of 5.5 cm/yr or is it a small fraction of this? The next great earthquake on the fault should answer this question or it may be answered by direct observations of displacements.

(2) What is the distribution of stress on the fault as a function of depth before and after a great earthquake? Is the failure stress in the near-surface brittle lithosphere larger or smaller than the yield stress in the deeper plastic lithosphere?

(3) What is the ratio of the stress drop during a great earthquake to the absolute stress transmitted across the fault? Is significant stress transmitted across the fault on the free sections? Does a great earthquake reduce the regional stress level?

(4) Is there a transfer of stress between the northern and southern locked sections of the fault? If there is a transfer, then seismic activity on the two sections is coupled, otherwise there is probably no dependence of the seismic risk in the northern section to the state of stress and seismicity on the southern section.

Acknowledgments

This research has been supported in part by the Division of Earth Sciences, National Science Foundation, NSF Grant EAR 74-03259 and the U.S. Geological Survey, Department of the Interior, under USGS Grant No. 14-08-001-G-385.

References

ATWATER, T. and MOLNAR, P., *Relative motion of the Pacific and North American plates deduced from sea floor spreading in the Atlantic, Indian, and South Pacific Oceans*, in *Proceedings of the Conference on Tectonic Problems of the San Andreas Fault System*, (eds. R. L. Kovach and A. Nur), (Stanford University Press, Stanford, California, 1973), pp. 136–148.

BOTT, M. H. P. and DEAN, D. S. (1973), *Stress diffusion from plate boundaries*, Nature *243*, 339–341.

BRUNE, J. N. and ALLEN, C. R. (1967), *A low stress-drop low-magnitude earthquake with surface faulting: The Imperial, California, earthquake of March 4, 1966*, Seis. Soc. Am. Bull. *57*, 501–514.

FRANK, F. C. (1966), *Deduction of earth strains from survey data*, Seis. Soc. Am. Bull. *56*, 35–42.

GRANTZ, A. and DICKINSON, W. R., *Indicated cumulative offsets along the San Andreas fault in the California Coast Ranges*, in *Proceedings of Conference on Geologic Problems of the San Andreas Fault System*, (eds. W. R. Dickinson and A. Grantz), (Stanford University Press, Stanford, California, 1968), pp. 117–120.

HENYEY, T. L. and WASSERBURG, G. J. (1971), *Heat flow near major strike-slip faults in California*, J. Geophys. Res. *76*, 7924–7946.

KNOPOFF, L. (1958), *Energy release in earthquakes*, Geophys. J. Roy. Astron. Soc. *1*, 44–52.

LACHENBRUCH, A. H. and SASS, J. H., *Thermo-mechanical aspects of the San Andreas fault system*, in *Proceedings of the Conference on Tectonic Problems of the San Andreas Fault System*, (eds. R. L. Kovach and A. Nur), (Stanford University Press, Stanford, California, 1973), pp. 192–214.

LAUDERBACK, G. D. (1947), *Central California earthquakes of the 1830's*, Seis. Soc. Am. Bull. *37*, 33–74.

MINSTER, J. B., JORDAN, T. H., MOLNAR, P. and HAINES, E. (1974), *Numerical modelling of instantaneous plate tectonics*, Geophys. J. Roy. Astron. Soc. *36*, 541–576.

SAVAGE, J. C. (1975), *Comment on 'An analysis of strain accumulation on a strike-slip fault' by D. L. Turcotte and D. A. Spence*, J. Geophys. Res. *80*, 4111–4114.

SAVAGE, J. C. (1976), private communication.

SAVAGE, J. C. and BURFORD, R. O. (1970), *Accumulation of tectonic strain in California*, Seis. Soc. Am. Bull. *60*, 1877–1896.

SAVAGE, J. C. and BURFORD, R. O. (1971), *Discussion of paper by C. H. Scholz and T. J. Fitch, 'Strain accumulation along the San Andreas fault'*, J. Geophys. Res. *76*, 6469–6479.

SAVAGE, J. C. and BURFORD, R. O. (1973), *Geodetic determination of relative plate motion in central California*, J. Geophys. Res. *78*, 832–845.

SPENCE, D. A. and TURCOTTE, D. L. (1976), *An elastostatic model of stress accumulation on the San Andreas fault*, Proc. Roy. Soc. London *A349*, 319–341.

THATCHER, W. (1975a), *Strain accumulation and release mechanism of the 1906 San Francisco earthquake*, J. Geophys. Res. *80*, 4862–4872.

THATCHER, W. (1975b), *Strain accumulation on the northern San Andreas fault zone since 1906*, J. Geophys. Res. *80*, 4873–4880.

TURCOTTE, D. L. and SPENCE, D. A. (1974), *An analysis of strain accumulation on a strike-slip fault*, J. Geophys. Res. *79*, 4407–4412.

WOOD, H. O. (1955), *The 1857 earthquake in California*, Seis. Soc. Am. Bull. *45*, 47–52.

WOODFORD, A. O. and MCINTYRE, D. B. (1976), *Curvature of the San Andreas fault, California*, Geology *4*, 573–575.

WYSS, M. and BRUNE, J. N. (1971), *Regional variations of source properties in southern California estimated from the ratio of short- to long-period amplitudes*, Seis. Soc. Am. Bull. *61*, 1153–1167.

(Received 1st November 1976)

Pageoph, Vol. 115 (1977), Birkhäuser Verlag, Basel

Shear Stress on the Base of a Lithospheric Plate

By Jay Melosh[1])

Abstract – This paper is a review of the theoretical and observational evidence bearing on the magnitude of the shear stress which acts on the base of a lithospheric plate. Estimates based on the viscosity of the upper mantle do not yield useful limits. Arguments based on the thermal stability of the upper mantle indicate that the basal shear stress is no larger than a few bars. An indirect measurement of the rheology and shear stress can be made by studying the diffusion of stress and displacement following a large decoupling earthquake. When applied to the 1965 Rat Island Earthquake, this method yields a basal shear stress of about 2 bars. These results indicate that for small plates the forces produced by basal shear stress are probably small in comparison with forces acting on plate boundaries. To a first approximation, the smaller plates act as if they were decoupled from the mantle below. These stress estimates lead to a model in which the motion of the smaller lithospheric plates is governed almost entirely by the forces acting on their edges. Forces due to basal shear stress may be comparable to forces acting on the edges of large lithospheric plates. Thus, complete decoupling may not be a good approximation for such plates.

Key words: Stress in lithosphere; Viscoelastic relaxation; Tectonics of plates.

1. *Introduction*

There are two kinds of surface force which may act on a lithospheric plate. One kind is due to shear stress applied to the base of the lithospheric plate and is thus proportional to the plate's area. The other is due to stress applied to the edges of the plate and is proportional to the plate's circumference. Suppose the basal shear stress is σ_b and let it act on a square plate of area L^2 and thickness h. If the edge stresses are of order σ_e, the circumferential forces ($4\ \sigma_e hL$) are equal to the areal forces ($\sigma_b L^2$) when $\sigma_b = (4h/L)\sigma_e$. For a typical large plate h may be 100 km while L is of order 10,000 km. If the basal shear stress σ_b is less than 1/25 the size of the average edge stress σ_e it can be neglected in the force balance on the plate. When this condition is satisfied, the plate may be considered as decoupled from the mantle below, since it moves primarily in response to forces applied at its edges, not on its base.

Estimates of the stress produced by ridges (Jacoby 1970) yield σ_e values of a few hundred bars. Stress drops of eathquakes on transform faults (Chinnery 1964) and in subduction zones (Kanamori and Anderson 1975) also yield σ_e values of a few hundred bars. Although there is some argument about whether these stress drops represent true tectonic stresses, we can regard such estimates of σ_e as lower limit, which is sufficient for our purpose. We thus conclude that the basal shear stress σ_b

---

[1]) California Institute of Technology, Pasadena, California, USA.

must be less than a few bars in order for the areal forces on a large lithospheric plate to be neglected with respect to the forces acting on its edges.

The purpose of this paper is to examine the current bounds on the basal shear stress σ_b and to determine how important areal forces are in the force balance on a plate. A crude estimate of σ_b can be made by using the viscosity of the asthenosphere. Unfortunately, most viscosity measurements are based on the assumption of Newtonian rheology, and we now have reason to believe that the asthenosphere is non-Newtonian (WEERTMAN and WEERTMAN 1976, POST and GRIGGS 1973, BRENNAN 1974, MELOSH 1976a). Moreover, the viscosities reported in the literature vary by two orders of magnitude, making any meaningful stress estimate virtually impossible. Table 1 contains a compilation of viscosities derived from the analysis of isostatic rebound and estimates of shear stress based on these viscosities. These estimates assume Newtonian rheology and also suppose that the flow in the asthenosphere is pure shear over a thickness d, across which the velocity variation Δv is 10 cm/yr. Thus, we take $\sigma_b = \eta \, \Delta v/d$, where d is assumed to be 150 km (roughly the thickness of the Low Velocity Zone) when it is not derived from the analysis itself. It is clear that these estimates of σ_b are not adequate to answer the question of whether lithospheric plates are decoupled from the mantle. The values of σ_b obtained vary from 0.3 bar (implying a decoupled lithosphere) to 600 bars (implying that basal shear stress dominates edge stress on the plate).

Table 1

Stress on the base of a lithospheric plate from estimates of the viscosity of the asthenosphere. All estimates assume Newtonian rheology

Source	Viscosity η, poise	Thickness of asthenosphere d, km	Basal shear stress σ_b,†) bars
A. Half space models			
HASKELL (1937)	9×10^{21}	150*)	190
VENING MEINESZ (1937)	3×10^{22}	150*)	630
GUTENBERG (1941)	3×10^{21}	150*)	63
CRITTENDEN (1967)	10^{20}–10^{21}	150*)	2–20
B. Thin channel models			
LLIBOUTRY (1971)	3×10^{19}	100	0.9
CATHLES (1975)	4×10^{20}	75	17
C. Depth dependent viscosity with an elastic lid (data are for lowest viscosity zone)			
McCONNELL (1968)	2.7×10^{21}	180	47
WALCOTT (1973)	10^{19}–10^{21}	100–500	0.3–6

*) These thicknesses are assumed.

†) σ_b is computed from $\sigma_b = \eta \, \Delta v/d$, where we assume a velocity differential $\Delta v = 10$ cm/yr occurs across a zone of thickness d, which deforms by simple shear.

In principle, σ_b could be measured directly if the compressive stress σ_{xx} at the top of the lithosphere were known. Since the shear stress on top of the lithosphere must be zero, shear stress acting on the base of the lithospheric plate must be balanced by compressive stresses within the plate. The shear stress in a thin plate with a free upper surface and shear stress σ_b on its base at a depth h is given by $\sigma_{xz} = \sigma_b(z/h)$, where z is depth in the plate (see Fig. 1). The stress equilibrium equations

$$\frac{\partial}{\partial x}\sigma_{xx} + \frac{\partial}{\partial z}\sigma_{xz} = 0 \tag{1}$$

(where we have neglected gravity) thus require that

$$\sigma_{xx} = \text{const} - (\sigma_b/h)x \tag{2}$$

The horizontal stress σ_{xx} is constant throughout the thickness of the plate (after the lithostatic pressure $\rho g z$ has been subtracted), and varies laterally along its length.

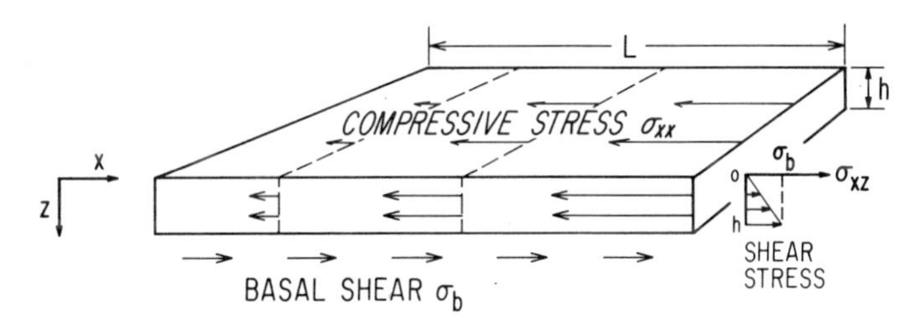

Figure 1
Shear stress σ_b acting on the base of a lithospheric plate of thickness h must fall to zero at the top of the plate (see inset graph of σ_{xz} versus depth). Stress equilibrium requires that the shear stress σ_{xz} must be balanced by a horizontal compressive stress σ_{xx} which increases laterally over the surface of the plate. In principle, measurements of σ_{xx} could determine σ_b.

Thus, if σ_{xx} could be measured at the top of the lithosphere, its horizontal rate of change could be used to compute σ_b directly. Unfortunately, the necessary measurements of σ_{xx} are not presently available. An approach which implicitly uses this method is that of SOLOMON *et al.* (1975) and RICHARDSON and SOLOMON (1976). These authors compare the observed pattern of intraplate stress with patterns computed on the basis of various assumptions about driving forces. The results indicate that the basal shear stress generates forces which are small in comparison with the forces acting on plate edges. FORSYTH and UYEDA (1975), using a direct inversion of plate velocities, reach the same conclusion. They also show that plate velocity is not correlated with plate area, consistent with the idea that basal shear stress does not contribute significantly to the force balance.

An indirect argument based on equation (2) shows that $\sigma_b \lesssim (2h/L)\sigma_e$. The difference in σ_{xx} between two portions of a plate separated by a distance L is $\Delta\sigma_{xx} = (L/h)\sigma_b$. If we suppose that $|\sigma_{xx}|$ cannot become larger than σ_e (otherwise, faulting would occur and a new plate boundary would be formed), then $\Delta\sigma_{xx}$ must be less than

$2\sigma_e$. Taking L to be the extent of the entire plate, we have $(L/h)\sigma_b \lesssim 2\sigma_e$ or $\sigma_b \lesssim (2h/L)\sigma_e$. If σ_e is a few hundred bars, then σ_b must be less than a few bars in order to satisfy this condition for all plates.

In the following two sections, we shall put forth further evidence that σ_b is less than a few bars. The first argument is based on the requirement that the asthenosphere be thermally stable and leads to an upper limit on σ_b of 1–5 bars. The second argument relates the rheology of the asthenosphere to the diffusion of stress and displacement following a large decoupling earthquake. The analysis of the 1965 Rat Island Earthquake indicates that the rheology is non-Newtonian and that the stress σ_b is about 2 bars at the local plate velocity of 7 cm/yr.

2. *Thermal stability of the asthenosphere*

By definition, the lithosphere behaves like an elastic solid over geologic periods of time. It can support applied shear stress by deforming elastically. The asthenosphere and mantle below it do not have this property. The materials of which they are composed creep steadily under applied shear stress. The shear stress σ_b acting on the base of the lithosphere must be maintained by velocity gradients in the asthenosphere and mantle. If the asthenosphere is a zone of especially low viscosity in the mantle, velocity gradients will be concentrated within it. This situation is potentially unstable to thermal runaways which can cause drastic modifications of the thermal structure of the asthenosphere.

The thermal instability of the asthenosphere is a direct result of the stress–strain rate relation for hot crystalline solids (WEERTMAN and WEERTMAN 1975). The shear strain rate $\dot{\varepsilon}_{xz}$ is given in terms of the applied shear stress σ_{xz} by

$$\dot{\varepsilon}_{xz} = B\,\sigma_{xz}^n \exp\left(-\frac{E^* + PV^*}{RT}\right) \tag{3}$$

where B is a constant and n is between 3 and 6 for most materials ($n = 1$ for a Newtonian material). In the exponential, E^* and V^* are activation energy and volume, respectively. P is pressure, T is absolute temperature and R is the gas constant. Equation (3) shows that for fixed σ_{xz} the strain rate $\dot{\varepsilon}_{xz}$ increases as the temperature T increases.

The 50–100 km thick lithospheric plate is an excellent insulator, having a thermal time constant of the order of 100 m yr. The velocity gradients in the asthenosphere generate shear heat in the amount of $\sigma_{xz}\dot{\varepsilon}_{xz}$ erg/cm^3 sec. Unless this heat is conducted away through the lithosphere, the temperature of the asthenosphere must rise. If it rises, $\dot{\varepsilon}_{xz}$ increases further: thermal runaway occurs, ceasing only when σ_{xz} is relieved or when temperatures reach the solidus and equation (3) no longer applies. This phenomenon was first studied by GRUNDFEST (1963) who showed that thermal runaway occurs only when a dimensionless parameter (now known as the 'Grundfest Parameter') exceeds unity.

MELOSH (1976b) studied this phenomenon in the geological context of a lithospheric plate moving over a low viscosity asthenosphere. This investigation took both radiogenic heat production and the pressure dependence of $\dot{\varepsilon}_{xz}$ into account. One of the results of this work is that thermally stable plate motion is possible only for a limited range of shear stress and velocity. Shear stress may range from zero to a value σ_c, the 'critical stress'. Steady solutions to the equations of motion do not exist for shear stresses larger than σ_c. Similarly, the velocity of the plate may range from zero up to a finite value v_c. Solutions ('supercritical solutions') with $\sigma_{xz} < \sigma_c$ *do* exist for plate velocities v larger than v_c, but they have rather bizarre properties (as the velocity increases, the shear stress decreases). These supercritical solutions are also unstable with respect to small temperature or stress perturbations, and so are unlikely to be of importance in the earth. None of the subcritical ($v < v_c$) solutions appreciably perturb the thermal structure of the asthenosphere (even at the largest possible stress σ_c and velocity v_c, the total dissipation amounts to only about 1/40 HFU). The subcritical solutions exhibit a normal relation between shear stress and velocity (Fig. 2).

There is currently no evidence that thermal runaway has occurred on a regional scale under any plate. A thermal runaway would generate unusually large heat flow values and perhaps widespread volcanic activity. We thus presume that all plates are currently moving in the subcritical regime. The basal shear stress on any plate is therefore less than the critical stress σ_c. If σ_c were known, its value would form an upper bound on the basal shear stress σ_b.

The critical stress σ_c itself is sensitive to a variety of poorly known rheological and thermal parameters. However, it can be shown (MELOSH 1976b) that the combination $\sigma_c v_c$ is nearly independent of these poorly known quantities (including the thickness of the asthenosphere d and the power n). To an accuracy of about 10%,

$$\sigma_c v_c = \frac{kT_*}{z_a} \tag{4}$$

where k is the thermal conductivity of the upper mantle (which we take to be 0.008 cal $\text{cm}^{-2}\ \text{sec}^{-1}\ ^\circ\text{K}^{-1}$) and T_* is a temperature parameter given by $RT_a^2/(E^* + P_a V^*)$. T_* is the number of °K rise in temperature needed to increase $\dot{\varepsilon}_{xz}$ in equation (3) by a factor e (σ_{xz} fixed). The quantities T_a, P_a and z_a are the temperature, pressure and depth of the zone of most intense shearing below the plate. They thus refer to some point in the asthenosphere.

Since we can place a lower bound on v_c($v_c \gtrsim$ 10 cm/yr, the velocity of the fastest plate), we obtain an upper bound on the critical stress

$$\sigma_c(\text{bar}) \leq 104 \frac{T_*(^\circ\text{K})}{z_a(\text{km}) v_c(\text{cm/yr})}. \tag{5}$$

Depending on the thermal and rheological model chosen, T_* and z_a may vary by as much as a factor of two. The ratio T_*/z_a is less sensitive to the model, however (for larger z_a, T_a is larger and hence T_* increases as well). If we take $T_* = 50°K$, $z_a =$ 150 km and $v_c = 10$ cm/yr, we obtain an upper limit

$$\sigma_c \leq 3.5 \text{ bar}. \tag{6}$$

Naturally, σ_b must be less than σ_c. If v_c were higher, say 50 cm/yr, σ_c would be still lower (0.7 bar in this case). The thermal stability argument thus shows that σ_b is certainly less than a few bars, and it may be much less than this.

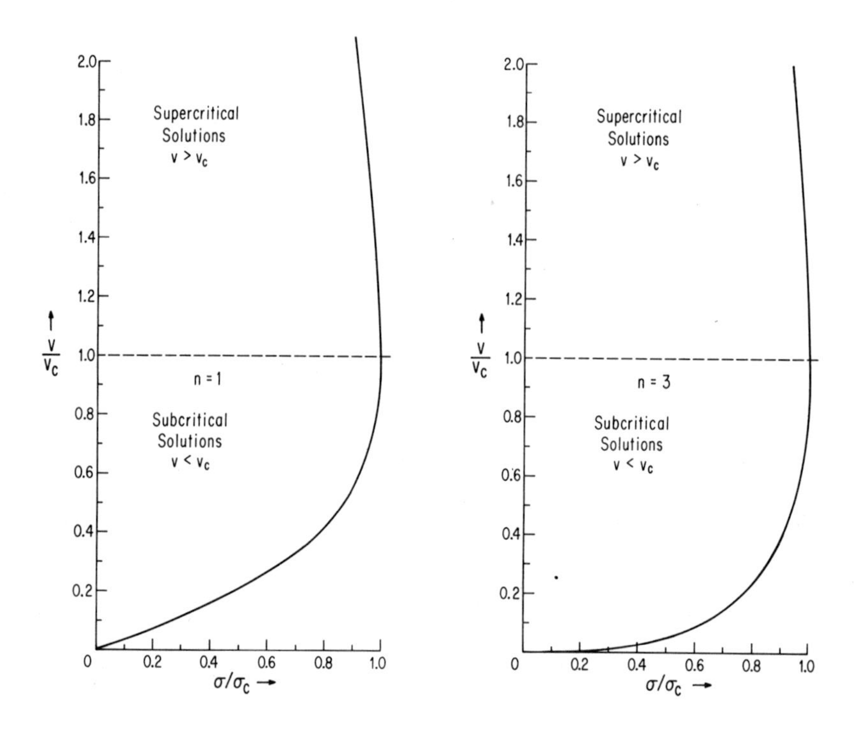

Figure 2

These plots display the relation between shear stress σ and velocity v acting on a plate. Shear heating in the asthenosphere has been taken into account, so that the shear stress cannot exceed a critical value σ_c without causing drastic modifications of the thermal structure of the asthenosphere. At the critical stress σ_c the plate velocity is finite, v_c. Although solutions exist for plate velocities higher than v_c (the 'supercritical solutions'), their negative slope on the v–σ plot coupled with their instability to small temperature and stress perturbations makes them difficult to realize in the earth. The relation between stress and velocity is normal for the 'subcritical solutions'. Moreover, at low stress the relation follows the usual law for no shear heating. Stress–velocity curves are presented for both Newtonian ($n = 1$) and a non-Newtonian ($n = 3$) stress–strain rate law. The asthenosphere's thickness is assumed to equal the thickness of the lithosphere.

3. *Stress diffusion following a large earthquake*

When the edge of a lithospheric plate is suddenly moved, as may occur during a decoupling earthquake in a subduction zone, a wave of stress and displacement propagates from the edge into the interior of the plate (ELSASSER 1967, BOTT and DEAN 1973). MELOSH (1976a) analyzed the propagation of such a wave in an elastic lithosphere overlying a thin Newtonian fluid asthenosphere governed by equation (3). It was found that the displacement field propagates away from the plate edge in such a way that the position of a constant displacement contour spreads as $t^{1/2n}$ where n is the power in $\dot{\varepsilon}_{xz} \sim \sigma^n_{xz}$ [equation (3)]. In the $n = 1$ (Newtonian) case, displacement spreads as $t^{1/2}$, a diffusion wave. For n larger than one, the spreading is initially rapid and later slows down much more quickly than for $n = 1$. The propagation of the displacement field following a single, sudden displacement of the plate edge is very sensitive to the rheology (that is, the value of the power n) of the asthenosphere. The observation of such a displacement field would thus constitute a measurement of the rheology. At present, geodetic measurements with the necessary accuracy of a few centimeters are not possible (although this may change in the near future). The displacement field, however, is accompanied by a stress field which is proportional to the derivative of the displacement with respect to the distance from the plate edge. It can be shown that the behavior expected for a stress *contour* is that it propagates rapidly into the plate after the plate edge is moved, then slows down and finally stops at a distance x_m from the plate edge. The stress contour then begins a slow retreat back to the plate edge, returning a time t_0 after the initial motion. The position of the stress contour as a function of time is strongly dependent upon the power n. Even relaxing the assumption of a thin asthenosphere seems to make little difference to the propagation pattern. While the shape of the curve of position of a stress contour versus time t_0 gives information about the rheology (n), the distance x_m and return time t_0 give information about stress levels and effective viscosity of the asthenosphere. The shear stress acting on the base of the plate is given by

$$\sigma_b = h\,E\left(\frac{v\,t_0}{c_\kappa u_0}\right)^{1/n}\frac{u_0}{x_m^2} \qquad (7)$$

where the average plate velocity v is assumed to drop from v to zero in the deep mantle (note, however, that the thickness of the zone over which this drop occurs does not enter the equation). E is the Young's modulus of the lithosphere, h is its thickness, u_0 is the amount of displacement at the plate edge and c_κ is a constant which is a function of the power n only ($c_\kappa(1) = 1.36$, $c_\kappa(3) = 26.2$; see MELOSH 1976a, Table 1 and equation (20) for further details). The return time t_0 and maximum distance x_m refer to *any* stress contour: the combination t_0/x_m^{2n} is independent of the value of the stress on that contour.

An indirect method can be used to measure the position of a stress contour as a function of time following a large decoupling earthquake. Suppose that aftershocks in the lithospheric plate are triggered by the changing stress field due to the diffusion of stress and displacement away from the plate edge. Suppose further that such aftershocks are frequent only when the stress exceeds some yield stress Y. Then the region over which most aftershocks occur at any given time is the region where the stress diffusing away from the plate edge exceeds Y. The edge of this region defines the contour of stress Y. Thus by observing the migration of the aftershock pattern away from the plate edge, the motion of the stress contour Y can be inferred. The motion of this contour is a function of the rheology of the asthenosphere.

The aftershocks in the oceanic plate following a large decoupling earthquake are expected to be confined initially to the rupture zone (or fault plane). As stress diffusion proceeds, the zone of aftershocks broadens and may extend a few hundred kilometers into the oceanic plate. Later, the aftershock zone narrows again, the edge of the zone retreating back to the plate edge. This is precisely the pattern observed for the aftershocks of the 4 February 1965, Rat Island Earthquake. Figure 3 shows the distance of the edge of the $M \geq 4.5$ aftershock zone from the initial rupture as a function of time. Also plotted are theoretical curves for different power laws (n) fit to the observed

$$x_m = 175 \pm 25 \text{ km}$$

$$t_0 = 2600 \pm 500 \text{ days}$$

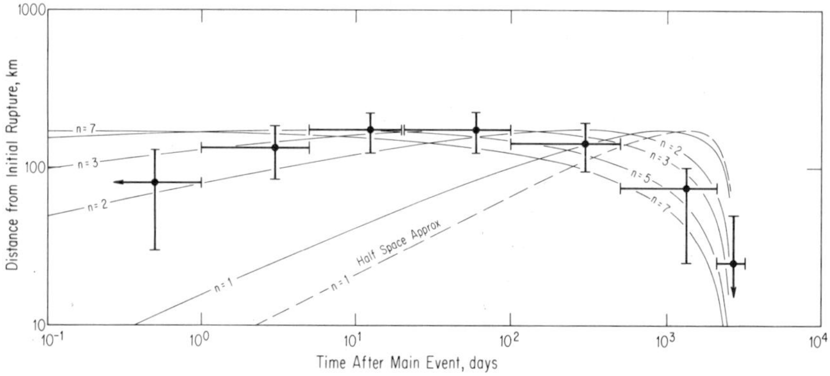

Figure 3

The width of the zone of $M \geq 4.5$ aftershocks in the oceanic plate following the February 4, 1965 Rat Island earthquake is plotted as a function of time after the event. The aftershocks are binned in logarithmically increasing time intervals in order to get roughly the same number of events in each bin. Errors on the width of the aftershock zone are based on detailed study of the aftershock locations. The solid curves are theoretical predictions of the motion of a contour of constant stress over a thin non-Newtonian asthenosphere. The curves are labeled by the power in $\dot{\varepsilon}_{xz} = A\sigma^n_{xz}$. The expected motion of a stress contour for a Newtonian thin asthenosphere model ($n = 1$, solid line) and a Newtonian viscous half space ($n = 1$, dashed line) are given for comparison. The fact that the width of the aftershock zone fails to follow the stress contours for a Newtonian model, yet does follow the contours for a non-Newtonian asthenosphere strongly suggests that the asthenosphere is non-Newtonian. The migration rate of the aftershock zone is consistent with a model asthenosphere based on measurements of the creep of olivine single crystals at low stress (KOHLSTEDT and GOETZE 1974). The maximum width of the aftershock zone is 175 ± 25 km and the time for a stress contour to return to the plate edge is 2600 ± 500 days, which leads to an estimate of the average basal shear stress on a lithospheric plate of about 2 bars.

A Newtonian stress–strain rate relation is clearly inconsistent with the observations (for either a thin asthenosphere or a viscous half space). The non-Newtonian models fit quite well for n between 2 and 6. Thus, although n is not well determined, it is certainly not 1. The observed values of x_m and t_0 are consistent with a rheological model based on KOHLSTEDT and GOETZE's (1974) low stress creep measurements on olivine. These values correspond to an effective viscosity in the asthenosphere of about 4×10^{19} poise (for all values of n) at the average shear stress

$$\sigma_b = 1.8 \pm 0.4 \text{ bar} \tag{8}$$

which has been calculated using equation (7). The average plate velocity in the Rat Island area is $v = 7$ cm/yr, $h = 70$ km, $E = 1.6 \times 10^6$ bar, and $u_0 = 2.5$ m. The initial displacement of the plate edge u_0 is the least well known quantity, being determined from the seismic moment of the 4 February 1965, event (WU and KANAMORI 1973). The errors assigned in (8) are based solely on the uncertainties assigned to x_m, t_0 and n (which is between 2 and 6). The value of σ_b in (8) thus depends somewhat upon the model assumptions, but the major result of this calculation cannot be altered: the basal shear stress is only a few bars.

4. *Conclusions*

The estimates presented in this paper strongly indicate that the shear stress acting on the base of a lithospheric plate is less than a few bars. The concordance of these estimates is remarkable, despite the diversity of the phenomena on which they are based. The viscosity estimates (of which at least some are consistent with the other estimates) are derived primarily from glacioisostatic rebound, while of the other two estimates one is theoretical, requiring only that the asthenosphere be thermally stable, and the second uses the migration pattern of aftershocks following a large earthquake.

The implications of such small basal shear stresses are not as clear as could be desired. The arguments presented in the introduction show that the net force acting on a plate due to the basal shear stress ($\sigma_b L^2$) is small in comparison to the forces acting on its edges ($4\,\sigma_e\, hL$) only when $\sigma_b < (4h/L)\,\sigma_e$. If we suppose that σ_e is a few hundred bars (based on stress drops of earthquakes at plate boundaries), then we require $\sigma_b \lesssim 2$ to 8 bars for a large ($L = 10{,}000$ km, $h = 100$ km) plate. This value of 2 to 8 bars is uncomfortably close to the 2 to 3 bar estimates we have found for σ_b, suggesting that basal shear stress *could* be important in the force balance for large plates. The case for negligible basal shear stress is stronger for the smaller plates.

We thus conclude that the motions of at least the smaller plates are governed primarily by forces acting on their edges. To a first approximation these plates may be considered to be mechanically decoupled from the mantle below. The low viscosity asthenosphere is the decoupling element in the system. It is probably this decoupling, rather than the intrinsic strength of the lithosphere, which allows the oceanic lithosphere to behave as rigid plates. These conclusions may also hold for large plates, although our knowledge of the rheology of the asthenosphere and the stresses acting on the edges of plates is not sufficient to answer this question unambiguously. It is clear, however, that areal forces acting on a plate are at most comparable with edge forces; the motion of the large plates cannot be dominated by forces due to basal shear stress, although such forces may make an important contribution to the overall force balance.

REFERENCES

BOTT, M. H. P. and DEAN, D. S. (1973), *Stress diffusion from plate boundaries*, Nature *243*, 339 341.

BRENNEN, C. (1974), *Isostatic recovery and strain rate dependent viscosity of the earth's mantle*, J. Geophys. Res. *79*, 3993–4001.

CATHLES, L. M., *The Viscosity of the Earth's Mantle* (Princeton Univ. Press, Princeton, N.J. 1975), p. 170.

CHINNERY, M. A. (1964), *The strength of the earth's crust under horizontal shear stress*, J. Geophys. Res. *69*, 2085–2089.

CRITTENDEN, M. D. (1967), *Viscosity and finite strength of the mantle as determined from water and ice loads*, Geophys. J. Roy. Astron. Soc. *14*, 261–279.

ELSASSER, W. M., *Convection and stress propagation in the upper mantle* in *The Application of Modern Physics to the Earth and Planetary Interiors* (ed. S. K. Runcorn) (Wiley Press, N.Y. 1967), pp. 223–246.

FORSYTH, D. W. and UYEDA, S. (1975), *On the relative importance of driving forces of plate motion*, Geophys. J. Roy. Astron. Soc. *43*, 163–200.

GRUNDFEST, I. J. (1963), *Thermal feedback in liquid flow: plane shear at constant stress*, Trans. Soc. Rheol. *7*, 195–207.

GUTENBERG, B. (1941), *Changes in sea level, postglacial uplift and mobility of the earth's interior*, Bull. Geol. Soc. Am. *52*, 721–772.

HASKELL, N. A. (1937), *The Viscosity of the asthenosphere*, Am. J. Sci. *33*, 22–28.

JACOBY, W. R. (1970), *Instability in the upper mantle and global plate movements*, J. Geophys. Res. *75*, 5671–5680.

KANAMORI, H. and ANDERSON, D. L. (1975), *Theoretical basis of some empirical relations in seismology*, Bull. Seis. Soc. Amer. *65*, 1073–1095.

KOHLSTEDT, D. L. and GOETZE, C. (1974), *Low-stress, high temperature creep in olivine single crystals*, J. Geophys. Res. *79*, 2045–2051.

LLIBOUTRY, L. A. (1971), *Rheological properties of the asthenosphere from Fennoscandian data*, J. Geophys. Res. *76*, 1433–1446.

MCCONNELL, R. K. (1968), *Viscosity of the earth's mantle from relaxation time spectra of isostatic adjustment*, J. Geophys. Res. *73*, 7089–7105.

MELOSH, H. J. (1976a), *Nonlinear stress propagation in the earth's upper mantle*, J. Geophys. Res. *81*, 5621–5632.

MELOSH, H. J. (1976b), *Plate motion and thermal instability in the asthenosphere*, Tectonophysics *35*, 363–390.

POST, R. L. and GRIGGS, D. T. (1973), *The earth's mantle: evidence of non-Newtonian flow*, Science *181*, 1242–1244.

RICHARDSON, R. M. and SOLOMON, S. C. (1976), *Intraplate stress as an indicator of plate tectonic driving forces*. J. Geophys. Res. *81*, 1847–1856.

SOLOMON, S. C., SLEEP, N. H. and RICHARDSON, R. M. (1975), *On the forces driving plate tectonics: inferences from absolute plate velocities and intraplate stress*. Geophys. J. Roy. Astron. Soc. *42*, 769–801.

VENING MEINESZ, F. A. (1937), *The determination of the earth's plasticity from the post-glacial uplift of Scandinavia: isostatic adjustment*, Proc. Kon. Ned. Akad. Wetensch. *40*, 654–662.

WALCOTT, R. I. (1973), *Structure of the earth from glacio-isostatic rebound*, Ann. Rev. Earth Planet. Sci. *1*, 15–37.

WEERTMAN, J. and WEERTMAN, J. R. (1975), *High temperature creep of rock and mantle viscosity*, Ann. Rev. Earth Planet. Sci. *3*, 293–315.

WU, F. T. and KANAMORI, H. (1973), *Source mechanism of February 4, 1965, Rat Island earthquake*, J. Geophys. Res. *78*, 6082–6092.

(Received 15th December 1976)

Pageoph, Vol. 115 (1977), Birkhäuser Verlag, Basel

Earthquake Stress Drops, Ambient Tectonic Stresses and Stresses That Drive Plate Motions

By THOMAS C. HANKS[1])

Abstract – A variety of geophysical observations suggests that the upper portion of the lithosphere, herein referred to as the elastic plate, has long-term material properties and frictional strength significantly greater than the lower lithosphere. If the average frictional stress along the non-ridge margin of the elastic plate is of the order of a kilobar, as suggested by the many observations of the frictional strength of rocks at mid-crustal conditions of pressure and temperature, the only viable mechanism for driving the motion of the elastic plate is a basal shear stress of several tens of bars. Kilobars of tectonic stress are then an ambient, steady condition of the earth's crust and uppermost mantle. The approximate equality of the basal shear stress and the average crustal earthquake stress drop, the localization of strain release for major plate margin earthquakes, and the rough equivalence of plate margin slip rates and gross plate motion rates suggest that the stress drops of major plate margin earthquakes are controlled by the elastic release of the basal shear stress in the vicinity of the plate margin, despite the existence of kilobars of tectonic stress existing across vertical planes parallel to the plate margin. If the stress differences available to be released at the time of faulting are distributed in a random, white fashion with a mean-square value determined by the average earthquake stress drop, the frequency of occurrence of constant stress drop earthquakes will be proportional to reciprocal faulting area, in accordance with empirically known frequency of occurrence statistics.

Key words: Earthquake stress drops; Plate tectonics; Stress in lithosphere.

1. Introduction

Since CHINNERY [1964] first concluded that 'the true stress relieved by [crustal earthquakes] is probably of the order of 10^7 dynes/cm^2' and noted that these stress drops were considerably less than the strength of the earth's crust and uppermost mantle as inferred by JEFFREYS [1959], the relationship of crustal earthquake stress drops to the ambient tectonic stress field giving rise to the earthquake in the first place has been a recurring theme in both observational and theoretical studies of the earthquake mechanism and in controlled laboratory experiments of rock failure. The advent of plate tectonics has only heightened geophysical interest in earthquake stress drops and ambient tectonic stresses, for it is generally agreed that the stress field that drives the motion of plates at the earth's surface should be related to earthquake stress drops along plate margins, the ambient tectonic stress across plate margins, or both. There is little agreement, however, on what such a relationship should be, chiefly because there is little agreement on the magnitude of tectonic stresses across

[1]) U.S. Geological Survey, Menlo Park, California 94025, USA.

plate margins or on the nature of the stress field driving plate motions across the earth's surface.

One point of view, first espoused by CHINNERY [1964], is that active crustal faults are throughgoing zones of weakness, with frictional or breaking strengths two orders of magnitude less than that for intact crustal rocks. Thus, while kilobars of deviatoric stress may exist in the crust beneath certain mountainous regions and perhaps locally on active crustal fault zones as well, the average tectonic stress in the earth's crust and uppermost mantle is probably not greater than 100 bars. This idea has been reinforced by BRUNE *et al.* [1969], who argued on the basis of the absence of a heat-flow anomaly localized to the San Andreas fault that the average frictional stress along the San Andreas fault could not exceed several hundred bars. In this framework, earthquake stress drops represent by and large a complete drop of the causative tectonic stresses, at least in the source region.

Laboratory experiments of rock failure have not, however, validated the fundamental proposition of this line of reasoning that crustal fault zones have strengths two orders of magnitude less than intact rocks. Across a wide range of rock types and surface preparations, kilobars of deviatoric stress are required to offset differentially rock masses subjected to mid-crustal conditions of pressure and temperature. Thus, if crustal faulting is governed by frictional processes, as it is generally agreed, kilobars of tectonic stress are required to induce fracture or frictional sliding of crustal materials through the seismogenic zone, unless fluid pressures essentially negate the lithostatic pressure.

The consequences of frictional stresses of kilobars on active crustal fault zones are far-reaching. It follows in a straightforward way that if kilobars of frictional stress resist plate motions across the upper several tens of kilometers of the plate margin, the only viable mechanism for driving plate motions at the earth's surface is a basal shear stress driving the motion of the plate in a direction parallel to itself, in a manner similar to the ductile model of LACHENBRUCH and SASS [1973]. Furthermore, this condition implies that tectonic stresses of kilobars across vertical planes parallel to the plate margin are an ambient, steady condition of the earth's crust and uppermost mantle. Less clear, however, is what relationship earthquake stress drops of tens of bars have to tectonic stresses of kilobars.

There are, nevertheless, several characteristics of earthquake stress drops in particular and faulting at plate margins in general that point to one such relationship. The first of these is the constancy of earthquake stress drops, in their average value, across the entire range of earthquakes for which instrumental recordings are available. More specifically, earthquakes along plate margins with fault lengths both much greater and much less than the seismogenic depth of approximately 15 km not only possess stress drops independent of source dimension but share the same average constant value of several tens of bars. A second characteristic is that the stress release accompanying even the largest earthquakes along plate margins is distinctly localized. It is well known from conventional geodetic observations that faulting displacements decay rapidly away from the fault surface, and both theory and observation are in accord that this decay is controlled by the depth or width of the fault. A third observation, probably related to the second, is that a variety of geophysical observations suggests that the frictional resistance to plate motions along their margins is not uniform across a 100 km thickness of the lithosphere but is probably concentrated in the upper several tens of kilometers.

The purpose of this study is to bring these observations, together with the proposition that kilobars of frictional stress exist on active crustal fault zones, into a framework which jointly relates them all, and several other matters of geophysical interest as well. Because it is not a consensus view of the geophysical community that an average frictional stress of kilobars exists on active crustal fault zones (e.g., CHINNERY 1964, BRUNE *et al.* 1969, FORSYTH and UYEDA 1975, CARTER 1976), the evidence for and against his proposition is presented. It is the point of view of this paper that the evidence supporting this proposition is, at the present time, persuasive although not irresistible. As such, we summarize the principal conclusions of this study in juxtaposition to the parallel consequences of the alternate (low-frictional stress) hypothesis at the end of this paper.

2. *Earthquake stress drops*

Figure 1 presents determinations of seismic moment (M_0), source dimension (r), and stress drop ($\Delta\sigma$) for 390 earthquakes, principally crustal earthquakes of the Southern California Region. These include 167 San Fernando aftershocks [TUCKER and BRUNE 1973], 138 southern California and northern Mexico earthquakes [THATCHER and HANKS 1973], 28 earthquakes in northern Baja California and the northern Gulf of California [THATCHER 1972], the 1968 Borrego Mountain [HANKS and WYSS 1972], the 1969 Coyote Mountain [THATCHER and HAMILTON 1973], the 1971 San Fernando [WYSS and HANKS 1972] and 1973 Pt. Mugu [ELLSWORTH *et al.* 1973] earthquakes. In addition, Fig. 1 includes 34 shallow focus earthquakes along the Tonga–Kermadec island arc [MOLNAR and WYSS 1972] and 19 intermediate and deep focus earthquakes in the Benioff zone beneath and behind the Tonga–Kermadec island arc [WYSS and MOLNAR 1972].

The distribution of stress drops, which is approximately logarithmically normal about the average value near 10 bars with two standard deviations corresponding to about an order of magnitude in stress drop, is generally but not completely independent of source strength. There are no stress drops less than 1 bar for earthquakes for which $M_0 \geq 10^{24}$ dyne-cm. THATCHER and HANKS [1973] suggested that r was likely to be overestimated near the lower range of M_0 for the earthquakes they studied (note the lower limit of $r \simeq 0.5$ km across two orders of magnitude in M_0 for their data in Fig. 1). A similar bias affects the stress drops of the smaller M_0 earthquakes considered by THATCHER [1972] and perhaps those of TUCKER and BRUNE

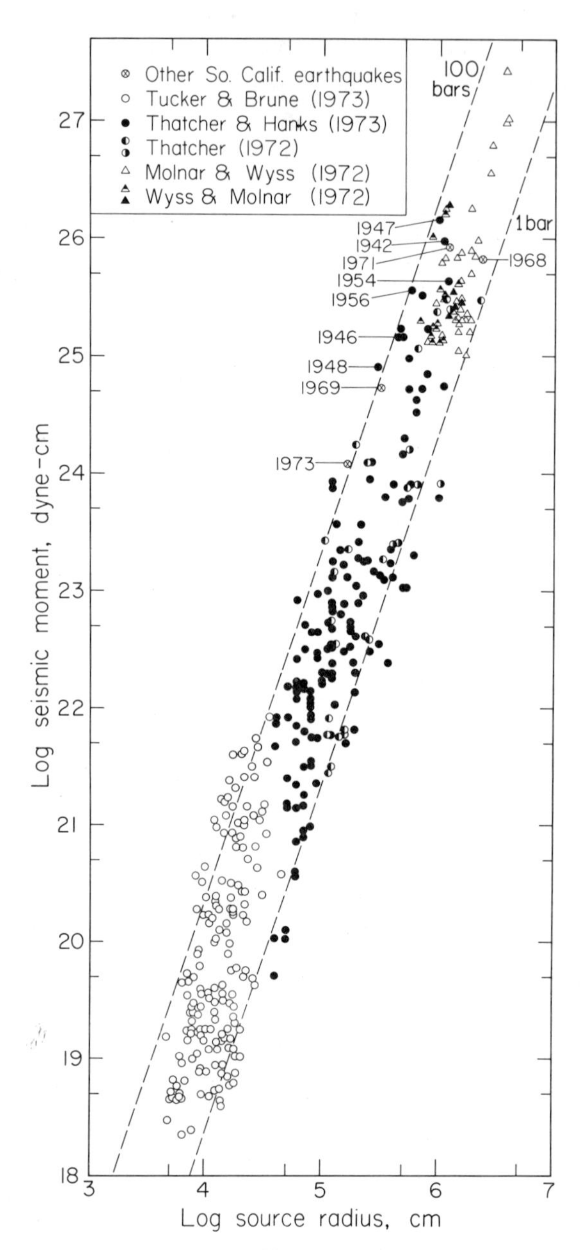

Figure 1

Earthquake stress drops. Open circles: San Fernando aftershocks; solid circles: earthquakes of the Southern California Region; half-filled circles: earthquakes in Baja California (left side filled) and in the Gulf of California (right side filled); triangles: earthquakes along the Tonga-Kermadec Island arc at shallow (open), intermediate (top filled), and deep (solid) depths. The dashed lines of slope 3 are lines of constant stress drop (1 bar and 100 bars as indicated).

[1973] as well. With allowance for these systematic effects, the average stress drop for the data in Fig. 1 would be slightly higher, perhaps 20–30 bars, with a corresponding reduction in the logarithmic standard deviation.

Because all of the data in Fig. 1 have been obtained by scaling spectral parameters of far-field body waves with the source model of BRUNE [1970, 1971], it is important to compare them with similar data for which source parameter estimates are available from geodetic data, field observations of fault length and offset, and aftershock distributions. With respect to the average earthquake stress drop, Fig. 1, quite remarkably, does little to change CHINNERY's [1964] conclusion, reproduced in the first sentence of this paper, reached on the basis of geodetic observations of faulting displacements for five major crustal earthquakes. Similarly, KANAMORI and ANDERSON [1975] (see also AKI 1972) have presented source parameter data for 41 moderate to great earthquakes ($8.3 \times 10^{24} \leq M_0 \leq 2 \times 10^{30}$ dyne-cm) for which source dimension data are principally determined by fault length observations and aftershock distributions. The resulting stress drops are similar to those for the larger earthquakes of Fig. 1, although KANAMORI and ANDERSON [1975] find that most of the values are between 10 and 100 bars.

It is worth noting that each of the stress drops in Fig. 1 and in KANAMORI and ANDERSON [1975] represents a value averaged over the entire faulting surface. They do not preclude the existence of larger, perhaps transient, stress differences across localized regions of the fault surface, as discussed for the San Fernando earthquake by HANKS [1974], the Borrego Mountain earthquake by BURDICK and MELLMAN [1976] and for a recent interpretation of peak accelerations [HANKS and JOHNSON, 1976]. Neither do these data preclude regional variations of stress drops within the observed range nor magnitude-dependent stress drops within a limited range of M_0 and r. But none of these considerations alter the principal conclusion to be extracted from the large number of observations presently available that crustal earthquake stress drops, in their average value $\overline{\Delta\sigma}$, are several tens of bars and that this average value is independent of source strength across twelve orders of magnitude in M_0.

It is curious that this is so, for the stress release of crustal earthquakes for which r is significantly less than the depth or width of faulting h is quite plainly a three-dimensional problem, whereas for earthquakes for which the fault length is significantly greater than h the stress release can be modelled well by two-dimensional geometry along most of the fault length [KASAHARA 1957, BYERLY and DENOYER 1958, CHINNERY 1961, THATCHER 1975, among many such studies]. In the following two sections, we explore the implications of this coincidence separately for earthquakes of small and large faulting dimensions.

3. *Stress drops of small dimension earthquakes*

An interesting way to explore the geophysical significance of the stress drops of small dimension earthquakes is in terms of the frequency of occurrence of constant stress drop events. The relations between the frequency of occurrence N of earthquake magnitude M

$$\log N = a - bM, \tag{1}$$

between M_0 and M

$$\log M_0 = cM + d, \tag{2}$$

and between M_0, r, and $\Delta\sigma$

$$M_0 = k\Delta\sigma r^3 \tag{3}$$

can be algebraically combined to obtain

$$\log N = \left(a + \frac{bd}{c}\right) - \frac{b}{c}\log(k\Delta\sigma r^3). \tag{4}$$

In these relations, a is a constant defined by the choice of region and time interval in which earthquakes are counted; d is an empirically determined constant, and k is a constant equal to 16/7 for a circular fault surface of radius r. It is empirically known that b is generally, but not always, very nearly equal to 1, irrespective of the choice of region and time interval in which earthquakes are counted. Also, c is empirically known to be 1.5 whether local magnitude [THATCHER and HANKS 1973] or surface wave magnitude [KANAMORI and ANDERSON 1975] is used in (2).

With $b = 1$ and $c = 1.5$, (4) reduces to

$$N = \frac{\text{const}}{(\Delta\sigma)^{2/3} r^2}. \tag{5}$$

If the earthquakes of the counted sample share the same $\Delta\sigma$, as they do on the average for all samples for which $\Delta\sigma$ has been determined, earthquake magnitude-frequency of occurrence statistics reduce to a simple matter of geometrical scaling in terms of the reciprocal faulting area.

This result can be interpreted in terms of a two-dimensional stress drop potential function on the fault surface, the functional form of which is specified by a mean-square value determined by $\overline{\Delta\sigma}$ and a spectral composition with constant amplitudes at all wavelengths $\leq h$; the stress drop potential in a region of incipient faulting is realized as the earthquake stress drop at the time of faulting. On the average, such a stress drop potential function will produce earthquakes with stress drop $\overline{\Delta\sigma}$, and their frequency of occurrence will scale as $1/r^2 (r \leq h)$, due to the constant spectral amplitudes at all wavelengths $\leq h$ in two dimensions; earthquakes with $\Delta\sigma$ both higher and lower than $\overline{\Delta\sigma}$ will occur, however, with certain probabilities. Whatever the origin of the stress differences recoverable in crustal faulting may be, then, it is more or less distributed as random, white noise on crustal fault zones. The stress drop potential function, moreover, retains its mean-square value and random, white characteristics until such time as the region of interest is faulted by a major earthquake, inasmuch as earthquakes with $r \ll h$ do not materially reduce the net stress on the fault surface.

4. Stress drops of major plate margin earthquakes

In the case of major plate margin earthquakes, a well-defined stress drop exists along most of the fault length through the seismogenic zone. Inasmuch as this represents a reduction of stresses opposing plate motions along the faulted portion of the plate margin, it must be balanced by a corresponding increase of resisting stresses elsewhere along the plate margin or a corresponding decrease in the stress-field driving the relative motion of the adjacent plates. There are, clearly, many options in placing this stress increment, but available geodetic observations of coseismic and postseismic displacements restrict the range of possibilities.

Geodetically measured surface displacements taken shortly (as much as years) after many major crustal earthquakes invariably reveal a rapid decay of earthquake-induced displacements away from the fault surface. When the fault length is much greater than the fault depth, such data are reasonably well-approximated by a two-dimensional elastostatic model of faulting [KASAHARA 1957, BYERLY and DENOYER 1958, CHINNERY 1961]. For an infinitely long, vertical fault in a uniform half space with constant offset U across $0 \leq z \leq h$ (Fig. 2a), *surface* displacements $u_y\,(x, z = 0)$ parallel to the fault are given by

$$u_y(x, z = 0) = \frac{U}{\pi} \tan^{-1}\left(\frac{h}{x}\right) \tag{6}$$

and the induced horizontal shear stress increment across any vertical plane $x =$ constant is

$$\sigma_{xy}(x, z = 0) = \mu \frac{U}{\pi h}\left(1 + \frac{x^2}{h^2}\right)^{-1} \tag{7}$$

where μ is the shear modulus.

The changes in σ_{xy} are thus strongly localized to the neighborhood of the fault, and the horizontal gradient of σ_{xy} is even more so. It requires little further analysis to conclude that $\partial\sigma_{xy}/\partial x$ can only be accompanied by corresponding gradients of stresses in the vertical direction across horizontal planes $z =$ constant. This requirement is, according to the equations of equilibrium in this two-dimensional problem,

$$\frac{\partial \sigma_{xy}}{\partial x} + \frac{\partial \sigma_{zy}}{\partial z} = 0. \tag{8}$$

In the geometry of Fig. 2b, we approximate $\partial\sigma_{xy}/\partial x$ in the region $0 \leq x \leq h$, $0 \leq z \leq h$ with

$$\overline{\frac{\partial \sigma_{xy}}{\partial x}} = -\frac{\Delta\sigma}{2h} \tag{9}$$

by taking the difference in σ_{xy} at $x = 0$ and $x = h$ along $z = 0$. In the same region, and again in an average sense without regard to localized stress changes induced by irregular faulting displacements, $\overline{\partial\sigma_{zy}/\partial z}$ must be a value comparable to (9). That is, the average stress difference $(\Delta\sigma)$ imposed on the fault $(x = 0, 0 \leq z \leq h)$ in the course of the earthquake is accompanied by a stress change of comparable magnitude $(\frac{1}{2}\Delta\sigma)$ and the same sign along $z = h, 0 \leq x \leq h$.

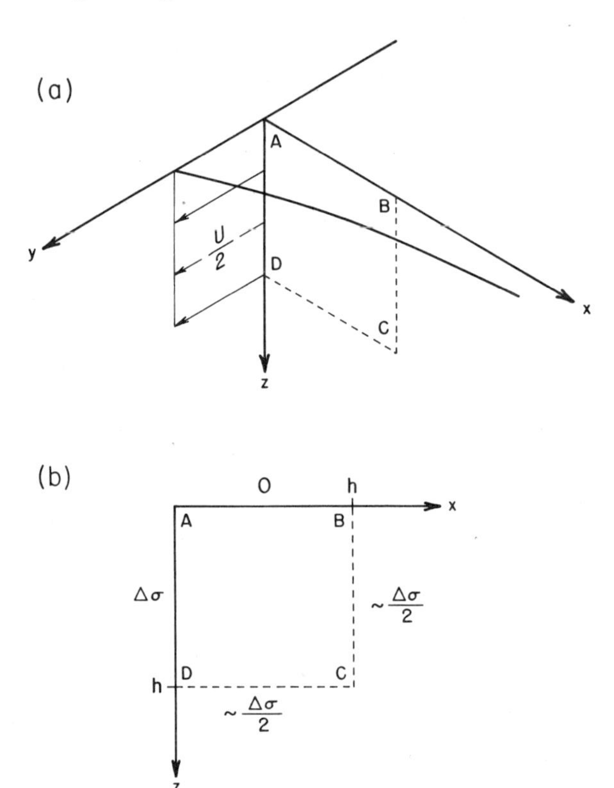

Figure 2
(a) Geometry of throughgoing faulting at a transform plate margin; the x-coordinate is into the plate interior, the y-coordinate is along the plate margin, and the z-coordinate measures depth. The curved line in the x–y plane is $u_y(x, z = 0) = U/\pi \tan^{-1}(h/x)$ which arises in a uniform elastic half space when a displacement $U/2$ is applied on (each side of) the fault through $0 \leq z \leq h$. (b) The approximate stress increments on the sides of ABCD in the x–z plane of (a), given that the faulting displacement $U/2$ through $0 \leq z \leq h$ corresponds to an average stress increment $\Delta\sigma$ through $0 \leq z \leq h$ on the fault plane $(x = 0)$.

To what extent the stress increments along $x = 0, z \leq h$ and $z = h, x \leq h$ cause postseismic displacements at greater depths on the fault is largely unknown. THATCHER [1975] has presented evidence that postseismic displacements in the several decades following the 1906 San Francisco earthquake between 10 and 30 km were of the same order as the coseismic displacements for $z < 10$ km. While the conventional geodetic observations are not well disposed to resolving relaxation over greater dimensions, there is no evidence within these data that significant displacement occurred at distances greater than several tens of kilometers from the fault.

The point of interest here is that the stress release following major plate margin earthquakes is distinctly localized within several fault depths (widths) of the corner of the lithospheric plate. Because of this, the stress change along the fault to depth h (or perhaps several h including postseismic relaxation) is accompanied by a comparable stress change of the same sign along horizontal planes at $z = h$ (or several h) within $x = h$ (or several h) of the plate margin. An interesting consequence of all of this, and the principal issue to be pursued in the remainder of this paper, is that the stress drops of earthquakes along plate margins *may* be controlled by recoverable stress differences of several tens of bars that exist along horizontal planes at a depth of an h or so and within an h or so of the plate margin, whatever the stresses on planes parallel to the plate margin may be.

5. *Strain accumulation and the basal shear stress*

The process of strain accumulation leading to major earthquakes along plate margins is not well understood, but a process that provides a roughly equal and opposite localization of strain energy to the manner in which it is released seems reasonable. Such a model, again for an infinitely long transform fault, is described by SAVAGE and BURFORD [1973] and is reproduced in Fig. 3a. Below the depth h, there is a uniform displacement $U/2$, giving a relative offset U at the fault surface for $z > h$, but for $z \leq h$ the fault is locked. The surface displacements parallel to the fault are then

$$u_y(x, z = 0) = \frac{U}{\pi} \tan^{-1}\left(\frac{x}{h}\right) \tag{10}$$

Following throughgoing faulting, we imagine the block motion indicated in Fig. 3b to ensue, as the addition of the coseismic surface displacements (6) to (10) suggests.

In this model of strain accumulation, it is reasonable to associate a shear stress across horizontal planes at depth $\sim h$ with the displacements $U/2$ below h, in a manner analogous to the ductile model of LACHENBRUCH and SASS [1973]. This possibility is envisioned in the simplest possible way: a single driving force, a basal shear stress driving the overlying plate in a direction parallel to itself, is in equilibrium with a single resisting force, the frictional stresses along the plate margin (Fig. 4). For a circular plate of radius R and thickness H, the condition of static equilibrium becomes

$$\sigma_b = 2\sigma_f \frac{H}{R} \tag{11}$$

where σ_b is the basal shear stress acting at depth H and σ_f is the frictional resisting stress along the plate margin.

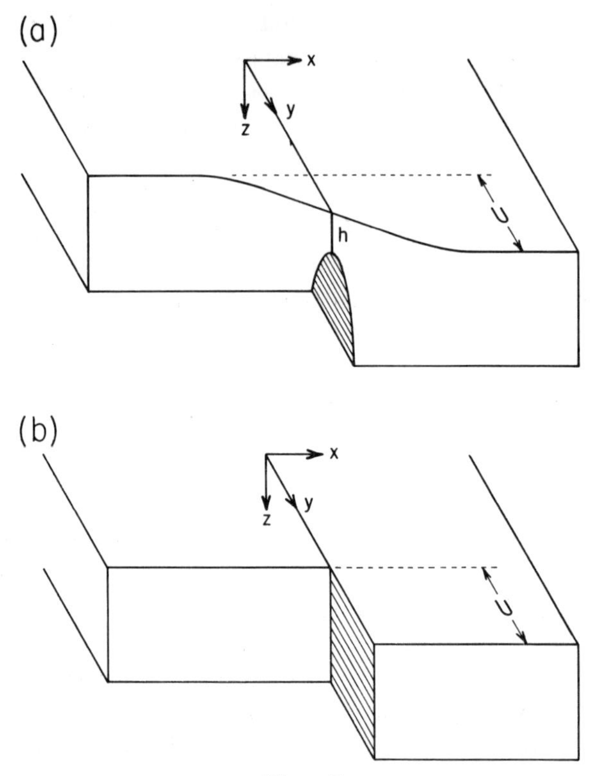

Figure 3
The SAVAGE and BURFORD (1973) model of strain accumulation along a transform plate margin; coordinate axes are the same as in Fig. 2. (a) Prior to throughgoing faulting, the seismogenic zone is essentially locked; strain accumulation in the vicinity of the plate margin arises from uniform, relative displacement U at depths greater than h. (b) After throughgoing faulting, slip on the seismogenic zone additive to aseismic displacements at greater depth results in an essentially undeformed block motion with relative offset U.

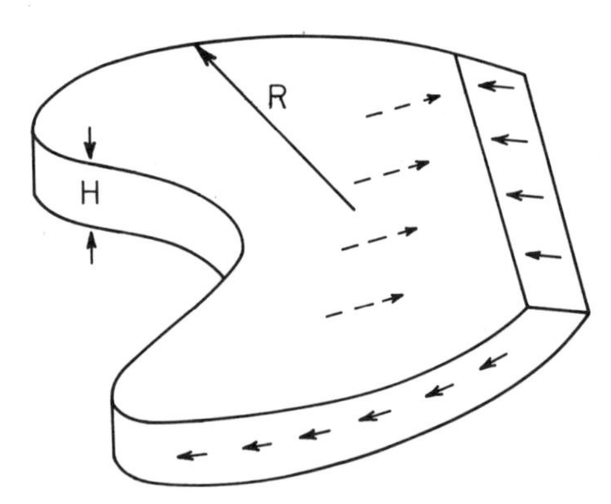

Figure 4
Schematic illustration of the mechanism driving the motion of the elastic plate: a single driving force, the shear stress acting at the base of the elastic plate (dashed arrows) is equilibrated by a single resisting force, the frictional stresses along the plate margin (solid arrows).

It is important to emphasize that the principal assumption of the model summarized in (11) lies not in the postulated existence and action of σ_b but in the values to be prescribed for σ_f, H, and R, particularly σ_f. This is simply because, if σ_f is of the order of a kilobar for H equal to several tens of kilometers or greater, the only viable mechanism for driving plate motions at the earth's surface is the basal shear stress, arising from thermal convection in the upper mantle. Neither the ridge push nor the net trench pull (for a discussion of the general classes of driving forces, see FORSYTH and UYEDA 1975) are sufficiently strong to offset resistance to plate motions of this magnitude (e.g., MCKENZIE 1969, ANDREWS 1972, SOLOMON *et al.* 1975, FORSYTH and UYEDA 1975).

We take this point of view here, postponing to the discussion a more complete statement of the evidence for and against the existence of kilobars of frictional stress on active crustal fault zones. On the basis of the many observations summarized in BRACE and BYERLEE [1970], BRACE [1972], STESKY and BRACE [1973], and ZOBACK and BYERLEE [1976], it seems likely that the maximum shear strength along active crustal fault zones is several kilobars near 15 km with a ductile strength decreasing below this depth due to increasing temperature and with a brittle strength decreasing above this depth due to decreasing confining pressure. The temperature dependence of the strength of rocks as well as the depth of earthquakes along plate margins suggest that the frictional resistance to plate motions along their margins is not uniform across 100 km thickness but is concentrated in the upper 20 to 30 km. These considerations motivate an important distinction made in this study between the *elastic plate thickness* of perhaps 20 to 30 km, at the base of which σ_b is presumed to act, and the lithospheric plate thickness, generally estimated to be 50 to 125 km on the basis of heat flow (e.g., MCKENZIE 1967, SCLATER and FRANCHETEAU 1970) and seismic (e.g., KANAMORI and PRESS 1970, FORSYTH 1977) studies. Additional support for the distinction of an elastic plate thickness considerably less than 100 km may be found in the model calculations of the deformation arising in the oceanic lithosphere offshore of many oceanic trenches [HANKS 1971, WATTS and TALWANI 1974].

With an average value of 1 kb for σ_f across $H = 30$ km along the plate margin and for $R = 2000$ km, $\sigma_b = 30$ bars. Since the chosen values of σ_f, H, and R are plainly uncertain to or variable over a factor of 2 or so, we conclude that σ_b is of the order of several tens of bars. We must, however, exclude those plates with R significantly less than 1000 km, and we do so on the grounds that their motions are more likely to be governed by forces transmitted through their boundaries rather than through their basal areas.

Depending on the scale of dimension and time, the basal shear stress plainly has two distinguishable effects. First, it induces a deformation field far from the plate margin that is unaffected by faulting at the margins and may well be steady on time scales of uniform plate motions, perhaps millions of years. This steady deformation, which is not shown in Fig. 3, is not measurable by any geodetic technique and gives rise to kilobars of tectonic stresses across vertical planes. In this case, it seem reasonable to associate σ_b with the creep strength of mantle materials at depths of several tens of kilometers, and several tens of bars is certainly a reasonable number [KIRBY 1977]. Additive to this steady deformation field, of course, is the simple translation (rotation on a spherical surface) of the plate determined by the plate velocity.

Within several h of the plate margin and on a time scale of minutes to years, σ_b plays a much different role: it is a recoverable, elastic stress available to be released at the time of faulting. In this context, the stress increment on the plate margin, to be released at the time of faulting, is equilibrated by σ_b within an h or so of the plate margin, also released at the time of faulting. If, at the time of faulting, the deformation deficit relative to $U/2$ away from the margin is fully recovered, the stress drop across horizontal planes within several h of the plate margin will be complete, and the earthquake stress drop will be approximately equal to σ_b. To assume that this deformation deficit is fully recovered is equivalent to assuming that, on time scales comparable to the repeat time of major plate margin earthquakes, all points in the elastic plate move the same amount. Given the approximate equality between slip rates at plate margins estimated from the cumulative seismic moments of earthquakes along them with gross plate motion velocities determined from magnetic anomalies [BRUNE 1968, DAVIES and BRUNE 1971], this is a reasonable enough assumption.

Thus, earthquake stress drops need not be controlled, and quite possibly are not controlled, by the shear stress existing across vertical planes parallel to the plate margins but instead are controlled by the tens of bars of basal shear stress across horizontal planes, which is released within an h or so of the plate margin at the time of throughgoing faulting. Before this occurs, we imagine the equilibrating stress increment on the elastic plate margin to be distributed in a random, white fashion, which leads to constant stress drop earthquakes at all dimensions $r \lesssim h$ for which the frequency of occurrence follows the geometric scaling described previously.

6. *Discussion*

In the preceding two sections, the stress drops of major plate margin earthquakes, the accumulation of strain energy leading to such earthquakes, and the connection between σ_b and $\overline{\Delta\sigma}$ have been discussed implicitly for the transform-type plate margin. The same ideas hold for a converging plate margin, although the problem is even simpler because the geometry is essentially one-dimensional. Figure 5 is a cross-section of the elastic plate in the vicinity of and normal to an oceanic trench. At the time of a major earthquake along the Benioff zone at shallow depth, the stress drop along the faulted region is simply balanced by a corresponding drop of σ_b to zero at the base of the elastic plate beneath the faulted region. The gross static equilibrium of the plate in this case, however, is maintained by a nearly horizontal compressive stress of several kilobars, acting normal to the trench axis and arising from frictional resistance to motion of the elastic plate along the Benioff zone at shallow depth.

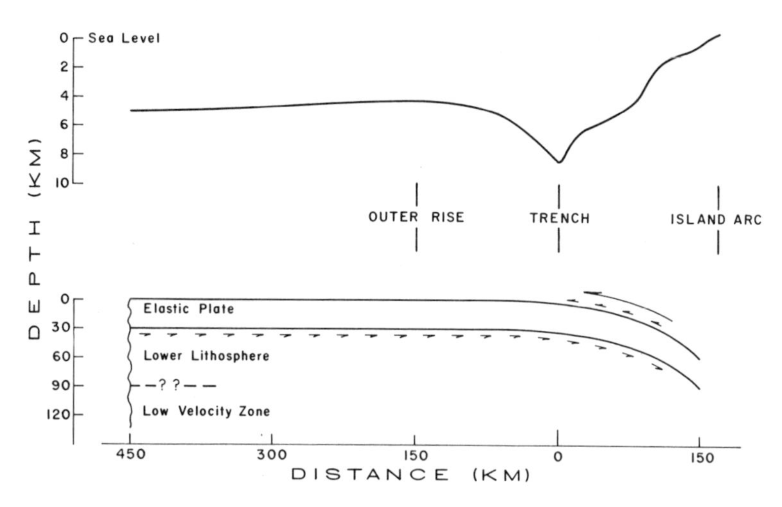

Figure 5

Cross-section of the oceanic lithosphere in the vicinity of and normal to a subducting plate margin. The top part of the figure displays typical bathymetry, at vertical exaggeration of 15:1. σ_b is denoted by small arrows at the bottom of the elastic plate, driving it to the right. On the upper surface of the subducted oceanic lithosphere between the trench and island arc (the crustal portion of the Benioff zone) the long arrow pointing to the left represents a frictional stress of kilobars, which maintains the gross static equilibrium of the plate and induces kilobars of nearly horizontal compression to deform the lithosphere offshore of the trench. The short arrows pointing to the left represent the potential $\Delta\sigma$ of a major earthquake in this region, which will be equilibrated by a drop of σ_b to zero at the base of the elastic plate in the same region.

Such a horizontal compressive stress has been inferred independently on the basis of the deformation of the oceanic lithosphere suggested by the bathymetry of the oceanic trench-outer rise systems by HANKS [1971] and WATTS and TALWANI [1974]. A bending moment applied onshore of the trench axis, however, provides much the same deflection as the horizontal load does (PARSON and MOLNAR 1976, CALDWELL *et al.* 1976). HANKS [1971] noted that horizontal compression is the more plausible loading mechanism in that it approximately negates the enormous tensional strains ($\sim 10^{-2}$) that develop on the surface of the oceanic lithosphere in the case of pure bending. Moreover, NAKAMURA and JACOB [1976] have found evidence in the alignment of flank eruptions on volcanoes along the Aleutian arc that the axis of maximum compression is parallel to the relative velocity vector of the converging Pacific plate. In any event, the gross difference between the predicted stress fields at the surface of the oceanic lithosphere deformed under these two loading conditions should allow them to be distinguished on the basis of hydrofracture measurements.

In either case, however, the deformed lithosphere in these regions must be capable of sustaining kilobars of deviatoric stress, if the upper several tens of kilometers behave in an elastic manner. If deformation in this region is principally anelastic, it must also be largely aseismic, for the oceanic lithosphere offshore of the trench is notably although not completely aseismic; major ($M \gtrsim 7$) earthquakes occur offshore of oceanic trenches only rarely. On the other hand, it seems to be of little value to invoke such an aseismic, anelastic mechanism for deformation of the oceanic lithosphere in its oldest, coldest, thickest and presumably most brittle condition known at the earth's surface when it is nevertheless capable of producing $M \simeq 7$ earthquakes at oceanic ridges in its youngest, hottest, thinnest and presumably most ductile condition. This matter returns us to the principal tenet of this study, that an average frictional stress of kilobars exists on active crustal fault zones and, by consequence, that tectonic stresses of kilobars are an ambient steady condition of the earth's crust and uppermost mantle. What further evidence is there for, and against, this proposition?

Certainly, this proposition is not a new one. JEFFREYS [1959] had concluded at an early date that 'stress differences of at least 1.5×10^9 dyne-cm^2 must exist within the outermost 50 km of the earth' in order to support high, uncompensated mountain ranges. CHINNERY [1964], explicitly assuming that crustal earthquake stress drops represented the breaking strength of crustal fault zones, was forced to conclude that crustal fault zones were throughgoing zones of weakness with breaking strengths approximately two orders of magnitude less than that for intact rock specimens. Laboratory experiments, however, are uniformly inconsistent with Chinnery's assumption. Across a wide range of rock types and surface preparations, kilobars of shear stress are required to offset differentially two rock masses at mid-crustal conditions of confining pressure and temperature (e.g., BRACE and BYERLEE 1970, BRACE 1972, STESKY and BRACE 1973, ZOBACK and BYERLEE 1976).

If internal fluid pressures are approximately equal to the confining pressures, however, the effective pressure and therefore the frictional strength are significantly reduced. The difference in density between rock and water, assuming water to be the pressurized fluid, requires the fluid pressure to be well above hydrostatic at all depths along the seismogenic zone, if this mechanism is to reduce frictional strength to a value significantly less than a kilobar. While superhydrostatic fluid pressures are known to occur in sedimentary basins, there is no evidence that this a prevailing condition across the seismogenic portion of active crustal fault zones, although our knowledge in this regard is hardly sufficient to rule this possibility out. In any event, holes drilled to even modest depth ($\sim$5 km) in active crustal fault zones should resolve this issue.

BRUNE *et al.* [1969] have argued that the absence of a heat flow anomaly localized to the San Andreas fault precluded the existence of frictional stresses along it in excess of several hundred bars. Their steady-state model, however, made no allowance for the heat required for transient temperature rises, the energy absorbed to create new fractures in fresh rock, or mechanisms of heat transport other than ordinary thermal conduction in the rock mass adjacent to the fault.

In view of the role water could play in reducing the frictional strength of active crustal fault zones, it is interesting to speculate on the role water could play in masking the heat flow anomaly estimated by BRUNE *et al.* [1969]. Although it is not localized to the fault zone, a substantial heat flow anomaly does exist in the Coast Ranges of central California (LACHENBRUCH and SASS 1973), but whether this excess heat flow is in fact due to the San Andreas fault, as opposed to residual thermal effects of late Cenozoic metamorphism, is unknown. The excess heat of this anomaly to ± 50 km away from the San Andreas fault, however, is enough to account for the heat generated by kilobars of frictional stress on the fault; the only problem is transporting this heat so far from the fault zone. A particularly efficient way of doing this would be to vaporize water in the fault zone with the frictional heat generated at the time of a major earthquake, convect it laterally away from the fault zone under the influence of thermally-induced pressure increases, and there release the heat carried in the latent heat of vaporization.

It is the point of view of this study that the argument of JEFFREYS [1959], the deformation of the lithosphere offshore of oceanic trenches, and especially the laboratory results of rock failure provide substantial evidence for the existence of ambient tectonic stresses of kilobars in the earth's crust and uppermost mantle and of frictional stresses of kilobars along active crustal faults; and that neither the possibility of nearly lithostatic fluid pressures to mid-crustal depths nor the absence of a heat flow anomaly localized to the San Andreas fault constitutes a compelling argument against these conditions. In view of the uncertainties associated with any line of evidence as well as the existence of contradictory possibilities and arguments, however, it is probably inappropriate to press this point too hard. As such, we summarize below the principal findings of this study, conditioned to the existence of frictional stresses of kilobars along active crustal fault zones, in juxtaposition with the parallel consequences of the alternate (low frictional stress) hypothesis.

7. *Summary and conclusions*

If kilobars of frictional stress resist plate motions along the upper several tens of kilometers of their non-ridge margins, it follows in a straightforward way that the only viable mechanism for driving the motion of the elastic plate is a basal shear stress of several tens of bars. Furthermore, tectonic stresses of kilobars are an ambient, steady condition of the earth's crust and uppermost mantle. The approximate equality of the basal shear stress and the average earthquake stress drop, the localization of strain release for major plate margin earthquakes, and the rough equivalence of plate margin slip rates and gross plate motion rates suggest that earthquake stress drops are governed by the elastic release of σ_b near the plate margin at the time of throughgoing faulting, despite the existence of kilobars of deviatoric stress existing across vertical planes parallel to the plate margin. Prior to throughgoing faulting, the stress increment $\overline{\Delta\sigma}$ is distributed in a random, white fashion along the elastic plate margin in order to provide the empirically known frequency of occurrence distribution of constant stress drop earthquakes.

σ_b must have its origin in some form of thermal convection beneath the elastic plate; its magnitude is several tens of bars, and its action is to drive the overlying plate in a direction parallel to itself. On a time scale of uniform plate motions, σ_b is presumably related to the creep strength of upper mantle materials at the appropriate pressure and temperature. On the much shorter time scales of throughgoing crustal faulting, however, σ_b is simply a recoverable elastic stress available to be released in the vicinity of the plate margin. On time scales between hundreds of seconds and hundreds of years, σ_b must be restored preparatory to the next major earthquake by some deformation mechanism largely confined to the neighborhood of the plate margin beneath the elastic plate. Whether or not σ_b in fact possesses these attributes is the major unresolved issue of this study, for little additional information is available with which this problem might be further constrained.

If the frictional stress on active crustal fault zones is of the order of 100 bars or less, there is no compelling reason to appeal to σ_b or to general thermal convection in the upper mantle as the principal mechanism for driving plate motions at the earth's surface. In this case, as SOLOMON *et al.* [1975] and FORSYTH and UYEDA [1975] have concluded, plate motions are probably governed by weak ridge pushes and trench pulls equilibrated by very weak basal drag forces. Ambient tectonic stresses will be comparable to the frictional stresses, and earthquake stress drops represent by and large a complete reduction of deviatoric stresses, at least in the source region. The most important consequence of this possibility, however, is that laboratory observations of rock failure at mid-crustal conditions of pressure and temperature have little if any relevance to the stress conditions governing crustal faulting, unless nearly lithostatic fluid pressures are a pervasive condition along crustal fault zones. The explanation of constant stress drop earthquakes for events with $r \lesssim h$ is unaffected by the choice of high or low frictional stress hypothesis.

The potential value of holes drilled in and near active crustal fault zones to depths of 5–10 km for resolving whether or not kilobars of frictional stress exist on active crustal fault zones cannot be overestimated. Measurements of the amount and fluid pressure of H_2O, the frictional strength of core samples, the permeability of the fault zone, the *in situ* stress, $\partial u_y/\partial z$, and the temperature field at these depths all bear directly on this central issue.

Acknowledgments

I have enjoyed the critical comments of many individuals in the development of the ideas presented in this manuscript, but those of D. J. Andrews, A. H. Lachenbruch, C. B. Raleigh, J. C. Savage, and W. R. Thatcher have been especially valuable.

REFERENCES

AKI, K. (1972), *Earthquake mechanism*, Tectonophysics *13*, 423–446.
ANDREWS, D. J. (1972), *Numerical simulation of sea-floor spreading*, J. Geophys. Res. *77*, 6470–6481.

BRACE, W. F. (1972), *Laboratory studies of stick-slip and their application to earthquakes*, Tectonophysics *14*, 189–200.

BRACE, W. F. and BYERLEE, J. D. (1970), *California earthquakes: why only shallow focus?* Science *168*, 1573–1575.

BRUNE, J. N. (1968), *Seismic moment, seismicity, and rate of slip along major fault zones*, J. Geophys. Res. *73*, 777–784.

BRUNE, J. N. (1970), *Tectonic stress and the spectra of seismic shear waves*, J. Geophys. Res. *75*, 4997–5009.

BRUNE, J. N. (1971), *Correction*, J. Geophys. Res. *76*, 5002.

BRUNE, J. N., HENYEY, T. L. and ROY, R. F. (1969), *Heat flow, stress, and rate of slip along the San Andreas fault, California*, J. Geophys. Res. *74*, 3821–3827.

BURDICK, L. J. and MELLMAN, G. R. (1976), *Inversion of the body waves from the Borrego Mountain earthquake to the source mechanism*, Bull. Seismol. Soc. Amer. *66*, 1485–1499.

BYERLY, P. and DE NOYER, J. *Energy in earthquakes as computed from geodetic observations* in *Contributions in Geophysics in Honor of Beno Gutenberg* (Pergamon Press, New York, 1958).

CALDWELL, J., HIXBY, W. F., KARIG, D. E. and TURCOTTE, D. L. (1976), *On the applicability of a universal elastic trench profile*, Earth Plan. Sci. Lett. *31*, 239–246.

CARTER, N. L. (1976), *Steady state flow of rocks*, Rev. Geophys. Space Phys. *14*, 301–360.

CHINNERY, M. A. (1961). *The deformation of the ground around surface faults*, Bull. Seismol. Soc. Amer. *51*, 355–372.

CHINNERY, M. A. (1964), *The strength of the Earth's crust under horizontal shear stress*, J. Geophys. Res. *69*, 2085–2089.

DAVIES, G. F. and BRUNE, J. N. (1971), *Regional and global fault slip rates from seismicity*, Nature *229*, 101–107.

ELLSWORTH, W. L., CAMPBELL, R. H., HILL, D. P., PAGE, R. A., ALEWINE, R. W., HANKS, T. C., HEATON, T. H., HILEMAN, J. A., KANAMORI, H., MINSTER, B. and WHITCOMB, J. H. (1973), *Point Mugu, California, earthquake of 21 February 1973 and its aftershocks*, Science *182*, 1127–1129.

FORSYTH, D. W. (1977), *The evolution of the upper mantle beneath mid-ocean ridges*, Tectonophysics, in press.

FORSYTH, D. and UYEDA, S. (1975), *On the relative importance of the driving forces of plate motion*, Geophys. J. Roy. Astron. Soc. *43*, 163–200.

HANKS, T. C. (1971), *The Kuril trench-Hokkaido rise system: large shallow earthquakes and simple models of deformation*, Geophys. J. Roy. Astron. Soc. *23*, 173–190.

HANKS, T. C. (1974), *The faulting mechanism of the San Fernando earthquake*, J. Geophys. Res. *79*, 1215–1229.

HANKS, T. C. and JOHNSON, D. A. (1976), *Geophysical assessment of peak accelerations*, Bull. Seismol. Soc. Amer. *66*, 959–968.

HANKS, T. C. and WYSS, M. (1972), *The use of body-wave spectra in the determination of seismic source parameters*, Bull. Seismol. Soc. Amer. *62*, 561–590.

JEFFREYS, H., *The Earth* (Cambridge University Press, Cambridge, 1959).

KANAMORI, H. and ANDERSON, D. L. (1975), *Theoretical basis of some empirical relations in seismology*, Bull. Seismol. Soc. Amer. *65*, 1073–1096.

KANAMORI, H. and PRESS, F. (1970), *How thick is the lithosphere?* Nature *226*, 330–331.

KASAHARA, K. (1957), *The nature of seismic origins as inferred from seismological and geodetic observations (1)*, Bull. Earthquake Res. Inst. Tokyo Univ. *35*, 475–532.

KIRBY, S. (1977), *State of stress in the lithosphere: inferences from the flow laws of olivine*, Pure appl. Phys. *115*, 245–258.

LACHENBRUCH, A. H. and SASS, J. H., *Thermo-mechanical aspects of the San Andreas fault system* in *Proceedings of the Conference on Tectonic Problems of the San Andreas fault system* (eds. R. L. Kovach and A. Nur) (Stanford University Publications, Stanford, 1973), pp. 192–205.

MCKENZIE, D. P. (1967), *Some remarks on heat flow and gravity anomalies*, J. Geophys. Res. *72*, 6261–6273.

MCKENZIE, D. P. (1969), *Speculations on the consequences and causes of plate motions*, Geophys. J. Roy. Astron. Soc. *18*, 1–32.

MOLNAR, P. and WYSS, M. (1972), *Moments, source dimensions and stress drops of shallow-focus earthquakes in the Tonga–Kermadec arc*, Phys. Earth Planet. Interiors *6*, 263–278.

NAKAMURA, K. and JACOB, K. (1976), *Volcanoes as indicators of tectonic stress orientation – Aleutians and Alaska* (abs.), EOS *57*, 1006.

PARSONS, B. and MOLNAR, P. (1976), *The origin of outer topographic rises associated with trenches*, Geophys. J. Roy. Astron. Soc. *45*, 707–712.

SAVAGE, J. C. and BURFORD, R. O. (1973), *Geodetic determination of relative plate motion in central California*, J. Geophys. Res. *78*, 832–845.

SCLATER, J. G. and FRANCHETEAU, J. (1970), *The implications of terrestrial heat flow observations on current tectonic and geochemical models of the crust and upper mantle of the earth*, Geophys. J. Roy. Astron. Soc. *20*, 509–542.

SOLOMON, S. C., SLEEP, N. H. and RICHARDSON, R. M. (1975), *On the forces driving plate tectonics: inferences from absolute plate velocities and intraplate stress*, Geophys. J. Roy. Astron. Soc. *42*, 769–801.

STESKY, R. M. and BRACE, W. F., *Estimation of frictional stress on the San Andreas fault from laboratory measurements* in *Proceedings of the Conference on Tectonic Problems of the San Andreas fault system* (eds. R. L. Kovach and A. Nur) (Stanford University Publications, Stanford, 1973), pp. 206–214.

THATCHER, W. (1972), *Regional variations of seismic source parameters in the northern Baja California area*, J. Geophys. Res. *77*, 1549–1565.

THATCHER, W. (1975), *Strain accumulation and release mechanism of the 1906 San Francisco earthquake*, J. Geophys. Res. *80*, 4862–4872.

THATCHER, W. and HAMILTON, R. M. (1973), *Aftershocks and source characteristics of the 1969 Coyote Mountain earthquake, San Jacinto fault zone, California*, Bull. Seismol. Soc. Amer. *63*, 647–661.

THATCHER, W. and HANKS, T. C. (1973), *Source parameters of southern California earthquakes*, J. Geophys. Res. *78*, 8547–8576.

TUCKER, B. E. and BRUNE, J. N., *Seismograms, S-wave spectra, and source parameters for aftershocks of San Fernando earthquake* in *San Fernando, California Earthquake of February 9, 1971*, Vol. III (Geological and Geophysical Studies, 69–122, U.S. Dept. of Commerce, 1973).

WATTS, A. B. and TALWANI, M. (1974), *Gravity anomalies seaward of deep-sea trenches and their tectonic implications*, Geophys. J. Roy. Astron. Soc. *36*, 57–90.

WYSS, M. and HANKS, T. C. (1972), *The source parameters of the San Fernando earthquake inferred from teleseismic body waves*, Bull. Seismol. Soc. Amer. *62*, 591–602.

WYSS, M. and MOLNAR, P. (1972), *Source parameters of intermediate and deep focus earthquakes in the Tonga arc*, Phys. Earth Planet. Interiors *6*, 279–292.

ZOBACK, M. D. and BYERLEE, J. D. (1976), *A note on the deformational behavior and permeability of crushed granite*, Int. J. Rock Mech. Min. Sci. Geomech. Abstr. *13*, 291–294.

(Received 14th January 1977)